Lecture Notes in Artificial Intelligence 8720

Subseries of Lecture Notes in Computer Science

LNAI Series Editors

Randy Goebel
University of Alberta, Edmonton, Canada
Yuzuru Tanaka
Hokkaido University, Sapporo, Japan
Wolfgang Wahlster
DFKI and Saarland University, Saarbrücken, Germany

LNAI Founding Series Editor

Joerg Siekmann
DFKI and Saarland University, Saarbrücken, Germany

Umberto Straccia   Andrea Calì (Eds.)

Scalable Uncertainty Management

8th International Conference, SUM 2014
Oxford, UK, September 15-17, 2014
Proceedings

 Springer

Volume Editors

Umberto Straccia
Istituto di Scienza e Tecnologie dell'Informazione (ISTI - CNR)
Pisa, Italy
E-mail: straccia@isti.cnr.it

Andrea Calì
Birkbeck, University of London
Department of Computer Science and Information Systems
London, UK
E-mail: andrea@dcs.bbk.ac.uk

ISSN 0302-9743   e-ISSN 1611-3349
ISBN 978-3-319-11507-8   e-ISBN 978-3-319-11508-5
DOI 10.1007/978-3-319-11508-5
Springer Cham Heidelberg New York Dordrecht London

Library of Congress Control Number: 2014948263

LNCS Sublibrary: SL 7 – Artificial Intelligence

Typesetting: Camera-ready by author, data conversion by Scientific Publishing Services, Chennai, India

Printed on acid-free paper

Springer is part of Springer Science+Business Media (www.springer.com)

Preface

Information systems are becoming increasingly complex, involving massive amounts of data coming from different sources. Information is often inconsistent, incomplete, heterogeneous, and pervaded with uncertainty.

The International Conference on Scalable Uncertainty Management (SUM) conferences series provides an international forum about the management of uncertain, incomplete, or inconsistent information.

This volume contains the papers presented at the 8th International Conference on Scalable Uncertainty Management (SUM 2014), which was held at St Anne's College, Oxford, UK, from the 15th to the 17th of September, 2014.

The call for papers solicited submissions in two categories: regular research papers and short papers, where the latter report on interesting work in progress or provide system descriptions. The call for papers resulted in 47 submissions. Based on the review reports and discussions, 26 papers were accepted for publication and presentation at the conference, among which 20 regular papers and 6 short papers. The conference program also included invited lectures by three leading researchers: Anthony Hunter (Department of Computer Science, University College London), Jens Lehmann (Department of Computer Science, University of Leipzig), and Dan Olteanu (Department of Computer Science, University of Oxford).

A conference such as this can only succeed as a team effort. We would like to thank: the authors of submitted papers, the invited speakers, and the conference participants; the members of the Program Committee and the external referees; Alfred Hofmann and Springer for providing assistance and advice in the preparation of the proceedings; the University of Oxford for providing local facilities; the creators and maintainers of the conference management system EasyChair. All of them made the success of SUM 2014 possible.

July 2014

Umberto Straccia
Andrea Calì

Organization

Program Committee

Leila Amgoud	IRIT - CNRS, France
Nahla Ben Amor	Institut Superieur de Gestion de Tunis, Tunisia
Leopoldo Bertossi	Carleton University, Canada
Isabelle Bloch	ENST - CNRS UMR 5141 LTCI, France
Fernando Bobillo	University of Zaragoza, Spain
Loreto Bravo	Universidad de Concepción, Chile
Andrea Calì	University of London, Birkbeck College, UK
Laurence Cholvy	ONERA - Toulouse, France
Jan Chomicki	University at Buffalo, USA
Fabio Cozman	Universidade de Sao Paulo, Brazil
Alfredo Cuzzocrea	ICAR - CNR, Italy
Michael Dekhtyar	Tver State University, Russia
Juergen Dix	Clausthal University of Technology, Germany
Didier Dubois	IRIT - CNRS, France
Thomas Eiter	Vienna University of Technology, Austria
Zied Elouedi	LARODEC, ISG de Tunis, Tunisia
Ronald Fagin	IBM Research - Almaden, USA
Minos Garofalakis	Technical University of Crete, Greece
Lluis Godo	Artificial Intelligence Research Institute, IIIA - CSIC, Spain
Nikos Gorogiannis	Middlesex University, UK
John Grant	Towson University, USA
Sergio Greco	University of Calabria, Italy
Stijn Heymans	SRI International, USA
Anthony Hunter	University College London, UK
Gabriele Kern-Isberner	Technische Universitaet Dortmund, Germany
Angelika Kimmig	KU Leuven, Belgium
Kathryn Laskey	George Mason University, USA
Jonathan Lawry	University of Bristol, UK
Sebastian Link	The University of Auckland, New Zealand
Peter Lucas	Radboud University Nijmegen, The Netherlands
Jianbing Ma	Bournemouth University, UK
Thomas Meyer	Centre for Artificial Intelligence Research, UKZN and CSIR Meraka, South Africa
Serafin Moral	University of Granada, Spain
Kedian Mu	School of Mathematical Sciences, Peking University, China

Jeff Z. Pan	University of Aberdeen, UK
Simon Parsons	University of Liverpool, UK
Rafael Peñaloza	TU Dresden, Germany
Olivier Pivert	IRISA - ENSSAT, France
Henri Prade	IRIT - CNRS, France
Andrea Pugliese	DEIS - University of Calabria, Italy
Guilin Qi	Southeast University, China
Fabrizio Riguzzi	University of Ferrara, Italy
Sebastian Rudolph	Technische Universität Dresden, Germany
Vítor Santos Costa	Universidade do Porto, Portugal
Steven Schockaert	Cardiff University, UK
Guillermo Ricardo Simari	Universidad Nacional del Sur in Bahia Blanca, Argentina
Giorgos Stoilos	National Technical University of Athens (NTUA), Greece
Umberto Straccia	ISTI - CNR, Italy
Heiner Stuckenschmidt	University of Mannheim, Germany
Vicenc Torra	IIIA - CSIC, Spain
Peter Vojtáš	Charles University, Czech Republic
Nic Wilson	4C, UCC, Cork, Ireland
Ronald Yager	Machine Intelligence Institute - Iona College, USA

Additional Reviewers

Britz, Arina
Casini, Giovanni
Garcia, Jhonatan
Klarman, Szymon
Meilicke, Christian

Molinaro, Cristian
Spina, Cinzia Incoronata
Tsalapati, Eleni
Weinzierl, Antonius

Table of Contents

Possibilistic Networks:
A New Setting for Modeling Preferences

Nahla BenAmor[1], Didier Dubois[2], Héla Gouider[1], and Henri Prade[2]

[1] LARODEC, Université de Tunis, ISG de Tunis, 41 rue de la Liberté, 2000 Le Bardo, Tunisia
[2] IRIT – CNRS, 118, route de Narbonne, Toulouse, France
nahla.benamor@gmx.fr, {dubois,prade}@irit.fr,
gouider.hela@gmail.com

Abstract. Possibilistic networks are the counterpart of Bayesian networks in the possibilistic setting. Possibilistic networks have only been studied and developed from a reasoning-under-uncertainty point of view until now. In this short note, for the first time, one advocates their interest in preference modeling. Beyond their graphical appeal, they can be shown to provide a natural encoding of preferences agreeing with the inclusion-based partial order applied to the subsets of preferences violated in the different situations. Moreover they do not encounter the limitations of CP-Nets in terms of representation capabilities. They also enjoy a logical counterpart that may be used for consistency checking. This short note provides a comparative discussion of the merits of possibilistic networks with respect to other existing preference modeling frameworks.

1 Introduction

Preferences are usually expressed by means of local pieces of information, rather than as a complete preorder between the different possible states of the world. This state of facts has led AI researchers to propose compact representation formats for preferences and procedures for computing a plausible ranking between completely described situations from such representations, in the last fifteen years. Conditional preference networks [6] (CP-Nets for short) have emerged as a popular reference setting for representing preferences, leading to different refinements [5,15], as well as some alternative approaches [4,8,13] (see [7] for a brief overview). Inspired from Bayesian networks, CP-Nets inherit their graphical nature, and besides, rely on a simple, apparently natural principle, named *ceteris paribus*, which allows to extend any contextual preference "in context c, I prefer a to $\neg a$" (denoted for short $c : a \succ \neg a$), to any particular specification b of the other variables used for describing the considered situations, i.e., the preference is understood as $\forall b, cab$ is preferred to $c \neg ab$. The CP-net approach perfectly exemplifies the ingredients needed for a satisfactory representation of preferences, stated in a conditional manner, into a partial order useful for a user: i) a simple representation setting, preferably having a graphical counterpart for elicitation ease, ii) a natural principle for making explicit the preferences between completely described situations, and iii) an algorithm for determining how to compare two complete situations according to the existence of a path of worsening flips linking them. In spite of their appealing features, CP-Nets have some limitations. First, there exist preorders that make sense and

U. Straccia and A. Calì (Eds.): SUM 2014, LNAI 8720, pp. 1–7, 2014.

for which there does not exist any CP-net that can be associated to them. They also tend to enforce some debatable priorities between the preferences associated to nodes in the CP-Nets, beyond what is really expressed by these preferences [11,12].

In this short paper, we advocate possibilistic networks as a valuable tool for representing preferences. First, possibilistic networks are the counterpart of Bayesian networks in possibility theory, based on a possibilistic Bayesian-like conditioning rule. Although they have been only used for uncertainty modeling until now, they can serve preference modeling purposes as well, as shown in the following, without having the CP-Nets limitations mentioned above. The paper is organized as follows. Section 2 provides a brief background on possibilistic networks. Then Section 3 proposes and explains their use in preference modeling and establishes some properties. The paper ends with a short discussion comparing CP-Nets and preference possibilistic networks.

2   Possibilistic Networks

Possibility theory [9,16] relies on the idea of a possibility distribution π, which is a mapping from a universe of discourse Ω to the unit interval $[0, 1]$, or to any bounded totally ordered scale. $\pi(\omega) = 0$ means that ω is fully impossible, while $\pi(\omega) = 1$ means that ω is fully possible. Nothing forbids to have $\omega \neq \omega', \pi(\omega) = \pi(\omega') = 1$. π is normalized if $\exists u, \pi(u) = 1$, which expresses that not all values in Ω are somewhat impossible, and thus consistency. Given a normalized possibility distribution π, the uncertainty about the occurrence of an event $A \subseteq \Omega$ is assessed via a possibility measure $\Pi(A) = \sup_{\omega \in A} \pi(\omega)$ and its dual necessity measure $N(A) = 1 - \Pi(\overline{A})$ (where $\overline{A}$ is the complement of A). $\Pi(A)$ (resp. $N(A)$) is the extent to which A is consistent with (resp. implied by) the information represented by π. Conditioning in possibility theory is defined from the Bayesian-like equation $\Pi(A \cap B) = \Pi(A|B) \otimes \Pi(B)$, where $\otimes$ stands for the product in a quantitative setting (using the full power of the unit interval $[0, 1]$), or for min in a qualitative setting where only the ordinal value of the grades makes sense. Possibilistic networks [2,3] are counterparts of Bayesian networks [14] which are based on the decomposition of a joint possibility distribution as a combination of conditional possibility distributions. Namely, given a set of variables $\{V_1, ..., V_n\}$, ordered arbitrarily, $\pi(V_1, ..., V_n) = \pi(V_n|V_1, ..., V_{n-1}) \otimes ... \otimes \pi(V_2|V_1) \otimes \pi(V_1)$. The conditional possibility distributions are normalized as soon as the joint possibility distribution is normalized. This decomposition can be further simplified by assuming conditional independence between variables [1]. For instance, if V_n is independent from $V_1, ..., V_i$ given $V_{i+1}, ..., V_{n-1}$ then $\pi(V_n|V_1, ..., V_{n-1}) = \pi(V_n|V_{i+1}, ..., V_{n-1})$.

Thus, a possibilistic network has (i) a graphical component which is a DAG (Directed Acyclic Graph) $\mathcal{G} = (\mathcal{V}, \mathcal{E})$ where $\mathcal{V}$ is a set of nodes representing variables and $\mathcal{E}$ a set of edges encoding conditional (in)dependencies between them; (ii) a data component associating a local normalized conditional possibility distribution to each variable $V_i \in \mathcal{V}$ in the context of its parents (denoted by $pa(V_i)$). The joint possibility distribution is then given by the chain rule: $\pi(V_1, ..., V_n) = \otimes_{i=1,...,n} \pi(V_i \mid pa(V_i))$ where $\otimes$ is either the min or the $product$ operator $*$ depending on the semantics underlying it. In the following, each variable V_i has a value domain $D(V_i)$, v_i denotes any value of V_i, and $\Omega = \{\omega_1, ..., \omega_m\}$ denotes the set of interpretations corresponding to the Cartesian product of all variable domains in $\mathcal{V}$.

3 Modeling Preferences with a Possibilistic Network

In this section, we introduce a new approach, briefly suggested in [10], based on product-based possibilistic networks, for representing preferences. The product has a greater discriminating power than the minimum operator. In this approach, possibility degrees may remain symbolic but stand for numbers. As we shall see, the representation is particularly faithful to the user's preferences. The ordering between interpretations obtained from this compact representation fully agrees with the inclusion ordering associated with the violation of preference statements, in the sense that if an interpretation ω violates all the preferences violated by another interpretation ω' plus some other(s), then ω' is strictly preferred to ω. Moreover, the relative importance of preferences can be easily taken into account when available. To illustrate the idea of representing preferences by means of possibilistic networks, we use the following example inspired from the CP-net literature [6].

Example 1 *Let us consider a simple example about a party suit with 4 variables standing for shirt (S), trousers (T), jacket (J) and shoes (H) s.t.* $D(S) = \{black(s),$ $red(\neg s)\}$, $D(T) = \{black(t), red(\neg t)\}$, $D(J) = \{red(j), white(\neg j)\}$ *and* $D(H) =$ $\{white(h), black(\neg h)\}$. *The preference conditional set is:*

The user prefers to wear a black shirt to a red one.
He prefers to wear black trousers to red ones.
If he wears a black shirt and black trousers, he prefers to wear a red jacket to a white one.
If he wears a black shirt and red trousers, he prefers to wear a white jacket.
If he wears a red shirt and black trousers, he prefers to wear a red jacket.
If he wears a red shirt and red trousers, he prefers to wear a white jacket.
If he wears a red jacket, he prefers to wear white shoes to black ones.
If he wears a white jacket, he prefers to wear black shoes.

The universe of discourse associated to this example is:
$\Omega = \{\omega_1 = tjsh, \omega_2 = tjs\neg h, \omega_3 = tj\neg sh, \omega_4 = tj\neg s\neg h, \omega_5 = t\neg jsh, \omega_6 = t\neg js\neg h,$
$\omega_7 = t\neg j\neg sh, \omega_8 = t\neg j\neg s\neg h, \omega_9 = \neg tjsh, \omega_{10} = \neg tjs\neg h, \omega_{11} = \neg tj\neg sh, \omega_{12} =$
$\neg tj\neg s\neg h, \omega_{13} = \neg t\neg jsh, \omega_{14} = \neg t\neg js\neg h, \omega_{15} = \neg t\neg j\neg sh, \omega_{16} = \neg t\neg j\neg s\neg h\}$.

The preference description is assumed to be given under the form of conditional statements of the form $c : a \succ \neg a$ where c stands for the specification of a context in terms of Boolean variable(s) and a is a Boolean variable. Unconditional preferences correspond to the case where c is the tautology $\top$. The graphical structure of the network is then directly determined from this description (as in the CP-net case). Namely each variable corresponds to a node and conditional preferences are expressed by means of edges. The possibilistic preference table (πP-table for short) associated to a node is defined in the following way. To each preference of the form $c : a \succ \neg a$, pertaining to a variable A whose domain is $\{a, \neg a\}$, is associated the conditional possibility distribution $\pi(a|c) = 1$ and $\pi(\neg a|c) = \alpha$ where α is a symbolic weight such that $\alpha < 1$. We write $\pi(\cdot|\top) = \pi(\cdot)$.

Figure 1 gives the possibilistic graph associated to the Example 1. For instance, the corresponding conditional possibility distribution of the variable H is $\pi(h|j) = 1$ and $\pi(\neg h|j) = \epsilon_1$, $\pi(\neg h|\neg j) = 1$ and $\pi(h|\neg j) = \epsilon_2$. Thanks to conditional independence relations as exhibited by the graph, and using the product-based chain rule, we have:

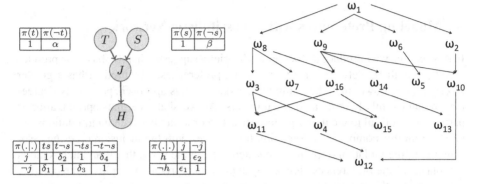

Fig. 1. A possibilistic network   **Fig. 2.** The Inclusion-based ordering

$\pi(TSJH) = \pi(H|J) * \pi(J|TS) * \pi(T) * \pi(S)$. We are then in position to compute the symbolic possibility degree expressing the satisfaction level of any interpretation. For instance, $\pi(\omega_4) = \pi(\neg h|j) * \pi(j|t\neg s) * \pi(t) * \pi(\neg s) = \epsilon_1\delta_2\beta$. Similarly, $\pi(\omega_3) = \pi(h|j) * \pi(j|t\neg s) * \pi(t) * \pi(\neg s) = \delta_2\beta$. Then, based on the fact that $\forall\, \alpha,\, \alpha < 1$, and $\forall \alpha,\, \beta,\, \alpha * \beta < \min(\alpha, \beta)$, we can define a *partial order* $\succ_\pi$ between interpretations under the form of a possibility distribution. In fact, given two interpretations $\omega_i, \omega_j \in \Omega$, $\omega_i \succ_\pi \omega_j$ iff $\pi(\omega_i) > \pi(\omega_j)$. Thus, for instance, $\omega_3 \succ_\pi \omega_4$. Besides, $\pi(\omega_6) = \delta_1$ and $\pi(\omega_{14}) = \alpha\delta_3$, thereby ω_6 and ω_{14} remain incomparable. However, if we further assume $\alpha < \delta_1$ expressing that the unconditional preference associated with node T is more important than the preference $ts : j \succ \neg j$, we become in position to establish that $\omega_6 \succ_\pi \omega_{14}$. Therefore, the approach leaves the freedom of specifying the *relative importance* of preferences.

Assume that for each node, i.e. each variable $V_i \in V$, two *distinct* symbolic weights are used, one for the context where the preferences associated with *each* parent nodes are satisfied, one *smaller* for all the other contexts. For instance, the symbolic weights of the variable J become $\delta_1 > \delta_2 = \delta_3 = \delta_4$ and those of the variable H become $\epsilon_1 > \epsilon_2$. The partial order induced from the possibilistic network (without adding other constraints between symbolic weights) is then faithful to the inclusion order associated to the violated constraints. It is, in fact, exactly the same ordering. This is due to the non comparability between some symbolic weights (following from the use of product). Figure 2 shows the inclusion-based order induced by the possibilistic graph with these additional assumptions.

4   Comparison with CP-Nets and Concluding Remarks

CP-Nets [6] are based on the ceteris paribus principle. As can be seen on the previous example (where ω_6 and ω_{14} are incomparable, while $\top : t \succ \neg t$), possibilistic networks do not obey that latter principle. The order induced by the CP-net is a refinement of the possibilistic order $\succ_\pi$, if no constraints about the relative importance of preferences are added. CP-Nets are, in some sense, too *bold* and too *cautious*. Too bold since, as a result of the systematic application of the ceteris paribus principle, some priority is given to preferences associated to parent nodes, which cannot be questioned nor modified, as already said. Too cautious since they usually lead to a partial order while a complete preorder may be more useful in practice. The basic ordering associated to a possibilistic network is just the inclusion-based ordering, which can then be completed by adding relative importance constraints. In particular, a complete ordering of the symbolic weights leads to a complete preordering of the interpretations. It is unknown whether CP-net orderings also respect the inclusion-based order, as it has apparently never been investigated.

Example 2 *Figures 3 and 4 show, respectively, the order induced by the CP-net and the possibilistic network of Figure 1. Here we assume $\alpha = \beta < \delta_1 < \delta_2 = \delta_3 = \delta_4 < \epsilon_1 < \epsilon_2$. For instance, let us consider the interpretations ω_7 and ω_{16}. In contrast to the possibilistic network, which gives a total preorder, the CP-net considers these two interpretations as incomparable. We notice that both interpretations violate two preferences: associated to a parent and to a grandchild for ω_7, and to two parents preferences for ω_{16}. As expected, ω_7 is preferred to ω_{16} in the possibilistic network as their possibility degrees are respectively $\pi(\omega_7) = \beta\epsilon_2$ and $\pi(\omega_{16}) = \alpha\beta$.*

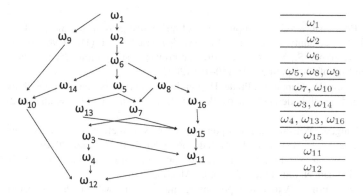

Fig. 3. The order induced by the CP-net **Fig. 4.** The order induced by the possibilistic network

Moreover, CP-Nets are sometimes unable to represent some user preferences.

Example 3 *Let us consider two binary variables A and B standing respectively for "vacations" and "good weather". Suppose that we have the following preference ordering (where one may have two variable switches between two successive interpretations in the ordering) : $ab \succ \neg a\neg b \succ a\neg b \succ \neg ab$. We observe that this complete preorder cannot be represented by a CP-net, while the possibilistic network can display it. Such preferences can be represented by a joint possibility distribution such that: $\pi(ab) > \pi(\neg a\neg b) > \pi(a\neg b) > \pi(\neg ab)$. Since any joint possibility distribution can be decomposed into conditional possibility distributions as shown by the possibilistic chain rule, any complete preorder can be represented by a possibilistic net. Here, we can take $\top : a \succ \neg a, a : b \succ \neg b$ and $\neg a : \neg b \succ b$. Note that encoding these preferences in a CP-Net way would lead to reverse some preferences, and to get $a\neg b \succ \neg a\neg b$. It* *corresponds to a network with two nodes with their corresponding conditional possibility distributions:* $\pi(a) = 1$, $\pi(\neg a) = \alpha$, $\pi(b|a) = 1$, $\pi(b|\neg a) = \gamma$, $\pi(\neg b|a) = \beta$ *and* $\pi(\neg b|\neg a) = 1$. *This yields* $\pi(ab) = 1 > \pi(\neg a \neg b) = \alpha > \pi(a \neg b) = \beta > \pi(\neg ab) = \alpha\gamma$ *taking* $\alpha > \beta$ *and* $\beta = \gamma$.

Lastly, it is important to mention that one of the advantages of the possibilistic graph is its ability to be translated into a possibility logic base [3,11,12] that can be used for executing the preference queries. This bridges the approach presented here with the *direct* representation of preferences by a possibilistic logic base, e.g. [11,12]. This short note has outlined a preliminary presentation of possibilistic networks as providing a convenient setting for acyclic preference representation. This setting remains close to the spirit of Bayesian networks since it relies on directed acyclic graphs, but is flexible enough, thanks to the introduction of symbolic weights, for capturing any ordering agreeing with the inclusion-based ordering. Further research is still needed for investigating their potential in greater detail.

References

1. Ben Amor, N., Benferhat, S.: Graphoid properties of qualitative possibilistic independence relations. Int. J. of Uncertainty, Fuzziness and Knowledge-based Sys. 13(1), 59–96 (2005)
2. Ben Amor, N., Benferhat, S., Mellouli, K.: Anytime propagation algorithm for min-based possibilistic graphs. Soft Computing 8(2), 150–161 (2003)
3. Benferhat, S., Dubois, D., Garcia, L., Prade, H.: On the transformation between possibilistic logic bases and possibilistic causal networks. Int. J. of Approximate Reasoning 29(2), 135–173 (2002)
4. Bienvenu, M., Lang, J., Wilson, N.: From preference logics to preference languages, and back. In: Lin, F., Sattler, U., Truszczynski, M. (eds.) Proc. KR 2010, Toronto (2010)
5. Brafman, R.I., Domshlak, C.: Introducing variable importance tradeoffs into CP-nets. In: Darwiche, A., Friedman, N. (eds.) Proc.UAI 2002, Alberta, pp. 69–76 (2002)
6. Boutilier, C., et al.: CP-nets: A tool for representing and reasoning with conditional ceteris paribus preference statements. JAIR 21, 135–191 (2004)
7. Domshlak, C., Hüllermeier, E., Kaci, S., Prade, H.: Preferences in AI: An overview. Artif. Intell. 175(7-8), 1037–1052 (2011)
8. Dubois, D., Kaci, S., Prade, H.: Approximation of conditional preferences networks "CP-nets" in possibilistic logic. In: Proc. FUZZ-IEEE 2006, Vancouver, pp. 16–21 (2006)
9. Dubois, D., Prade, H.: Possibility Theory: An Approach to Computerized Processing of Uncertainty. Plenum Press (1988)
10. Dubois, D., Prade, H.: Qualitative possibility theory in information processing. In: Nikravesh, M., Kacprzyk, J., Zadeh, L.A. (eds.) Forging New Frontiers: Fuzzy Pioneers II. STUDFUZZ, vol. 218, pp. 53–83. Springer, Heidelberg (2008)
11. Dubois, D., Prade, H., Touazi, F.: Conditional Preference-nets, possibilistic logic, and the transitivity of priorities. In: Bramer, M., Petridis, M. (eds.) Research and Development in Intelligent Systems XXX, pp. 175–184 (2013)
12. Kaci, S., Prade, H.: Mastering the processing of preferences by using symbolic priorities in possibilistic logic. In: Ghallab, M., Spyropoulos, C.D., Fakotakis, N., Avouris, N.M. (eds.) Proc. ECAI 2008, Patras, pp. 376–380. IOS Press (2008)
13. Kaci, S., van der Torre, L.: Reasoning with various kinds of preferences: logic, non-monotonicity, and algorithms. Annals of Operations Research 163(1), 89–114 (2008)

14. Pearl, J.: Probabilistic Reasoning in Intelligent Systems: Networks of Plausible Inference. Morgan Kaufmann Publishers Inc., San Francisco (1988)
15. Wilson, N.: Computational techniques for a simple theory of conditional preferences. Artif. Intell. 175(7-8), 1053–1091 (2011)
16. Zadeh, L.A.: Fuzzy sets as a basis for a theory of possibility. Fuzzy Sets & Sys. 1, 3–28 (1978)

Min-based Assertional Merging Approach for Prioritized DL-Lite Knowledge Bases

Salem Benferhat[1], Zied Bouraoui[1], Sylvain Lagrue[1], and Julien Rossit[2]

[1] Univ Lille Nord de France, F-59000 Lille, France
UArtois,CRIL - CNRS UMR 8188, F-62300 Lens,France
{benferhat,bouraoui,lagrue}@cril.fr
[2] Université Paris Descartes, LIPADE-France
julien.rossit@parisdescartes.fr

Abstract. *DL-Lite* is a powerful and tractable family of description logics specifically tailored for applications that use huge volumes of data. In
many real world applications, data are often provided by several and
potentially conflicting sources of information having different levels of
priority. Possibility theory offers a very natural framework to deal with
ordinal and qualitative uncertain beliefs or prioritized preferences. Thus,
to encode prioritized assertional facts, a possibility *DL-Lite* logic is more
suited.

We propose in this paper a *min*-based assertional merging operator
for possibilistic *DL-Lite* knowledge bases. We investigate in particular
the situation where the sources share the same terminological base. We
present a syntactic method based on conflict resolution which has a meaningful semantic counterpart when merging possibility distributions. We
finally provide an analysis in the light of a new set of postulates dedicated
to uncertain *DL-Lite* merging.

1 Introduction

Description Logics (DLs) provide a powerful formalism for representing and reasoning on ontologies [2]. A DL knowledge base is formed by a terminological
base, called TBox, and an assertional base, called ABox. The TBox contains
intentional (or generic) knowledge of the application domain whereas the Abox
stores data (or individuals or constants) that instantiate terminological knowledge. In the last years, there has been an increasingly interest in Ontology-based
Data Access (OBDA), in which a TBox is used to reformulate posed queries to
offer a better access to the set of data encoded in the ABox [16]. Recently, a lot
of attention was given to *DL-Lite*, a family of lightweight DLs specifically fitted
towards OBDA [9]. *DL-Lite* is especially dedicated for applications that use huge
volumes of data, in which query answering is the most important reasoning task.
DL-Lite offers a very low computational complexity for the reasoning process.
In particular query answering is in *LogSpace* for spatial complexity (w.r.t. the
overall size of the ABox). Moreover knowledge base consistency test and all DLs
standard reasoning services are polynomial for combined complexity (w.r.t. the
overall size of the knowledge base) [1].

U. Straccia and A. Calì (Eds.): SUM 2014, LNAI 8720, pp. 8–21, 2014.
© Springer International Publishing Switzerland 2014

In many real OBDA applications, assertional facts are often provided by several and potentially conflicting sources of information having different reliability levels. Moreover, a given source may provide its set of data with different confidence levels. Possibilistic theory (*e.g.* [11]) offers a very natural framework to deal with ordinal and qualitative uncertain beliefs or prioritized preferences. This framework allows to deal with non-probabilistic information and is particularly appropriate when the uncertainty (or priority) scale only reflects a priority relation between different pieces of information. An important problem that arises in such a situation is how to aggregate these different sets of data. This problem is closely related to the *belief merging problem* (*e.g.* [8,14]), largely studied when knowledge bases are encoded in propositional logic framework. Belief merging focuses on aggregating pieces of information issued from distinct, and possibly conflicting or inconsistent, sources of information. This process produces a global point of view over considered problems by taking advantage of pieces of information provided by each source. Within the possibility theory framework, several merging operators (*e.g.* [10,6,7]) have been proposed for merging pieces of information. These merging operators lead to combine multiple possibility distributions, that encode information provided by different sources, to obtain a unique possibility distribution which represents the global point of view from available information. Syntactic counterparts have been introduced for most of them.

Recently, a possibilistic extension of *DL-Lite*, denoted *DL-Lite$^\pi$*, was proposed in [3]. In particular, *DL-Lite$^\pi$* guarantees a computational complexity that is identical to the one of standard *DL-Lite*. In this paper, we use *DL-Lite$^\pi$* to encode and reason with available knowledge. Merging possibilistic DLs knowledge bases has been recognized as an important issue [17]. Recently, in [4], a *min*-based merging operator dedicated to possibilistic *DL-Lite* knowledge bases was proposed as an adaptation of the well-known idempotent conjunctive operator lastly introduced within possibilistic logics setting. This latter, suitable when sources are assumed to be dependent, is very cautious in the sense where it leads to ignore too many axioms in order to ensure the consistency of the resulting knowledge base.

In this paper, we go one step further in the definition of merging operators for *DL-Lite$^\pi$* knowledge bases by investigating the aggregation of assertional bases (ABox) which are linked to the same terminological base. The rest of this paper is organized as follows. Section 2 gives brief preliminaries on *DL-Lite$^\pi$* as extension of *DL-Lite* within possibility theory setting. In Section 3, we first introduce a syntactic merging operator, namely a *min*-based assertional operator based on conflict resolution. We show that such a merging operator gives a more satisfactory result compared with the one proposed in [4]. We then study, in Section 4, merging at a semantic level, and we show that our operator has a natural counterpart when combining several possibility distributions. We also rephrase within *DL-Lite* framework the set of postulates proposed in [15] to characterize the logical behavior of belief bases merging operators. Thus, we provide a postulates-based logical analysis of the *min*-based assertional operator in the light of this new set of postulates dedicated to the uncertain *DL-Lite* framework. Section 5 concludes the paper. Two important results of this study are: (i) our merging approach based on conflict resolution can be easily extended to define others merging operators and (ii) the computational complexity of *min*-based assertional fusion outcome is polynomial.

2   Possibilistic DL-Lite

In this section, we recall main notions of possibilistic *DL-Lite* framework [3], denoted by $DL\text{-}Lite^\pi$, as an extension of *DL-Lite* within a possibility theory setting. $DL\text{-}Lite^\pi$ provides an excellent mechanism to deal with uncertainty and to ensure reasoning under inconsistency while keeping a computational complexity identical to the one used in standard *DL-Lite*.

2.1   A Brief Refresh on *DL-Lite*

For the sake of simplicity, we only present $DL\text{-}Lite_{core}$ the core fragment of all the *DL-Lite* family [9]. However, results of this paper are valid for $DL\text{-}Lite_R$ and $DL\text{-}Lite_F$, the two main fragments of the *DL-Lite* family.

A *DL-Lite* knowledge base $\mathcal{K}=\langle \mathcal{T}, \mathcal{A} \rangle$ is composed of a set of atomic concepts (*i.e.* unary predicates), a set of atomic roles (*i.e.* binary predicates) and a set of individuals (*i.e.* constants). Complex concepts and roles are built as follows:

$$B \longrightarrow A|\exists R \quad R \longrightarrow P|P^- \quad C \longrightarrow B|\neg B$$

where A (*resp.* P) is an atomic concept (*resp.* role). B (*resp.* C) is called basic (*resp.* complex) concept and role R is called basic role. The TBox $\mathcal{T}$ includes a finite set of inclusion assertions of the form $B \sqsubseteq C$ where B and C are concepts. The ABox $\mathcal{A}$ contains a finite set of assertions on atomic concepts and roles of the form $A(a)$ and $P(a,b)$ where a and b are two individuals.

The semantics of *DL-Lite* is given by an interpretation $I=(\Delta, .^I)$ which consists of an infinite and non-empty domain, denoted Δ, and an interpretation function, denoted $.^I$. The function $.^I$ associates with each individual a an element a^I of Δ^I, to each concept C a subset C^I of Δ^I and to each role R a binary relation R^I over Δ^I. Furthermore, the interpretation function $.^I$ is extended in a straightforward way for complex concepts and roles as follows: $(\neg B)^I=\Delta^I \backslash B^I$, $(P^-)^I=\{(y,x) \in \Delta^I \times \Delta^I |(x,y) \in P^I\}$ and $(\exists R)^I=\{x \in \Delta^I|\exists y \in \Delta^I \ such\ that\ (x,y) \in R^I\}$.

An interpretation I is said to be a model of an inclusion assertion $B \sqsubseteq C$, denoted by $I \vDash B \sqsubseteq C$, iff $B^I \subseteq C^I$. Similarly, we say that an interpretation I is a model of a membership assertion $A(a)$ (*resp.* $P(a,b)$), denoted by $I \vDash A(a)$ (*resp.* $I \vDash P(a,b)$), iff $a^I \in A^I$ (*resp.* $(a^I,b^I) \in P^I$). I is a model of $\mathcal{K}=\langle \mathcal{T}, \mathcal{A} \rangle$, denoted by $I \vDash \mathcal{K}$, iff $I \vDash \mathcal{T}$ and $I \vDash \mathcal{A}$ where $I \vDash \mathcal{T}$ (*resp.* $I \vDash \mathcal{A}$) means that I is a model of all axioms in $\mathcal{T}$ (*resp.* $\mathcal{A}$). A knowledge base $\mathcal{K}$ is said to be consistent if it admits at least one model, otherwise $\mathcal{K}$ is said to be inconsistent. A *DL-Lite* TBox $\mathcal{T}$ is said to be incoherent if there exists at least a concept C such that for each interpretation I which is a model of $\mathcal{T}$, we have $C^I=\emptyset$. Note that within a *DL-Lite* setting, the inconsistency problem is always defined with respect to some ABox since a TBox may be incoherent but never inconsistent.

2.2   Possibility Distribution over *DL-Lite* Interpretation

Let Ω be a universe of discourse composed by a set of *DL-Lite* interpretations $(I=(\Delta, .^I) \in \Omega)$. The semantic counterpart of a *DL-Lite*$^\pi$ knowledge base is given by a possibility distribution, denoted by π, which is a mapping from Ω to the unit interval $[0, 1]$ that assigns to each interpretation $I \in \Omega$ a possibility degree $\pi(I) \in [0, 1]$ that represents its compatibility or consistency with respect to the set of available knowledge. When $\pi(I)=0$, we say that I is impossible and it is fully inconsistent with the set of available knowledge, whereas when $\pi(I)=1$, we say that I is totally possible and it is fully consistent with the set of available knowledge. For two interpretations I and I', when $\pi(I) > \pi(I')$ we say that I is more consistent or more preferred than I' *w.r.t* available knowledge. Lastly, π is said to be normalized if there exists at least one totally possible interpretation, namely $\exists I \in \Omega, \pi(I)=1$, otherwise, we say that π is sub-normalized. The concept of sub-normalization reflects the presence of conflicts in the set of available information.

Given a possibility distribution π defined on a set of interpretations Ω, one can define two measures on a *DL-Lite* axiom φ: A possibility measure $\Pi(\varphi)=\max\limits_{I\in\Omega}$ $\{\pi(I) : I \vDash \varphi\}$ that evaluates to what extent an axiom φ is compatible with the available knowledge encoded by π and a necessity measure $N(\varphi)=1 - \max\limits_{I\in\Omega}\{\pi(I) :$ $I \nvDash \varphi\}$ that evaluates to what extent φ is certainty entailed from available knowledge encoded by π.

2.3   *DL-Lite*$^\pi$ Knowledge Base

Let $\mathcal{L}$ be a *DL-Lite* description language, a *DL-Lite*$^\pi$ knowledge base is a set of possibilistic axioms of the form (φ, α) where φ is an axiom expressed in $\mathcal{L}$ and $\alpha \in \,]0, 1]$ is the degree of certainty of φ. Namely, a *DL-Lite*$^\pi$ knowledge base $\mathcal{K}$ is such that $\mathcal{K}=\{(\varphi_i, \alpha_i) : i = 1, ..., n\}$. Only somewhat certain information are explicitly represented in a *DL-Lite*$^\pi$ knowledge base. Namely, axioms with a null degree ($\alpha = 0$) are not explicitly represented in the knowledge base. The weighted axiom (φ, α) means that the certainty degree of φ is at least equal to α (namely $N(\varphi) \geq \alpha$). A *DL-Lite*$^\pi$ knowledge base $\mathcal{K}$ will also be represented by a couple $\mathcal{K}=\langle \mathcal{T}, \mathcal{A} \rangle$ where both elements in $\mathcal{T}$ and $\mathcal{A}$ may be uncertain. It is important to note that, if we consider all $\alpha_i = 1$ then we found a classical *DL-Lite* knowledge base: $\mathcal{K}^*=\{\varphi_i : (\varphi_i, \alpha_i) \in \mathcal{K}\}$.

Given $\mathcal{K}=\langle \mathcal{T}, \mathcal{A} \rangle$ a *DL-Lite*$^\pi$ knowledge base, we define the α-cut of $\mathcal{K}$ (*resp.* $\mathcal{T}$ and $\mathcal{A}$), denoted by $\mathcal{K}_{\geq\alpha}$ (*resp.* $\mathcal{T}_{\geq\alpha}$ and $\mathcal{A}_{\geq\alpha}$), the subbase of $\mathcal{K}$ (*resp.* $\mathcal{T}$ and $\mathcal{A}$) composed of axioms having weights at least greater than α. We say that $\mathcal{K}$ is consistent if the standard knowledge base obtained from $\mathcal{K}$ by ignoring the weights associated with axioms is consistent. In case of inconsistency, we attach to $\mathcal{K}$ an inconsistency degree. The inconsistency degree of a *DL-Lite*$^\pi$ knowledge base $\mathcal{K}$, denoted by $Inc(\mathcal{K})$, is syntactically defined as follow: $Inc(\mathcal{K})=max\{\alpha:\mathcal{K}_{\geq\alpha}$ is inconsistent$\}$.

Given a *DL-Lite*$^\pi$ knowledge base $\mathcal{K}$, one can associate to $\mathcal{K}$ a joint possibility distribution, denoted by $\pi_\mathcal{K}$, defined over the set of all interpretations $I=(\Delta, .^I)$ by associating to each interpretation its level of consistency with the set of available knowledge, that is, with $\mathcal{K}$. Namely:

Definition 1. *The possibility distribution induced from a DL-Lite$^\pi$ is defined as follows:* $\forall I \in \Omega : \pi_\mathcal{K}(I) = \begin{cases} 1 & \text{if } \forall \ (\varphi_i, \alpha_i) \in \mathcal{K}, I \models \varphi_i \\ 1 - max\{\alpha_i : (\varphi_i, \alpha_i) \in \mathcal{K}I \nvDash \varphi_i\} & \text{otherwise} \end{cases}$

A *DL-Lite$^\pi$* knowledge base $\mathcal{K}$ is said to be consistent if its joint possibility distribution $\pi_\mathcal{K}$ is normalized. If not, $\mathcal{K}$ is said to be inconsistent and its inconsistency degree is defined semantically as follow: $Inc(\mathcal{K})=1 - \max_{I \in \Omega}\{\pi_\mathcal{K}(I)\}$.

It was shown in [3] that computing the inconsistency degree of a *DL-Lite$^\pi$* knowledge base comes from the extension of the algorithm presented in [9] by modifying it to query for individuals with a given certainty degree.

Example 1. Let $\mathcal{K}=\langle\mathcal{T}, \mathcal{A}\rangle$ be a *DL-Lite$^\pi$* knowledge base where $\mathcal{T}=\{(A \sqsubseteq B, 1), (B \sqsubseteq \neg C, .9)\}$ and $\mathcal{A}=\{(A(a), .6), (C(b).5)\}$. The possibility distribution $\pi_\mathcal{K}$ associated to $\mathcal{K}$ is computed using Definition 1 as follows where $\Delta=\{a, b\}$:

Table 1. Example of a possibility distribution induced from a *DL-Lite$^\pi$* KB

I	$\cdot^I$	$\pi_\mathcal{K}$
I_1	A={a},B={},C={b}	0
I_2	A={a},B={a},C={b}	1
I_3	A={},B={},C={a,b}	.4
I_4	A={a,b},B={a,b},C={}	.5

One can observe that $\pi_\mathcal{K}(I_2)=1$ meaning that $\pi_\mathcal{K}$ is normalized, and thus, $\mathcal{K}$ is consistent.

3 Syntactic Merging of *DL-Lite$^\pi$* Assertional Bases

Let us consider $\mathcal{A}_1,...,\mathcal{A}_n$ a set of assertional bases (ABox) where each $\mathcal{A}_i$ represents assertional facts provided by a single source of information. We assume that we have a well-formed and coherent terminological base (TBox) $\mathcal{T}$ where each $\mathcal{A}_i$ is consistent with $\mathcal{T}$. This is not a restriction. This particular case can be handled outside the fusion problem considered in this paper. Note that this choice is motivated by the fact that such situation is widely occurring in Ontology-Based Data Access. Throughout the rest of this paper, we cast available information within the *DL-Lite$^\pi$* framework. For the sake of simplicity, we omit the weights notation attached to the TBox axioms considered as the ones having the highest certainty level, namely, an axiom in $\mathcal{T}$ is of the form $(\varphi, 1)$. We only represent explicitly weights attached to $\mathcal{A}_i$ assertions. An assertion f in $\mathcal{A}_i$ is of the form $f=(\varphi, \alpha)$ where $\alpha \in [0, 1]$. Note that copies of the same assertions φ are allowed in several $\mathcal{A}_i$ and they are considered as different in the sense of priorities or certainty and not in terms of interpretations since we use the *unique name assumption*. In this section, we study syntactic merging of n assertional bases $\mathcal{A}_1,...,\mathcal{A}_n$ that are linked to the same TBox $\mathcal{T}$.

Let us consider $S_1, ..., S_n$ be the signatures of $\mathcal{A}_1,...,\mathcal{A}_n$ and $\mathcal{T}$. Recall that a signature S of a knowledge base $\mathcal{K}$ is the set of concept names and role names used in $\mathcal{K}$. We assume that all $\mathcal{A}_i$'s and $\mathcal{T}$ share the same signature. Namely if a concept name (*resp.* role name) A appears in S_1 and S_2 then A is assumed to be the same. We look to identify a syntactical merging operator on the $\mathcal{A}_i$'s *w.r.t* a TBox $\mathcal{T}$ which will be semantically meaningful. Merging at semantic level will be presented in Section 4.

3.1   Merging using the Classical *min*-based Operator

In this section, we perform merging of $\mathcal{A}_1,...,\mathcal{A}_n$ a set of ABox *w.r.t* a TBox $\mathcal{T}$ using the classical *min*-based merging operator proposed in [4] to aggregate *DL-Lite*$^\pi$ knowledge bases. This operator is a direct extension of the well-known idempotent conjunctive operator (*e.g.* [6]) within possibilistic *DL-Lite* setting. It is recommended when distinct sources that provide information are assumed to be dependent.

Let $\mathcal{T}$ be a TBox and $\mathcal{A}_1,...\mathcal{A}_n$ be a set of ABox provided by n distinct sources of information to be linked to $\mathcal{T}$. The *min*-based merging operator, denoted by $\oplus$ considers the union of all ABox. Namely:

$$\mathcal{A}_\oplus=\mathcal{A}_1 \cup \mathcal{A}_2 \cup \ldots \cup \mathcal{A}_n.$$

The merging of two consistent knowledge bases is not guaranteed to be consistent. Namely, the resulting knowledge base $\mathcal{K}_\oplus=\langle\mathcal{T},\mathcal{A}_\oplus\rangle$ may be inconsistent. To restore the consistency of the resulting knowledge base a normalization step is required. The following definition gives the formal logical representation of the normalized knowledge base.

Definition 2. *Let $\mathcal{T}$ be a TBox and $\mathcal{A}_\oplus$ be the aggregation of $\mathcal{A}_1,...\mathcal{A}_n$, n ABox using classical min-based operator. Let $x=Inc(\langle\mathcal{T},\mathcal{A}_\oplus\rangle)$. Then, the normalized knowledge base, denoted, $\mathcal{K}_{N\oplus}$ is such that:*

$$\mathcal{K}_{N\oplus}=\langle\mathcal{T},\{(\varphi,\alpha):(\varphi,\alpha)\in\mathcal{A}_\oplus \text{ and } \alpha>x\}\rangle$$

Example 2 (continued). Let us continue with the TBox $\mathcal{T}=\{A \sqsubseteq B,\ B \sqsubseteq \neg C\}$ presented in Example 1 while assuming that the certainty degree of each axioms is set to 1. Let us consider the following set of ABox to be linked to $\mathcal{T}$: $\mathcal{A}_1=\{(A(a),.6),\quad (C(b),.5)\}$, $\mathcal{A}_2=\{(C(a),.4),\quad (B(b),.8),\quad (A(b),.7)\}$ and $\mathcal{A}_3=\{(A(b),.2),(A(c),.5),(B(c),.4)\}$. We have $\mathcal{A}_\oplus=\{(A(a),.6),(C(b),.5),(C(a),.4),(B(b),.8),(A(b),.7),(A(b),.2),(A(c),.5),(B(c),.4)\}$ where $Inc(\langle\mathcal{T},\mathcal{A}_\oplus\rangle)=.5$. Then $\mathcal{K}_{N\oplus}=\mathcal{T}\cup\{(A(a),.6),(B(b),.8),(A(b),.7)\}$.

According to Definition 2, merging operation does not modify the certainty degrees of the *DL-Lite*$^\pi$ knowledge base. It just permits to ignore the presence of contradictions (or conflicts) and maintain all the assertions of $\mathcal{A}_\oplus$ whose certainty degrees are higher than the inconsistency degree of $\langle\mathcal{T},\mathcal{A}_\oplus\rangle$. It is clear that the formal expression of the normalized *DL-Lite*$^\pi$ knowledge base $\mathcal{K}_\oplus$ given in Definition 2 provides a consistent knowledge base. However, this result is not very satisfactory, since many assertions in $\mathcal{A}_1,...,\mathcal{A}_n$, which are not involved in any conflict are thrown out. As pointed in [5], restoring consistency in possibilistic logics suffers generally from an important drawback problem in the sense that some axiom from $\mathcal{A}_\oplus$-$\mathcal{A}_{\oplus>Inc(\mathcal{T}\cup\mathcal{A}_\oplus)}$ that are not involved in any conflict are inhibited as we can see in the above example.

Example 3 (continued). One can see that the assertions $(A(c), .5)$ and $(B(c), .4)$ are not involved in any conflict, but they are nor integrated in the merging result.

In the next section, we investigate a new approach to merge assertional base based on conflict detection. This approach allows recovering of all elements, non involved in any conflict and inhibited when restoring consistency using the classical *min*-based merging operator.

3.2   Min-based Assertional Merging Using Conflict Resolution

Let $\mathcal{K}=\langle\mathcal{T},\mathcal{A}\rangle$ be a *DL-Lite$^\pi$* knowledge base. In [3] it was shown that computing the inconsistency degree of $\mathcal{K}$ comes down to compute the one of $\langle\pi - neg(\mathcal{T}), \mathcal{A}\rangle$ where π-$neg(\mathcal{T})$ is the negated closure of $\mathcal{T}$. The negated closure will contain all the possibilistic negated axioms of the form $(B_1\sqsubseteq\neg B_2, \alpha)$ that can be derived from $\mathcal{T}$. The set π-$neg(\mathcal{T})$ is obtained by applying a set of three rules that extend the ones defined in standard *DL-Lite*. For instance after adding all NI of $\mathcal{T}$ to π-$neg(\mathcal{T})$ a rule said that If $(B_1\sqsubseteq B_2, \alpha_1) \in \mathcal{T}$ and $(B_2\sqsubseteq\neg B_3, \alpha_2)$ in π-$neg(\mathcal{T})$ then add $(B_1\sqsubseteq\neg B_3, min(\alpha_1, \alpha_2))$ to π-$neg(\mathcal{T})$. See [3] for a more detailed description of *DL-Lite$^\pi$*. Indeed, computing inconsistency degree of $\mathcal{K}$ consists on calculating the maximal weight attached to minimal inconsistent subsets involved in inconsistency. More formally, a minimal inconsistent set is defined as follows.

Definition 3. *A minimal inconsistent subset $\mathcal{M}\subseteq\mathcal{K}$ is a subset of $\langle\pi - neg(\mathcal{T}),$ $\mathcal{A}\rangle$ of the form:* $\{(B_1\sqsubseteq\neg B_2, \alpha_1), (B_1(a), \alpha_2), (B_2(a), \alpha_3)\}$ *where* $(B_1\sqsubseteq\neg B_2, \alpha_1)\in$ π-$neg(\mathcal{T})$, $(B_1(a), \alpha_2)\in\mathcal{A}$ *and* $(B_2(a), \alpha_3)\in\mathcal{A}$.

Clearly, a minimal inconsistent subset is a subset of information involving three elements: an axiom of π-$neg(\mathcal{T})$ and two assertions of $\mathcal{A}$ up to a particular case where $B_1=B_2$ belongs to π-$neg(\mathcal{T})$. This corresponds to the situation of insatisfiable concept. Namely, no way to find an individual that belongs to B. In this case $B_1=B_2$ is minimal inconsistent subset composed only of two elements: an axiom of π-$neg(\mathcal{T})$ and an assertions of $\mathcal{A}$. Within a *DL-Lite* setting, the inconsistency problem is always defined with respect to some ABox, since a TBox may be incoherent but never inconsistent. Recall that in this paper, we assume that $\mathcal{T}$ is coherent. So, from the definition of minimal inconsistent subset, we define the notion of conflict as a minimal inconsistent subset of assertions that contradict a negative inclusion axiom. More formally:

Definition 4. *Let $\mathcal{K}=\langle\mathcal{T},\mathcal{A}\rangle$ be an inconsistent DL-Lite$^\pi$ knowledge base where axioms in $\mathcal{T}$ are set to 1. A sub-base $\mathcal{C}\subseteq\mathcal{A}$ is said to be an assertional conflict set of $\mathcal{K}$ iff*

- *$Inc(\langle\mathcal{T},\mathcal{C}\rangle) > 0$ and*
- *$\forall\ f\in\mathcal{C},\ Inc(\langle\mathcal{T},\mathcal{C} - \{f\}\rangle)=0$ with $f=(\varphi, \alpha)$*

It is clear that in Definition 4, removing any assertion φ from $\mathcal{C}$ restores the consistency of $\langle\mathcal{T},\mathcal{C}\rangle$. Recall that when the TBox is coherent, a conflict involves exactly two assertions.

Example 4 (Example continued). Let us consider $\mathcal{T}$ and $\mathcal{A}_\oplus$ from the above example. The $\pi\text{-}neg(\mathcal{T})=\{A\sqsubseteq\neg C,\ B\sqsubseteq\neg C\}$. One can compute the following conflict sets: $\mathcal{C}_1=\{(A(a),.6),(C(a),.4)\}$, $\mathcal{C}_2=\{(C(b),.5),(B(b),.8)\}$, $\mathcal{C}_3=\{(C(b),.5),(A(b),.7)\}$ and $\mathcal{C}_4=\{(C(b),.5),(A(b),.2)\}$.

Let us assume that $\mathcal{A}_1,...,\mathcal{A}_n$ are assertional bases provided by n sources of information to be linked to the same TBox $\mathcal{T}$ and they use the same scale to represent uncertainty. Let denote by $f=(\varphi,\alpha)$ an assertion or a fact in $\mathcal{A}_i$, we define the notion of conflict vector as follows:

Definition 5. *Let $\mathcal{T}$ be a TBox and $\mathcal{A}_1,...,\mathcal{A}_n$ be a set of ABox provided by n distinct sources of information to be linked to $\mathcal{T}$. Then $\forall f\in\mathcal{A}_i$ we define a conflict vector associated with $f=(\varphi,\alpha)\in\mathcal{A}_i:\mathcal{V}(f)=\langle\nu_1,\nu_2,...,\nu_n\rangle$ such that*

$$\forall j=1..n: \mathcal{V}_j(f) = \begin{cases} 1\ if\ \langle\mathcal{T},\{(\varphi,1)\cup\mathcal{A}_i\}\rangle\ is\ consistent \\ Inc(\langle\mathcal{T},\{(\varphi,1)\cup\mathcal{A}_i\}\rangle)\ otherwise \end{cases}$$

Where $\mathcal{V}_i$ represents the ith component of the vector $\mathcal{V}$.

Intuitively, for each assertion provided by an information source we built upon a vector that represents to what extend this latter contradicts other ones provided by other source. To this end, we add first the assertion with a highest prescribed level in each source and then we compute the inconsistency degree of this one. It is obvious that the conflict vector of a non conflicting assertion is equal to $\mathcal{V}(f)=\langle1,1,...,1\rangle$. However assertions that are involved in conflict will have at least a ν_i strictly less than 1.

Example 5 (continued). One can obtain the following conflict vectors: $\mathcal{V}((A(a),.6))=\langle1,.6,1\rangle,\mathcal{V}((A(b),.7))=\langle.5,1,1\rangle,\mathcal{V}((A(b),.2))=\langle.5,1,1\rangle,\mathcal{V}((A(c),.5))=\langle1,1,1\rangle,\mathcal{V}((B(b),.8))=\langle.5,1,1\rangle,\ \mathcal{V}((B(c),.4))=\langle1,1,1\rangle,\mathcal{V}((C(a),.4))=\langle.4,1,1\rangle$ and $\nu((C(b),.5))=\langle1,.2,.8\rangle$

From now on, we give the way to aggregate assertional bases using conflict vectors attached to each assertion. Let denote by Σ the set of conflict vectors, we define the *min*-based assertional merging operators, denoted by Λ as follows:

Definition 6. *Let $\mathcal{T}$ be a TBox and $\mathcal{A}_1,\mathcal{A}_2,...,\mathcal{A}_n$ be a set of ABox provided by n sources to be linked to $\mathcal{T}$. Let Σ be the collection of conflict vectors associated to each assertion on $\mathcal{A}_i$. Then the min-based assertional merging operator, denoted by Λ, is defined on Σ as follows:*

$$\forall\mathcal{V}(f)\in\Sigma:\ \Lambda(f)=min\{\nu_i(f)\}$$

Let us denote by Σ_Λ, the vector resulting by *min* aggregation of conflict vectors.

Example 6 (Example continued). Σ_Λ contains the following elements: $\Lambda((A(a),.6))=.6,\ \Lambda((A(b),.7))=.5,\ \Lambda((A(b),.2))=.5,\ \Lambda((A(c),.5))=1,\ \Lambda((B(b),.8))=.5,\ \Lambda((B(c),.4))=1,\Lambda((C(a),.4))=.4$ and $\Lambda((C(b),.5))=.2$

According to conflict vectors, one can associate to the set of assertions a new pre-order by attaching to each of them a new weight (*i.e.* $\forall (\varphi, \alpha) \in \mathcal{A}_i : (\varphi, \alpha) = (\varphi, \Lambda(f))$). According to this new pre-order, we define the knowledge base resulting from fusion operation as follows.

Definition 7. *Let $\mathcal{T}$ be a TBox and $\mathcal{A}_1, ..., \mathcal{A}_n$ be a set of n ABox to be linked to $\mathcal{T}$. Let $\mathcal{A}_\Lambda = \{(\varphi, \Lambda(f)) : f = (\varphi, \alpha) \in \mathcal{A}_i$ and $\Lambda(f) \in \Sigma_\Lambda\}$. Let $x = Inc(\langle \mathcal{T}, \mathcal{A}_\Lambda \rangle)$. Then the resulting knowledge base $\mathcal{K}_\Lambda$ is such that:*

$$\mathcal{K}_\Lambda = \langle \mathcal{T}, \{(\varphi, \alpha) : (\varphi, \alpha) \in \mathcal{A}_\Lambda \text{ and } \alpha > x\} \rangle$$

Example 7 (continued). One can obtain $\mathcal{A}_\Lambda = \{(A(a), .6), (A(b), .5), (A(b), .5), (A(c), 1), (B(b), .5), (B(c), 1), (C(a), .4), (C(b), .2)\}$ where $Inc(\langle \mathcal{T}, \mathcal{A}_\Lambda \rangle = .4$. Then $\mathcal{K}_\Lambda = \mathcal{T} \cup \{(A(a), .6), (A(b), .5), (A(b), .5), (A(c), 1), (B(b), .5), (B(c), 1)\}$.

According to Definition 7, it is clear that method based on conflict vectors is more productive that the classical definition of the *min*-based merging operator proposed in Definition 2. Note that this approach can easily propose others aggregation modes such as product-based merging or sum-based merging. The definition of this merging operator is based on a notion of conflict measure between sources of information. However, one can observe that original weights attached to assertions are lost. Regarding for instance assertion $B(c)$, it is provided by only one where its initial weight was .4. This means that $B(c)$ is not a totally reliable information. In the new knowledge base its weight is raised to 1. This can be justified by the fact that such assertion is not involved in any conflict. However when we need to an iteration process this approach may not be very useful. To overcome such limitation while preserving the same productivity of the fusion result, we propose the following definition.

Definition 8. *Let $\mathcal{T}$ be a TBox and $\mathcal{A}_1, ..., \mathcal{A}_n$ be a set of n ABox to be linked to $\mathcal{T}$. Let $\mathcal{A}_\Lambda = \{(\varphi, \Lambda(f)) : (\varphi, \alpha) \in \mathcal{A}_i\}$. Let $x = Inc(\langle \mathcal{T}, \mathcal{A}_\Lambda \rangle)$. Then the resulting knowledge base $\mathcal{K}'_\Lambda$ is such that:*

$$\mathcal{K}'_\Lambda = \langle \mathcal{T}, \{f = (\varphi, \alpha) \in \mathcal{A}_i : i \in \{1, ..., n\}, (\varphi, \Lambda(f)) \in \mathcal{A}_\Lambda \text{ and } \Lambda(f) > x\} \rangle$$

4   Semantic Counterpart

Let us consider $\mathcal{A}_1, ..., \mathcal{A}_n$ a set of assertional bases (ABox) where each $\mathcal{A}_i$ represents data of a single source of information. We assume that we have a well-formed and coherent terminological base (TBox) $\mathcal{T}$ where each $\mathcal{A}_i$ is consistent with the $\mathcal{T}$. Let $\pi_1, ..., \pi_n$ be the set of possibility distributions associated with $\mathcal{K}_1, ..., \mathcal{K}_n$ where each $\mathcal{K}_i = \langle \mathcal{T}, \mathcal{A}_i \rangle$. Namely each $DL\text{-}Lite^\pi$ knowledge base $\mathcal{K}_i$ is associated with a possibility distribution π_i which is its semantic counterpart. In this section, we investigate fusion of weighted $DL\text{-}Lite^\pi$ assertional bases at semantic level. We show that such merging operation is the natural semantic counterpart of the Λ merging operators (presented in Section 3) used to merge $DL\text{-}Lite^\pi$ ABox $\mathcal{A}_1, ..., \mathcal{A}_n$ w.r.t a $\mathcal{T}$. More formally, given $(\pi_1, ..., \pi_n)$ possibility distributions associated with $(\mathcal{K}_1, ..., \mathcal{K}_n)$ $DL\text{-}Lite^\pi$ knowledge bases, then for the proposed operator Λ applied to aggregate $\mathcal{A}_1, ..., \mathcal{A}_n$ w.r.t $\mathcal{T}$, we look for a $DL\text{-}Lite^\pi$ possibility distribution π_Λ constructed from the aggregation of $(\pi_1, ..., \pi_n)$ with the semantic counterpart of Λ that corresponds to the possibility distribution $\pi_{\mathcal{K}_\Lambda}$ induced from $\mathcal{K}_\Lambda$. Namely $\pi_\Lambda = \pi_{\mathcal{K}_\Lambda}$.

4.1   Min-based Assertional Merging of Possibility Distributions

Let us assume that $\pi_1,...,\pi_n$ share the same domain of interpretations (namely $\Delta_1=...=\Delta_n$), and that all possibility distributions use the same scale to represents uncertainty. In [4], the semantic counterpart of the classical min-based operator or idempotent conjunctive operator, denoted by $\oplus$, was defined as a mapping from a vectors of possibility values $(\forall I \in \Omega, \nu(I) = \langle \pi_1(I), ..., \pi_n(I) \rangle)$ to an interval $[0,1]$ as follows: $\pi_\oplus(I) = min\{\nu_i(I)\}$. Generally merging two normalized possibility distributions gives an sub-normalized one. Normalizing $\pi_\oplus$ consists in maintaining only axioms having certainty degrees higher than the inconsistency degree deduced from $\pi_\oplus$. In this section, we deal with assertional bases merging at semantic level. We propose the natural semantic counterpart of the min-based assertional merging operator, denote Λ, presented in Section 3.2 which is based on conflict resolution. The following definition introduces the semantic definition of conflict vectors.

Definition 9. *Let $\mathcal{A}_1,...,\mathcal{A}_n$ be a set of ABox and $\pi_1,...,\pi_n$ be a the set of possibility distributions induced from $\mathcal{K}_1,...,\mathcal{K}_n$ where each $\mathcal{K}_i = \langle \mathcal{T}, \mathcal{A}_i \rangle$. Then $\forall f \in \mathcal{A}_i$ with $f = (\varphi, \alpha)$, we define semantically a conflict vector, denoted by $\mathcal{V}(f)$, as follows:*

$$\mathcal{V}(f) = \langle \Pi_{\pi_1}(\varphi), \Pi_{\pi_2}(\varphi), ..., \Pi_{\pi_n}(\varphi) \rangle$$

where $\forall i = 1..n : \Pi_{\pi_i}(f)$ denotes the possibility measure of φ induced from the possibility distribution π_i

Intuitively, a conflict vector associated to any ABox assertion represents to what extent this latter is compatible with available knowledge provided by each source.

Example 8 (continued). Assuming that $\Delta = \{a, b, c\}$, let us consider the following possibility distributions π_1, π_2 and π_3 to be merged. Note that we have only considered interpretations model of $\mathcal{T}$.

Table 2. Possibility distributions induced from three knowledge bases

I		π_1	π_2	π_3
I_1	A={a},B={a},C={b,c}	1	.2	.5
I_2	A={b},B={b},C={a,c}	.4	1	.5
I_3	A={c},B={c},C={a,b}	.4	.2	.8
I_4	A={a,b},B={a,b},C={c}	.5	.6	.5
I_5	A={a,c},B={a,c},C={b}	1	.2	.8
I_6	A={b,c},B={b,c},C={a}	.4	1	1
I_7	A={a,b,c},B={a,b,c},C={}	.5	.6	1
I_8	A={},B={},C={a,b,c}	.4	.2	.5

One can compute the following conflict vectors for each assertion:
$\mathcal{V}(A(a)) = \langle max(1,.5,1,1), max(.2,.6,.2,.6), max(.5,.6,.8,1) \rangle = \langle 1,.6,1 \rangle$,
$\mathcal{V}(A(b)) = \langle max(.4,.5,.5,5), max(1,.6,.1,.6), max(.5,.5,1,1) \rangle = \langle .5,1,1 \rangle$,
$\mathcal{V}(A(c)) = \langle max(.4,1,.4,.5), max(.2,.2,1,.6), max(.8,.8,1,1) \rangle = \langle 1,1,1 \rangle$,
$\mathcal{V}(B(b)) = \langle .5,1,1 \rangle$, $\mathcal{V}(B(c)) = \langle 1,1,1 \rangle$, $\mathcal{V}(C(a)) = \langle .4,1,1 \rangle$ and $\mathcal{V}(C(b)) = \langle 1,.2,.8 \rangle$ which are equal the ones computed syntactically in Example 5.

Let us denote by Σ the collection of conflict vectors associated to each assertion of $\mathcal{A}_i$. The next definition introduces min-based assertional merging operator, denoted Λ, on the conflict vectors of Σ.

Definition 10. *Let $\mathcal{A}_1,...,\mathcal{A}_n$ be a set of ABox and $\pi_1,...,\pi_n$ be a the set of possibility distributions induced from $\mathcal{K}_1,...,\mathcal{K}_n$ where each $\mathcal{K}_i=\langle \mathcal{T},\mathcal{A}_i\rangle$. Let Σ the collection of conflict vectors associated to each assertion on $\mathcal{A}_i$ computed using Definition 9. Then the min-based assertional merging operator, denoted by Λ, is defined on Σ as follows: $\forall \mathcal{V}(f)\in\Sigma:\mathcal{V}(f)=\langle \Pi_{\pi_1}(\varphi), \Pi_{\pi_2}(\varphi), ..., \Pi_{\pi_n}(\varphi)\rangle$,*

$$\Lambda(f)= min\{\nu_i(f) \in \mathcal{V}(f)\}$$

Let us denote by Σ_Λ, the vector resulting by min-based aggregation of conflict vectors.

Example 9 (Example continued). One can compute the set Σ_Λ as follow: $\Lambda((A(a), .6))=.6$, $\Lambda((A(b),.7))=.5$, $\Lambda((A(b),.2))=.5$, $\Lambda((A(c),.5))=1$, $\Lambda((B(b),.8))=.5$, $\Lambda((B(c), .4))=1$, $\Lambda((C(a),.4)) =.4$ and $\Lambda((C(b),.5))=.2$

From Definition 10, one can associate to each assertion a new weight that represents its compatibility with others assertions provided other sources.

Definition 11. *Let $\mathcal{A}_1,...,\mathcal{A}_n$ be a set of ABox and $\pi_1,...,\pi_n$ be a the set of possibility distributions induced from $\mathcal{K}_1,...,\mathcal{K}_n$ where each $\mathcal{K}_i=\langle \mathcal{T},\mathcal{A}_i\rangle$. Then the possibility distribution π_Λ as follows:*

$$\forall I \in \Omega : \pi_\Lambda(I) = \begin{cases} 1 \ if \forall \ (\varphi,\alpha) \in \mathcal{A}_i, I \models \varphi \\ 1 - max\{\Lambda((\varphi,\alpha)) : (\varphi,\alpha) \in \mathcal{A}_i, and \ I \nvDash \varphi\} \ otherwise \end{cases}$$

where $\Lambda(\varphi_i)$ is the compatibility measure of φ_i computed using definition 10

Example 10. From Example 3, we have $(A(c),.1)$, $(B(c),1)$, $(A(a),.6)$, $(A(b),.5)$, $(B(b),.5)$, $(C(a),.4)$, $(C(a),.2)$. Then:

Table 3. Possibility distributions resulting from assertional min-based merging of possibility distributions

I	I_1	I_2	I_3	I_4	I_5	I_6	I_7	I_8
π_Λ	0	0	.4	0	.5	.4	.6	0

One can check that merging normalized possibility distributions may lead to sub-normalized possibility distribution. This is the case with our example. Indeed, we focus on the normalization problem when the use of min-based assertional operators min provides a subnormal possibility distribution.

Definition 12. *Let us consider: $h(\pi_\Lambda)=\max\limits_{I\in\Omega}\{\pi_\Lambda(I)\}$. Then for every $I\in\Omega$ and*

$$h(\pi_\Lambda)>0, \ \pi_{N\Lambda}(I) = \begin{cases} 1 \ if \ \pi_\Lambda(I) = h(\pi_\Lambda) \\ \pi_\Lambda(I) \ otherwise \end{cases}$$

Example 11 (continued). From previous Example, we have:

Table 4. Normalized possibility distributions resulting from assertional *min*-based merging

I	I_1	I_2	I_3	I_4	I_5	I_6	I_7	I_8
π_Λ	0	0	.4	0	.5	.4	.6	0
π_Λ	0	0	.4	0	.5	.4	1	0

The following proposition states the equivalence between the semantic and syntactic approaches.

Proposition 1. *Let $\mathcal{A}_1,...,\mathcal{A}_n$ be a set of ABox and $\pi_1,...,\pi_n$ be a the set of possibility distributions induced from $\mathcal{K}_1,...,\mathcal{K}_n$ where each $\mathcal{K}_i = \langle \mathcal{T}, \mathcal{A}_i \rangle$. Then the possibility distribution*

$$\pi_{N\Lambda}(I) = \begin{cases} 1 \ if \ \pi_\Lambda(I) = h(\pi_\Lambda) \\ \pi_\Lambda(I) \ otherwise \end{cases}$$

is associated with

$$\mathcal{K}_\Lambda = \langle \mathcal{T}, \{(\varphi, \Lambda(f)) : (\varphi, \Lambda(f)) \in \mathcal{A}_\Lambda \ and \ \Lambda(f) > x \} \rangle$$

4.2 Logical Properties

Let us use $E = \{\mathcal{K}_1, ..., \mathcal{K}_n\}$ to denote a multi-set, called belief profile, that represents the knowledge bases to be merged (where each $\mathcal{K}_i$ is associated with a possibility distribution π_i). Let us use $\triangle$ to denote a merging operator. This merging operator can be parametrized by an integrity constraint, being a konwledge base $\mathcal{K}$, and $\triangle_\mathcal{K}(E)$ denotes the result of the merging operator under this constraint $\mathcal{K}$. A logical characterization of integrity constraint merging operators has been proposed in [14] through a set of rational postulates extended from the ones proposed for belief revision [12]. The following postulates rephrase the ones proposed in [14] within *DL-Lite* framework.

$(\mathbf{M_0^\pi})$ $\triangle_\mathcal{K}(E) \models \mathcal{K}$

$(\mathbf{M_1^\pi})$ if $\mathcal{K}$ is consistent, then $\triangle_\mathcal{K}(E)$ is consistent

$(\mathbf{M_2^\pi})$ if $\mathcal{K} \cup \bigcup_{\mathcal{K}_i \in E} \mathcal{K}_i$ is consistent, then $\triangle_\mathcal{K}(E) = \mathcal{K} \cup \bigcup_{\mathcal{K}_i \in E} \mathcal{K}_i$

$(\mathbf{M_3^\pi})$ if $E_1 \approx E_2$ and $\mathcal{K}_1 \equiv \mathcal{K}_2$, then $\triangle_{\mathcal{K}_1}(E_1) \equiv \triangle_{\mathcal{K}_2}(E_2)$.

$(\mathbf{M_4^\pi})$ if $\mathcal{K}_1 \models \mathcal{K}$ and $\mathcal{K}_2 \models \mathcal{K}$, then $\triangle_\mathcal{K}(\mathcal{K}_1 \cup \mathcal{K}_2)$ is consistent implies that $\triangle_\mathcal{K}(\mathcal{K}_1 \cup \mathcal{K}_2) \cup \mathcal{K}_2$ is consistent

$(\mathbf{M_5^\pi})$ $\triangle_\mathcal{K}(E_1) \cup \triangle_\mathcal{K}(E_2) \models \triangle_\mathcal{K}(E_1 \uplus E_2)$

$(\mathbf{M_6^\pi})$ if $\triangle_\mathcal{K}(E_1) \cup \triangle_\mathcal{K}(E_2)$ is consistent, then $\triangle_\mathcal{K}(E_1 \uplus E_2) \models \triangle_\mathcal{K}(E_1) \cup \triangle_\mathcal{K}(E_2)$

$(\mathbf{M_7^\pi})$ $\triangle_\mathcal{K}(E) \cup \mathcal{K}' \models \triangle_{\mathcal{K} \cup \mathcal{K}'}(E)$

$(\mathbf{M_8^\pi})$ if $\triangle_\mathcal{K}(E) \cup \mathcal{K}'$ is consistent, then $\triangle_{\mathcal{K} \cup \mathcal{K}'}(E) \models \triangle_\mathcal{K}(E) \cup \mathcal{K}'$

$(\mathbf{M_{maj}^\pi})$ $\exists n \ \triangle_\mathcal{K}(E_1 \uplus E_2^n) \models \triangle_\mathcal{K}(E_2)$

$(\mathbf{M_I^\pi})$ $\forall n \ \triangle_\mathcal{K}(E_1 \uplus E_2^n) \equiv \triangle_\mathcal{K}(E_1 \uplus E_2)$

With:

1. $\mathcal{K}_1 \models \mathcal{K}_2$ iff $\arg\max_I \pi_{\mathcal{K}_1}(I) \subseteq \arg\max_I \pi_{\mathcal{K}_2}(I)$
2. $\mathcal{K}_1 \equiv \mathcal{K}_2$ iff $\mathcal{K}_1 \models \mathcal{K}_2$ and $\mathcal{K}_2 \models \mathcal{K}_1$
3. $E_1 \approx E_2$ if and only if there exists a bijection g from E_1 to E_2 such that $\forall \mathcal{K} \in E_1 : \pi_{\mathcal{K}} = \pi_{g(\mathcal{K})}$
4. $\uplus$ is the union of multisets [13]
5. $E^n = \underbrace{E \uplus ... \uplus E}_{n \text{ times}}$

Note that in the special case where we only consider only one TBox $\mathcal{T}_1$ for E, these postulates are equivalent with the ones proposed in [18], by considering the revision of $\mathcal{T}_1$ by the shared TBox $\mathcal{T}$. Hence, our postulates extend (with very few adaptations) the notion of Revision of [18].

For the merging process considered in the present paper, the integrity constraint is $\mathcal{K} = \langle \mathcal{T}, \emptyset \rangle$ where $\mathcal{T}$ is the set of TBox axioms of each $\mathcal{K}_i \in E$ and $\mathcal{K}_i = \langle \mathcal{T}, \mathcal{A}_i \rangle$.

Proposition 2. *Our min-based assertional merging merging satisfies* $(\mathbf{M_0^\pi})$, $(\mathbf{M_1^\pi})$, $(\mathbf{M_2^\pi})$, $(\mathbf{M_3^\pi})$, $(\mathbf{M_5^\pi})$, $(\mathbf{M_6^\pi})$, $(\mathbf{M_7^\pi})$, $(\mathbf{M_8^\pi})$, $(\mathbf{M_I^\pi})$ *and falsifies* $(\mathbf{M_4^\pi})$, $(\mathbf{M_{maj}^\pi})$.

For the counter-examples, let us consider $\mathcal{K} = \langle \mathcal{T}, A \rangle$ where $\mathcal{T} = \{A \sqsubseteq \neg B\}$, $\mathcal{A}_1 = \{(\mathcal{A}(a), .9)\}$ and $\mathcal{A}_2 = \{(B(a), .5)\}$. In this case, $\mathcal{K}_\Lambda = \langle \{A \sqsubseteq \neg B\}, \{(\mathcal{A}(a), .9)\} \rangle$. $\mathcal{K}_\Lambda$ is consistent, contrary to $\mathcal{K}_\Lambda \cup \{(B(a), .5)\}$, which falsifies $(\mathbf{M_4^\pi})$. Moreover, repeating $\mathcal{A}_2$ will not change the result: $(\mathbf{M_{maj}^\pi})$ is also falsified.

5   Conclusion

We propose in this paper a new operator for merging multiple sources ABoxes sharing a same terminology in the context of *DL-Lite$^\pi$*. We propose a syntactic version of this operator and its semantic counterpart. This operator turns out to be more productive than the operator previously proposed in [4], without increasing the complexity of the merging process. In particular, it picks any pieces of information that is not in contradiction with other bases: it is not affected by the drowning effect. We finally provide an analysis in the light of a new set of postulates dedicated to uncertain *DL-Lite* merging.

This paper opens several perspectives. For instance, we focus on a *min* operator for aggregating conflict vectors, in order to preserve possibilistic semantics. Nevertheless, other aggregation operators can be considered (e.g. the product operator) or direct comparisons from vectors (e.g. G-max based operator). From a postulate point of view, other postulates dedicated to DL knowledge bases could be studied and adapted (e.g. arbitration [15]).

Acknowledgement. This work has been supported by the french Agence Nationale de la Recherche for the ASPIQ project ANR-12-BS02-0003.

References

1. Artale, A., Calvanese, D., Kontchakov, R., Zakharyaschev, M.: The dl-lite family and relations. J. Artif. Intell. Res. (JAIR) (2009)
2. Baader, F., Calvanese, D., McGuinness, D.L., Nardi, D., Patel-Schneider, P.F.: The Description Logic Handbook: Theory, Implementation and Applications, 2nd edn. Cambridge University Press, New York (2010)
3. Benferhat, S., Bouraoui, Z.: Possibilistic dl-lite. In: Liu, W., Subrahmanian, V.S., Wijsen, J. (eds.) SUM 2013. LNCS, vol. 8078, pp. 346–359. Springer, Heidelberg (2013)
4. Benferhat, S., Bouraoui, Z., Loukil, Z.: Min-based fusion of possibilistic dl-lite knowledge bases. In: Web Intelligence, pp. 23–28. IEEE Computer Society (2013)
5. Benferhat, S., Dubois, D., Prade, H.: Argumentative inference in uncertain and inconsistent knowledge bases. In: Heckerman, D., Mamdani, E.H. (eds.) UAI, pp. 411–419. Morgan Kaufmann (1993)
6. Benferhat, S., Dubois, D., Prade, H.: Syntactic combination of uncertain information: A possibilistic approach. In: Gabbay, D.M., Kruse, R., Nonnengart, A., Ohlbach, H.J. (eds.) ECSQARU-FAPR 1997. LNCS, vol. 1244, pp. 30–42. Springer, Heidelberg (1997)
7. Benferhat, S., Kaci, S.: Fusion of possibilistic knowledge bases from a postulate point of view. Int. J. Approx. Reasoning 33(3), 255–285 (2003)
8. Bloch, I., Hunter, A., Appriou, A., Ayoun, A., Benferhat, S., Besnard, P., Cholvy, L., Cooke, R.M., Cuppens, F., Dubois, D., Fargier, H., Grabisch, M., Kruse, R., Lang, J., Moral, S., Prade, H., Saffiotti, A., Smets, P., Sossai, C.: Fusion: General concepts and characteristics. Int. J. Intell. Syst. 16(10), 1107–1134 (2001)
9. Calvanese, D., Giacomo, G.D., Lembo, D., Lenzerini, M., Rosati, R.: Tractable reasoning and efficient query answering in description logics: The dl-lite family. J. Autom. Reasoning 39(3), 385–429 (2007)
10. Dubois, D., Lang, J., Prade, H.: Dealing with multi-source information in possibilistic logic. In: ECAI, pp. 38–42 (1992)
11. Dubois, D., Prade, H.: Possibility theory. Plenum Press, New-York (1988)
12. Katsuno, H., Mendelzon, A.O.: Propositional knowledge base revision and minimal change. Artificial Intelligence 52(3), 263–294 (1991)
13. Knuth, D.E.: The Art of Computer Programming. Seminumerical Algorithms, vol. 2, pp. 694–695. Addison Wesley (1998)
14. Konieczny, S., Pérez, R.P.: Merging information under constraints: A logical framework. J. Log. Comput. 12(5), 773–808 (2002)
15. Konieczny, S., Pino Pérez, R.: Merging information under constraints: a logical framework. Journal of Logic and Computation 12(5) (2002)
16. Poggi, A., Lembo, D., Calvanese, D., Giacomo, G.D., Lenzerini, M., Rosati, R.: Linking data to ontologies. J. Data Semantics 10, 133–173 (2008)
17. Qi, G., Ji, Q., Pan, J.Z., Du. Extending, J.: description logics with uncertainty reasoning in possibilistic logic. Int. J. Intell. Syst. 26(4), 353–381 (2011)
18. Qi, G., Liu, W., Bell, D.A.: Knowledge base revision in description logics. In: Fisher, M., van der Hoek, W., Konev, B., Lisitsa, A. (eds.) JELIA 2006. LNCS (LNAI), vol. 4160, pp. 386–398. Springer, Heidelberg (2006)

On the Revision of Prioritized DL-Lite Knowledge Bases

Salem Benferhat, Zied Bouraoui, and Karim Tabia

Univ Lille Nord de France, F-59000 Lille, France
UArtois, CRIL - CNRS UMR 8188, F-62300 Lens, France
{benferhat,bouraoui,tabia}@cril.fr

Abstract. *DL-Lite* is a tractable family of description logics particularly suitable for query answering. One of the fundamental issues in this
area is the dynamics of the knowledge base which is a problem closely
related to the belief revision one. This paper investigates revision of prioritized *DL-Lite* knowledge bases when a new input piece of information,
possibly conflicting or uncertain, becomes available. To encode the prioritized knowledge, we use a possibility theory-based *DL-Lite* logic. We first
study revision at the semantic level consisting in directly conditioning
possibility distributions. In particular, we show that such conditioning
provides in some situations some counterintuitive results compared with
the ones of conditioning directly the knowledge base syntactically. We
then study revision at the syntactic level of possibilistic *DL-Lite* knowledge bases. Finally, we show that such revision process has a meaningful
semantic counterpart.

1 Introduction

There is an increasing use of ontologies in many application areas in the last
years. Description Logics (DLs) represent a powerful formalism for encoding and
reasoning on ontologies. Recently, a lot of attention is given to *DL-Lite*, a family
of lightweight DLs specifically designed for applications using huge volumes of
data such as Web applications where query answering is the most important
reasoning task [6]. *DL-Lite* guarantees an efficient computational complexity of
the reasoning process. In many applications, the available knowledge is often
affected by uncertainty especially when it is provided by several and potentially conflicting sources. Generally, concatenating them gives a prioritized or a
stratified knowledge base [3]. In [9], it was shown that handling priorities is in
a complete agreement with possibility theory. This latter offers a very natural
framework to deal with ordinal and qualitative uncertainty or preferences and
priorities. Recently, a particular attention was given to the extension of DLs and
DL-Lite within the possibility theory setting (*e.g.* [12,1]). One of the interesting
aspects of possibilistic knowledge bases and more generally weighted knowledge
bases is the ability of reasoning with partially inconsistent knowledge.

Originally DLs have been proposed to represent the static knowledge of a
domain of interest. However in some applications (like Web-based ones), the

U. Straccia and A. Calì (Eds.): SUM 2014, LNAI 8720, pp. 22–36, 2014.

knowledge may be non static and may evolve and change from one situation to another in order to take into account and integrate the changes that occur over time [16]. Dynamics of a DL-based knowledge base gave rise to increasing interest (*e.g.* [15,14,13]) and often concerns the situation where a new information should be incorporated while ensuring the consistency of the results. This issue is closely related to the belief revision problem where old beliefs are revised to take into account the newly available pieces of information. Revision here is often seen as knowledge change and is characterized for instance by the well-known AGM postulates in the propositional logic setting, or by the Hansson's postulates for revising belief bases.

Recently, several works have dealt with revising *DL-Lite* knowledge bases [16,7,2]. Unfortunately, there is to the best of our knowledge no approach for revising prioritized DLs or *DL-Lite* knowledge bases when a new uncertain information is available. This paper fills this gap and investigates revising prioritized *DL-Lite* knowledge bases. In order to encode and reason with the available prioritized knowledge, a possibilistic *DL-Lite* logic [1], denoted π-*DL-Lite*, is more appropriate. In particular, this extension guarantees a computational complexity identical to the one of standard *DL-Lite*. We first study revision of π-*DL-Lite* knowledge bases semantically by conditioning the possibility distribution associated to *DL-Lite* interpretations by the new information. We start by adapting the standard conditioning proposed in the possibilistic setting to the π-*DL-Lite* setting. We show in particular that conditioning the possibility distribution within *DL-Lite* differs from the one proposed by [4] within the standard possibilistic setting in the sense that a direct adaptation of conditioning to π-*DL-Lite* framework is not satisfactory. Roughly speaking, according to the interaction between the new information and the knowledge base, we identify situations where conditioning in *DL-Lite* differs from the one of the standard possibilistic setting. To this end, we study revision at syntactic level of π-*DL-Lite* knowledge bases. We propose two other definitions that generalize and refine the classical one. An important result is that revision operation is done efficiently without additional extra computational costs.

2 Possibilistic DL-Lite

In this section, we recall the main notions of possibilistic *DL-Lite* logic [1], denoted π-*DL-Lite*. This formalism is an extension of *DL-Lite* within the possibility theory setting. π-*DL-Lite* provides a powerful and natural mechanism to deal with uncertainty and to ensure reasoning under inconsistency while keeping a computational complexity identical to the one used in standard *DL-Lite*.

2.1 DL-Lite Logic

We briefly recall *DL-Lite$_{core}$* fragment which is the core fragment for all the *DL-Lite* family [6] in order to introduce possibilistic *DL-Lite*.

Syntax A *DL-Lite* knowledge base (KB) $\mathcal{K}=\langle \mathcal{T}, \mathcal{A} \rangle$ is built upon a set of atomic concepts (*i.e.* unary predicates), a set of atomic roles (*i.e.* binary predicates) and a set of individuals Complex concepts and roles are formed as follows:

$$B \longrightarrow A|\exists R \quad C \longrightarrow B|\neg B \quad R \longrightarrow P|P^-$$

where A (*resp.* P) is an atomic concept (*resp.* role). B (*resp.* C) are called basic (*resp.* complex) concepts and role R is called basic role. The TBox $\mathcal{T}$ consists of a finite set of inclusion axioms between concepts of the form: $B \sqsubseteq C$. The ABox $\mathcal{A}$ consists of a finite set of membership assertions on atomic concepts and on atomic roles of the form: $A(a_i)$, $P(a_i, a_j)$, where a_i and a_j are two individuals.

Semantics The *DL-Lite* semantics is given by an interpretation $I=(\Delta, .^I)$ which consists of a nonempty domain Δ and an interpretation function $.^I$. The function $.^I$ assigns to each individual a an element $u^I \in \Delta^I$, to each concept C a subset $C^I \subseteq \Delta^I$ and to each role R a binary relation $R^I \subseteq \Delta^I \times \Delta^I$ over Δ^I. The interpretation function $.^I$ is extended for all the constructs of the *DL-Lite$_R$*. Namely: $(\neg B)^I = \Delta^I \setminus B^I$, $(\exists R)^I = \{x \in \Delta^I | \exists y \in \Delta^I$ such that $(x, y) \in R^I\}$ and $(P^-)^I = \{(y, x) \in \Delta^I \times \Delta^I | (x, y) \in P^I\}$. For the TBox, we say that I satisfies a concept inclusion axiom, denoted by $I \models B \sqsubseteq C$ iff $B^I \subseteq C^I$. For the ABox, we say that I satisfies a concept (*resp.* role) membership assertion, denoted by $I \models A(a_i)$ (*resp.* $I \models P(a_i, a_j)$), iff $a_i^I \in A^I$ (*resp.* $(a_i^I, a_j^I) \in P^I$). Note that we only consider *DL-Lite* with unique name assumption. Lastly, an interpretation I is said to satisfy a KB $\mathcal{K}=\langle \mathcal{T}, \mathcal{A} \rangle$ iff I satisfies every axiom in $\mathcal{T}$ and every axiom in $\mathcal{A}$. Such interpretation is said to be a model of $\mathcal{K}$.

2.2   Possibility Theory and *DL-Lite*

Let $\mathcal{L}$ be a *DL-Lite* description language, Ω be a universe of discourse consisting of a set of *DL-Lite* interpretations ($I=(\Delta, .^I) \in \Omega$). An epistemic state is represented by a possibility distribution π which is a mapping from Ω to the unit interval $[0, 1]$ that assigns to each interpretation $I \in \Omega$ a possibility degree $\pi(I) \in [0, 1]$. $\pi(I)$ represents the compatibility or consistency of I with respect to the set of available knowledge about the real world. When $\pi(I)=0$, I is said impossible and it is fully inconsistent with the set of available knowledge, whereas when $\pi(I)=1$, I is said totally possible and it is fully consistent with the available knowledge (namely nothing prevents I from being the real world). For two interpretations I and I', when $\pi(I) > \pi(I')$ we say that I is more consistent or more preferred than I' w.r.t the available knowledge. Lastly, π is said normalized if there exists at least one totally possible interpretation, namely $\exists I \in \Omega$, $\pi(I)=1$, otherwise, we say that π is sub-normalized. Note that the concept of sub-normalization reflects the presence of conflicts in the set of available information. Given a possibility distribution π defined on a set of interpretations Ω, two dual measures are generally used to assess the uncertainty of any event of interest $\phi \subseteq \Omega$: the possibility Π and the necessity N measures such that $\Pi(\phi)=\max_{\omega \in \phi}(\pi(\omega))$ and $N(\phi)=1-\Pi(\neg \phi)$. These two measures are extended for a *DL-Lite* axiom ϕ as follows:

Possibility Measure: $\Pi(\phi)=\max\limits_{I\in\Omega}\{\pi(I) : I \vDash \phi\}$ evaluates to what extent an axiom ϕ is compatible with the available knowledge encoded by π.

Necessity Measure: The necessity degree $N(\phi)=1-\max\limits_{I\in\Omega}\{\pi(I) : I \nvDash \phi\}$ evaluates to what extent ϕ is certainty entailed from the available knowledge encoded by π where $I \nvDash \phi$ means that I is not a model of ϕ.

2.3 π-*DL-Lite* Knowledge Bases

Syntactic representation Let $\mathcal{L}$ be a *DL-Lite* description language, a π-*DL-Lite* KB is a set of possibilistic axioms of the form (ϕ, α) where ϕ is an axiom expressed in $\mathcal{L}$ and $\alpha \in \,]\,0, 1]$ is the degree of certainty of ϕ. Formally, $\mathcal{K}=\{(\phi_i, \alpha_i): i=1...n\}$. Only somewhat certain information $(\alpha>0)$ is explicitly represented in a π-*DL-Lite* KB. A weighted axiom (ϕ, α) means that the certainty degree of ϕ is at least equal to α (namely, $N(\phi)\geq\alpha$). A π-*DL-Lite* KB $\mathcal{K}$ will also be represented by a couple $\mathcal{K}=\langle\mathcal{T}, \mathcal{A}\rangle$ where both elements in $\mathcal{T}$ and $\mathcal{A}$ may be uncertain. It is important to note that if for every axiom ϕ_i, we have $\alpha_i=1$ then this gives a classical *DL-Lite* KB denoted $\mathcal{K}^*=\{\phi_i : (\phi_i, \alpha_i) \in \mathcal{K}\}$.

Given $\mathcal{K}=\langle\mathcal{T}, \mathcal{A}\rangle$ a π-*DL-Lite* KB, we define the α-cut of $\mathcal{K}$ (*resp.* $\mathcal{T}$ and $\mathcal{A}$), denoted by $\mathcal{K}_{\geq\alpha}$ (*resp.* $\mathcal{T}_{\geq\alpha}$, $\mathcal{A}_{\geq\alpha}$), the sub-base of $\mathcal{K}$ (*resp.* $\mathcal{T}$ and $\mathcal{A}$) composed of axioms having weights α_i that are at least equal to α and the strict α-cut of $\mathcal{K}$ (*resp.* $\mathcal{T}$ and $\mathcal{A}$), denoted by $\mathcal{K}_{>\alpha}$ (resp. $\mathcal{T}_{>\alpha}$, $\mathcal{A}_{>\alpha}$), as a sub-base of $\mathcal{K}$ (*resp.* $\mathcal{T}$ and $\mathcal{A}$) composed of axioms having weights α_i strictly greater than α. We say that $\mathcal{K}$ is consistent if the standard base obtained from $\mathcal{K}$ by ignoring the weights associated with axioms is consistent. In case of inconsistency, we associate to $\mathcal{K}$ an inconsistency degree defined as follows:

Definition 1. *The inconsistency degree of a π-DL-Lite KB $\mathcal{K}$, denoted $Inc(\mathcal{K})$, is syntactically defined as follows:* $Inc(\mathcal{K})=\max\{\alpha:\mathcal{K}_{\geq\alpha}$ *is inconsistent*$\}$.

In [1], the computation of the inconsistency degree of a π-*DL-Lite* KB is performed using an extension of the algorithm proposed in [6]. This extension consists first on computing the negated closure of the KB, denoted π-$neg(\mathcal{T})$, using the rules presented in [1]. This π-$neg(\mathcal{T})$ is transformed to weighted queries performed over the set of individuals in $\mathcal{A}$ in order to compute the inconsistency degree. The inconsistency associated with a query and a given tuple of assertions provided as an answer for the query is the maximum weight among all the certainty degrees of the query and this tuple. The maximum among these inconsistency degrees is the inconsistency degree associated with the KB.

Semantics Given a π-*DL-Lite* KB $\mathcal{K}$, one can associate to $\mathcal{K}$ a joint possibility distribution, denoted $\pi_\mathcal{K}$, defined over the set of all interpretations $I=(\Delta, .^I)$ by associating to each interpretation I its level of consistency with the set of available knowledge encoded in $\mathcal{K}$.

Definition 2. *The possibility distribution $\pi_\mathcal{K}$ induced from a π-DL-Lite KB $\mathcal{K}$ is defined as follows: $\forall I \in \Omega$:*

$$\pi_\mathcal{K}(I) = \begin{cases} 1 & if \forall (\phi_i, \alpha_i) \in \mathcal{K}, I \models \phi_i \\ 1 - max\{\alpha_i : (\phi_i, \alpha_i) \in \mathcal{K}, I \nvDash \phi_i\} & otherwise \end{cases}$$

A π-*DL-Lite* KB $\mathcal{K}$ is said consistent if its joint possibility distribution $\pi_\mathcal{K}$ is normalized, otherwise $\mathcal{K}$ is said inconsistent and its inconsistency degree is defined semantically as follows:

Definition 3. *The inconsistency degree of a π-DL-Lite KB $\mathcal{K}$, denoted $Inc(\mathcal{K})$, is semantically defined as follows: $Inc(\mathcal{K}) = 1 - \max\limits_{I \in \Omega}\{\pi_\mathcal{K}(I)\}$.*

Example 1. Let $\mathcal{K} = \langle \mathcal{T}, \mathcal{A} \rangle$ be a π-*DL-Lite* KB where $\mathcal{T} = \{(A \sqsubseteq B, .4)\}$ and $\mathcal{A} = \{(A(a), .5), (C(a), .7)\})$. One can compute $\pi_\mathcal{K}$ the possibility distribution induced from $\mathcal{K}$ using Definition 2.

Table 1. Example of a possibility distribution $\pi_\mathcal{K}$ computed using Definition 2

I	$.^I$	$\pi_\mathcal{K}$	I	$.^I$	$\pi_\mathcal{K}$
I_1	$A = \{\}, B = \{\}, C = \{\}$	.3	I_2	$A = \{a\}, B = \{\}, C = \{\}$	.3
I_3	$A = \{\}, B = \{a\}, C = \{\}$	.3	I_4	$A = \{\}, B = \{\}, C = \{a\}$	.5
I_5	$A = \{a\}, B = \{a\}, C = \{\}$	.3	I_6	$A = \{a\}, B = \{\}, C = \{a\}$	.6
I_7	$A = \{\}, B = \{a\}, C = \{a\}$	.5	I_8	$A = \{a\}, B = \{a\}, C = \{a\}$	1

One can observe that $\pi_\mathcal{K}(I_8) = 1$ meaning that the KB is consistent. Note that we have chosen a simple example in order to enumerate all interpretations. This will be helpful to illustrate the conditioning of a π-*DL-Lite* possibility distribution.

3 Revising the $\pi_\mathcal{K}$ Distribution

Let $\mathcal{K} = \langle \mathcal{T}, \mathcal{A} \rangle$ be a π-*DL-Lite* KB where $\pi_\mathcal{K}$ is its joint possibility distribution computed according to Definition 2. For the sake of simplicity, we assume that $\mathcal{K}$ is consistent (namely $\pi_\mathcal{K}$ is normalized). Let us denote by (φ, μ) the new information to be accepted. Within the π-*DL-Lite* setting, φ may be an assertion of the form $A(a)$ or $P(a, b)$, a positive inclusion axiom (PI) of the form $B_1 \sqsubseteq B_2$ or a negative inclusion axiom (NI) of the form $B_1 \sqsubseteq \neg B_2$ and $\mu \in]0, 1]$. The new input can be a totally reliable information (i.e. $\mu = 1$) or uncertain (i.e. $0 < \mu < 1$). In π-*DL-Lite*, revision comes down to add the new information with its prescribed level of certainty while ensuring the consistency of the revision results.

In the following, we investigate revision at the semantic level. It consists in conditioning the original possibility distribution $\pi_\mathcal{K}$ by the new information (φ, μ). This operation takes as input a possibility distribution $\pi_\mathcal{K}$ and the new information (φ, μ) and transforms $\pi_\mathcal{K}$ to a revised possibility distribution $\pi' = \pi_\mathcal{K}(. | (\varphi, \mu))$. Here, the input (φ, μ) is considered as a constraint that must be satisfied in π'. More precisely, the revised distribution is such that $\Pi'(\varphi) = 1$ (in the possibilistic setting, in order for an event φ to have a certainty degree greater than zero, it must be totally possible, hence $\Pi'(\varphi)=1$) and $N'(\varphi)\geq\mu$ meaning that the axiom φ is certain at least to the degree μ. Here Π' (*resp.* N') is the possibility (*resp.* necessity) measure induced by the revised possibility distribution π'.

3.1 Logical Properties

In [4], conditioning in the possibilistic logic setting is characterized with the following properties. A revised possibility distribution π' is considered eligible for revising the initial distribution $\pi_\mathcal{K}$ with the new input (φ,μ) if it satisfies the following properties.

    (A1) $\max_{I\in\Omega}(\pi'(I))=1$.
    (A2) $\Pi'(\varphi)=1$ and $N'(\varphi)\geq\mu$.
    (A3) $\forall I_1\nvDash\varphi,\ I_2\nvDash\varphi$, if $\pi_\mathcal{K}(I_1)\leq\pi_\mathcal{K}(I_2)$ then $\pi'(I_1)\leq\pi'(I_2)$.
    (A4) $\forall I_1\vDash\varphi,\ I_2\vDash\varphi$, if $\pi_\mathcal{K}(I_1)\leq\pi_\mathcal{K}(I_2)$ then $\pi'(I_1)\leq\pi'(I_2)$.
    (A5) If $N_\mathcal{K}(\varphi)>0$ then $\forall I\vDash\varphi$: $\pi_\mathcal{K}(I)=\pi'(I)$
    (A6) If $\pi_\mathcal{K}(I)=0$ then $\pi'(I)=0$.

Property **A1** ensures the consistency of the revised possibility distribution by guaranteeing a normalized distribution π'. **A2** guarantees that the added information should be inferred from the revised distribution π' with a weight at least equal to its prescribed priority level. **A3** ensures that the relative order between the interpretations that falsify φ is preserved. **A4** states that the new possibility distribution π' should preserve the previous pre-order between interpretations which are models of φ. **A5** means that the revision process does not affect models of φ when φ is a priori fully accepted. **A6** states that every impossible interpretation remains impossible after conditioning. In order to satisfy properties **A3** and **A4**, it is clear that the revision operation should condition both the interpretations satisfying φ and those falsifying φ. According to properties **A1-A6**, two different types of possibility distribution conditioning when $\Pi(\varphi)>0$ are proposed in [8], namely in an ordinal setting and in a quantitative setting. These conditionings are extended to the case where the new input is uncertain in [10] and studied in [5]. In this paper, we only focus on conditioning in the ordinal setting, well-known as min-based conditioning [4].

Belief revision with uncertain information was studied in many works and its close relation to Jeffrey's rule [11] (generalizing probability theory's conditioning) is pointed out. In [4] the possibilistic counterpart was given for belief revision with uncertain inputs when dealing with belief bases encoded in possibilistic logics. The authors show that the revision process comes down syntactically to adding the new information with a prescribed level of certainty while maintaining the consistency of the resulting base and semantically to conditioning the possibility distribution representing the current epistemic state in order to add the new input.

3.2   Min-based π-*DL-Lite* Possibility Distribution Conditioning

In order to define conditioning of possibility distribution $\pi_{\mathcal{K}}$, let us first recall that in standard propositional possibilistic logic, the necessity measure is the dual of the possibility measure and it is defined by $N(\phi)=1\text{-}\Pi(\neg\phi)$ where ϕ is a propositional formula. In possibilistic DL-Lite, a necessity measure cannot be defined as the dual of the possibility measure because the negation of an axiom in *DL-Lite* is not allowed. Instead, we define $\Pi_n(\varphi)=\max\limits_{I\in\Omega}\{\pi(I):I\nvDash\varphi\}$.

One can see that $\Pi_n(\varphi)$ is intuitively similar to $\Pi(\neg\phi)$ where $\omega\vDash\neg\phi$ with ω is a propositional logic interpretation (and it denotes in *DL-Lite* an interpretation I falsifying φ, denoted $I \nvDash \varphi$). The following definition rephrases conditioning within the possibilistic *DL-Lite* setting.

Definition 4. *Let $\mathcal{K}=\langle\mathcal{T},\mathcal{A}\rangle$ be a π-DL-Lite KB and $\pi_{\mathcal{K}}$ be its joint possibility distribution. Let (φ,μ) be the new information. The min-based conditioning is extended to the π-DL-Lite setting as follows:*

$$- \ \forall I\vDash\varphi, \ \pi_{\mathcal{K}}(.|_m(\varphi,\mu))=\begin{cases}1 & if \ \pi_{\mathcal{K}}(I)=\Pi(\varphi)\\ \pi(I) & otherwise\end{cases}$$

$$- \ \forall I\nvDash\varphi, \ \pi_{\mathcal{K}}(.|_m(\varphi,\mu))=\begin{cases}1\text{-}\mu & if \ \pi(I)=\Pi_n(\varphi)\\ 1\text{-}\mu & if \ \pi_{\mathcal{K}}(I)>1\text{-}\mu\\ \pi(I) & otherwise\end{cases}$$

According to Definition 4, accepting the input consists in raising the degree of the most plausible model of φ to 1. This allows to deal only with axioms that are consistent with the input. For the counter-models, it is clear that the most plausible is set to $1\text{-}\mu$ and all the interpretations that are more compatible than $1\text{-}\mu$ should be shifted down to $1\text{-}\mu$.

Proposition 1. *Let $\mathcal{K}=\langle\mathcal{T},\mathcal{A}\rangle$ be a π-DL-Lite KB and $\pi_{\mathcal{K}}$ be its joint possibility possibility distribution. Let (φ, μ) be the new information. Then $\pi'=\pi_{\mathcal{K}}(.|(\varphi,\mu))$ computed using Definition 4 satisfies postulates (A1)-(A6).*

Example 2. Let us consider $\pi_{\mathcal{K}}$ presented in Example 1. Assume that we have in this example separately two cases of new information pieces to be accepted. The first one is $(B\sqsubseteq\neg C,.9)$ and the second one is $(B\sqsubseteq\neg C,.2)$. Using Definition 4, the min-based revised possibility distribution $\pi'=\pi_{\mathcal{K}}(I|_m(B\sqsubseteq\neg C,.9))$ (*resp.* $\pi'=\pi_{\mathcal{K}}(I|_m(B\sqsubseteq\neg C,.2))$) is as follows:

In this example, the first scenario is revising $\pi_{\mathcal{K}}$ associated to $\mathcal{K}$ with the input $(B\sqsubseteq\neg C,.9)$. Given that in $\pi_{\mathcal{K}}$, we have a priori $\Pi(B\sqsubseteq\neg C)=.6$ (hence it's necessity is 0) then the new input requires to be satisfied to increase the necessity of the axiom $B\sqsubseteq\neg C$ until .9. In the second scenario, the necessity of the axiom $B\sqsubseteq\neg C$ has to be shifted down to .2. One can observe in $\pi_{\mathcal{K}}$ that the interpretations $\{I_1,I_2,I_3,I_4,I_5,I_6\}\vDash B\sqsubseteq\neg C$ where $\Pi(B\sqsubseteq\neg C)=.6$ while $\{I_7,I_8\}\nvDash B\sqsubseteq\neg C$ where $\Pi_n(B\sqsubseteq\neg C)=1$. □

Definition 4 is a direct adaptation of conditioning in possibilistic logic [8] to π-DL-Lite framework. As it will be shown in the following example, conditioning of

Definition 4 is not satisfactory as it provides somehow counterintuitive results. More precisely, conditioning of Definition 4 works when the new information is inconsistent with the KB or it is a priori inferred with a weight less than its prescribed level μ. Hence revision here consists in simply adding the new information to the old knowledge (it is a kind of knowledge expansion). However, conditioning of Definition 4 does not work properly when the input is a priori inferred with a weight greater than its prescribed level μ. The following example illustrates this situation.

Example 3. Assume that we have a π-*DL-Lite* KB $\mathcal{K}$ where the TBox $\mathcal{T}=\{(A\sqsubseteq B, .4), (B\sqsubseteq C,.7)\}$ and the ABox $\mathcal{A}=\{(A(a),.3)\}$. One can easily check that we have a priori $\mathcal{K}\vDash_\pi(A\sqsubseteq C,.4)$ (indeed, as it is shown in Table 3, the axiom $A\sqsubseteq C$ has a necessity degree of .4 in the possibility distribution $\pi_\mathcal{K}$ associated to $\mathcal{K}$). Now assume the two following situations: In the first one, the information piece to be accepted by $\mathcal{K}$ is $(A\sqsubseteq C,.9)$ while in the second situation $\mathcal{K}$ is revised with $(A\sqsubseteq C,.2)$. Let $\pi'=\pi_\mathcal{K}(I|_m(A\sqsubseteq C,.9))$ (*resp.* $\pi''=\pi_\mathcal{K}(I|_m(A\sqsubseteq C,.2)))$ the conditioned min-based possibility distribution using Definition 4. The interpretations $\{I_1,I_2,I_3,I_4,I_5,I_6\}$ satisfy the input axiom $A\sqsubseteq C$ and we have a priori $\Pi(A\sqsubseteq C)=1$ and $\Pi_n(A\sqsubseteq C)=.6$. The possibility degrees of the interpretations $\{I_7,I_8\}$ are set to $(1-.9)=.1$ in order to ensure that $N'(A\sqsubseteq C)=.9$. It is easy to check that properties **(A1)-(A6)** are satisfied by the distribution π' computed according to Definition 4. However when the input is $(A\sqsubseteq C,.2)$, there is a prob-

Table 2. Example of possibility distribution revision by two information pieces

I	$.^I$	$\pi_\mathcal{K}$	$\pi_\mathcal{K}(I|_m(B\sqsubseteq\neg C,.9))$	$\pi_\mathcal{K}(I|_m(B\sqsubseteq\neg C,.2))$
I_1	$A=\{\},B=\{\},C=\{\}$	.3	.3	.3
I_2	$A=\{a\},B=\{\},C=\{\}$	.3	.3	.3
I_3	$A=\{\},B=\{a\},C=\{\}$	.3	.3	.3
I_4	$A=\{\},B=\{\},C=\{a\}$	.5	.5	.5
I_5	$A=\{a\},B=\{a\},C=\{\}$	.3	.3	.3
I_6	$A=\{a\},B=\{\},C=\{a\}$	.6	1	1
I_7	$A=\{\},B=\{a\},C=\{a\}$	.5	.1	.5
I_8	$A=\{a\},B=\{a\},C=\{a\}$	1	.1	.8

Table 3. Second example of possibility distribution revision by two information pieces

I	$.^I$	$\pi_\mathcal{K}$	$\pi'=\pi_\mathcal{K}(I|_m(A\sqsubseteq C,.9))$	$\pi''=\pi_\mathcal{K}(I|_m(A\sqsubseteq C,.2))$
I_1	$A=\{\}, B=\{a\}, C=\{\}$	.3	.3	.3
I_2	$A=\{a\}, B=\{\}, C=\{a\}$	.6	.6	.6
I_3	$A=\{\}, B=\{\}, C=\{\}$	.7	.7	.7
I_4	$A=\{\}, B=\{\}, C=\{a\}$	.7	.7	.7
I_5	$A=\{\}, B=\{a\}, C=\{a\}$	.7	.7	.7
I_6	$A=\{a\}, B=\{a\}, C=\{a\}$	1	1	1
I_7	$A=\{a\}, B=\{\}, C=\{\}$	.6	.1	.8
I_8	$A=\{a\},B=\{a\},C=\{\}$	.3	.1	.3

lem regarding the possibility degree associated to I_2 in π''. Indeed, we have $A \sqsubseteq C$ is implied by the fact $A \sqsubseteq B$ and $B \sqsubseteq C$. Hence, in order to have a necessity degree of $A \sqsubseteq C$ of .2 then one has to shift down at least the necessity degree of the axiom $A \sqsubseteq B$ down to .2 as it has a lower priority than $B \sqsubseteq C$. However, if the necessity of $A \sqsubseteq B$ is shifted down to .2 then the corresponding $\pi_\mathcal{K}$ after this modification will not be equivalent to the one given in Table 3. For instance, the interpretation I_2 will be associated with a degree of .8 instead of .6 currently. Clearly revision with conditioning of Definition cannot fully capture syntactic revision detailed in the following section. $\square$

It is important to note that in the *DL-Lite* framework, it is not guaranteed that any set of interpretations represents a *DL-Lite* axiom [1]. In the next section, we analyze revision at syntactic level. We then provide a definition of conditioning possibility distributions that refines Definition 4.

4 Syntactic Revision

In this section, we study revision with the new information (φ, μ) at the syntactic level. Revision here consists in obtaining from a π-*DL-Lite* KB $\mathcal{K} = \langle \mathcal{T}, \mathcal{A} \rangle$ associated to a possibility distribution $\pi_\mathcal{K}$ and an uncertain input information (φ, μ), a new π-*DL-Lite* KB $\mathcal{K}' = \langle \mathcal{T}', \mathcal{A}' \rangle$. As in possibilistic logic, in π-*DL-Lite*, revision comes down to add the new information with its prescribed level of certainty while ensuring the consistency of the revision results. When adding the new information to the KB, several situations may be encountered, namely when the input is consistent or inconsistent with the the original knowledge.

4.1 The Input (φ, μ) is Inconsistent with $\mathcal{K}$

We address here the situation where the new information (φ, μ) is inconsistent with the KB $\mathcal{K}$, namely $\Pi_\mathcal{K}(\varphi) < 1$ (recall that in possibility theory, if $\Pi(\varphi) < 1$ then $N(\varphi) = 0$). There are two situations to be considered. The first one is when (φ, μ) is implicitly inhibited by higher priority TBox or ABox axioms that contradict it. The second one is when (φ, μ) is not inhibited by higher priority axioms that contradict it. For these two cases, the construction of the augmented π-*DL-Lite* KB $\mathcal{K}'$ is performed according to the following steps: (1) Add the input φ to the KB $\mathcal{K}$ with the highest prescribed level (i.e. $\mu = 1$). (2) Compute the inconsistency degree $\beta = Inc(\mathcal{K}_1)$ with $\mathcal{K}_1 = \mathcal{K} \cup \{(\varphi, 1)\}$. (3) Drop every axiom in $\mathcal{K}_1$ having a priority less than or equal to the inconsistency degree β. Let $\mathcal{K}_2$ the obtained consistent KB. (4) Add φ with its prescribed level μ to $\mathcal{K}_2$. Let $\mathcal{K}' = \mathcal{K}_2 \cup \{(\varphi, \mu)\}$.

These steps ensure the consistency of the resulting KB after adding the input (φ, μ) with its prescribed level. The following proposition relates the resulting KB $\mathcal{K}'$ with the possibility distribution $\pi_{\mathcal{K}'}$ associated to $\mathcal{K}'$ with the results of conditioning at the semantic level using Definition 4.

Proposition 2. *Let $\mathcal{K} = \langle \mathcal{T}, \mathcal{A} \rangle$ be a π-DL-Lite KB and $\pi_\mathcal{K}$ be its joint possibility distribution. Let (φ, μ) be the added uncertain input information and $\beta = Inc(\mathcal{K}_1)$*

where $\mathcal{K}_1 = \mathcal{K} \cup \{(\varphi, 1)\}$. Let $\mathcal{K}' = \langle \mathcal{T}', \mathcal{A}' \rangle$ such that $\mathcal{K}' = \{(\varphi, \mu)\} \cup \{(\phi, \alpha) : (\phi, \alpha) \in \mathcal{K} \text{ and } \alpha > \beta\}$ and let $\pi_{\mathcal{K}'}$ be the possibility distribution associated to $\mathcal{K}'$. Then,

$$\forall I \in \Omega, \pi_{\mathcal{K}'}(I) = \pi_{\mathcal{K}}(I|_m(\varphi, \mu)),$$

where $\pi_{\mathcal{K}}(I|_m(\varphi,\mu))$ denotes the revised possibility distribution $\pi_{\mathcal{K}}$ computed using min-based conditioning defined in Definition 4.

Example 4. (examples 1 and 2 continued) Let us first assume a new input $(B \sqsubseteq \neg C, .9)$ and then another input $(B \sqsubseteq \neg C, .2)$. One can easily check that $Inc(\mathcal{K} \cup \{(B \sqsubseteq \neg C, 1)\}) = .4$. So, $(B \sqsubseteq \neg C, .2)$ (*resp.* $(B \sqsubseteq \neg C, .9)$) is inhibited (*resp.* not inhibited) by higher priority axioms that contradict it. For the first case, it is easy to check that $\mathcal{K}' = \{(B \sqsubseteq \neg C, .2), (A(a), .5), (C(a), .7)\}$ is such that $\pi_{\mathcal{K}'}(I) = \pi_{\mathcal{K}}(I|_m(B \sqsubseteq \neg C, .2))$ presented in Example 2. For the second case however, $\mathcal{K}' = \{(B \sqsubseteq \neg C, .9), (A(a), .5), (C(a), .7)\})$ such that $\pi_{\mathcal{K}'}(I) = \pi_{\mathcal{K}}(I|_m(B \sqsubseteq \neg C, .9))$ presented in Example 2.   □

4.2   The Input (φ, μ) is Consistent with $\mathcal{K}$

When the input (φ, μ) is consistent with the KB $\mathcal{K}$ (namely $\Pi(\varphi) = 1$), two situations are to be considered: The first one is when (φ, μ) is a priori inferred from the KB $\mathcal{K}$, namely $\mathcal{K} \models_\pi \phi$, the second one is when (φ, μ) cannot be inferred from $\mathcal{K}$, namely $\mathcal{K} \not\models_\pi \phi$. Here, revision is performed with a simple expansion of $\mathcal{K}$ with the input (φ, μ), namely $\mathcal{K}' = \mathcal{K} \cup (\varphi, \mu)$.

Let us first discuss the situation where the input (φ, μ) is a priori inferred from the KB $\mathcal{K}$. In this situation, two scenarios can hold depending on the a priori necessity measure of φ (denoted $N(\varphi) = \nu$), and its prescribed posterior necessity $N'(\varphi) = \mu$. Namely: (i) When $\nu \le \mu$ meaning that the new information is inferred with a certainty degree ν less than its prescribed one μ. Note that this situation is similar to the case of revising with a certain input (namely case where $\mu = 1$). (ii) When $\nu > \mu$ meaning that the new information is inferred with a certainty degree ν that is greater than its prescribed one μ.

In π-*DL-Lite*, to determine to what extent the input (φ) is inferred from the KB, namely $\mathcal{K} \models_\pi (\varphi, \nu)$ with $\nu \ge \mu$ or $\nu < \mu$, we first add to $\mathcal{K}$ the assumption that φ is false encoded by the following statements: $\{(Y \sqsubseteq C_1, 1), (Y \sqsubseteq \neg C_2, 1), (Y(y), 1)\}$ if $\varphi = C_1 \sqsubseteq C_2$ and $\{(Y \sqsubseteq \neg C_1, 1), (Y(a), 1)\}$ if $\varphi = C_1(a)$ where Y (*resp.* y) is a new concept (*resp.* individual) not appearing in $\mathcal{K}$. Then we compute the inconsistency degree of the augmented KB. This inconsistency degree corresponds to ν. Namely $\mathcal{K} \models_\pi (\varphi, \nu)$ iff $Inc(\mathcal{K}_1) = \nu$ where $\mathcal{K}_1 = \langle \mathcal{T}_1, \mathcal{A}_1 \rangle$ with $\mathcal{T}_1 = \mathcal{T} \cup \{(Y \sqsubseteq C_1, 1), (Y \sqsubseteq \neg C_2, 1)\}$ and $\mathcal{A}_1 = \{(Y(y), 1)\}$ or $\mathcal{T}_1 = \mathcal{T} \cup \{(Y \sqsubseteq \neg C_1, 1)\}$ and $\mathcal{A}_1 = \mathcal{A} \cup \{(Y(a), 1)\}$. Now, the construction of the augmented π-*DL-Lite* KB $\mathcal{K}'$ is performed using the following steps: (1) Add the assumption that φ is false to $\mathcal{K}$ with the highest prescribed level (i.e. $\mu = 1$). (2) Compute the inconsistency degree of the augmented KB (*i.e.* $Inc(\mathcal{K}_1) = \nu$). (3) If $\mu \ge \nu$, then the revision outcome is $\mathcal{K}' = \mathcal{K} \cup \{(\varphi, \mu)\}$. (4) if $(\mu < \nu)$ two solutions can be proposed. (4.1) The first one is to shift down the weights of axioms in $\mathcal{K}$ which are between μ and ν to μ. (4.2) The second solution is to compute first the set $\mathcal{X} \subseteq \mathcal{K}$ of axioms in $\mathcal{K}$ that imply φ. Then we shift down the weights of axioms in $\mathcal{X}$ which are between μ and ν to μ.

These steps ensure inferring the new input φ from the resulting KB $\mathcal{K}'$ with its prescribed level μ. Following these steps, it is clear that the revision process does not change the initial weights attached to axioms of $\mathcal{K}$ if $\mathcal{K} \vDash_\pi (\varphi, \nu)$ with $\nu \leq \mu$. However it changes the initial weights attached to some axioms responsible or not for inferring φ from $\mathcal{K}$ with the weight μ when $\nu > \mu$. According to the Example 3 presented in the previous section, conditioning proposed by Definition 4 is counterintuitive when $(\mu < \nu)$. To this end, we fit Definition 4 before giving the formal representation of $\mathcal{K}'$.

4.3   Semantic Counterpart

Let us start with the case where $\nu > \mu$. The following definition gives a min-based conditioning of π-DL-$Lite$ possibility distribution generalizing Definition 4.

Definition 5. *Let $\mathcal{K} = \langle \mathcal{T}, \mathcal{A} \rangle$ be a π-DL-$Lite$ KB and $\pi_\mathcal{K}$ be its joint possibility possibility distribution. Let (φ, μ) be the new information. The min-based conditioning is extended to the π-DL-$Lite$ setting as follows:*

$$- \forall I \vDash \varphi, \ \pi_\mathcal{K}(.|_m(\varphi, \mu)) = \begin{cases} 1 & if \ \pi_\mathcal{K}(I) = \Pi(\varphi) \\ 1 - \mu & if \ \Pi_n(\varphi) \leq \pi_\mathcal{K}(I) \leq 1\text{-}\mu \\ \pi(I) & otherwise \end{cases}$$

$$- \forall I \nvDash \varphi, \ \pi_\mathcal{K}(.|_m(\varphi, \mu)) = \begin{cases} 1\text{-}\mu & if \ \pi(I) = \Pi_n(\varphi) \\ 1\text{-}\mu & if \ \pi_\mathcal{K}(I) > 1\text{-}\mu \\ \pi(I) & otherwise \end{cases}$$

According to Definition 5, accepting the input consists in raising the degree of the most plausible model of φ to 1. Moreover when $N(\varphi) > \mu$, some models of φ will all be set to 1-μ. For the counter-models, the most plausible is set to 1-μ and all interpretations that are more compatible than 1-μ should be shifted down to 1-μ. Moreover, when $N(\varphi) = \nu > \mu$ the interpretations that falsify less priority axioms inferring φ will be revised.

Proposition 3. *Let $\mathcal{K} = \langle \mathcal{T}, \mathcal{A} \rangle$ be a π-DL-$Lite$ KB and $\pi_\mathcal{K}$ be its joint possibility possibility distribution. Let (φ, μ) be the new information. Then $\pi' = \pi_\mathcal{K}(.|(\varphi, \mu))$ computed using Definition 5 satisfies postulates (A1), (A2), (A3), (A4), (A6).*

The following proposition relates the resulting KB $\mathcal{K}'$ with the possibility distribution $\pi_{\mathcal{K}'}$ associated to $\mathcal{K}'$ with the results of conditioning at the semantic level using Definition 5

Proposition 4. *Let $\mathcal{K} = \langle \mathcal{T}, \mathcal{A} \rangle$ be a π-DL-$Lite$ KB and $\pi_\mathcal{K}$ be its joint possibility distribution. Let (φ, μ) be the added uncertain input information and $\nu = Inc(\mathcal{K}_1)$ where $\mathcal{K}_1$ is the augmented KB by the assumption that φ is false. Then the revised π-DL-$Lite$ KB $\mathcal{K}' = \langle \mathcal{T}', \mathcal{A}' \rangle$ such that:*

$$\mathcal{K}'=\{(\varphi,\mu)\}\cup\{(\phi,\alpha):(\phi,\alpha)\in\mathcal{K}\,and\,\alpha>\nu\}\cup\{(\phi,\alpha):(\phi,\alpha)\in\mathcal{K}\,and\,\alpha<\mu\}\,\cup$$
$$\{(\phi,\mu):(\phi,\alpha)\in\mathcal{K}\,and\,\mu\leq\alpha\leq\nu\}$$

The possibility distribution $\pi_{\mathcal{K}'}$ associated to $\mathcal{K}'$ is such that:

$$\forall I\in\Omega,\pi_{\mathcal{K}'}(I)=\pi_{\mathcal{K}}(I|_m(\varphi,\mu)),$$

where $\pi_{\mathcal{K}}(I|_m(\varphi,\mu))$ denotes the revised possibility distribution of $\pi_{\mathcal{K}}$ using the min-based conditioning of Definition 5.

Example 5 (Examples 3 continued). We have $\mathcal{T}=\{(A\sqsubseteq B,.4),\,(B\sqsubseteq C,.7)\}$ and $\mathcal{A}=\{(A(a),.3)\}$. Let us consider $(A\sqsubseteq C,.9)$ and $(A\sqsubseteq C,.2)$. One can easily check that $Inc(\mathcal{K}_1)=.4$ where $\mathcal{K}_1=\langle\mathcal{T}\cup\{(Y\sqsubseteq A,1),(Y\sqsubseteq\neg C,1)\},\{(Y(y),1)\}\rangle$. So $\mathcal{K}\vDash_\pi(A\sqsubseteq C,.4)$. When the input is $(A\sqsubseteq C,.9)$, then $\mathcal{K}'=\{(A\sqsubseteq B,.4),\,(B\sqsubseteq C,.7),$ $(A\sqsubseteq C,.9),\,(A(a),.3)\}$ such that $\pi_{\mathcal{K}'}(I)=\pi_{\mathcal{K}}(I|_m(A\sqsubseteq C,.9)$ presented in Example 3. Now, when the input is $(A\sqsubseteq C,.2)$, then $\mathcal{K}'=\{(A\sqsubseteq B,.2),\,(B\sqsubseteq C,.7),\,(A\sqsubseteq C,.2),$ $(A(a),.2)\}$ such that $\pi_{\mathcal{K}'}(I)=\pi_{\mathcal{K}}(I|_m(A\sqsubseteq C,.2)$ presented in Example 3 becomes as follows:$\pi_{\mathcal{K}'}(I_1)=.3,\pi_{\mathcal{K}'}(I_2)=.8,\,\pi_{\mathcal{K}'}(I_3)=.8,\,\pi_{\mathcal{K}'}(I_4)=.8,\,\pi_{\mathcal{K}'}(I_5)=.8,\,\pi_{\mathcal{K}'}(I_6)=1,$ $\pi_{\mathcal{K}'}(I_7)=.8$ and $\pi_{\mathcal{K}'}(I_8)=.3$.   □

Proposition 4 leads to shift down the weights of axioms in $\mathcal{K}$ which are between μ and ν to μ. However, one can improve the result with a minimal change consisting in revising only the weights of some axioms responsible of implying the new information. Given the set $\mathcal{X}\subseteq\mathcal{K}$ of axioms in $\mathcal{K}$ that infer φ, we distinguish semantically four sets of interpretations when the new information φ is satisfied: (1) Interpretations that are models of $\mathcal{X}$ and $\mathcal{K}$-$\mathcal{X}$, (2) Interpretations that are models of $\mathcal{X}$ but are not models of $\mathcal{K}$-$\mathcal{X}$, (3) Interpretations that are models of $\mathcal{K}$-$\mathcal{X}$ but are not models of $\mathcal{X}$ and (4) Interpretations that are neither models of $\mathcal{K}$-$\mathcal{X}$ nor $\mathcal{X}$. The following definition provides another min-based conditioning of π-DL-Lite possibility distribution that also adapts Definition 4.

Definition 6. *Let $\mathcal{K}=\langle\mathcal{T},\mathcal{A}\rangle$ be a π-DL-Lite KB and $\pi_{\mathcal{K}}$ be its joint possibility possibility distribution. Let (φ,μ) be the new information. Let $\mathcal{X}\subseteq\mathcal{K}$ be the set of axioms inferring φ. Let $\mu'=max\{\alpha:(\phi,\alpha)\in\mathcal{K}-\mathcal{X}\,and\,I\nvDash\phi\}$. In an ordinal setting, we define the min-based conditioning as follows:*

$$-\ \forall I\vDash(\varphi\cup\mathcal{X}),\ \pi(.|_m(\varphi,\mu))=\begin{cases}1 & if\,\pi(I)=\Pi(\varphi)\\ \pi(I) & otherwise\end{cases}$$

$$-\ \forall I\vDash\varphi\cup(\mathcal{K}\text{-}\mathcal{X}),\ I\nvDash\mathcal{X},\ \pi(.|_m(\varphi,\mu))=\begin{cases}1\text{-}\mu & if\,\pi(I)=\Pi_n(\varphi)\\ \pi(I) & otherwise\end{cases}$$

$$-\ \forall I\vDash\varphi,I\nvDash\mathcal{X},\ I\nvDash\mathcal{K}-\mathcal{X},\pi(.|_m(\varphi,\mu))=\begin{cases}1\text{-}\mu & if\,\pi(I)=\Pi_n(\varphi)\,and\,1-\mu'\geq1\text{-}\mu\\ 1\text{-}\mu' & if\,\pi(I)=\Pi_n(\varphi)\,and\,1-\mu'\leq1\text{-}\mu\\ \pi(I) & otherwise\end{cases}$$

$$-\ \forall I\nvDash\varphi,\pi(.|_m(\varphi,\mu))=\begin{cases}1\text{-}\mu & if\,\pi(I)=\Pi_n(\varphi)\\ 1\text{-}\mu & if\,\pi(I)>1\text{-}\mu\\ \pi(I) & otherwise\end{cases}$$

Proposition 5. *Let $\mathcal{K}=\langle\mathcal{T},\mathcal{A}\rangle$ be a π-DL-Lite KB and $\pi_\mathcal{K}$ be its joint possibility possibility distribution. Let (φ,μ) be the new information. Then $\pi'=\pi_\mathcal{K}(.|(\varphi,\mu))$ computed using Definition 6 satisfies postulates (A1),(A2),(A3) and (A6).*

The following proposition relates the resulting KB $\mathcal{K}'$ with the possibility distribution $\pi_{\mathcal{K}'}$ associated to $\mathcal{K}'$ with the results of conditioning at the semantic level using Definition 6.

Proposition 6. *Let $\mathcal{K}=\langle\mathcal{T},\mathcal{A}\rangle$ be a π-DL-Lite KB and $\pi_\mathcal{K}$ be its joint possibility distribution. Let (φ,μ) be the added uncertain input information and $\nu=Inc(\mathcal{K}_1)$ where $\mathcal{K}_1$ is the augmented KB by the assumption that φ is false. Then the revised π-DL-Lite KB $\mathcal{K}'=\langle\mathcal{T}',\mathcal{A}'\rangle$ such that*

$$\mathcal{K}'-\{(\varphi,\mu)\}\cup\{\mathcal{K}\quad\mathcal{X}\}\cup\{(\phi,\alpha):(\phi,\alpha)\in\mathcal{X}\,und\,\alpha>\nu\}\cup\{(\phi,\mu):(\phi,\nu)\in\mathcal{X}\,and\,\nu=\alpha\}$$

The possibility distribution $\pi_{\mathcal{K}'}$ associated to $\mathcal{K}'$ is such that:

$$\forall I\in\Omega,\pi_{\mathcal{K}'}(I)=\pi_\mathcal{K}(I|_m(\varphi,\mu)),$$

where $\pi_\mathcal{K}(I|_m(\varphi,\mu))$ denotes the revised possibility distribution of $\pi_\mathcal{K}$ using the min-based conditioning defined in Definition 6.

Example 6 (Examples 3 continued). When the input is $(A\sqsubseteq C,.9)$, then $\mathcal{K}'=\{(A\sqsubseteq B,.4),(B\sqsubseteq C,.7),(A\sqsubseteq C,.9),(A(a),.3)\}$ such that $\pi_{\mathcal{K}'}(I)=\pi_\mathcal{K}(I|_m(A\sqsubseteq C,.9)$ presented in Example 3. Now, when the input is $(A\sqsubseteq C,.2)$, then $\mathcal{K}'=\{(A\sqsubseteq B,.2),(B\sqsubseteq C,.7),(A\sqsubseteq C,.9),(A(a),.3)\}$ such that $\pi_{\mathcal{K}'}(I)=\pi_\mathcal{K}(I|_m(A\sqsubseteq C,.2)$ becomes as follows: $\pi_{\mathcal{K}'}(I_1)=.3,\pi_{\mathcal{K}'}(I_2)=.8,\pi_{\mathcal{K}'}(I_3)=.7,\pi_{\mathcal{K}'}(I_4)=.7,\pi_{\mathcal{K}'}(I_5)=.7,\pi_{\mathcal{K}'}(I_6)=1,\pi_{\mathcal{K}'}(I_7)=.8$ and $\pi_{\mathcal{K}'}(I_8)=.3$. □

Let us now discuss the case where $\mu\geq\nu$. It is similar to the revision by a totally reliable information (i.e. $\mu=1$). In this case, it is natural that all the interpretations that are models of φ must be preserved and all the interpretations that falsify φ must be set as impossible (the necessity degree of the input equals 0). In this case the conditioning operation follows from Definitions 5 and 6. Moreover conditioning according Definitions 5 and 6 agrees with Definition 4. Finally when (φ,μ) cannot be inferred from $\mathcal{K}$, this means that the revision process is performed simply with an expansion of $\mathcal{K}$ with the input. In such situation, conditioning follows trivially according to Definitions 5 and 6 and coincides with Definition 4. It is similar to the case where the input is inconsistent with $\mathcal{K}$.

5 Discussions and Concluding Remarks

This paper addressed revision of π-DL-Lite KBs when a new piece of information (φ,μ), possibly conflicting or uncertain, becomes available. We first studied revision at the semantic level by adapting conditioning of possibility distributions proposed within the possibilistic setting. We have shown that such conditioning provides a counterintuitive results. We then investigated revision at the syntactic level of π-*DL-Lite* KBs. Finally, we proposed two others definitions of π-*DL-Lite* possibility distribution conditioning that generalize the first one.

According to the new definition, conditioning of π-*DL-Lite* possibility distribution with (φ, μ) establishes a new pre-order between counter-models and models of φ. This new ranking depends on the a priori necessity measure of φ, and the prescribed posterior necessity measure of φ. Roughly speaking, if $N(\varphi){\leq}\mu$ then with a min-based conditioning every interpretation that falsifies φ and that is more compatible than 1-μ is shifted down to 1-μ. This means that some a priori pre-order on these interpretations will be lost. Moreover, the fact that within π-*DL-Lite* framework, the necessity measure is not the dual of the possibility measure, some a priori pre-order on interpretations which are models of φ will also be lost. This is a consequence of shifting down to 1-μ some more compatible counter-models of φ when $N(\varphi){\leq}\mu$. Regarding the computational complexity of the syntactic revision, it is obvious that it is polynomial since computing the inconsistency degree of a π-*DL-Lite* KB is polynomial using the algorithm proposed in [1]. To compute the revision outcome, we need one step further when (φ, μ) is inferred from the KB. Namely, we need to compute the set of axioms responsible for deducing the input. The computational complexity of this subset is also polynomial. A future works will focus on the assertional-based revision as well as terminological-based revision within the π-*DL-Lite* setting.

Acknowledgement. This work has received support from the French National Research Agency, ASPIQ project reference ANR-12-BS02-0003.

References

1. Benferhat, S., Bouraoui, Z.: Possibilistic DL-Lite. In: Liu, W., Subrahmanian, V.S., Wijsen, J. (eds.) SUM 2013. LNCS, vol. 8078, pp. 346–359. Springer, Heidelberg (2013)
2. Benferhat, S., Bouraoui, Z., Papini, O., Würbel, E.: Assertional-based removed set revision of DL-Liter belief bases. In: ISAIM (2014)
3. Benferhat, S., Dubois, D., Prade, H.: How to infer from inconsisent beliefs without revising? In: IJCAI, pp. 1449–1457. Morgan Kaufmann (1995)
4. Benferhat, S., Dubois, D., Prade, H., Williams, M.A.: A practical approach to revising prioritized knowledge bases. Studia Logica 70(1), 105–130 (2002)
5. Benferhat, S., Tabia, K., Sedki, K.: Jeffrey's rule of conditioning in a possibilistic framework. Annals of Mathematics and Artificial Intelligence 61(3), 185–202 (2011)
6. Calvanese, D., Giacomo, G.D., Lembo, D., Lenzerini, M., Rosati, R.: Tractable reasoning and efficient query answering in description logics: The dl-lite family. J. Autom. Reasoning 39(3), 385–429 (2007)
7. Calvanese, D., Kharlamov, E., Nutt, W., Zheleznyakov, D.: Evolution of DL-Lite knowledge bases. In: Patel-Schneider, P.F., Pan, Y., Hitzler, P., Mika, P., Zhang, L., Pan, J.Z., Horrocks, I., Glimm, B. (eds.) ISWC 2010, Part I. LNCS, vol. 6496, pp. 112–128. Springer, Heidelberg (2010)

8. Didier, D., Henri, P.: Possibility theory: Qualitative and quantitative aspects. In: Smets, P. (ed.) Quantified Representation of Uncertainty and Imprecision, Handbook of Defeasible Reasoning and Uncertainty Management Systems, vol. 1, pp. 169–226. Springer Netherlands (1998), http://dx.doi.org/10.1007/978-94-017-1735-9_6

9. Dubois, D., Prade, H.: Epistemic entrenchment and possibilistic logic. Artif. Intell. 50(2), 223–239 (1991)

10. Dubois, D., Prade, H.: A synthetic view of belief revision with uncertain inputs in the framework of possibility theory. International Journal of Approximate Reasoning 17(2-3), 295–324 (1997)

11. Jeffrey, J.: The logic of decision. McGraw-Hill Press (1965)

12. Qi, G., Ji, Q., Pan, J.Z., Du, J.: Extending description logics with uncertainty reasoning in possibilistic logic. Int. J. Intell. Syst. 26(4), 353–381 (2011)

13. Qi, G., Du, J.: Model-based revision operators for terminologies in description logics. In: IJCAI, pp. 891–897 (2009)

14. Qi, G., Haase, P., Huang, Z., Ji, Q., Pan, J.Z., Völker, J.: A kernel revision operator for terminologies — algorithms and evaluation. In: Sheth, A.P., Staab, S., Dean, M., Paolucci, M., Maynard, D., Finin, T., Thirunarayan, K. (eds.) ISWC 2008. LNCS, vol. 5318, pp. 419–434. Springer, Heidelberg (2008)

15. Qi, G., Liu, W., Bell, D.A.: Knowledge base revision in description logics. In: Fisher, M., van der Hoek, W., Konev, B., Lisitsa, A. (eds.) JELIA 2006. LNCS (LNAI), vol. 4160, pp. 386–398. Springer, Heidelberg (2006)

16. Wang, Z., Wang, K., Topor, R.W.: A new approach to knowledge base revision in DL-Lite. In: AAAI (2010)

Interval-Based Possibilistic Networks

Salem Benferhat, Sylvain Lagrue, and Karim Tabia

Univ Lille Nord de France, F-59000 Lille, France
UArtois, CRIL UMR CNRS 8188, F-62300 Lens, France
{benferhat,lagrue,tabia}@cril.fr

Abstract. In this paper, we study foundations of interval-based possibilistic networks where possibility degrees associated with nodes are no longer singletons but sub-intervals of [0,1]. This extension allows to compactly encode and reason with epistemic uncertainty and imprecise beliefs as well as with multiple expert knowledge. We propose a natural semantics based on compatible possibilistic networks. The last part of the paper shows that computing the uncertainty bounds of any event can be computed in interval-based networks without extra computational cost with respect to standard possibilistic networks.

1 Introduction

Graphical models are powerful tools for modeling and reasoning with uncertain and complex information [7]. They are compact and expressive representations of beliefs that can be elicited from an agent or automatically learnt from empirical data. However, the difficulty for an agent to provide precise and reliable crisp belief degrees has led researchers to the development of alternative and flexible formalisms for representing and managing ill-known beliefs. An example of flexible frameworks is the one of interval-based representations of uncertainty which is justified in many situations by the availability of few information pieces and knowledge, the existence of multiple and potentially contradictory information sources, the impreciseness of sensors' outputs, etc. Several alternative theories are now developed to encode ill-known beliefs like imprecise probabilities [18], belief functions and evidence theory [17], etc. Interval-based representations are widely adopted to encode ill-known beliefs but there is no consensus on the semantics and interpretations underlying belief intervals. Existing works on interval-based graphical models deal only with probabilistic networks[1] and they are mostly interested in inference issues [8] [9] [10]. One of the major problems of interval-based probabilistic networks is their exorbitant computational complexity for inferring posterior probability intervals while in practice the obtained intervals are often **too large to be exploited** [13]. As it will be shown in the inference section, min-based possibilistic networks have different behaviors due to the idempotence property of the min and max operators of possibility theory.

While the possibilistic setting is more appropriate for dealing with partial ignorance and incomplete information, interval-based possibilistic networks have not been investigated. In [4], we proposed three-valued possibilistic networks where only three

---

[1] Note that in the probabilistic setting, interval-based networks are called credal networks [6] [14].

U. Straccia and A. Calì (Eds.): SUM 2014, LNAI 8720, pp. 37–50, 2014.
© Springer International Publishing Switzerland 2014

possibility levels (namely, *fully accepted*, *fully rejected* or *unknown*) are used to encode incomplete information. In the logical setting, there is only our work dealing with
interval-based logic bases proposed in the possibility theory framework [2]. However,
as it will be shown later, this work cannot be used for the purpose of our paper. Besides,
different extensions of standard possibilistic logic have been proposed to deal with multiple source information or with temporal information using intervals [12]. Again, these
works cannot be used for the purpose of this paper.

Interval-based possibilistic networks (IPNs) generalize standard possibilistic networks [5][1] to encode ill-know beliefs where these latter are encoded by means of
sub-intervals of [0, 1]. More precisely, IPNs allow to compactly encode families of
standard joint possibility distributions. The motivation of interval-based possibilistic
representations is to encode and reason with ill-known and imprecise beliefs, confidence intervals, multi-source information, etc. In particular, standard possibility theory
does not allow incomparability since the possibility of any two events are comparable.
This may appear as a strong assumption in some applications. Besides, in practice it is
not always possible to have necessity/possibility bounds for every proposition. Intervals
offer more flexibility to represent and to handle incomparable events. The objective of
this paper is to analyze foundational issues of IPNs and its main contributions are:

- We propose a definition of IPNs which extend standard possibilistic ones. This
  flexible model has not been addressed before.
- We extend the definition of chain rule for IPNs. We show that contrary to standard
  possibility theory, an IPN can neither be faithfully represented by an interval-based
  possibility distribution nor by an interval-based possibilistic logic knowledge base.
- We propose a natural semantics for IPNs based on compatible standard possibilistic networks. We provide precise relations between a set of compatible standard
  possibilistic networks and an interval-based possibility distribution induced by the
  extended chain rule.
- Lastly, we show that computing the uncertainty bounds of any event, given in terms
  of guaranteed possibility measure Δ and possibility measure Π can be computed
  in IPNs without extra computational cost with respect to standard possibilistic
  networks.

2   A Brief Refresher on Possibility Theory and Possibilistic Networks

2.1   Possibility Theory

Possibility theory [19] provides a powerful and simple alternative to probability theory
in particular for dealing with qualitative uncertainty where only the plausibility ordering
between events is important. One of the fundamental concepts of possibility theory
is the concept of possibility distribution π which is a mapping from the universe of
discourse Ω to the unit interval [0, 1]. A possibility degree $\pi(w_i)$ expresses to what
extent $w_i \in \Omega$ can be the real world. $\pi(w_i)=1$ means that w_i is totally possible and
$\pi(w_i)=0$ denotes an impossible event. The relation $\pi(w_i)>\pi(w_j)$ means that w_i is more
possible than w_j. A possibility distribution π is normalized if $max_{w_i \in \Omega}\pi(w_i) = 1$. A

possibility measure $\Pi(\phi)$ evaluates the possibility degree relative to an event $\phi \subseteq \Omega$. It is defined as follows:

$$\Pi(\phi) = \max_{w_i \in \phi}(\pi(w_i)). \tag{1}$$

Another important concept, introduced in [11][3], is the one of guaranteed possibility measure $\Delta(\phi)$ which evaluates to which extent the possibility degree of an event is guaranteed:

$$\Delta(\phi) = \min_{w_i \in \phi}(\pi(w_i)). \tag{2}$$

The two measures $\Delta(\phi)$ and $\Pi(\phi)$ allow to define the minimal and maximal possibility degrees associated with the event ϕ given a joint possibility distribution π. When π encodes a set of preferences, $\Pi(\phi)$ represents the most satisfactory solution while $\Delta(\phi)$ represents the least satisfactory solution where ϕ is true.

The other fundamental notion in possibility theory is the one of conditioning defined as follows [15]: $\forall \phi \subseteq \Omega$,

$$\pi(w_i|\phi) = \begin{cases} 1 & \text{if } \pi(w_i)=\Pi(\phi) \text{ and } w_i \in \phi; \\ \pi(w_i) & \text{if } \pi(w_i)< \Pi(\phi) \text{ and } w_i \in \phi; \\ 0 & \text{otherwise.} \end{cases} \tag{3}$$

2.2 Possibilistic Networks

A possibilistic network $\mathcal{G}=<G,\Theta>$ is specified by:

i) A *graphical component* G consisting of a directed acyclic graph (DAG) where vertices represent the variables and edges represent direct *dependence* relationships between variables. Each variable A_i is associated with a domain D_i containing the values a_i taken by a variable A_i.

ii) A *numerical component* Θ allowing to assess the uncertainty relative to each variable using local possibility tables. The possibilistic component consists in a set of local possibility tables $\Theta_i=\{\theta_{a_i|u_i}\}$ where $a_i \in D_i$ and u_i is an instance of U_i denoting the parent variables of A_i in the network $\mathcal{G}$.

We will use the notations $\pi(A_i)$ and θ_{A_i} interchangeably to denote the local possibility table of variable A_i whenever no confusion arises. Note that all the local possibility distributions Θ_i must be normalized, namely $\forall i=1..n$, $\forall u_i \in D_{U_i}$, $\max_{a_i \in D_i}(\theta_{a_i|u_i})=1$.

Example 1. Figure 1 gives an example of a possibilistic network over four Boolean variables A, B, C and D. The structure of $\mathcal{G}$ encodes a set of independence relationships $I=\{I(A_i,U_i,Y)\}$ where each variable A_i in the context of its parents U_i is independent of its non descendants Y. For example, in the network of Figure 1, variable C is independent of B and D in the context of A.

In the qualitative possibilistic setting, the joint possibility distribution is factorized using the min-based chain rule:

$$\pi(a_1, a_2, .., a_n) = \min_{i=1}^{n}(\pi(a_i|u_i)). \tag{4}$$

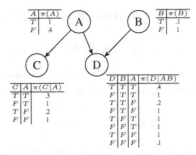

Fig. 1. Example of a possibilistic network

3 Interval-Based Possibilistic Networks: Definitions

In order to encode ill-known beliefs, interval-based representations rely on sub-intervals $[\alpha, \beta]$ of $[0, 1]$. When the plausibility of an event ϕ is encoded by the interval $[\alpha, \beta]$, this is interpreted as *the actual but ill-known plausibility of ϕ is a unique value within the interval $[\alpha, \beta]$*. Here β represents the upper bound of $\Pi(\phi)$ (denoted $\overline{\Pi}(\phi)$) and α its lower bound (denoted $\underline{\Pi}(\phi)$). The underlying interpretation is disjunctive in the sense that $\Pi(\phi) \in [\alpha, \beta]$. This is in opposite interpretation of intervals used in [12] in the context of handling temporal information where $\Pi(\phi) \in [\alpha, \beta]$ means that $\Pi(\phi)$ is true in the period between time instants α and β.

Definition 1. *An IPN $\mathcal{I}=<G,\Theta^I>$ is a network where the uncertainty is represented by intervals. Namely, $\mathcal{I}$ consists of*

1. a directed acyclic graph G encoding direct independence relationships between variables $I(A_i, U_i, Y)$ and
2. a set of local interval-based possibility tables Θ^I where $\forall \theta^I_{a_i|u_i} \in \Theta^I$, $\theta^I_{a_i|u_i} \subseteq [0, 1]$.

It is clear that in case where all the parameters of the network are singletons (pointwise-based possibilities), then the network is a standard (pointwise-based) network. Hence, an IPN $\mathcal{I}$ is a possibilistic network where the graphical component has the same representation while local possibility tables contain intervals allowing to encode some imprecision on the encoded beliefs.

Example 2. Figure 2 is an example of an IPN over two Boolean variables A and B.

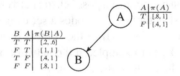

Fig. 2. $\mathcal{I}$: Example of an IPN

In the following, we introduce on the concepts of compatible and coherent possibilistic networks.

3.1 Compatible Possibilistic Networks

Let us define the concept of compatible network.

Definition 2. *Let $\mathcal{I}=<G,\Theta^I>$ be an IPN. A pointwise-based possibilistic network $\mathcal{G}=<G,\Theta>$ is compatible with $\mathcal{I}$ iff*

1. $\mathcal{I}$ and $\mathcal{G}$ have exactly the same graph and
2. $\forall \theta_{a_i|u_i} \in \Theta$, $\theta_{a_i|u_i} \in \theta^I_{a_i|u_i}$ with $\theta^I_{a_i|u_i} \in \Theta^I$.

In Definition 2, a possibilistic network $\mathcal{G}$ is compatible with an IPN $\mathcal{I}$ if they have the same structure (hence encode the same conditional independence relationships) and every local possibility distribution $\theta_{a_i|u_i}$ of $\mathcal{G}$ is compatible with its corresponding interval-based distribution $\theta^I_{a_i|u_i}$ in the IPN $\mathcal{I}$.

Example 3. Let us consider the interval-based network of Figure 2 over two Boolean variables A and B. One can easily check that the network of Figure 3 is normalized and compatible with the interval-based network of Figure 2.

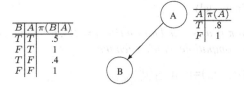

B	A	$\pi(B\|A)$
T	T	.5
F	T	1
T	F	.4
F	F	1

A	$\pi(A)$
T	.8
F	1

Fig. 3. Example of a possibilistic network compatible with the network $\mathcal{I}$ of Figure 2

3.2 Coherent Interval-Based Possibilistic Networks

In standard possibilistic networks, the normalization condition prevents inconsistencies. In IPNs, we refer to the existence of compatible networks with the concept of coherent IPN defined as follows:

Definition 3. *An IPN $\mathcal{I}=<G,\Theta^I>$ is coherent iff*

- *there is at least one pointwise-based possibilistic network $\mathcal{G}$ which is compatible with $\mathcal{I}$ and*
- *all the values composing the parameters $\theta^I_{a_i|u_i}$ are feasible. Namely $\forall \theta^I_{a_i|u_i} \in \Theta^I_i$, $\forall \alpha \in \theta^I_{a_i|u_i}$, there exists a compatible network $\mathcal{G}=<G,\Theta>$ such that $\theta_{a_i|u_i}=\alpha$.*

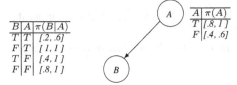

B	A	$\pi(B\|A)$
T	T	[.2, .6]
F	T	[1, 1]
T	F	[.4, 1]
F	F	[.8, 1]

A	$\pi(A)$
T	[.8, 1]
F	[.4, .6]

Fig. 4. Example of an incoherent IPN

Example 4. The IPN of Figure 2 is coherent since there exists at least the pointwise-based network of Figure 3 which is compatible with it. However, the network of Figure 4 is incoherent since it is impossible to build a normalized pointwise-based network where the possibility degree of $A=T$ is 0.8.

In the following, we only consider coherent IPNs.

4   IPNs: 2 Alternative Semantics

The previous section dealt with the syntax of IPNs and the coherence concept. This section proposes two semantics for coherent IPNs.

4.1   Semantics Based on Compatible Models

The semantics views a coherent IPN $\mathcal{I}$ as a family of compatible pointwise-based networks. Each of these compatible networks encodes a joint pointwise-based distribution we call simply c-model (for compatible model).

Definition 4. *A possibility distribution π is a c-model for an IPN $\mathcal{I}=<G, \Theta^I>$ if there exists a standard network $\mathcal{G}=<G, \Theta>$ compatible with $\mathcal{I}$ such that*

$$\forall a_1..a_n \in \Omega, \ \pi(a_1..a_n)=\min_{i=1}^{n}(\theta^G_{a_i|u_i}),$$

where u_i is the parent of a_i in $a_1..a_n$.

Definition 4 states that the c-models π are those possibility distributions associated with the compatible possibilistic networks $\mathcal{G}$ using the min-based chain rule of Equation 4.

Example 5. The possibility distribution π of Table 1 is the c-model corresponding to the network of Figure 3, obtained using the min-based chain rule.

Table 1. A c-model corresponding to network of Figure 3

B	A	$\pi(AB)$
T	T	.5
F	T	.8
T	F	.4
F	F	1

Let us use $\mathcal{F}_C$ to denote the set of c-models of the IPN $\mathcal{I}$. Clearly, if $\mathcal{F}_C$ is empty then the network $\mathcal{I}$ is incoherent.

There is another way to define a compatible joint distribution based on the concept of π-model, defined as follows:

Definition 5. *A pointwise-based distribution π is a π-model of the IPN $\mathcal{I}$ iff it retrieves all the parameters of $\mathcal{I}$ and satisfies all the independence relations encoded by $\mathcal{I}$, namely:*

- *Condition 1: $\forall \theta^I_{a_i|u_i} \in \Theta^I$, $\Pi(a_i|u_i) \in \theta^I_{a_i|u_i}$.*

- *Condition 2:* $\forall I(A_i, U_i, Y) \in I,\ \Pi(A_i | U_i, Y) = \Pi(A_i | U_i)$.

Definition 5 states that a pointwise-based distribution π is a π-model of an $\mathcal{I}$ if the possibility of any event a_i in the context of its parents u_i computed from π (namely $\Pi(a_i | u_i)$) is in the interval $\theta^I_{a_i | u_i}$ (Condition 1). In addition to retrieving[2] the network's parameters, a distribution is a π-model if it satisfies also all the independence relationships encoded by the network (Condition 2). Then we have the following finding:

Proposition 1. *A distribution π is a π-model of an IPN $\mathcal{I} = <G, \Theta^I>$ iff π is a c-model of $\mathcal{I}$, namely, $\pi \in \mathcal{F}_C$.*

Proof. The proof of Proposition 1 is straightforward. Indeed, recall that in the standard possibilistic setting, given a possibilistic network G, the possibility distribution obtained using the min-based chain rule of Equation 4 allows to retrieve all the parameters of G and satisfies the independence relations. Hence, if one starts with a compatible network with the IPN $\mathcal{I}$, then its associated distribution π satisfies the two conditions of Definition 5, which means that π is also π-model of the IPN. The converse is also true. Assume that π_1 is a π-model of the IPN $\mathcal{I}$. Let us construct a standard network G where G and $\mathcal{I}$ have the same graph and where the parameters $\theta_{a_i | u_i}$ are those computed from π_1, namely $\theta_{a_i | u_i} = \Pi_1(a_i | u_i)$. Clearly, the construction of such a network gives a compatible network and since applying the min-based chain rule exactly gives π_1 we conclude that $\pi_1 \in \mathcal{F}_C$. $\qquad\square$

4.2 Semantics Based on Extending the Chain Rule

In the above section, we presented a semantics for IPNs based on a family of possibility distributions $\mathcal{F}_C$. Here we discuss if an IPN can be seen as an interval-based joint possibility distribution. This latter is defined as follows:

Definition 6. *An interval-based joint distribution π_I is a mapping from the universe of discourse Ω to the set of sub-intervals of $[0, 1]$ which associates with each interpretation $\omega_i \in \Omega$ an interval $[\alpha_i, \beta_i] \subseteq [0, 1]$.*

Let us now define compatible possibility distributions:

Definition 7. *A pointwise-based possibility distribution π is compatible with the interval-based one π_I iff $\forall \in \omega_i \Omega,\ \pi(\omega_i) \in \pi_I(\omega_i)$.*

By definition a compatible distribution must be normalized.

Let $\mathcal{F}_C^I$ denote the set of pointwise-based joint possibility distributions that are compatible with the interval-based distribution π_I. Given an interval-based distribution π_I, the plausibility of an event ϕ is assessed as follows:

Definition 8. *Let $\phi \subseteq \Omega$ be an arbitrary event and let π_I be an interval-based distribution. Then the possibility interval associated with ϕ, computed from π_I, is the interval defined as follows:*

$$\Pi_I(\phi) = [\min_{\pi \in \mathcal{F}_C^I} (\Pi(\phi)),\ \max_{\pi \in \mathcal{F}_C^I} (\Pi(\phi))]. \tag{5}$$

---

[2] A possibility distribution π retrieves some parameter α associated with some conditional event $a|b$, if $\Pi(a|b) = \alpha$ where Π is computed from π. If α is an interval then the property requires that $\Pi(a|b)$ belongs to that interval.

Similarly, the conditional possibility of $\phi \subseteq \Omega$ given an evidence $\psi \subseteq \Omega$ is defined as an interval as follows:

$$\Pi_I(\phi \mid \psi) = [\min_{\pi \in \mathcal{F}_C^I} (\Pi(\phi \mid \psi)), \max_{\pi \in \mathcal{F}_C^I} (\Pi(\phi \mid \psi))]. \tag{6}$$

A natural question raises here about how to induce from an IPN $\mathcal{I}$ an interval-based joint distribution π_I. Namely, what is the counterpart of the min-based chain rule of Equation 4 in the interval-based setting? Two methods can be considered:

1. Extending the min-based chain rule of Equation 4 to the interval-based setting directly using the associativity property of the minimum operator, stated as follows:

$$\min([\alpha_1, \beta_1], [\alpha_2, \beta_2]) = [\min(\alpha_1, \alpha_2), \min(\beta_1, \beta_2)]. \tag{7}$$

Using Equation 7, the min-based chain rule of Equation 4 is extended to the interval-based setting as follows:

$$\pi_{\mathcal{I}}(a_1, a_2, .., a_n) = [\min_{i=1}^{n} \underline{\pi}(a_i \mid u_i), \min_{i=1}^{n} \overline{\pi}(a_i \mid u_i)]. \tag{8}$$

2. Using the compatible networks, a joint interval-based distribution can be directly computed from the compatible networks. Interestingly enough, this is equivalent to the extended chain rule of Equation 8. Namely,

Proposition 2. *Let $\mathcal{F}_C^I$ denote the set of compatible pointwise-based networks $\mathcal{G}$ with the interval-based one $\mathcal{I}$. Then $\forall \omega \in \Omega$,*

$$\pi_I(\omega) = [\min_{\pi \in \mathcal{F}_C^I} (\Pi(\omega)), \max_{\pi \in \mathcal{F}_C^I} (\Pi(\omega))]. \tag{9}$$

It is easy to show that the joint distributions computed from Equations 8 and 5 are equivalent (just apply Equation 5, the min-based chain rule of Equation 4 and the associativity property of Equation 7). It is important to note that in the joint interval-based distribution π_I, for the upper bound distribution $\overline{\pi}_I$ there exists a compatible pointwise-based network while there may exist situations where there does not exist a compatible pointwise-based network for the lower bound distribution $\underline{\pi}_I$ as shown in the following example.

Example 6. Let $\mathcal{I}$ be IPN having the unique node A with the domain $D_A = \{a_1, a_2\}$. Let the distribution associated with A be $\pi_I(a_1) = [.3, 1]$ and $\pi_I(a_2) = [.2, 1]$. Using Equations 8 and 5, the lower bound distribution is $\underline{\pi}_I(a_1) = .3$ and $\underline{\pi}_I(a_2) = .2$. It is clear that the distribution $\underline{\pi}_I = \{.3, .2\}$ is sub-normalized and there is no compatible network with $\mathcal{I}$ that encodes $\underline{\pi}_I$.

Now, given an interval-based joint distribution π_I over a set of variables $A_1..A_n$, is it possible to encode π_I by an IBN $\mathcal{I}$? This issue is addressed in the following section.

5   Analysis of IPNs Semantics

5.1   Relating the Extended Chain Rule with the Compatible Distributions

The following proposition relates the set of pointwise-based distributions $\mathcal{F}_C$ induced by the compatible standard networks and the set $\mathcal{F}_C^I$ of pointwise-based distributions compatible with the interval-based possibility distribution π_I.

Proposition 3. *Let $\mathcal{I}=<G,\Theta^I>$ be an IPN and let π_I be the induced interval-based distribution from $\mathcal{I}$. Then*

$$\mathcal{F}_C \subseteq \mathcal{F}_C^I \text{ but } \mathcal{F}_C^I \nsubseteq \mathcal{F}_C.$$

Let us provide an example showing that considering compatible pointwise-based distributions with the interval-based distribution π_I induced from the network $\mathcal{I}$ does not necessarily retrieve the network's parameters.

Example 7. Let A and B be two Boolean variables and let $\mathcal{I}$ be the IPN of Figure 5.

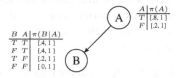

B	A	$\pi(B\|A)$
T	T	[.4, 1]
F	T	[.4, 1]
T	F	[.2, 1]
F	F	[0, 1]

A	$\pi(A)$
T	[.8, 1]
F	[.2, 1]

Fig. 5. Counter example IPN $\mathcal{I}$

The interval-based distribution $\pi_{\mathcal{I}}$ induced from the IPN $\mathcal{I}$ of Figure 5 using the interval-based chain rule of Equation 8 is the following (left side table):

A	B	$\pi_{\mathcal{I}}(AB)$	A	B	$\pi(AB)$
T	F	[.4, 1]	T	F	.5
T	T	[.4, 1]	T	T	.5
F	F	[.2, 1]	F	F	1
F	T	[0, 1]	F	T	1

From the distribution π (right side table), it follows that $\Pi(A=T)=.5$. One can check that π is compatible with π_I but it is impossible to build a compatible network with the IPN $\mathcal{I}$ of Figure 5 that encodes π since in the IPN $\mathcal{I}$ the possibility degree of $A=T$ is at least .8.

From this example, one can claim that the semantics of an IPN is not an interval-based joint distribution. The semantics that can be associated with an IPN is **the set of possibility distributions induced from the pointwise-based possibilistic networks $\mathcal{G}_1..\mathcal{G}_n$ compatible with $\mathcal{I}$.**

5.2   Can an Interval-Based Network Encode a Joint Interval-Based Distribution?

The above section shows that an IPN can be represented by a family of compatible distributions $\mathcal{F}_C$ but not by an interval-based joint distribution π_I. This section addresses the converse problem, namely:

(1) Given an interval-based possibility distribution π_I, is it possible to build from π_I an IPN $\mathcal{I}$ such that $\forall\omega\in\Omega$, $\pi_I(\omega)=\pi_{\mathcal{I}}(\omega)$?
(2) Given a family of compatible possibility distributions $\mathcal{F}$, can we build an IPN $\mathcal{I}$ such that $\mathcal{F}=\mathcal{F}_C^I$?

The interesting point is that we get the converse of the results obtained in the previous section. Indeed, the statement (2) is false as shown in the following example.

Example 8. Consider one binary variable A and assume that $\mathcal{F}$ is composed of the following two possibility distributions π_1 and π_2.

A	$\pi_1(A)$
T	1
F	.5

A	$\pi_2(A)$
T	1
F	.7

Clearly, it is impossible to build an IPN $\mathcal{I}$ such that $\mathcal{F}=\mathcal{F}_C^I$. Indeed, assume that such an IPN $\mathcal{I}$ exists. Then we have two cases: i) either the intervals associated with the variable A are singletons then there exist a single compatible possibility distribution and this contradicts the fact that $\mathcal{F}$ contains two distributions, ii) or the intervals associated with A are not singletons, then the set of compatible possibility distributions is infinite and this contradicts the fact that $\mathcal{F}$ is finite.

However, statement (1) is true as it is shown in the following proposition:

Proposition 4. *An interval-based possibility distribution π_I can be represented by an IPN $\mathcal{I}$. Namely,*

$$\forall \omega_i \Omega,\ \pi_I(\omega_i)=\pi_{\mathcal{I}}(\omega_i),$$

where $\pi_{\mathcal{I}}(\omega_i)$ is the interval computed from the IPN $\mathcal{I}$ using the interval-based chain rule of Equation 8.

Proof. The proof is immediate since one can use the well-known chain rule in graphical models to factorize any belief distribution. In fact, one can always build an IPN $\mathcal{I}$ that encodes the interval-based distribution π_I. This can be done as follows:

i) For the structure of the IPN $\mathcal{I}$, select the complete graph G where for each variable A_i, set its parents $U_i=\{A_{i+1},..,A_n\}$. One can easily show that the obtained network is a DAG.

ii) For the parameters of $\mathcal{I}$, one has just to associate with $\theta^I_{a_i|u_i}$ the set of values $\pi_C^I(a_i|u_i)$ where π_C^I is a possibility distribution compatible with π_I. One can check that for any a_i and u_i, $\theta^I_{a_i|u_i}$ is an interval.   □

Example 9. Table 2 is an example of interval-based joint possibility distribution π_I over two binary variables A_1 and A_2. Following our proof, we can build the interval-based network $\mathcal{I}$ given in Figure 6 which encodes the distribution π_I of Table 2.

Table 2. Example of interval-based joint distribution π_I

A_1	A_2	$\pi_I(A_1 A_2)$
T	T	[.3, 1]
F	T	[.2, .3]
T	F	[.2, .3]
F	F	[.6, 1]

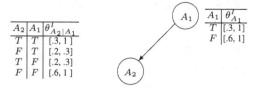

A_2	A_1	$\theta^I_{A_2|A_1}$
T	T	$[.3, 1]$
F	T	$[.2, .3]$
T	F	$[.2, .3]$
F	F	$[.6, 1]$

A_1	$\theta^I_{A_1}$
T	$[.3, 1]$
F	$[.6, 1]$

Fig. 6. IPN encoding π_I of Table 2

One can easily check that the interval-based joint distribution $\pi_\mathcal{I}$ computed from the IPN $\mathcal{I}$ of Figure 6 using the interval-based chain rule of Equation 8 is equivalent to the interval-based joint distribution π_I of Table 2.

6 Inferring Uncertainty Bounds from IPNs

Inference is an important task for reasoning with belief networks. It is well-known that it is a hard task even in standard probabilistic networks [8]. This task becomes harder for an IPN since we consider the family of standard networks compatible with the IPN. In this section, we show that it is possible to compute two uncertainty bounds of any event of interest from an IPN $\mathcal{I}$ without extra complexity with respect to standard networks thanks to the ordinal nature of qualitative possibility theory[3]. More precisely, we show that for any event $\phi \subseteq \Omega$, the upper bound of the possibility degree $\overline{\Pi}_I(\phi)$ and the guaranteed possibility degree $\Delta_I(\phi)$ can be computed from two special standard networks, hence **the uncertainty interval is computed using only two calls for an inference algorithm in standard networks**.

Recall that we are interested here in computing the interval $[\Delta_I(\phi), \overline{\Pi}_I(\phi)]$ for a given event $\phi \subseteq \Omega$. Recall also that $\Delta(\phi)$ represents the lowest possibility degree associated with ϕ (see Equation 2). When we reason with an IPN $\mathcal{I}$, we define $\Delta_I(\phi) = \{\min_{\pi \in \mathcal{F}_C}(\pi(\phi))\}$. It turns out that computing $\Delta_I(\phi)$ can be performed using a particular pointwise-based network, namely the lower bound $\Delta_I(\phi)$ can be computed from the standard network G_L defined as follows:

Proposition 5. *Let $\mathcal{I} = <G, \Theta^I>$ be an IPN. Let $\mathcal{G}_L = <G, \Theta>$ be the standard possibilistic network such that*

1. *$\mathcal{I}$ and $\mathcal{G}_L$ have the same graphical structure and*
2. *$\forall a_i \in D_{A_i}, \theta^{G_L}_{a_i|u_i} = \underline{\theta}^I(a_i|u_i)$.*
   Then $\forall \phi \subseteq \Omega, \Delta_I(\phi) = \Pi^{G_L}(\phi)$.

Proof sketch. The proof is based on the fact that if $\Pi^{G_L}(\phi) = \alpha$ then there always exists a compatible network $\mathcal{G}$ such that $\Pi^G(\phi) = \alpha$ and there is no way to build another compatible network $\mathcal{G}'$ such that $\Pi^{G'}(\phi) < \alpha$. □

Similarly to computing $\Delta_I(\phi)$, given an IPN $\mathcal{I}$, the upper bound $\overline{\Pi}_I(\phi)$ can be computed from the standard network G_U defined as follows:

---

[3] Namely, because of the idempotence property of the max and min operators used in the qualitative possibilistic setting.

Proposition 6. *Let* $\mathcal{I}=<G,\,\Theta^I>$ *denote an IPN. Let* $\mathcal{G}_U=<G,\,\Theta>$ *be the standard possibilistic network such that*

1. $\mathcal{I}$ *and* $\mathcal{G}_U$ *have the same graph and*
2. $\forall a_i \in D_{A_i},\, \theta^{G_U}_{a_i|u_i} = \overline{\theta}^I(a_i \mid u_i).$

$$\text{Then, } \forall \phi \subseteq \Omega, \overline{\Pi}_I(\phi)=\Pi^{G_U}(\phi).$$

Proof sketch. It is easy to show from the interval-based chain rule of Equation 8 that the upper bound is computed only by considering the upper bounds of $a_i|u_i$ over all the compatible networks $\mathcal{G}$ with $\mathcal{I}$. It is also straightforward to show that the network $\mathcal{G}_U$ is a compatible network. $\qquad\square$

This section shows that computing the uncertainty bounds $\overline{\Pi}_I(\phi)$ and $\Delta_I(\phi)$ from interval possibilistic networks has the same computational complexity as the inference process in standard possibilistic networks. Hence, the expressive power of possibilistic networks is extended without increasing their computational complexity.

7   Discussions and Concluding Remarks

This paper provided foundations of interval-based possibilistic networks. It addressed both semantics and inference issues. More precisely, it provided two semantics for IPNs where the first one is based on compatible possibilistic networks while the second is based on interval-based joint possibility distributions. The paper related the two semantics and showed that they are not equivalent. The semantics that can be associated with an IPN is the family of possibility distributions induced from the pointwise-based compatible possibilistic networks. As a first consequence, inference algorithms like the well-known junction tree algorithm [16] cannot directly be adapted for the interval-based setting since such algorithms rely on local distributions unless they are adapted to consider the set of possibility distributions. Indeed, given that an interval-based network $\mathcal{I}$ cannot be represented by an interval-based joint distribution π_I but with $\mathcal{F}_C$, the junction tree algorithm will associate with each clique an interval-based potential (a kind of joint distribution over a subset of variables) while the associated semantics should be a family of compatible distributions. For instance, if we consider an interval-based network with two variables A and B where the graph contains the unique dependence relation $A \rightarrow B$. Using the junction tree algorithm, the potential associated with the clique (A, B) is the interval-based joint distribution over A and B. Clearly, existing propagation algorithms cannot be directly adapted using interval-based joint distributions.

Like possibilistic graphical models, possibilistic logic offers an efficient and compact framework of representing and reasoning with partial ignorance. A possibilistic knowledge base Σ is a set of weighted formulas $\{(\phi_i, \alpha_i),\ i=1,.,n\}$ where ϕ_i is a propositional formula and $\alpha_i \in]0, 1]$ is the necessity degree associated with ϕ_i. In [2] we proposed interval-based possibilistic logic where each propositional logic formula ϕ_i is associated with an interval $[\alpha_i, \beta_i]$ containing the actual necessity degree of ϕ_i. It is shown that inference in interval-based possibilistic knowledge bases does not imply extra complexity in comparison with inference in standard possibilistic logic. A natural question then is whether it is possible to transform an IPN $\mathcal{I}$ into an equivalent interval-based possibilistic logic knowledge base such that inference can be performed efficiently. There are basically two issues making it impossible to rely on transformations from graphical representations to possibilistic logic knowledge bases in order to encode and reason with IPNs. The first issue is that in interval-based possibilistic logic, formulas associated with the necessity degree 0 are ignored which is not the case if we transform an IPN where we have upper bounds equal to 1 (because of the normalization condition). The second issue is related to the semantics underlying the two interval-based representations. While an IPN is seen as a collection of normalized pointwise-based networks, hence *consistent*, an interval-based possibilistic knowledge base is seen as a collection of potentially *inconsistent* pointwise-based possibilistic logic knowledge bases.

To sum up, the main findings of this paper are:

- We viewed an interval-based possibilistic network as a family of compatible and coherent standard possibilistic networks. We provided two semantics for IPNs:
  i) The first one is based on standard possibility distributions induced by the coherent possibilistic networks.
  ii) The second one is based on the interval-based possibility distribution induced by the extended chain rule.
- We provided precise relationships between these two semantics. In particular, we showed that the semantics behind IPNs cannot be an interval-based joint possibility distribution. Hence, interval-based possibility distributions cannot be used for answering queries from IPNs.
- We pointed out that interval-based possibilistic logic cannot be used to perform inference in IPNs since an interval-based possibilistic logic base cannot encode an interval-based possibilistic network.
- Interestingly enough, computing the uncertainty bounds in terms of possibility degree and guaranteed possibility degree of any event can be done using two special standard networks without any extra complexity.

As future works, in addition to real applications with IPNs, we will extend our framework to penalty logic where each conditional event is attached with a weight representing a penalty to pay if it is violated. In some applications, this penalty may be provided with intervals.

References

1. Ben Amor, N., Benferhat, S., Mellouli, K.: A two-steps algorithm for min-based possibilistic causal networks. In: Benferhat, S., Besnard, P. (eds.) ECSQARU 2001. LNCS (LNAI), vol. 2143, pp. 266–277. Springer, Heidelberg (2001)
2. Benferhat, S., Hué, J., Lagrue, S., Rossit, J.: Interval-based possibilistic logic. In: Proceedings of the 22nd International Joint Conference on Artificial Intelligence, Barcelona, Catalonia, Spain, July 16-22, pp. 750–755 (2011)
3. Benferhat, S., Kaci, S.: Logical representation and fusion of prioritized information based on guaranteed possibility measures: application to the distance-based merging of classical bases. Artif. Intell. 148, 291–333 (2003)

4. Benferhat, S., Tabia, K.: Three-valued possibilistic networks. In: 20th European Conference on Artificial Intelligence, Montpellier, France, August 27-31, pp. 157–162. IOS Press (2012)
5. Kruse, R., Borgelt, C., Gebhardt, J.: Possibilistic graphical models. In: International School for the Synthesis of Expert Knowledge (ISSEK 1998), Udine (Italy), pp. 51–68 (1998)
6. Cozman, F.G.: Credal networks. Artif. Intell. 120(2), 199–233 (2000)
7. Darwiche, A.: Modeling and Reasoning with Bayesian Networks. Cambridge University Press (2009)
8. de Campos, C.P.: New complexity results for map in bayesian networks. In: Proceedings of the 22nd International Joint Conference on Artificial Intelligence, IJCAI 2011, Barcelona, Catalonia, Spain, July 16-22, pp. 2100–2106 (2011)
9. de Campos, C.P., Cozman, F.G.: The inferential complexity of bayesian and credal networks. In: IJCAI 2005, Proceedings of the Nineteenth International Joint Conference on Artificial Intelligence, Edinburgh, Scotland, UK, July 30-August 5, pp. 1313–1318 (2005)
10. de Campos, L., Huete, J., Moral, S.: Uncertainty management using probability intervals. In: Bouchon-Meunier, B., Yager, R.R., Zadeh, L.A. (eds.) IPMU 1994. LNCS, vol. 945, pp. 190–199. Springer, Heidelberg (1995)
11. Dubois, D., Hájek, P., Prade, H.: Knowledge-driven versus data-driven logics. J. of Logic, Lang. and Inf. 9, 65–89 (2000)
12. Dubois, D., Lang, J., Prade, H.: Timed possibilistic logic. Fundam. Inf. 15, 211–234 (1991)
13. Guo, P., Tanaka, H.: Decision making with interval probabilities. European Journal of Operational Research 203(2), 444–454 (2010)
14. Ha, V.A., Doan, A., Vu, V.H., Haddawy, P.: Geometric foundations for interval-based probabilities. Annals of Mathematics and Artificial Intelligence 24, 1–21 (1998)
15. Hisdal, E.: Conditional possibilities independence and non interaction. Fuzzy Sets and Systems, 283–297 (1978)
16. Jensen, F.V.: An Introduction to Bayesian Networks. UCL Press, London (1996)
17. Shafer, G.: A Mathematical Theory of Evidence. Princeton University Press, Princeton (1976)
18. Walley, P.: Statistical reasoning with imprecise probabilities (1991)
19. Zadeh, L.: Fuzzy sets as a basis for a theory of possibility. Fuzzy Sets and Systems 100 (suppl. 1), 9–34 (1999)

Tractable vs. Intractable Cases of Query Answering under Matching Dependencies

Leopoldo Bertossi[1] and Jaffer Gardezi[2]

[1] Carleton University, SCS ,Ottawa, Canada
[2] University of Ottawa, SITE., Ottawa, Canada

Abstract. Matching Dependencies (MDs) are a recent proposal for declarative entity resolution. They are rules that specify, on the basis of similarities satisfied by values in a database, what values should be considered duplicates, and have to be matched. On the basis of a chase-like procedure for MD enforcement, we can obtain clean (duplicate-free), and possibly several, *resolved* instances. The *resolved answers* to a query are invariant under the class of resolved instances. Previous work identified classes of queries and sets of MDs for which resolved query answering is tractable, with special emphasis on cyclic sets of MDs. In this work we further investigate the complexity of this problem, identifying intractable cases, and exploring the frontier between tractability and intractability. We concentrate mostly on acyclic sets of MDs. For a special case we obtain a dichotomy result relative to *NP*-hardness.

1 Introduction

A database may contain several representations of the same external entity, i.e. "duplicates", which may be undesirable; and the database has to be cleaned. The problem of *duplicate- or entity-resolution* (ER) is about (a) detecting duplicates, and (b) merging duplicate representations into single representations. It is a classic and complex problem in data management and data cleaning, in particular [7, 9, 3]. In this work we deal with the merging part of the problem, in a relational context.

The problem can be approached by specifying what attribute values have to be matched (made identical) under what conditions. For this, a declarative language with a precise semantics can be used. In this direction, matching dependencies (MDs) have been recently introduced [10, 11]. They represent rules for resolving pairs of duplicate representations (two tuples at a time). When certain similarity relationships between attribute values hold, an MD indicates what attribute values have to be made the same.

Example 1. The similarities of phone and address indicate that the tuples refer to the same person, and the names should be matched. Here, 723-9583 $\approx$ (750) 723-9583 and 10-43 Oak St. $\approx$ 43 Oak St. Ap. 10.

People	Name	Phone	Address
	John Smith	723-9583	10-43 Oak St.
	J. Smith	(750) 723-9583	43 Oak St. Ap. 10

This MD captures this resolution policy: (with P standing for predicate *People*): $P[Phone] \approx P[Phone] \wedge P[Address] \approx P[Address] \rightarrow P[Name] \doteq P[Name]$. It involves only one database predicate, but an MD may involve two different relations. We can also have several, interacting MDs on the schema. □

U. Straccia and A. Cal` (Eds.): SUM 2014, LNAI 8720, pp. 51–65, 2014.
© Springer International Publishing Switzerland 2014

The framework for MD-based ER we use was introduced in [12], with a precise, chase-based semantics for the MDs originally introduced in [11]. The problem of *resolved query answering* (RQA) was introduced in [12]. For a fixed set of MDs, and a fixed query, it is about deciding, given an "unresolved" instance, and a candidate query answer $\bar{a}$, whether $\bar{a}$ is an answer to the query under all admissible ways of resolving the duplicates as dictated by the MDs. This problem is generally intractable [12].

The RQA problem was studied further in [14, 13]. A class of tractable cases of RQA was identified [14], for which a technique based on query rewriting into stratified Datalog with aggregation was developed [13]. In those tractable cases, we find conjunctive queries with certain restrictions on joins, and sets of MDs that *cyclically* depend on each other. These are the (cyclic) *HSC sets* identified in [14]. It was shown that, in general, cyclic dependencies on MDs make the problem tractable, because the requirement of chase termination imposes a relatively simple structure on the clean instances [14].

We concentrate here on *acyclic sets* of MDs, which completely change the picture wrt. previous work. As just mentioned, for HSC sets, tractability of RQA holds [14]. This is the case, e.g., for the cyclic $M = \{R[A] \approx R[A] \to R[B] \doteq R[B], R[B] \approx R[B] \to R[A] \doteq R[A]\}$. However, as we will show, for the following acyclic, somehow syntactically similar example, $M' = \{R[A] \approx R[A] \to R[B] \doteq R[B], R[B] \approx R[B] \to R[C] \doteq R[C]\}$, RQA can be intractable. This example, and our general results, show that, possibly counter-intuitively, the presence of cycles in sets of MDs tends to make resolved query answering easier in comparison with the acyclic case.

We further explore the complexity of RQA. Instead of considering isolated intractable cases as in [12, 14], we take a more systematic approach, developing syntactic criteria on sets of two MDs that, when satisfied by a given pair of MDs, implies intractability of RQA. We show, under an additional assumption about the similarity operators, that RQA is tractable for sets of MDs not satisfying these criteria, leading to a dichotomy result. We extend these results also considering (in)tractability of sets of more than two MDs. All the results apply to acyclic sets of MDs, and are complementary to those in [14, 13], providing a broader picture of the complexity of RQA.

Summarizing, in this paper, we undertake a systematic investigation of the data complexity of the problems of deciding and computing resolved answers to conjunctive queries under MDs. This sheds light on the intrinsic computational limitations of retrieving, from a database with unresolved duplicates, the information that is invariant under the ER process as captured by MDs. Our contributions are the following:

1. We identify a class of conjunctive queries that are relevant for the investigation of tractability vs. intractability of RQA. Intuitively, these queries return data that can be modified by application of the MDs. We call them *changeable attribute queries*.

2. Having investigated in [13, 14] cases of cyclic sets of MDs, we complement these results by studying the complexity of RQA for sets of MDs that do not have cycles.

3. For certain *pairs* of MDs that satisfy a syntactic condition, we establish an intractability result, proving that deciding resolved answers to changeable attribute queries is *NP*-hard in data.

4. For similarity relations that are transitive (a special case), we establish that the conditions for hardness mentioned in the previous item, lead to a dichotomy result: pairs of MDs that satisfy them are always *hard*, otherwise they are always *easy* (for RQA). This shows, in particular, that the result mentioned in item 3. cannot be extended to a wider

class of MDs for arbitrary similarity relations. We also prove that the dichotomy result does not hold when the hypothesis on similarity is not satisfied.

5. Relying on the results for pairs of MDs, we consider acyclic *sets* of MDs of arbitrary size. In particular, we prove intractability of the RQA problem for certain acyclic sets of MDs that have the syntactic property of *non-inclusiveness*.

The structure of the paper is as follows. Section 2 introduces notation, terminology, and previous results. Section 3 identifies classes of MDs, queries and assumptions that are relevant for this research. Sections 3.1 and 4 investigate the complexity of the problem of computing resolved answers for sets of two MDs. Section 5 extends those results to sets of MDs of arbitrary size. Section 6 summarizes results, and makes comparisons with *consistent query answering*. Full proofs of our results can be found in [6].

2 Preliminaries

In this work we consider relational database schemas and instances. Schemas are usually denoted with $\mathcal{S}$, and contain relational predicates. Instances are usually denoted with D. Matching dependencies (MDs) are symbolic rules of the form:

$$\bigwedge_{i,j} R[A_i] \approx_{ij} S[B_j] \rightarrow \bigwedge_{k,l} R[A_k] \doteq S[B_l], \tag{1}$$

where R, S are relational predicates in $\mathcal{S}$, and the $A_i, \ldots$ are attributes for them. The LHS captures similarity conditions on a pair of tuples belonging to the extensions of R and S in an instance D. We abbreviate (1) as: $R[\bar{A}] \approx S[\bar{B}] \rightarrow R[\bar{C}] \doteq S[\bar{E}]$.

The similarity predicates (or operators) $\approx$ are domain-dependent and treated as built-ins. For them we assume symmetry and reflexivity, but not transitivity.

MDs have a *dynamic interpretation* requiring that those values on the RHS should be updated to some arbitrary value in common (of the database domain). Attributes on a RHS of an MD are called *changeable*. MDs are expected to be "applied" iteratively until duplicates are solved, through *chase sequences*. In order to keep track of the changes and comparing tuples and instances, we use global tuple identifiers, a non-changeable surrogate key for each database predicate. The auxiliary, extra attribute (when shown) appears as the first attribute in a relation, e.g. t is the identifier in $R(t, \bar{x})$. A *position* is a pair (t, A) with t a tuple id, and A an attribute (of the relation where t is an id). The *position's value*, $t[A]$, is the value for A in tuple (with id) t.

2.1 MD Semantics

The semantics of MDs acting on database instances [12] is based on a *chase procedure* that is iteratively applied to the original instance D. A *resolved instance* D' is obtained from a finitely terminating sequence of database instances:

$$D =: D_0 \mapsto D_1 \mapsto D_2 \mapsto \cdots \mapsto D_n =: D'. \tag{2}$$

D' satisfies the MDs as *equality generating dependencies* [1], i.e. replacing $\doteq$ by $=$.

The semantics specifies the one-step transitions or updates applied to go from D_{i-1} to D_i, i.e. "$\mapsto$" in (2). Only *modifiable positions* within the instance are allowed to change their values in such a step, and as forced by the MDs. Actually, the modifiable positions syntactically depend on the set M of MDs and the instance at hand; and can be recursively defined (see [12, 14] for the details).[1] Intuitively, a position (t, A) is

---

[1] As a consequence, a *changeable* attribute may not necessarily give rise to a corresponding *modifiable* position for a given instance at hand.

modifiable iff: (a) There is a t' such that t and t' satisfy the similarity condition of an MD with A on the RHS; and (b) $t[A]$ has not already been resolved (it is different from one of its other duplicates).

Example 2. For schema $R(A, B)$, consider the MD $R[A] = R[A] \rightarrow R[B] \doteq R[B]$, and the instance $R(D)$ below. The positions of the underlined values in D are modifiable, because their values are unresolved (wrt the MD and instance $R(D)$).

$R(D)$	A	B
t_1	a	$\underline{b}$
t_2	a	$\underline{c}$

$\mapsto$

$R(D')$	A	B
t_1	a	d
t_2	a	d

D' is a resolved instance since it satisfies the MD interpreted as the FD $R : A \rightarrow B$. Here, the update value d is arbitrary.

D' has no modifiable positions with unresolved values: the values for B are already the same, so there is no reason to change them (and we don't). $\qquad\square$

More formally, the *single step semantics* ($\mapsto$ in (2)) is as follows. Each pair (D_i, D_{i+1}) in an update sequence (2), i.e. a chase step, must *satisfy* the set M of MDs *modulo unmodifiability*, denoted $(D_i, D_{i+1}) \models_{um} M$, which holds iff: (a) For every MD in M, say $R[\bar{A}] \approx S[\bar{B}] \rightarrow R[\bar{C}] \doteq S[\bar{D}]$ and pair of tuples t_R and t_S, if $t_R[\bar{A}] \approx t_S[\bar{B}]$ in D_i, then $t_R[\bar{C}] = t_S[\bar{D}]$ in D_{i+1}; and (b) The value of a position can only differ between D_i and D_{i+1} if it is modifiable wrt D_i. Accordingly, in (2) we also require that $(D_i, D_i) \not\models_{um} M$, for $i < n$, and $(D_n, D_n) \models_{um} M$ (the *stability* condition).[2]

This semantics captures as close as possible the spirit of MDs as originally, and rather informally introduced in [11], and also *uncommitted* in the sense that the MDs do not specify how the matchings have to be realized (also as in [11]).

Example 3. Consider the instance $R(D)$ below and the set of MDs: $\{R[A] = R[A] \rightarrow R[B] \doteq R[B]; R[B] = R[B] \rightarrow R[C] \doteq R[C]\}$. Attribute $R(C)$ is changeable. Position (t_2, C) is not modifiable wrt. M and D: There is no justification to change its value *in one step* on the basis of an MD and D. Position (t_1, C) is modifiable. D has two resolved instances, $R(D_1)$ and $R(D_2)$. $R(D_1)$ cannot be obtained in a single (one step) update (the underlined value is for a non-modifiable position). But $R(D_2)$ can.

$R(D)$	A	B	C
t_1	a	b	d
t_2	a	c	$\underline{e}$
t_3	a	b	e

$R(D_1)$	A	B	C
t_1	a	b	d
t_2	a	b	d
t_3	a	b	d

$R(D_2)$	A	B	C
t_1	a	b	e
t_2	a	b	e
t_3	a	b	e

$\qquad\square$

For arbitrary sets of MDs, some (admissible) chase sequences may not terminate. However, it can be proved that there are always terminating chase sequences. As a consequence, for some sets of MDs, there are both terminating and non-terminating chase sequences. In any case, the class of resolved instances is always well-defined.

We prefer *resolved instances* that are the closest to the original instance. A *minimally resolved instance* (MRI) of D is a resolved instance D' whose *number of changes of attribute values* wrt. D is a minimum. In Example 3, instance D_2 is an MRI, but not D_1 (2 vs. 3 changes). We denote with $Res(D, M)$ and $MinRes(D, M)$ the classes of resolved, resp. minimally resolved, instances of D wrt M.

Infinite chase sequences may occur when the MDs *cyclically* depend on each other, in which case updated instances in a such a sequence may alternate between two or more states [14, Example 6]. However, for the chase sequences that do terminate in

---

[2] The case $D' = D_0$ occurs only when D is already resolved.

a minimally resolved instance, the chase imposes a relatively easily characterizable structure [14, 13], allowing us to obtain a query rewriting methodology. So, cycles help us achieve tractability for some classes of queries [13] (cf. Section 2.2).

On the other side, it has been shown that if a set of MDs satisfies a certain *acyclicity* property, then all chase sequences terminate after a number of iterations that depends only on the set of MDs and not on the instance [12, Lemma 1] (cf. Theorem 1 below). But the number of resolved instances may still be "very large". Sets of MDs considered in this work are acyclic.

2.2 Resolved Query Answers

Given a conjunctive query $\mathcal{Q}$, a set of MDs M, and an instance D, the *resolved answers* to $\mathcal{Q}$ from D are invariant under the entity resolution process, i.e. they are answers to $\mathcal{Q}$ that are true in all MRIs of D:

$$ResAns_M(\mathcal{Q}, D) := \{\ \bar{a}\ |\ D' \models \mathcal{Q}[\bar{a}], \text{ for every } D' \in MinRes(D, M)\}. \quad (3)$$

The *resolved query answering* (RQA) corresponding decision problem is $RA(\mathcal{Q}, M) :$ $= \{(D, \bar{a}) \mid \bar{a} \in ResAns_M(\mathcal{Q}, D)\}$.

In [13, 14], a query rewriting methodology for computing resolved answers to queries under MDs was presented. In this case, the rewritten queries turn out to be Datalog queries with counting, and can be obtained for two main kinds of sets of MDs: (a) MDs do not depend on each other, i.e. *non-interacting* sets of MDs [12]; (b) MDs that depend cyclically on each other, e.g. $R[A] \approx R[A] \rightarrow R[B] \doteq R[B]$ and $R[B] \approx R[B] \rightarrow R[A] \doteq R[A]$

For these sets of MDs, a conjunctive query can be rewritten to retrieve, in polynomial time in data, the resolved answers, provided the queries have no joins on existentially quantified variables corresponding to changeable attributes. The latter form the class of *unchangeable attribute join conjunctive* (UJCQ) queries [14].

For example, for the MD $R[A] = R[A] \rightarrow R[B, C] \doteq R[B, C]$ on schema $R[A, B, C]$, $\mathcal{Q}$: $\exists x \exists y \exists z (R(x, y, c) \wedge R(z, y, d))$ is *not* UJCQ; whereas $\mathcal{Q}'$: $\exists x \exists z (R(x, y, z) \wedge R(x, y', z'))$ is UJCQ. For queries outside UJCQ, the resolved answer problem can be intractable even for one MD [14].

The set of MDs (4), which is neither non-interacting nor cyclic, is not covered by the positive cases for Datalog rewriting above.

$$R[A] \approx R[A] \rightarrow R[B] \doteq R[B], \quad (4)$$
$$R[B] \approx R[B] \rightarrow R[C] \doteq R[C],$$

Actually, for this set RQA becomes intractable for very simple queries, like $\mathcal{Q}(x, z) :$ $\exists y R(x, y, z)$, that is UJCQ [12]. Sets of MDs like (4) are the main focus of this work.

3 Intractability of RQA

We just briefly described classes of queries and MDs for which RQA can be done in polynomial time in data (via the Datalog rewriting). We showed that there are intractable cases, by pointing to a specific query and set of MDs. Natural questions that we start addressing are: (a) What is the complexity of RQA outside the Datalog rewritable cases? (b) Do the exhibited query and MDs fall into a more general intractable class?

For all sets M of MDs we consider below, *we assume that at most two relational predicates, say R, S, appear in M*, e.g. as in $M = \{R[A] \approx S[B] \rightarrow R[C] \doteq S[E]\}$. In same cases we assume that there are exactly two predicates. The purpose of this restriction is to simplify the presentation. All results can be generalized to sets of MDs with more than two predicates. To do this, definitions and conditions concerning the two relations in the MDs can be extended to cover the additional relations as well.

At the other extreme, when a single predicate occurs in M, say R, as in Example 3, the results for at most two predicates can be reformulated and applied by replacing S with R'. Although R and R' are the same relation in this case, the prime is used to distinguish between the two tuples to which the MD refers.

All the sets of MDs considered below are both *interacting* (non-interaction does not bring complications) and *acyclic*. Both notions and others can be captured in terms of the MD *graph*, $MDG(M)$, of M. It is a directed graph, such that, for $m_1, m_2 \in M$, there is an edge from m_1 to m_2 if there is an overlap between $RHS(m_1)$ and $LHS(m_2)$ (the right- and left-hand sides of the arrows as sets of attributes) [12]. Accordingly, M is *acyclic when $MDG(M)$ is acyclic*. In fact, the sets of MDs in this work satisfy a stronger property, defined below, which we call *strong acyclicity*.

Definition 1. [12] 1. Let M be a set of MDs on schema S. (a) The symmetric binary relation $\doteq_r$ relates attributes $R[A]$, $S[B]$ of S whenever there is $m \in M$ in which $R[A] \doteq S[B]$ occurs. (b) The *attribute closure* of M is the reflexive and transitive closure of $\doteq_r$. (c) $E_{R[A]}$ denotes the equivalence class of attribute $R[A]$ in the attribute closure of M.
2. The *augmented MD-graph* of M, denoted $AMDG(M)$, is a directed graph with a vertex labeled with m for each $m \in M$, and with an edge from m to m' iff there is an attribute, say $R[A]$, with $R[A] \in RHS(m)$ and $E_{R[A]} \cap LHS(m') \neq \emptyset$.
3. M is *strongly acyclic* if $AMDG(M)$ has no cycles.   □

Because $R[A] \in E_{R[A]}$, for any set M of MDs, all edges in $MDG(M)$ are also edges in $AMDG(M)$. So, strong acyclicity implies acyclicity, but the converse is not true.

Example 4. The set M of MDs $m_1: R[F] \approx S[G] \rightarrow R[A] \doteq S[H]$; $m_2: R[A] \approx S[B] \rightarrow R[C] \doteq S[E]$; $m_3: R[C] \approx S[E] \rightarrow R[I] \doteq S[H]$ is acyclic but not strongly acyclic. $MDG(M)$ has three vertices, m_1, m_2, m_3, and edges (m_1, m_2) and (m_2, m_3). $AMDG(M)$ has the additional edge (m_3, m_2), because $E_{R[I]} = \{R[I], S[H], R[A]\} \cap LHS(m_2) = \{R[A]\}$.   □

In this work, we consider strongly acyclic sets of MDs. In particular, two interesting and common kinds that form large classes of sets M of MDs: *linear pairs*, which consist of two MDs such that $MDG(M)$ contains a single edge from one to the other (c.f. Definition 5); and acyclic sets that are *pair-preserving* (c.f. Definition 7). From the definitions of these two kinds of sets of MDs it will follow that they are strongly acyclic.

Theorem 1. [12] Let M be a strongly acyclic set of MDs on schema S, and D an instance for S. Every sequence of M-based updates to D as in (2) terminates with a resolved instance after at most $d + 1$ steps, where d is the maximum length of a path in $AMDG(M)$.   □

As mentioned previously, there may be infinite chase sequences when M is not acyclic. For the cyclic case, Theorem 1 only tells us about the chase termination and lengths, but not about the data; and does not guarantee tractability for RQA. This leaves room for tractable and intractable cases. Actually, it can still be the case that there are exponentially many MRIs. A reason for this is that the application of an MD to an instance may produce new similarities among the values of attributes in $RHS(m_1)$ that are not strictly required by the chase, but result from a particular choice of update values. Such "accidental similarities" affect subsequent updates, resulting in exponentially many possible update sequences. This is illustrated in the next example.

Example 5. Consider the strongly acyclic set M: $R[A] \approx R[A] \rightarrow R[B] \doteq R[B]$; $R[B] \approx R[B] \rightarrow R[C] \doteq R[C]$. When the following instance is updated according to M, the sets of value positions $\{t_1[B], t_2[B]\}$ and $\{t_3[B], t_4[B]\}$ must be merged.

$R(D_1)$	A	B	C		$R(D_1)$	A	B	C
t_1	a	m	e		t_1	a	m	e
t_2	a	d	f	$\mapsto$	t_2	a	m	f
t_3	b	c	g		t_3	b	m	g
t_4	b	k	h		t_4	b	m	h

One possible update is as above. The similarities between the attribute B values of the top and bottom pairs of tuples are accidental, because they result from the choice of update values. In the absence of accidental similarities, there is only one possible set of sets of values that are merged in the second update, namely $\{\{t_1[C], t_2[C]\}, \{t_3[C], t_4[C]\}\}$.

Accidental similarities increase the complexity of query answering over the instance by adding another possible set of sets of merged values, $\{\{t_1[C], t_2[C], t_3[C], t_4[C]\}\}$. More generally, for an instance with n sets of merged value positions in the B column, the number of possible sets of sets of value positions in the C column that are merged in the second update is $\Omega(2^{n^2})$.   □

We want to investigate the frontier between tractability and intractability. For this reason, we make the assumption that, *for each similarity relation, $\approx$, there is an infinite set of mutually dissimilar values*. Actually, without this assumption, the resolved answer problem becomes immediately tractable for certain similarity operators (e.g. transitive similarity operators). This is because, for these operators, the whole class of minimal resolved instances of an instance can be computed in polynomial time.

Proposition 1. For strongly acyclic sets of MDs, if the similarity predicates are transitive and there is no infinite set of mutually dissimilar values, the set of minimal resolved instances for a an instance D can be computed in polynomial time in the size of D. □

Our next results require some terms and notation that we now introduce.

Definition 2. For a set M of MDs with predicates R and S, a *changeable attribute query* $\mathcal{Q}$ is a (conjunctive) query in UJCQ, containing a conjunct of the form $R(\bar{x})$ or $S(\bar{y})$ all whose variables are free and none occur in another conjunct of the form $R(\bar{x})$ or $S(\bar{y})$. Such a conjunct is a *join-restricted free occurrence* of the predicate R or S. □

By definition, the class of *changeable attribute queries* (CHAQ) is a subclass of UJCQ. Both depend on the set of MDs at hand. For example, for the MDs in (4), $\exists y R(x, y, z) \in$ UJCQ $\setminus$ CHAQ, but $\exists w \exists t (R(x, y, z) \land S(x, w, t)) \in$ CHAQ. We confine our attention to UJCQ and subsets of it, because, as mentioned in the previous section, intractability limits the applicability of the duplicate resolution method for queries outside UJCQ.

The requirement that the query contains a join-restricted free occurrence of R or S eliminates from consideration certain queries in UJCQ for which the resolved answer problem is immediately tractable. For example, for the MDs in (4), the query $\exists y \exists z R(x, y, z)$ is not CHAQ, and is tractable simply because it does not return the values of a changeable attribute (the resolved answers are the answers in the usual sense). The restriction on joins simplifies the analysis while still including many useful queries.

In order to eliminate queries like $\exists y \exists z R(x, y, z)$ wrt M in (4), CHAQ imposes a strong condition. Actually, the condition can be weakened, requiring to have *at least one* of the variables satisfying the condition in the definition for CHAQ. Weakening the condition makes the presentation much more complex since a finer interaction with the MDs has to be brought into the picture. (We leave this issue for an extended version.)

Definition 3. A set M of MDs is *hard* if, for every CHAQ $\mathcal{Q}$, $RA(\mathcal{Q}, M)$ is *NP*-hard. M is *easy* if, for every CHAQ $\mathcal{Q}$, $RA(\mathcal{Q}, M)$ is in *PTIME*.   □

Of course, a set of MDs still may not be hard or easy. For the resolved answer problem, membership of *NP* is open. However, for strongly acyclic sets, the bound on the length of the chase implies an upper bound of Π_2^P [12, Theorem 5].

In the following we give some syntactic conditions that guarantee hardness for classes of MDs. To state them we need to introduce some useful notions first.

Definition 4. For an MD m, the symmetric binary relation $LRel(m)$ ($RRel(m)$) relates pairs of attributes $R[A]$ and $S[B]$ when $R[A] \approx S[B]$ (resp. $R[A] \doteq S[B]$) appears in m. An *L-component* (*R-component*) of m is an equivalence class of the reflexive, transitive closure $LRel(m)^{=+}$ (resp. $RRel(m)^{=+}$) of $LRel(m)$ (resp. $RRel(m)$).   □

Example 6. For $R[A] \approx S[B] \land R[A] \approx S[C] \rightarrow R[E] \doteq S[F] \land R[G] \doteq S[H]$, there is only one L-component: $\{R[A], S[B], S[C]\}$; and two R-components: $\{R[E], S[F]\}$ and $\{R[G], S[H]\}$.   □

3.1 Hardness of Linear Pairs of MDs

Most of the results that follow already hold for pairs of MDs.

Definition 5. A set $M = \{m_1, m_2\}$ of MDs is a *linear pair*, denoted (m_1, m_2), if its graph $MDG(M)$ consists of vertices m_1 and m_2 with only an edge from m_1 to m_2. □
Notice that if (m_1, m_2) is a generic linear pair, say

$$m_1: \ R[\bar{A}] \approx_1 S[\bar{B}] \rightarrow R[\bar{C}] \doteq S[\bar{E}], \tag{5}$$
$$m_2: \ R[\bar{F}] \approx_2 S[\bar{G}] \rightarrow R[\bar{H}] \doteq S[\bar{I}],$$

then, from the definition of the MD graph, it follows that $(R[\bar{C}] \cup S[\bar{E}]) \cap (R[\bar{F}] \cup S[\bar{G}]) \neq \emptyset$, whereas $(R[\bar{H}] \cup S[\bar{I}]) \cap (R[\bar{A}] \cup S[\bar{B}]) = \emptyset$. In the following we have to analyze other different forms of (non-)interaction between the attributes in linear pairs.

Definition 6. Let (m_1, m_2) be a linear pair as in (5). (a) B_R is a binary (reflexive and symmetric) relation on attributes of R: $(R[U_1], R[U_2]) \in B_R$ iff $R[U_1]$ and $R[U_2]$ are in the same R-component of m_1 or the same L-component of m_2. Similarly for B_S. (b) An *R-equivalent set* (*R-ES*) of attributes of (m_1, m_2) is an equivalence class of $TC(B_R)$, the transitive closure of B_R, with at least one attribute in the equivalence class belonging to $LHS(m_2)$. The definition of an *S-equivalent set* (*S-ES*) is similar, with R replaced by S. (c) An (R or S)-ES E of (m_1, m_2) is *bounded* if $E \cap LHS(m_1)$ is non-empty.   □

Example 7. Consider the schema $R[A, C, F, H, I, M]$, $S[B, D, E, G, N]$, and the linear pair (m_1, m_2) with:

m_1: $R[A] \approx S[B] \rightarrow R[C] \doteq S[D] \land R[C] \doteq S[E] \land R[F] \doteq S[G] \land R[H] \doteq S[G]$,
m_2: $R[F] \approx S[E] \land R[I] \approx S[E] \land R[A] \approx S[E] \land R[F] \approx S[B] \rightarrow R[M] \doteq S[N]$.

It holds: (a) $B_R(R[F], R[H])$ due to the occurrence of $R[F] \doteq S[G]$, $R[H] \doteq S[G]$. (b) $B_R(R[F], R[I])$ due to $R[F] \approx S[E]$, $R[I] \approx S[E]$. (c) $B_R(R[I], R[A])$ due to $R[I] \approx S[E]$, $R[A] \approx S[E]$. (d) $\{R[A], R[F], R[I], R[H]\}$ is an R-ES, and since $\{R[A], R[F], R[I], R[H]\} \cap LHS(m_1) = \{R[A]\} \neq \emptyset$, it is also bounded.   □

Theorem 2. Let (m_1, m_2) be a linear pair, with relational predicates R and S. Let E_R, E_S be the sets of R-ESs and S-ESs, resp. The pair (m_1, m_2) is hard if $RHS(m_1) \cap RHS(m_2) = \emptyset$, and at least one of (a) and (b) below holds: (a) All of the following hold: (i) $Attr(R) \cap (RHS(m_1) \cap LHS(m_2)) \neq \emptyset$. (ii) There are unbounded ESs in E_R. (iii) For some L-component L of m_1, $Attr(R) \cap (L \cap LHS(m_2)) = \emptyset$. (b) Same as (a), but with R replaced by S.   □

Theorem 2 says that a linear pair of MDs is hard unless the syntactic form of the MDs is such that there is a certain association between changeable attributes in $LHS(m_2)$ and attributes in $LHS(m_1)$ as specified by conditions (ii) and (iii).

For pairs of MDs satisfying the negation of (a)(ii) or that of (a)(iii) (or the negation of (b)(ii) or that of (b)(iii)) in Theorem 2, the similarities resulting from applying m_2 are restricted to a subset of those that are already present among the values of attributes in $LHS(m_1)$, making the problem tractable. However, when condition (ii) or (iii) is satisfied, accidental similarities among the values of attributes in $RHS(m_1)$ cannot be passed on to values of attributes in $RHS(m_2)$.

Example 8. The linear pair (m_1, m_2), with m_1 : $R[A] \approx S[B] \rightarrow R[C] \doteq S[D]$; m_2: $R[C] \approx S[D] \rightarrow R[E] \doteq S[F]$, is hard. In fact, first: $RHS(m_1) \cap RHS(m_2) = \emptyset$.

Now, it satisfies condition (a): Condition (a)(i) holds, because $R[C] \in RHS(m_1) \cap LHS(m_2)$. Conditions (a)(ii) and (a)(iii) are trivially satisfied, because there are no attributes of $LHS(m_1)$ in $LHS(m_2)$.   □

As mentioned above, Theorem 2 generalizes to the case of more or fewer than two database predicates. It is easy to verify, for the former case, that if there are more than two predicates in a linear pair, then there must be exactly three of them, one of which appears in both MDs. In this case, hardness is implied by condition (a) in Theorem 2 alone, with R the predicate in common.

Example 9. The linear pair (m_1, m_2) with three predicates m_1 : $R[A] \approx S[B] \rightarrow R[C] \doteq S[E]$; m_2: $R[C] \approx P[B] \rightarrow R[F] \doteq P[G]$ is hard if it satisfies condition (a) in Theorem 2, which it does: (i) $Attr(R) \cap (RHS(m_1) \cap LHS(m_2)) = \{R[C]\}$. (ii) The ES $\{R[C]\}$ is unbound. (iii) Part (iii) holds with $L = \{R[A], S[B]\}$.   □

With only one predicate R in the linear pair, in order to apply Theorem 2, we need to derive from it a special result, Corollary 1 below. It is obtained by first labeling the different occurrences of the (same) predicate in M, and then generating conditions (four of them, analogous to (a) and (b) in Theorem 2) for the labeled version, M'. When M' satisfies those conditions, the original set M is hard. An algorithm, *Conditions*, (described in detail in [6]) does both the labeling and the condition generation to be checked on

M'. After the labeling, there is still only one predicate in M'. The labeling simply provides a convenient way to refer to different sets of attributes. Example 10 demonstrates in informal terms the use of the algorithm and the application of the corollary.

Corollary 1. A linear pair containing one predicate is hard if it satisfies $RHS(m_1) \cap RHS(m_2) = \emptyset$ and at least one of the four sets of three conditions (i)-(iii) generated by Algorithm *Conditions*. $\square$

Example 10. Consider the linear pair M: m_1 : $R[A] \approx R[B] \wedge R[C] \approx R[E] \rightarrow R[F] \doteq R[G] \wedge R[B] \doteq R[G]$; m_2: $R[G] \approx R[H] \wedge R[B] \approx R[I] \wedge R[L] \approx R[I] \rightarrow R[J] \doteq R[K]$. Algorithm *Conditions* produces the following labeling:

$$m_1': \ R_1^1[A] \approx R_1^2[B] \wedge R_1^1[C] \approx R_1^2[E] \rightarrow R_1^1[F] \doteq R_1^2[G] \wedge R_1^1[B] \doteq R_1^2[G],$$
$$m_2': \ R_2^1[G] \approx R_2^2[H] \wedge R_2^1[B] \approx R_2^2[I] \wedge R_2^1[L] \approx R_2^2[I] \rightarrow R_2^1[J] \doteq R_2^2[K].$$

With the above labeling, R^1 (R^2)-equivalent sets can be defined analogously to R (S)-equivalent sets in the two relation case, except that they generally include attributes from two "relations", R_1^1 and R_2^1 (R_1^2 and R_2^2), instead of one. For example, in $\{m_1', m_2'\}$, one R^1-ES is $\{R_1^1[F], R_1^1[B], R_2^1[B], R_2^1[L]\}$.

The conditions output by *Conditions* for the combination $X = 1$, $Y = 2$ is the following: (i) $Attr(R_1^2) \cap Attr(R_2^1) \cap (RHS(m_1) \cap LHS(m_2)) \neq \emptyset$, (ii) There are R^1-equivalent sets that do not contain attributes in $Attr(R_1^2) \cap LHS(m_1)$, and (iii) For some L-component L of m_1, $Attr(R_1^2) \cap Attr(R_2^1) \cap (L \cap LHS(m_2)) = \emptyset$.

These conditions are satisfied by M'. In fact, for (i) this set is $\{R_2^1[G]\}$; for (ii) the R^1-ES $\{R_2^1[G]\}$ satisfies the condition; and for (iii) $L = \{R_1^1[C], R_1^2[E]\}$ satisfies the condition. Thus, by Corollary 1, M is hard. $\square$

Example 11. (example 5 cont.) M is hard by Corollary 1. In fact, Algorithm *Conditions* produces the following labeled set M': $R_1^1[A] \approx R_1^2[A] \rightarrow R_1^1[B] \doteq R_1^2[B]$; $R_2^1[B] \approx R_2^2[B] \rightarrow R_2^1[C] \doteq R_2^2[C]$, which satisfies the conditions (i)-(iii) for the choice $X = 1$, $Y = 2$: for (i) this set is $\{R[B]\}$; for (ii) the R^1-ES $\{R_1^1[B], R_2^1[B]\}$ satisfies the property; and for (iii) we use $L = \{R_1^1[A], R_1^2[A]\}$.

As mentioned in Section 2.2, for the given M and the query $\mathcal{Q}(x, z)$: $\exists y R(x, y, z)$, RQA is intractable [12]. This query is in UJCQ $\setminus$ CHAQ. Now, we have just obtained that RQA for that M is also intractable for *all* CHAQ queries. $\square$

Example 12. Consider M consisting of m_1 : $R[A] \approx R[A] \rightarrow R[B] \doteq R[B]$; m_2 : $R[A] \approx R[A] \wedge R[B] \approx R[B] \rightarrow R[C] \doteq R[C]$. It does not satisfy the conditions of Theorem 2 (actually, Corollary 1). The sole L-component of m_1 is $\{R[A]\}$, and all attributes of this set occur in $LHS(m_2)$. M is easy, because the non-interacting set $\{R[A] \approx R[A] \rightarrow R[B] \doteq R[B], \ R[A] \approx R[A] \rightarrow R[C] \doteq R[C]\}$ is equivalent to it in the sense that, for any instance, the MRIs are the same for either set. This is because applying m_1 to the tuples of R and S results in an instance such that all pairs of tuples satisfying the first conjunct to the left of the arrow in m_2 satisfy the entire similarity condition. $\square$

Theorem 2 gives a syntactic condition for hardness. It is an important result, because it applies to simple sets of MDs such as that in Example 5 that we expect to be commonly encountered in practice. Moreover, in Section 5, we use Theorem 2 to show that similar sets involving more than two MDs are also hard. The conditions for hardness in Theorem 2 are not necessary conditions. Actually, the set of MDs in Example 13 below is hard, but does not satisfy the conditions of this theorem.

4   A Dichotomy Result

All syntactic conditions/constructs on attributes above, in particular, the transitive closures on attributes, are "orthogonal" to semantic properties of the similarity relations. When similarity predicates are transitive, every linear pair not satisfying the hardness criteria of Theorem 2 is easy.

Theorem 3. Let (m_1, m_2) be a linear pair with $RHS(m_1) \cap RHS(m_2) = \emptyset$. If the similarity operators are transitive, then (m_1, m_2) is either easy or hard. More precisely, if the conditions of Theorem 2 hold, M is hard. Otherwise, M is easy. $\square$

This result does not hold in general when similarity is not transitive (c.f. Proposition 2 below). The possibilities for accidental similarities are reduced by disallowing that two dissimilar values are similar to a same value. Actually, the complexity of the problem is reduced to the point where the resolved answer problem becomes tractable.

Example 13. The linear pair M consisting of $m_1: R[A] \approx S[B] \wedge R[I] \approx S[J] \rightarrow R[E] \doteq S[F]$; $m_2: R[E] \approx S[F] \wedge R[A] \approx S[J] \wedge R[I] \approx S[B] \rightarrow R[G] \doteq S[H]$ does not satisfy the conditions of Theorem 2, because m_1 has two L-components, $\{R[A], S[B]\}$ and $\{R[I], S[J]\}$. Since $LHS(m_2)$ includes one attribute of R and S from each of these L-components, conditions (a)(iii) and (b)(iii) are not satisfied. Then, by Theorem 3, M is easy when $\approx$ is transitive. $\square$

Example 12 showed that a pair of MDs is easy for arbitrary $\approx$ by exhibiting an equivalent non-interacting set. This method cannot be applied in Example 13, because the similarity condition of m_1 is not included in that of m_2. The set of MDs in Example 13 can be hard for non-transitive similarity relations, as the following proposition shows.

Proposition 2. There exist (non-transitive) similarity operators $\approx$ for which the set of MDs in Example 13 is hard. $\square$

5   Hardness of Acyclic Sets of MDs

We consider now acyclic sets of MDs of arbitrary finite size, concentrating on a class of them that is common in practice.

Definition 7. A set M of MDs is *pair-preserving* if for every attribute appearing in M, say $R[A]$, there is exactly one attribute appearing in M, say $S[B]$, such that $R[A] \approx S[B]$ or $R[A] \doteq S[B]$ (or the other way around) occurs in M. $\square$

It is easy to verify that pair-preserving, acyclic sets of MDs are strongly acyclic. Pair-preservation typically holds in ER, because values for pairs of attributes are compared only if they hold the same kind of information (e.g. both addresses or both names).

Example 14. M in Example 12 is pair-preserving. The set of MDs in Example 13 is not pair-preserving, because $S[B]$ is paired with both $R[A]$ and $R[C]$ in m_1. It is also possible for cyclic sets of MDs to be pair-preserving. For example, the set $R[A] \approx R[A] \rightarrow R[B] \doteq R[B]$; $R[B] \approx R[B] \rightarrow R[A] \doteq R[A]$ is pair-preserving. $\square$

Now, recall from the previous section that syntactic conditions on linear pairs (m_1, m_2), like the absence of certain attributes in $LHS(m_1)$ from $LHS(m_2)$ (c.f. conditions (a)(iii) or (b)(iii)), imply hardness. *Non-inclusiveness* wrt. *subsets* of M is a syntactic condition on acyclic, pair-preserving sets M of MDs that generalizes those conditions that ensure hardness for linear pairs.

Definition 8. Let M be acyclic and pair-preserving, B an attribute in M, and $M' \subseteq M$. B is *non-inclusive* wrt. M' if, for every $m \in M \smallsetminus M'$ with $B \in RHS(m)$, there is an attribute C such that: (a) $C \in LHS(m)$, (b) $C \notin \bigcup_{m' \in M'} LHS(m')$, and (c) C is *non-inclusive* wrt. M'. □

This is a recursive definition, with base case when C is not in $RHS(m)$ for any m (then is inclusive, i.e. not non-inclusive). Since $C \in LHS(m)$ in the definition, for any m_1 with $C \in RHS(m_1)$, there is an edge from m_1 to m. Therefore, we are traversing an edge backwards with each recursive step, and the recursion terminates by acyclicity.

Example 15. In the acyclic, pair-preserving set $\{m_1 : R[I] \approx S[J] \to R[A] \doteq S[E];$ $m_2 : R[A] \approx S[E] \to R[C] \doteq S[B];$ $m_3 : R[G] \approx S[H] \to R[I] \doteq S[J]\}$, $R[A]$ is non-inclusive wrt. $\{m_2\}$ because $R[A] \in RHS(m_1)$ and there is an attribute, $R[I]$, in $LHS(m_1)$ that satisfies conditions (a), (b), and (c) of Definition 8. Conditions (a) and (b) are obviously satisfied. Condition (c) is satisfied, because $R[G]$ is non-inclusive wrt. $\{m_1\}$. This is trivially true, since $R[G] \notin RHS(m_1) \cup RHS(m_3)$. □

Non-inclusiveness is a generalization of conditions (a) (iii) and (b) (iii) in Theorem 2 to finite *sets* of MDs. It expresses a condition of inclusion of attributes in the LHS of one MD in the LHS of another. In particular, suppose $M = (m_1, m_2)$ is a pair-preserving linear pair, and take $M' = \{m_2\}$. It is easy to verify that the requirement that there is an attribute in $RHS(m_1)$ that is non-inclusive wrt. M' is equivalent to conditions (a)(iii) and (b)(iii) of Theorem 2. Theorem 4 tells us that a non-inclusive set of MDs is hard.

Theorem 4. Let M be acyclic and pair-preserving. Assume there is $\{m_1, m_2\} \subseteq M$, and attributes $C \in RHS(m_2)$, $B \in RHS(m_1) \cap LHS(m_2)$ with: (a) C is non-inclusive wrt $\{m_1, m_2\}$, and (b) B is non-inclusive wrt $\{m_2\}$. Then, M is hard. □

Example 16. (example 15 cont.) The set of MDs is hard. This follows from Theorem 4, with $\{m_1, m_2\}$ and C, B in Theorem 4 being $\{m_1, m_2\}$ and $R[C], R[A]$ in the example, resp. Part (b) of Theorem 4 was shown in the first part of this example. Part (a) holds trivially, since $R[C] \notin RHS(m_3)$. □

Example 17. Consider $M = \{m_1, m_2, m_3\}$ with $m_1 : R[G] \approx S[H] \to R[I] \doteq S[J];$ $m_2 : R[G] \approx S[H] \wedge R[I] \approx S[J] \to R[A] \doteq S[E];$ $m_3 : R[G] \approx S[H] \wedge R[A] \approx S[E] \to R[C] \doteq S[B]$. It does not satisfy the condition of Theorem 4. The only candidates for $\{m_1, m_2\}$ in Theorem 4 are $\{m_1, m_2\}$ and $\{m_2, m_3\}$ in this example, because of the requirement that $RHS(m_1) \cap LHS(m_2) \neq \emptyset$. In the first case, B in the Theorem 4 is $R[I]$ (or $S[J]$), which does not satisfy (b) because $LHS(m_1) \backslash LHS(m_2) = \emptyset$. In the second case, B in Theorem 4 is $R[A]$ (or $S[E]$). Because $R[G]$ and $S[H]$ are in $LHS(m_3)$, $R[A]$ can only satisfy (b) if $R[I]$ does. $R[I]$ does not satisfy (b), since $LHS(m_1) \backslash LHS(m_3) = \emptyset$.

Actually, M is easy, because it is equivalent to the non-interacting set $m'_1 : R[G] \approx S[H] \to R[I] \doteq S[J];$ $m'_2 : R[G] \approx S[H] \to R[A] \doteq S[E];$ $m'_3 : R[G] \approx S[H] \to R[C] \doteq S[B]$. This can be shown as in Example 12. □

Our dichotomy result applies to linear pairs (and transitive similarities). However, tractability can be obtained in some cases of larger sets of MDs for which hardness cannot be obtained via Theorem 4 (because the conditions do not hold). The following is a general result concerning sets such as M in Example 17.

Theorem 5. Let M be an acyclic, pair-preserving set of MDs. If, for each $m \in M$, all changeable attributes $A \in LHS(m)$ are inclusive wrt $\{m\}$, then M is easy. $\qquad \square$

Example 18. (example 17 cont.) As expected, the set M of MDs $\{m_1, m_2, m_3\}$ satisfies the requirement of Theorem 5. To show this, the only attributes to be tested for inclusiveness wrt. an MD are $R[A]$ and $R[I]$. Specifically, it must be determined whether $R[I]$ is inclusive wrt $\{m_2\}$ and whether $R[A]$ is inclusive wrt $\{m_3\}$. $R[I]$ is inclusive wrt $\{m_2\}$, because all attributes in $LHS(m_1)$ are in $LHS(m_2)$. $R[A]$ is inclusive wrt $\{m_3\}$, since $R[G] \in LHS(m_3)$ and $R[I]$ is inclusive wrt $\{m_3\}$. $\qquad \square$

Example 19. (example 16 cont.) $\{m_1, m_2, m_3\}$ in Example 15 was shown to be hard in Example 16. As expected, it does not satisfy the requirement of Theorem 5. This is because $R[A]$ is changeable, $R[A] \in LHS(m_2)$, and $R[B]$ is non-inclusive wrt $\{m_2\}$ since $R[I] \in LHS(m_1)$, $R[I] \notin LHS(m_2)$, and $R[I]$ is non-inclusive wrt $\{m_2\}$. $\qquad \square$

The conditions of Theorems 4 and 5 are mutually exclusive: B in Theorem 4 is changeable (since $B \in RHS(m_1)$), $B \in LHS(m_2)$, and B is non-inclusive wrt $\{m_2\}$. Together, they do not provide a dichotomy result, as the following example shows.

Example 20. The set formed by $m_1: R[E] \approx R[E] \to R[B] \doteq R[B]$; $m_2: R[B] \approx R[B] \to R[C] \doteq R[C]$; $m_3: R[E] \approx R[E] \to R[C] \doteq R[C]$ does not satisfy the condition of Theorem 5, because $R[B]$ is changeable and non-inclusive wrt. $\{m_2\}$. Nor condition (a) of Theorem 4, because C is inclusive wrt. $\{m_1, m_2\}$ ($R[E] \in LHS(m_1)$).

The tractability of this case cannot be determined through the theorems above, but it is easy, because, for any update sequence that leads to an MRI, each set of merged duplicates must be updated to a value in the set (to satisfy minimality of change). It is easily verified that, with this restriction, the second update to the values of $R[C]$ is subsumed by the first, and therefore this update has no effect on the instance. Thus, sets of duplicates can be computed in the same way as with non-interacting sets. $\qquad \square$

Notice that the condition of Theorem 2 that there exists an ES that is not bounded does not appear in Theorem 4. This is because, for pair-preserving, acyclic sets of MDs, this condition is always satisfied by any subset of the set that is a linear pair. Indeed, for such a subset (m_1, m_2), if all its ESs are bounded, then by the pair-preserving requirement, $LHS(m_2) \subseteq LHS(m_1)$. Since (m_1, m_2) is a linear pair, $LHS(m_2) \cap RHS(m_1) \neq \emptyset$. This implies $LHS(m_1) \cap RHS(m_1) \neq \emptyset$, contradicting the acyclicity assumption.

For linear pairs, Theorem 4 becomes Theorem 2. For such pairs, condition (a) of Theorem 4 is always satisfied. If the (acyclic) linear pair is also a pair-preserving, as required by Theorem 4, the conditions of Theorem 2 reduce to conditions (a)(iii) and (b)(iii), which, as noted previously, are equivalent to condition (b) of Theorem 4.

6 Conclusions

We have shown that RQA is typically intractable when the MDs have non-cyclic dependencies on each. Our results depend on our chase-based semantics. Alternatives to this chase have been considered [6]. A quite different chase, which applies one MD at a time and uses *matching functions*, is presented in [5, 2].

The definition of resolved answer reminds us of *consistent query answering* (CQA) [4], where much research has been about (polynomial-time) query rewriting methodologies. In all the cases identified in the literature (see [4, 18] for recent surveys), the rewritings have been first-order. For MDs, the exhibited rewritings are in Datalog [13].

RQA brings many new challenges in comparison to CQA, and results for the latter cannot be applied (at least not in an obvious manner): (a) MDs contain usually non-transitive similarity relations. (b) Enforcing consistency of updates requires computing the transitive closure of such relations. (c) *Tuple-based repairs* are usually considered in CQA [4]. The minimality of *value changes*, not always used in CQA, has not been considered for consistent rewritings. (d) The semantics of resolved query answering for MD-based ER is given, in the end, in terms of a chase procedure.[3] However, the semantics of CQA is model-theoretic, given in terms of non-operationally defined repairs that arise from set-theoretic conditions. For additional discussions of differences and connections between CQA and RQA, see [12, 14].

We have presented the first dichotomy result for the complexity of RQA. Its cases depend on the set of MDs, for a fixed class of queries. In CQA with FDs, dichotomy results have been obtained for limited classes of conjunctive queries [15, 16, 18]. However, in CQA the cases depend mainly on the queries, as opposed to FDs.

Some open problems for ongoing and future research are: (a) Extending the class of CHAQ queries, considering additional projections, and also boolean queries. (b) Deriving a dichotomy result for acyclic, pair-preserving sets analogous to the one for linear pairs. (c) Since, FDs (and other equality generating dependencies) can be expressed as MDs, with equality as a transitive symmetry relation, applying the dichotomy result in Theorem 3 to CQA under FDs under a *value-based* repair semantics [4].

References

[1] Abiteboul, S., Hull, R., Vianu, V.: Foundations of Databases. Addison-Wesley (1995)
[2] Bahmani, Z., Bertossi, L., Kolahi, S., Lakshmanan, L.: Declarative entity resolution via matching dependencies and answer set programs. In: Proc. KR 2012 (2012)
[3] Benjelloun, O., Garcia-Molina, H., Menestrina, D., Su, Q., Euijong Whang, S., Widom, J.: Swoosh: A generic approach to entity resolution. VLDB Journal 18(1), 255–276 (2009)
[4] Bertossi, L.: Database Repairing and Consistent Query Answering. Morgan & Claypool, Synthesis Lectures on Data Management (2011)
[5] Bertossi, L., Kolahi, S., Lakshmanan, L.: Data cleaning and query answering with matching dependencies and matching functions. Theory of Computing Systems 52(3), 441–482 (2013)
[6] Bertossi, L., Gardezi, J.: Tractable vs. Intractable Cases of Matching Dependencies for Query Answering under Entity Resolution. Corr ArXiv: 1309.1884 (2013)
[7] Bleiholder, J., Naumann, F.: Data fusion. ACM Computing Surveys 41(1), 1–41 (2008)
[8] Cali, A., Lembo, D., Rosati, R.: On the decidability and complexity of query answering over inconsistent and incomplete databases. In: Proc. PODS 2003, pp. 260–271 (2003)
[9] Elmagarmid, A., Ipeirotis, P., Verykios, V.: Duplicate record detection: A survey. IEEE Trans. Knowledge and Data Eng. 19(1), 1–16 (2007)

---

[3] See [17] for some implicit connections between repairs and chase procedures, e.g. as used in data exchange; and [8] for connections with the chase used for database completion with ICs.

[10] Fan, W.: Dependencies revisited for improving data quality. In: Proc. PODS 2008 (2008)

[11] Fan, W., Jia, X., Li, J., Ma, S.: Reasoning about record matching rules. In: Proc. VLDB 2009 (2009)

[12] Gardezi, J., Bertossi, L., Kiringa, I.: Matching dependencies: semantics, query answering and integrity constraints. Frontiers of Computer Science 6(3), 278–292 (2012)

[13] Gardezi, J., Bertossi, L.: Query rewriting using datalog for duplicate resolution. In: Barceló, P., Pichler, R. (eds.) Datalog 2.0 2012. LNCS, vol. 7494, pp. 86–98. Springer, Heidelberg (2012)

[14] Gardezi, J., Bertossi, L.: Tractable cases of clean query answering under entity resolution via matching dependencies. In: Hüllermeier, E., Link, S., Fober, T., Seeger, B. (eds.) SUM 2012. LNCS, vol. 7520, pp. 180–193. Springer, Heidelberg (2012)

[15] Kolaitis, P., Pema, E.: A dichotomy in the complexity of consistent query answering for queries with two atoms. Information Processesing Letters 112(3), 77–85 (2012)

[16] Koutris, P., Suciu, D.: A dichotomy on the complexity of consistent query answering for atoms with simple keys. In: Proc. ICDT 2014, pp. 165–176 (2014)

[17] ten Cate, B., Fontaine, G., Kolaitis, P.: On the data complexity of consistent query answering. In: Proc. ICDT 2012, pp. 22–33 (2012)

[18] Wijsen, J.: A survey of the data complexity of consistent query answering under key constraints. In: Beierle, C., Meghini, C. (eds.) FoIKS 2014. LNCS, vol. 8367, pp. 62–78. Springer, Heidelberg (2014)

Analogical Classification: Handling Numerical Data

Myriam Bounhas[1], Henri Prade[2], and Gilles Richard[2]

[1] LARODEC Laboratory, ISG de Tunis, 41 rue de la Liberté, 2000 Le Bardo, Tunisia
& Emirates College of Technology, P.O. Box: 41009, Abu Dhabi, United Arab Emirates
[2] IRIT – CNRS, 118, route de Narbonne, Toulouse, France
& QCIS, University of Technology, Sydney, Australia
myriam_bounhas@yahoo.fr, {prade,richard}@irit.fr

Abstract. The formal modeling of analogical proportions (i.e., statements of the form "a is to b as c is to d") has led in the last past years to different proposals for classification algorithms, which have been quite successful on benchmarks where data are described by binary or nominal features. As far as we know and up to one exception, numerical data have never been considered. We propose here a new algorithm for handling numerical data. Starting from multiple-valued logical modelings of analogical proportions, more or less strongly encoding the idea that the change from a to b is the same as the change from c to d, we investigate different implementations leading to very good results on classical benchmarks.

1 Introduction

The study and the use of analogical reasoning has a long history in artificial intelligence [7,8]; see for a recent survey the introductory chapter of [18]. In the last decade, a renewal of the interest for the modeling of *analogical proportions* [9,23,22], i.e., statements of the form "a is to b as c is to d", often denoted $a : b :: c : d$, has motivated a series of successful applications to classification, and more generally machine learning [11,13,21], among other works.

The use of analogical proportions in classification has only been investigated for *binary* attributes until recently, which corresponds to the case where analogical proportions involve Boolean variables. Two approaches to the case of *nominal* attributes have been proposed recently [4,3], while the case of *numerical* attributes was the topic of a preliminary study a few years ago [20].

Indeed, the handling of numerical data by means of analogical proportions raises specific problems. While the question of deciding if an analogical proportion holds for a Boolean, or a nominal attribute between four situations, may receive a yes-or-no answer, it should become a matter of degree for a numerical attribute. Indeed, if one considers that for instance $0.30 : 0.80 :: 0.30 : 0.80$ is a perfect analogical proportion, one may expect that $0.30 : 0.80 :: 0.30 : 0.79$ is still an analogical proportion to a high degree, rather than being completely false. It turns out that there are several potentially meaningful ways for computing this degree [15].

In this paper, we propose a new algorithm for the classification problem, in the case of numerical data, based on graded analogical proportions. The paper is organized as follows. The next section presents a background on the logical modeling of analogical proportions, with a special emphasis on its extension to intermediary degrees of truth.

U. Straccia and A. Calì (Eds.): SUM 2014, LNAI 8720, pp. 66–79, 2014.

The third main section presents an overview of the existing analogical proportion-based algorithms for classification of binary, nominal and numerical data. Then in the next section a new approach and two algorithms are proposed for numerical data. The last main section reports experimental results obtained on benchmarks for binary or multiple classes problems, and provides a detailed, comparative discussion of the results that show the interest of the new approach.

2   Background on Analogical Proportions

Arithmetical (resp. geometrical) numerical proportions assert equality between two differences: $a - b = c - d$ (resp. ratios: $\frac{a}{b} = \frac{c}{d}$), where a, b, c, d are numbers. They are at the root of the idea of analogical proportions. More generally, a (symbolic) analogical proportion is a statement of the form "a is to b as c is to d" (denoted $a : b :: c : d$ and where the type of a, b, c, d is not specified for now), expressing informally that "a differs from b as c differs from d" and vice versa. As it is the case for numerical proportions, this "analogical" statement is supposed to still hold when the pairs (a, b) and (c, d) are exchanged, or when the mean terms b and c are permuted (see [16] for a detailed discussion). In the following subsections, we shall first focus on the case where a, b, c, d are Boolean truth variables, i.e. taking their values in $\mathbb{B} = \{0, 1\}$. Then, we will recall how analogical proportions can be extended to graded truth values when variables take their value in $[0, 1]$.

2.1   Boolean Case

When considering Boolean variables, a simple way to abstract the symbolic counterpart of numerical proportions has been given in [14] by focusing on indicators that capture the ideas of "similarity" and "dissimilarity". Namely, for a pair (a, b) of Boolean variables, four indicators are associated to such a pair, namely the Boolean functions:

    $- a \wedge b$ and $\neg a \wedge \neg b$: they are respectively *positive similarity* and *negative similarity* indicators ; $a \wedge b$ (resp. $\neg a \wedge \neg b$) is true iff only both a and b are true (resp. false);

    $- a \wedge \neg b$ and $\neg a \wedge b$: they are *dissimilarity* indicators ; they are true iff only one of a or b is true and the other is false.

Then, a logical proportion (see [17] for a thorough investigation of this notion) is a conjunction of two equivalences between such indicators, and the best "clone" of the numerical proportion is the analogical proportion defined as:

$$(a \wedge \neg b \equiv c \wedge \neg d) \wedge (\neg a \wedge b \equiv \neg c \wedge d) \quad (1)$$

This logical expression of an analogical proportion, using only dissimilarities, could be informally read as *what is true for a and not for b is exactly what is true for c and not for d, and vice versa*. It perfectly fits with the reading "a differs from b as c differs from d and vice versa". As such, a logical proportion is a Boolean formula involving 4 variables and it can be easily checked on its truth table (Table 1) that the logical expression of $a : b :: c : d$ satisfies symmetry ($a : b :: c : d \Rightarrow c : d :: a : b$) and central permutation ($a : b :: c : d \Rightarrow a : c :: b : d$), which are key properties of an analogical proportion, acknowledged for a long time, while $a : b :: a : b$ (and $a : a :: b : b$) always hold true. Thus, in terms of generic patterns, we see that analogical proportion always holds for

Table 1. Valuations where $a : b :: c : d$ is true

a b c d	a : b :: c : d
0 0 0 0	1
1 1 1 1	1
0 0 1 1	1
1 1 0 0	1
0 1 0 1	1
1 0 1 0	1

the three following patterns: $s : s :: s : s$, $s : s :: t : t$ and $s : t :: s : t$ where s and t are distinct values.

Thanks to properties of Boolean algebra, it can be easily seen that definition (1) is equivalent to:

$$((a \to b) \equiv (c \to d)) \wedge ((b \to a) \equiv (d \to c)) \quad (2)$$

but also, which is less obvious, that (1) is equivalent to:

$$((a \wedge d) \equiv (b \wedge c)) \wedge ((\neg a \wedge \neg d) \equiv (\neg b \wedge \neg c))$$

and thus to

$$((a \wedge d) \equiv (b \wedge c)) \wedge ((a \vee d) \equiv (b \vee c)) \quad (3)$$

where there is no negation operator [12]. Note that the last two expressions now refer to similarity indicators. Moreover, the analogical proportion "a is to b as c is to d" now reads "what a and d have in common, b and c have it also (both positively and negatively)", which is a less straightforward reading of the idea of analogy than the one associated with (1) or (2).

One of the side product of geometrical proportions is the well-known "rule of three" allowing to compute a suitable 4th item $x = d$ in order to complete a proportion $\frac{a}{b} = \frac{c}{x}$. This property has a counterpart in the Boolean case where the problem can be stated as follows. Given a triple (a, b, c) of Boolean values, does it exist a Boolean value x such that $a : b :: c : x = 1$, and in that case, is this value unique? It is easy to see that there are cases where the equation has no solution since the triple a, b, c may take $2^3 = 8$ values, while $a : b :: c : d$ is true only for 6 distinct 4-tuples. Indeed, the equations $1 : 0 :: 0 : x = 1$ and $0 : 1 :: 1 : x = 1$ have no solution. It is easy to prove that the analogical equation $a : b :: c : x = 1$ is solvable iff $(a \equiv b) \vee (a \equiv c)$ holds true. In that case, the *unique* solution is given by $x = a \equiv (b \equiv c)$.

Representing objects with a single Boolean value is not generally sufficient and we have to consider situations where items are represented by *vectors* of Boolean values, each component being the value of a binary attribute. A simple extension of the previous definitions to Boolean vectors in $\mathbb{B}^n$ of the form $\overrightarrow{a} = (a_1, \cdots, a_n)$ can be done as follows:

$$\overrightarrow{a} : \overrightarrow{b} :: \overrightarrow{c} : \overrightarrow{d} \text{ iff } \forall i \in [1, n], \ a_i : b_i :: c_i : d_i$$

Obviously, all the basic properties (symmetry, central permutation) still hold for vectors. On top of that, the equation solving process is still valid and provides a new insight about analogical proportion: analogical proportions are *creative*. Indeed, let us consider the following example where $\overrightarrow{a} = (1, 0, 0)$, $\overrightarrow{b} = (0, 1, 0)$ and $\overrightarrow{c} = (1, 0, 1)$. Solving the analogical equation $\overrightarrow{a} : \overrightarrow{b} :: \overrightarrow{c} : \overrightarrow{x}$ yields $\overrightarrow{x} = (0, 1, 1)$, which is a new vector *different* from $\overrightarrow{a}$, $\overrightarrow{b}$ and $\overrightarrow{c}$.

2.2 Multiple-valued Models

If we consider the Boolean expression of the analogical proportion given by formula (1), one may think of many possible multiple-valued extensions, depending on the choice of the connectives associated to $\neg$, $\wedge$ and $\equiv$. Then, it is important to make proper choices that are in agreement with the intended meaning of analogical proportion. Some properties seem very natural to preserve, such as

- i) the independence with respect to the positive or negative encoding of properties (one may describe a price as the extent to which it is cheap, as well as it is not cheap), which leads to require that $\neg a : \neg b :: \neg c : \neg d$ holds if $a : b :: c : d$ holds (as it is already the case with Boolean truth values);
- ii) the definition should agree with the Boolean definitions in the limit case where a, b, c, d take their values in $\{0, 1\}$.

In [19], a careful analysis of these requirements, ensuring that the value of b can be retrieved from the value of a and the ones of $a \wedge \neg b$ and $\neg a \wedge b$, leads to choose (for the sake of simplicity, we keep the same notation for a variable and its truth value):

- Łukasiewicz implication: $a \rightarrow b = \min(1, 1 - a + b)$ (or equivalently $a \wedge \neg b = \max(0, a - b)$);
- conjunction: $a \wedge b = \min(a, b)$;
- equivalence: $a \equiv b = \min(a \rightarrow b, b \rightarrow a) = 1 - |a - b|$.

When starting from definition (2), this leads to the following expression which both generalizes the Boolean case to multiple-valued entries and introduces a graded view of the analogical proportion:

$$a : b :: c : d = 1 - |(a - b) - (c - d)| \text{ if } a \geq b \text{ and } c \geq d, \text{ or } a \leq b \text{ and } c \leq d$$

$$a : b :: c : d = 1 - \max(|a - b|, |c - d|) \text{ otherwise} \quad (A)$$

As can be seen, $a : b :: c : d = 1$ if and only if $(a - b) = (c - d)$. We thus recognize the arithmetical proportion.[1]

---

[1] The geometrical proportion could be retrieved as well, choosing Goguen implication $s \rightarrow t = \min(1, t/s)$ and $s \rightarrow t = 1$ if $s = 0$, product conjunction $s \wedge t = s \cdot t$, and equivalence $s \equiv t = \min(s/t, t/s)$ [12]. It yields

$$a : b :: c : d = \min\left(\frac{\min(1, \frac{b}{a})}{\min(1, \frac{d}{c})}, \frac{\min(1, \frac{d}{c})}{\min(1, \frac{b}{a})}\right) \cdot \min\left(\frac{\min(1, \frac{a}{b})}{\min(1, \frac{c}{d})}, \frac{\min(1, \frac{c}{d})}{\min(1, \frac{a}{b})}\right).$$

with the convention 0/0 = 1. Clearly, $a : b :: c : d = 1$ if and only if $a/b = c/d$. In spite of this nice property, it can be checked that $1/2 : 0 :: 1 : 0 = 1$, which is not satisfactory, since we expect here a value less than 1 [19]. For this reason, we do not consider this option further in this paper.

When starting from definition (3), involving the operator $\vee$ defined as

$$a \vee b = \neg(\neg a \wedge \neg b) = 1 - min(1 - a, 1 - b) = max(a, b)$$

we get another expression:

$$a : b :: c : d = min(1 - |min(a, d) - min(b, c)|, 1 - |max(a, d) - max(b, c)|)$$

which can be rewritten as:

$$a : b :: c : d = 1 - max(|min(a, d) - min(b, c)|, |max(a, d) - max(b, c)|) \quad (A^*)$$

Unfortunately, the equivalence between the 2 definitions A and A^* is no longer valid in the multiple-valued case. Roughly speaking, for A^* to hold (i.e. to have value 1), the pair (a, d) and the pair (b, c) should have the same min and max, so $A^*(0, 0.5, 0.5, 1)$ has the value 0.5 and then does not hold at degree 1. But applying the definition of A shows that $A(0, 0.5, 0.5, 1) = 1$.

In fact, A^* provides a more restrictive view of analogical proportion than A: for instance, in the case of a tri-valued semantics where the domain of the variables is reduced to $\{0, 0.5, 1\}$, the truth table of A has 19 lines leading to 1 but A^* has only 15 such lines [15]. Moreover, as discussed in detail in [6], A agrees with the ideas of linear interpolation and extrapolation, since, in particular, the solution of $A(a, x, x, b) = 1$ is $x = \frac{a+b}{2}$, which is not the case of A^*.

Lastly, we can extend the notion of "true" proportion to vectors in $[0, 1]^n$ as in the Boolean model with (where P denotes A or A^*):

$$P(\overrightarrow{a}, \overrightarrow{b}, \overrightarrow{c}, \overrightarrow{d}) = 1 \text{ iff } \forall i \in [1, n], \; P(a_i, b_i, c_i, d_i) = 1$$

In the frequent case where $P(a_i, b_i, c_i, d_i) \in]0, 1[$ for some index i, $P(\overrightarrow{a}, \overrightarrow{b}, \overrightarrow{c}, \overrightarrow{d})$ is not defined and we have different options to allocate a truth value to the whole proportion. Obviously, one option is to compute the min of the truth values, another option may be to compute their mean $\frac{\sum_{i=1}^{n} P(a_i, b_i, c_i, d_i)}{n}$. We will further discuss these options in the next section.

3   Analogical Proportions and Classification

Numerical proportions are closely related to the ideas of extrapolation and of linear regression, i.e., to the idea of predicting a new value on the ground of existing values. Analogical proportions may serve similar purposes. The equation solving property recalled above is at the root of a brute force method for classification. It is based on a kind of "proportional continuity principle": if the attributes of 4 objects are componentwise in analogical proportion, then this should still be the case for their classes. Let us briefly recall the Boolean case, before considering the numerical attribute case.

3.1 Boolean and Discrete Cases

Having a 2-class classification problem, and 4 Boolean objects $\overrightarrow{a}$, $\overrightarrow{b}$, $\overrightarrow{c}$, $\overrightarrow{d}$ over $\mathbb{B}^n$, the proportional continuity principle can be stated as follows:

$$\frac{\overrightarrow{a} : \overrightarrow{b} :: \overrightarrow{c} : \overrightarrow{d}}{cl(\overrightarrow{a}) : cl(\overrightarrow{b}) :: cl(\overrightarrow{c}) : cl(\overrightarrow{d})}$$

In the case where $\overrightarrow{a}$, $\overrightarrow{b}$, $\overrightarrow{c}$ are in the training set with known classes and $\overrightarrow{d}$ being the object to be classified, if the equation $cl(\overrightarrow{a}) : cl(\overrightarrow{b}) :: cl(\overrightarrow{c}) : x = 1$ is solvable, we can allocate its solution to $cl(\overrightarrow{d})$ just by applying the continuity principle.

The case of attributes on discrete domains and of a number of classes larger than 2 can be handled as easily as the binary case. Indeed, consider a finite attribute domain $\{v_1, \cdots, v_m\}$ (note that the attribute may also be the class itself). This attribute, say $\mathcal{A}$, can be straightforwardly binarized by means of the m properties "having value v_i, or not". Consider the partial description of objects $\overrightarrow{a}$, $\overrightarrow{b}$, $\overrightarrow{c}$, and $\overrightarrow{d}$ wrt $\mathcal{A}$. Assume, for instance, that objects $\overrightarrow{a}$ and $\overrightarrow{c}$ have value v_1, while objects $\overrightarrow{b}$ and $\overrightarrow{d}$ have value v_2. This situation is summarized in Table 2 where the respective truth-values of the four objects wrt each binary property "having value v_i" are indicated. As can be seen on this table, an analogical proportion holds true between the four objects for each binary property, and in the example, can be more compactly encoded as an analogical proportion between the attribute values themselves, namely here: $v_1 : v_2 :: v_1 : v_2$. More generally, x and y denoting possible values of a considered attribute $\mathcal{A}$, the analogical proportion between objects $\overrightarrow{a}$, $\overrightarrow{b}$, $\overrightarrow{c}$, and $\overrightarrow{d}$ holds for $\mathcal{A}$ iff the 4-tuple $(\mathcal{A}(\overrightarrow{a}), \mathcal{A}(\overrightarrow{b}), \mathcal{A}(\overrightarrow{c}), \mathcal{A}(\overrightarrow{d}))$ is equal to one 4-tuple having one of the three forms (s, s, s, s), (s, t, s, t), or (s, s, t, t). This continuity principle has led to diverse implementations that we recall now.

- In [1], the authors use a measure of *analogical dissimilarity* between 4 objects. It estimates how far 4 objects are from being in analogical proportion. Roughly speaking, the analogical dissimilarity ad between 4 Boolean values is the minimum number of bits that have to be switched to get a proper analogy. Thus $ad(1, 0, 1, 0) = 0$, $ad(1, 0, 1, 1) = 1$ and $ad(1, 0, 0, 1) = 2$. Thus, $a : b :: c : d$ holds if and only if $ad(a, b, c, d) = 0$. Moreover ad differentiates two cases where analogy does not hold, namely the 8 cases with an odd number of 0 and an odd number of 1 among the 4 Boolean values, such as $ad(0, 0, 0, 1) = 1$ or $ad(0, 1, 1, 1) = 1$, and the two cases $ad(0, 1, 1, 0) = ad(1, 0, 0, 1) = 2$. When we deal with 4 Boolean vectors in $\mathbb{B}^n$, adding the ad evaluations componentwise generalizes the analogical dissimilarity to Boolean vectors, and leads to an integer belonging to the interval $[0, 2n]$.

Table 2. Handling non binary attributes

	v_1	v_2	v_3	$\cdots$	v_m	
$\overrightarrow{a}$	1	0	0	$\cdots$	0	v_1
$\overrightarrow{b}$	0	1	0	$\cdots$	0	v_2
$\overrightarrow{c}$	1	0	0	$\cdots$	0	v_1
$\overrightarrow{d}$	0	1	0	$\cdots$	0	v_2

It is used in [1] in the implementation of a classification algorithm where the input parameters are a set TS of classified items, an integer k, and a new item d to be classified. It proceeds as follows:

Step 1: Compute the analogical dissimilarity ad between d and all the triples in TS^3 that produce a solution for the class of d.

Step 2: Sort these n triples by the increasing value of ad wrt with d.

Step 3: Let p be the value of ad for the k-th triple, then find k' as being the greatest integer such that the k'-th triple has the value p.

Step 4: Solve the k' analogical equations on the label of the class. Take the winner of the k' votes and allocate this winner as the class of d.

This approach provides remarkable results and, in several cases, outperforms the best known algorithms [11].

– In the algorithm proposed in [14], there is no use of a dissimilarity measure but a straightforward implementation of the continuity principle, keeping flexibility by allowing to have some components where analogy does not hold. Triples of Boolean vectors (a, b, c) are considered such that the class equation $cl(a) : cl(b) :: cl(c) : x$ is solvable and such that the number of componentwise analogies $card(\{i \in \{1, n\}| a_i : b_i :: c_i : d_i \text{ holds}\})$ is maximal. Then the label solution of the corresponding class equation is allocated to d, implementing a majority vote in case of multiple candidate triples.

– In [13], flexibility is introduced via an integer p indicating how many components we tolerate with a failure of the analogical proportion. In that case, a candidate voter is just a triple of Boolean vectors a, b, c such that the class equation $cl(a) : cl(b) :: cl(c) : x$ is solvable and

$$card(\{i \in \{1, n\}| a_i : b_i :: c_i : d_i \text{ holds}\}) \geq n - p$$

Still a majority vote is applied among the candidate voters. The main difference with the approach of [1] is that there is no distinction between the cases where analogy does not hold. In terms of complexity, all these approaches are cubic in the size of the training set, leading to high runtime. In terms of accuracy, there is no significant differences between the three implementation techniques.

– A completely different approach has been recently proposed where the building blocks are *pairs* rather than triples [4]. To each pair of examples is associated a "change" (resp. "no change") rule if the two examples are in different classes (resp. are in the same class). Then the observed change in some of the attribute values (while the other attributes keep the same value in the two examples of the pair) is considered as responsible for the change in classes, or as having no influence (if the two examples are in the same class). This gives birth to rules that are then applied to pairs made of one example and of the item to classify in such a way that an analogical proportion holds componentwise between the pair associated with the rule and the pair involving the item. It turns out that many rules are applicable in this way, and a majority vote is applied between the different predictions obtained for the class of the item. Again good results are obtained for the case of binary and *nominal* attributes. Although in this paper we shall not extend such a rule-based view to numerical attributes, we shall keep in mind the idea of working with pairs.

– Let us also mention the work of [22] although it does not deal with Boolean vectors strictly speaking, since they are the first authors to suggest the proportional continuity principle as an underlying mechanism for building analogical learners. They propose an algebraic framework for defining analogical proportions between structured data: lattices, free monoids and trees. In these universes, they introduce the notion of *factorization* allowing the decomposition of items into smaller parts bound to satisfy diverse constraints. For instance, in the case of *words*, this allows to consider "viewing : reviewer :: searching : researcher" as a valid proportion, generalizing the work of [9]. To solve analogical equations in such universes, diverse heuristic approaches are used, but with the advantage of making analogical inference tractable even for large databases. Their approach provide satisfactory results [21].

All the previous works are focused on discrete data and none of them tackles the issue of dealing with numerical values.

3.2   Numerical Case

In that situation, we consider 4 real-valued vectors $\vec{a}, \vec{b}, \vec{c}, \vec{d}$ over $[0,1]^n$ (the numerical values are previously normalized in the unit interval through a linear transformation, if necessary), but still with discrete classes. Then, we interpret the value of an attribute as a truth value (corresponding to the extent to which the property underlying the attribute holds): thus the value of $P(a_i, b_i, c_i, d_i)$ can always be computed and belongs also to $[0,1]$. Actually, this truth value is rarely equal to 1 but can be close to 1. So the following inference principle, cloning the Boolean case:

$$\frac{P(\vec{a}, \vec{b}, \vec{c}, \vec{d}) = 1}{cl(\vec{a}) : cl(\vec{b}) :: cl(\vec{c}) : cl(\vec{d})}$$

has to be adapted for a proper implementation.

As far as we know, the work presented in [20] is the only one to deal with numerical data. Starting from datasets coming from UCI repository [10], the data are normalized in order to get values in $[0, 1]$ considered as truth degrees which allows the application of the graded semantics previously described in this paper. Given a new data $\vec{d}$ to be classified, the main idea is to consider *all* the triples $(\vec{a}, \vec{b}, \vec{c})$ such that the corresponding class equation is solvable. We denote $Solv(TS^3)$ this set. Actually, these triples are the only ones able to provide a prediction for the unknown label of $\vec{d}$. We compute for each of these triples the vector $(P(a_1, b_1, c_1, d_1), \ldots, P(a_i, b_i, c_i, d_i), \ldots, P(a_n, b_n, c_n, d_n))$. Then we order these vectors of truth values using the leximin[2] as a total order. The best triple, i.e. the one maximizing $(P(a_1, b_1, c_1, d_1), \ldots, P(a_i, b_i, c_i, d_i), \ldots, P(a_n, b_n, c_n, d_n))$ w.r.t. leximin is then chosen to allocate a label to the new item $\vec{d}$. As highlighted in [20], the accuracy results of the corresponding classifier are quite good, and in some cases, outperform well-known algorithms.

---

[2] $(u_1, \ldots, u_i, \ldots, u_n) >_{leximin} (v_1, \ldots, v_i, \ldots, v_n)$, *once the components of each vector have been increasingly ordered, iff* $\exists j < n \, \forall i = 1, j \, u_i = v_i$ *and* $u_{j+1} > v_{j+1}$.

4 Implementation

In this paper, we suggest another viewpoint in order to drastically reduce the number of triples to be investigated. Instead of systematically surveying $Solv(TS^3)$, we first consider the nearest neighbours $\vec{c}$ of $\vec{d}$ (w.r.t. the 1-norm distance). This idea of first building a pair with the new item $\vec{d}$ and one of its nearest neighbors has also been successfully investigated in the case of binary and nominal attributes in a companion paper [3].

We then apply exactly the same decision rule restricted to the subset of $Solv(TS^3)$ where $\vec{c}$ is one of the nearest neighbours of $\vec{d}$. More formally, let us assume we have computed a numerical value for $P(\vec{a}, \vec{b}, \vec{c}, \vec{d})$. In our case, we use for P the definition as the mean of the truth values obtained in componentwise manner (given above). This use (in place of leximin) is inspired from practice in information retrieval (see [2] for comparisons and possible refinements). Then we can associate with $\vec{c}$ the finite subset of $[0, 1]$:

$$\{P(\vec{a}, \vec{b}, \vec{c}, \vec{d})|(\vec{a}, \vec{b}, \vec{c}) \in Solv(TS^3)\}$$

At this stage, we have 2 options:

1. The first option is to consider the union of all these finite subsets, getting a new finite subset of $[0, 1]$, then having a unique maximum element m_0 corresponding to a triple $(\vec{a_0}, \vec{b_0}, \vec{c_0})$ such that the class equation $cl(\vec{a_0}) : cl(\vec{b_0}) :: cl(\vec{c_0}) : x = 1$ is solvable. Still keeping the same philosophy as with the Boolean case and considering $(\vec{a_0}, \vec{b_0}, \vec{c_0})$ as the "best" triple to predict the class, we allocate to $\vec{d}$ the solution of this class equation. In the quite unlikely case (we are in a numerical setting) where we would have more that one triple corresponding to this maximum element m_0, we use a majority vote among the corresponding triples.

2. The second option is to consider, for each finite subset, the triple $(\vec{a_0}, \vec{b_0}, \vec{c_0})$ corresponding to its maximum element $m_{0,\vec{c}}$. This triple generates a candidate label for $\vec{d}$. Finally we get a set of candidate labels, one per neighbour $\vec{c}$: we implement a majority vote to get the final label.

It is quite straightforward to implement the previous options leading to the algorithms 1 and 2. In the latter algorithm, we use a function $vote(E, \vec{c}, \vec{d})$ whose input is a set of pairs of numerical vectors E and 2 numerical vectors, returning as output the label given by the triple maximizing the number $P(\vec{a}, \vec{b}, \vec{c}, \vec{d})$, with $(\vec{a}, \vec{b}) \in E$. $candidate(\vec{d})$ is just the list of class vote for $cl(\vec{d})$: obviously, the class having the maximum number of occurrences ($nbocc$) is the allocated label for $\vec{d}$ as expressed by the last command $cl(\vec{d}) = argmax_l\{nbocc(l)\ in\ candidate(\vec{d})\}$.

Note that, as we have seen with [20], it is not absolutely necessary to define a global truth value for $P(\vec{a}, \vec{b}, \vec{c}, \vec{d})$: the only thing we need is a total order on the set $Solv(TS^3)$ (or more generally TS^3).

Algorithm 1. Numerical Analogical Classifier without majority vote

Input: $k > 1$, $\overrightarrow{d} \notin TS$ a new instance to be classified
$cl(\overrightarrow{d}) = undefined;\ maxTruthValue = 0$
for each $\overrightarrow{c} \in TS$ **do** compute $||\overrightarrow{c} - \overrightarrow{d}||_1$
sort by increasing order the list L of values $\{||\overrightarrow{c} - \overrightarrow{d}||_1 | \overrightarrow{c} \in TS\}$
build up the set $NN_k(\overrightarrow{d}) = \{\overrightarrow{c} \in TS$ s.t. $rank(||\overrightarrow{c} - \overrightarrow{d}||_1)$ in $L \leq k\}$
for each $\overrightarrow{c}$ in $NN_k(\overrightarrow{d})$ **do**
  for each pair $(\overrightarrow{a}, \overrightarrow{b})$ such that: $cl(\overrightarrow{a}) : cl(\overrightarrow{b}) :: cl(\overrightarrow{c}) : x$ has solution l **do**
    if $maxTruthValue < P(\overrightarrow{a}, \overrightarrow{b}, \overrightarrow{c}, \overrightarrow{d})$ **then**
      $maxTruthValue = P(\overrightarrow{a}, \overrightarrow{b}, \overrightarrow{c}, \overrightarrow{d})$;
      $cl(\overrightarrow{d}) = l$
    end if
  end for
end for
return $cl(\overrightarrow{d})$

Algorithm 2. Numerical Analogical Classifier with majority vote among the nearest neighbors

Input: $k > 1$, $\overrightarrow{d} \notin TS$ a new instance to be classified
$candidate(\overrightarrow{d}) = nil$;
for each $\overrightarrow{c} \in TS$ **do** compute $||\overrightarrow{c} - \overrightarrow{d}||_1$
sort by increasing order the list L of values $\{||\overrightarrow{c} - \overrightarrow{d}||_1 | \overrightarrow{c} \in TS\}$
build up the set $NN_k(\overrightarrow{d}) = \{\overrightarrow{c} \in TS$ s.t. $rank(||\overrightarrow{c} - \overrightarrow{d}||_1)$ in $L \leq k\}$
for each $\overrightarrow{c}$ in $NN_k(\overrightarrow{d})$ **do**
  build $E = (\overrightarrow{a}, \overrightarrow{b})$ s.t. $cl(\overrightarrow{a}) : cl(\overrightarrow{b}) :: cl(\overrightarrow{c}) : x$ has solution
  $candidate(\overrightarrow{d}) = vote(E, \overrightarrow{c}, \overrightarrow{d}).candidate(\overrightarrow{d})$
end for
$cl(\overrightarrow{d}) = argmax_l\{nbocc(l)$ in $candidate(\overrightarrow{d})\}$
return $cl(\overrightarrow{d})$

5 Experimental Results and Comparison

In order to evaluate the efficiency of analogical proportion-based classifiers with numerical data, we have tested the two algorithms on 8 data sets taken from the UCI machine learning repository [10]. A brief description of these data sets is given in Table 3 where we focus on classification problems involving numerical attributes only. In terms of classes, we deal with a maximum number of 8 classes. In order to apply our multiple-valued semantics framework, all attribute values are normalized in a standard way to get numbers in [0,1]: a real value is thus changed into a number that may be understood as a truth value. In terms of protocol, we apply a standard 10 fold cross-validation technique and we run our tests both for A and A^* definitions of the graded analogical proportion.

Table 3. Description of datasets

Datasets	Instances	Attributes	Classes
Diabetes	768	8	2
W. B. Cancer	699	9	2
Heart	270	13	2
Iris	150	4	3
Wine	178	13	3
Satellite Image	346	36	6
Glass	214	9	7
Ecoli	336	7	8

Table 4. Classification accuracies given as mean and with standard deviation

Datasets	Algo.	A				A*			
value of k		1	3	5	11	1	3	5	11
Diabetes	Algo 1	65.4±4.4	64.8± 4.1	65.7± 4.4	65.2±4.5	65.4±4.6	64.8±5.3	65.0±5.2	64.3± 5.0
	Algo 2		68.5±4.6	71.0±4.3	**73.0±4.8**		67.5±5.0	69.7±4.7	71.7± 5.2
W. B. Cancer	Algo 1	96.0±1.9	95.2±2.0	95.1 ±1.9	94.7±2.3	96.2±1.8	96.0±2.0	95.8±2.1	95.5 ±2.4
	Algo 2		96.7±2.0	96.7±1.9	96.6±2.3		**97.0±2.0**	96.8±1.9	96.8±2.1
Heart	Algo 1	73.3±7.1	71.7±8.7	72.2±7.9	72.4±7.3	72.9±7.9	71.4±8.5	70.9±7.9	70.6±7.6
	Algo 2		77.1±6.8	78.2±6.9	**82.1±6.1**		77.3±6.9	78.7±6.7	79.8±6.1
Iris	Algo 1	94.2±5.3	95.7±4.6	94.5±5	93.0±5.5	94.2±5.0	93.4±5.7	93.1±5.8	93.2±4.9
	Algo 2		95.8±4.8	95.3±5.1	**96.9 ±4.5**		95.7±4.5	95.2±4.9	94.9±4.9
Wine	Algo 1	95.3±4.0	96.1± 3.6	96.2± 4.2	95.8±4.3	95.8± 4	95.8± 3.9	95.3± 4.3	95.9±3.8
	Algo 2		96.6±3.2	96.9 ± 3.3	**98.2±2.7**		97.1±3.5	97.3±3.4	97.1±3.5
Sat. Image	Algo 1	94.1± 3.6	**95.3±3.4**	95.1±3.2	94.8±2.9	93.5±3.8	94.2±3.8	94.4±3.8	94.7±4.0
	Algo 2		95.1±3.9	94.4±4.1	94.5±3.9		94.8±3.7	94.1±4.2	93.5±4.3
Glass	Algo 1	71.7±8.9	70.2±8.6	70.7±8.6	71.3±9.1	73.7±8.9	73.4±8	73.8±7.8	74.4±8.2
	Algo 2		72.0±8.2	74.1±8	72.1±9.8		74.2±8.4	**74.6±8.7**	73.6±9.3
Ecoli	Algo 1	79.6±6.8	77.4±7.8	77.2±7.4	76.7±5.8	79.7±5.5	78.9±6.2	78.2±6.5	78.6±6.2
	Algo 2		82.3±6.6	84.6±5.6	**86.8±6.0**		81.7±5.7	83.1±5.9	83.9±5.7

Table 5. Comparison with classification results of some well-known classifiers

Datasets	SVM	JRip	IBK(K=1,k=10)	[20]	Algo2 with A and k=11
Diabetes	77.3	76.0	70.0, 71.1	71	73.0
Cancer	97.1	96.0	96.2 , 96.9	-	96.6
Heart	83.7	81.1	74.8, 81.4	-	82.1
Iris	96.0	95.3	95.3, 96.0	99	96.9
Wine	98.3	92.7	94.9, 95.5	-	98.2
Sat. Image	94.2	93.9	94.2, 92.2	90	94.5
Glass	57.9	69.1	70.5, 64.5	-	72.1
Ecoli	84.2	81.2	80.3, 86.0	-	86.8

In Table 4, we provide mean accuracies and standard deviations for diverse values of k ($k = 1$ means that we apply the algorithm without vote, k being the number of nearest neighbors of $\vec{d}$ used: in that case, Algo 1 is just Algo 2).

From Table 4, we can notice that:

– Algorithm 2 with $k = 11$ and using definition A performs generally better than Algorithm 1: in that case, the voting procedure to get a label for $\vec{d}$ includes triples

$(\overrightarrow{a}, \overrightarrow{b}, \overrightarrow{c})$ where $\overrightarrow{c}$ could be quite far from $\overrightarrow{d}$. It appears that it is better to consider a majority vote, allowing each neighbor to provide a candidate label, than to only consider the neigbor providing the triple $(\overrightarrow{a}, \overrightarrow{b}, \overrightarrow{c})$ with the highest P.

- It is quite clear that we get (except in 1 case) better results when using A definition than when using A^* definition for analogical proportion. It could be the case that A^*, providing a more restrictive view of analogical proportion than A, removes too many candidates from the voting triples. It remains to investigate if there is a way to qualify a target dataset as being more suitable for classification using A or using A^*. This is an open problem.

In Table 5, we provide the known accuracy results for standard machine learning algorithms: the SVM, JRip an optimized propositional rule learner and the k-Nearest Neighbors IBk (with $k = 1$ and $k = 10$). Accuracy results for SVM, JRip and IBk are obtained by applying the free implementation of Weka software.

- This table highlights the fact that our analogical classifiers perform more or less in the same way as the best known algorithms, but without using any optimization trick. Especially, Algorithm2, using A, with $k = 11$ outperforms all other classifiers for data sets "Iris", "Sat.Image", "Glass" and "Ecoli" and have performances similar to SVM for datasets "Cancer" and "Wine".
- The classification success of Algorithm 2 for "Iris", "Sat.Image", "Glass" and "Ecoli" (which have multiple classes) demonstrates its ability to deal with multiple class data sets.
- The analogy-based classifiers seem to be efficient when classifying data sets with a large number of attributes as in the case of "Sat.Image" for example.
- Contrary to multiple class data sets, the analogy-based classifiers seems to be less efficient when classifying some binary data sets such as "Diabetes" specially with small values of k. For this reason we tested the behavior of Algorithm 2 with definition A when using larger number of neighbors, we get an accuracy of 74.9 (instead of 73.0 for $k = 11$) just by increasing k to 13. We can conclude that, for binary data sets, the classifier requires to investigate more neighbors to be able to distinguish between classes. It is still an open question to identify the optimal k suitable for each data set.
- Although, our algorithms do not use the same order on TS^3 as the approach developed in [20], we note that our results are better than the results obtained by the previous approach for numerical data, for the tested data sets "Diabetes" and "Sat. Image".
- To compare our analogical classifier to the IBk classifier, we used the Wilcoxon Matched-Pairs Signed-Ranks Test as proposed by Demsar [5]. It is a non-parametric alternative to the paired t-test that enables us to compare the two classifiers over multiple data sets. In our case, the null hypothesis, states that our algorithm performs as IBk. This hypothesis leads to a p-value = 0.0179 and then has to be rejected. Algorithm 2 thus performs significantly better than IBk classifier.
- Lastly, we also computed the frequency that examples $\overrightarrow{a}$ or $\overrightarrow{b} \in (\overrightarrow{a}, \overrightarrow{b})$ are among k nearest neighbors $\overrightarrow{c}$ of $\overrightarrow{d}$. We get a very low average frequency (1.2% on 8 benchmarks). Thus we can conclude that $\overrightarrow{a}$ and $\overrightarrow{b}$ are usually far from the

example $\overrightarrow{d}$ to be classified. This shows once again that analogical classifiers are not just another view of k-nn algorithm.

6 Conclusion

In this paper, we have presented a new way to classify data with numerical attributes using analogical proportions. The procedure can be summarized in the following way. Inspired by the k-nn algorithm, we avoid a brute-force investigation of all the triples $\overrightarrow{a}, \overrightarrow{b}, \overrightarrow{c}$ to build up a valid proportion with the new item $\overrightarrow{d}$ to be classified. We first look for a neighbor $\overrightarrow{c}$ of $\overrightarrow{d}$. Then, we compute for all pairs $(\overrightarrow{a}, \overrightarrow{b})$, the n truth values $P(a_i, b_i, c_i, d_i), i \in [1, n]$, P being one of two multiple-valued definitions of an analogical proportion w.r.t. each attribute, getting a n-dimensional real vector of truth values. Such pairs, associated with $\overrightarrow{c}$ constitute the candidate voters provided that the corresponding class equation is solvable. For such triple, we compute a global value for P by adding componentwise the value of P, thus associating to each triple a unique real number. This provides a complete ordering of the triples. One option is to choose, among the candidate voters, the triple with the greatest P value as the ultimate voter. Another option is to extend the number of candidate voters by allowing each nearest neighbor to provide a unique voter and to implement a majority vote for the final label. Our implementation of the two options, using each of the 2 multiple-valued definitions of an analogical proportion exhibits very good results on 8 UCI benchmarks. Ultimately, the second option provides better results than the first one.

While classifiers like k-nn focus on the neighborhood of the target item, analogical classifiers go beyond this neighborhood. Rather than just "copying" what emerges among close neighbors of $\overrightarrow{c}$, they "take inspiration" of relevant information that is usually far from this immediate neighborhood. Indeed this information comes from pairs of examples $(\overrightarrow{a}, \overrightarrow{b})$ that are not among nearest neighbors. We think that this way to proceed with analogical proportions is paving the way to what could be called "creative machine learning".

References

1. Bayoudh, S., Miclet, L., Delhay, A.: Learning by analogy: A classification rule for binary and nominal data. In: Proc. Inter. Join. Conf. on Artificial Intelligence, IJCAI 2007, pp. 678–683 (2007)
2. Boughanem, M., Loiseau, Y., Prade, H.: Refining aggregation functions for improving document ranking in information retrieval. In: Prade, H., Subrahmanian, V.S. (eds.) SUM 2007. LNCS (LNAI), vol. 4772, pp. 255–267. Springer, Heidelberg (2007)
3. Bounhas, M., Prade, H., Richard, G.: Analogical classification: A new way to deal with examples. Tech. Rep. RR–2014-03–FR, Institut de Recherche en Informatique de Toulouse (IRIT), March, to appear in Proc. 21st Europ. Conf. on Artificial Intelligence, Prague, August 18-22 (2014)
4. Bounhas, M., Prade, H., Richard, G.: Analogical classification: A rule-based view. In: Laurent, A., Strauss, O., Bouchon-Meunier, B., Yager, R.R. (eds.) IPMU 2014, Part II. CCIS, vol. 443, pp. 485–495. Springer, Heidelberg (2014)

5. Demsar, J.: Statistical comparisons of classifiers over multiple data sets. Journal of Machine Learning Research 7, 1–30 (2006)
6. Dubois, D., Prade, H., Richard, G.: Multiple-valued extensions of analogical proportions - A discussion (submitted)
7. Gentner, D., Holyoak, K.J., Kokinov, B.N.: The Analogical Mind: Perspectives from Cognitive Science. Cognitive Science, and Philosophy. MIT Press, Cambridge (2001)
8. Helman, D.H. (ed.): Analogical Reasoning: Perspectives of Artificial Intelligence. Cognitive Science, and Philosophy. Kluwer, Dordrecht (1988)
9. Lepage, Y.: Analogy and formal languages. Electr. Notes Theor. Comput. Sci. 53 (2001)
10. Mertz, J., Murphy, P.: UCI repository of machine learning databases (2000), ftp://ftp.ics.uci.edu/pub/machine-learning-databases
11. Miclet, L., Bayoudh, S., Delhay, A.: Analogical dissimilarity: definition, algorithms and two experiments in machine learning. JAIR 32, 793–824 (2008)
12. Miclet, L., Prade, H.: Handling analogical proportions in classical logic and fuzzy logics settings. In: Sossai, C., Chemello, G. (eds.) ECSQARU 2009. LNCS, vol. 5590, pp. 638–650. Springer, Heidelberg (2009)
13. Moraes, R.M., Machado, L.S., Prade, H., Richard, G.: Classification based on homogeneous logical proportions. In: Bramer, M., Petridis, M. (eds.) Proc. of AI-2013, The Thirty-third SGAI International Conference on Innovative Techniques and Applications of Artificial Intelligence, Cambridge, England, UK, pp. 53–60. Springer, Heidelberg (2013)
14. Prade, H., Richard, G.: Reasoning with logical proportions. In: Lin, F.Z., Sattler, U., Truszczynski, M. (eds.) Proc. 12th Int. Conf. on Principles of Knowledge Representation and Reasoning, KR 2010, Toronto, May 9-13, pp. 545–555. AAAI Press (2010)
15. Prade, H., Richard, G.: Analogical proportions and multiple-valued logics. In: van der Gaag, L.C. (ed.) ECSQARU 2013. LNCS, vol. 7958, pp. 497–509. Springer, Heidelberg (2013)
16. Prade, H., Richard, G.: Homogeneous logical proportions: Their uniqueness and their role in similarity-based prediction. In: Brewka, G., Eiter, T., McIlraith, S.A. (eds.) Proc. 13th Int. Conf. on Principles of Knowledge Representation and Reasoning (KR 2012), Roma, June 10-14, pp. 402–412. AAAI Press (2012)
17. Prade, H., Richard, G.: From analogical proportion to logical proportions. Logica Universalis 7(4), 441–505 (2013)
18. Prade, H., Richard, G. (eds.): Computational Approaches to Analogical Reasoning: Current Trends. SCI, vol. 548. Springer, Heidelberg (2014)
19. Prade, H., Richard, G.: Multiple-valued logic interpretations of analogical, reverse analogical, and paralogical proportions. In: Proc. 40th IEEE Int. Symp. on Multiple-Valued Logic (ISMVL 2010), Barcelona, pp. 258–263 (2010)
20. Prade, H., Richard, G., Yao, B.: Enforcing regularity by means of analogy-related proportions-a new approach to classification. International Journal of Computer Information Systems and Industrial Management Applications 4, 648–658 (2012)
21. Stroppa, N., Yvon, F.: An analogical learner for morphological analysis. In: Online Proc. 9th Conf. Comput. Natural Language Learning (CoNLL 2005), pp. 120–127 (2005)
22. Stroppa, N., Yvon, F.: Analogical learning and formal proportions: Definitions and methodological issues. Tech. rep. (June 2005)
23. Yvon, F., Stroppa, N., Delhay, A., Miclet, L.: Solving analogical equations on words. Tech. rep., Ecole Nationale Supérieure des Télécommunications (2004)

Lazy Analogical Classification:
Optimization and Precision Issues

William Correa Beltran, Hélène Jaudoin, and Olivier Pivert

University of Rennes 1 – Irisa
Technopole Anticipa 22305 Lannion Cedex France
william.correa_beltran@irisa.fr, {jaudoin,pivert}@enssat.fr

Abstract. This paper presents a novel approach for lazy classification based on the notion of analogical proportions. Our starting point is a method from the literature based on a measure of analogical dissimilarity. Based on some observations about the effectiveness of different analogical proportion situations for classification purposes, we optimize this method, considerably reducing the size of the training set. These results raise some questions about the reasons of the effectiveness of the analogical approach, which are briefly discussed at the end of the paper.

1 Introduction

Several lazy classification algorithms based on the notion of an analogical proportion [6] have been proposed in the literature, see for instance [7,4]. An analogical proportion is a statement of the form "A is to B as C is to D" which (implicitly) states that the way two objects A and B, otherwise similar, differ is the same way as the two objects C and D, which are similar in some respects, differ.

In this paper, we consider the method of "classification by analogy" introduced in [7] where the authors describe an algorithm named FADANA. This algorithm uses a measure of *analogical dissimilarity* between four objects, which estimates how far these objects are from being in analogical proportion. We show that a modification of the algorithm aimed to favor a certain situation of analogical proportion makes it possible to considerably reduce the size of the training set while preserving (and sometimes slightly improving) the accuracy of the approach, even in the case where the dataset involves numerical attributes (whereas the basic FADANA algorithm is limited to Boolean features).

The remainder of the paper is structured as follows. Section 2 provides a refresher on the notion of analogical proportion and presents the principle of FADANA. Section 3 describes our optimized algorithm. Experimental results are presented and discussed in Section 4. Section 5 deals with the relationship between the optimized version of FADANA and a classification method based on the k nearest neighbors technique (kNN). Section 6 recalls the main contributions and outlines perspectives for future work.

U. Straccia and A. Calì (Eds.): SUM 2014, LNAI 8720, pp. 80–85, 2014.

2   Analogical Proportion

The following presentation is mainly drawn from [3]. An analogical proportion is a statement of the form "A is to B as C is to D". This will be denoted by $(A : B :: C : D)$. In the context considered, A, B, C, and D are assumed to correspond to descriptions of objects such as sets, multisets, vectors, strings or trees. In the following, A, B, C, and D are tuples having n attributes, i.e., $A = \langle a_1, \ldots, a_n \rangle$, $\ldots$, $D = \langle d_1, \ldots, d_n \rangle$, and we shall say that A, B, C, and D are in analogical proportion if and only if for each component i an analogical proportion "a_i is to b_i as c_i is to d_i" holds.

We now have to specify what kind of relation an analogical proportion may mean. Intuitively speaking, we have to understand how to interpret "is to" and "as" in "A is to B as C is to D". A may be similar (or identical) to B in some respects, and differ in other respects. The way C differs from D should be the same as A differs from B, while C and D may be similar in some other respects, if we want the analogical proportion to hold. This view is enough for justifying three postulates that date back to Aristotle's time, i.e., i) $(A : B :: A : B)$, ii) $(A : B :: C : D) \Leftrightarrow (C : D :: A : B)$, and iii) $(A : B :: C : D) \Leftrightarrow (A : C :: B : D)$, where the first and second property express reflexivity and symmetry for the comparison "as" respectively, while the latter allows for central permutation.

A *logical proportion* [5] is a particular type of Boolean expression $T(a, b, c, d)$ involving four variables a, b, c, d, whose truth values belong to $\mathbb{B} = \{0, 1\}$. It is made of the conjunction of two distinct equivalences, involving a conjunction of variables a, b on one side, and a conjunction of variables c, d on the other side of $\equiv$, where each variable may be negated. Analogical proportion is a special case of a logical proportion, and its expression is [3]: $(a\overline{b} \equiv c\overline{d}) \wedge (\overline{a}b \equiv \overline{c}d)$. The six valuations yielding *true* are (0, 0, 0, 0), (1, 1, 1, 1), (0, 0, 1, 1), (1, 1, 0, 0), (0, 1, 0, 1) and (1, 0, 1, 0). Using the terminology from [4], one may group these situations in three classes: (i) similarity $(a : a :: a : a)$, (ii) pairwise identity $(a : a :: b : b)$, and (iii) identity of change $(a : b :: a : b)$.

As noted in [6], the idea of proportion is closely related to the idea of extrapolation, i.e., to guess/compute a new value on the ground of existing values. In other words, if it is known that a logical proportion holds between four binary elements, three being known, then one may try to infer the value of the fourth one (corresponding to its class in the context considered in the following).

Analogical dissimilarity [7] indicates how far four objects are from being in analogical proportion. In the Boolean case, ad indicates the minimum number of bits that have to be switched to get a proper analogy. For instance, $\mathsf{ad}(1, 0, 1, 0) = 0$, $\mathsf{ad}(1, 0, 1, 1) = 1$, and $\mathsf{ad}(1, 0, 0, 1) = 2$. Thus, denoting by $\mathcal{A}$ the relation of analogical proportion, we have $\mathcal{A}(a, b, c, d) \Leftrightarrow \mathsf{ad}(a, b, c, d) = 0$.

In [1], we showed how analogical proportions relaxed by considering an *approximate equality* relation between the values could be applied to the predict missing numerical attribute values. Analogical proportion in this case is defined as $a : b :: c : d \Leftrightarrow (((a \approx b) \wedge (c \approx d)) \vee ((a \approx d) \vee (b \approx d)))$, where $x \approx b$ is interpreted as $|x - y| \leq \lambda$, λ being a threshold in [0, 1] (the numerical values have been normalized into [0, 1] beforehands).

3   Making Analogical Classification more Efficient

Our work is inspired from the algorithm FADANA [7], which handles Boolean vectors, adding the **ad** evaluations componentwise to get the analogical dissimilarity, leading to an integer in the interval $[0, 2n]$. This algorithm takes as input a training set S of classified items, a new item d to be classified, and an integer k. It proceeds as follows: 1) for every triple (a, b, c) of S^3, compute $\mathbf{ad}(a, b, c, d)$; 2) sort these n triples by increasing value of their **ad** when associated with d; 3) if the k-th triple has the integer value p for **ad**, then let k' be the greatest integer such that the k'-th triple has the value p; 4) solve the k' analogical equations on the label of the class. Take the winner of the k' votes as the class of d.

3.1   Some Remarks about Fadana

The computation of **ad** for each triplet $\in S^3$ completed by a given d_i is the most expensive part of the algorithm — it takes around 80% of the overall processing time with the datasets we used. In [2], it is shown that a training set comprising 40 items is in general sufficient to reach quasi optimal precision. Taking into account the symmetry and central permutation properties (cf. Section 2), the number of triples formed is $c * (c - 1) * (c - 2)/2$ where c is the cardinality of the training set. For instance, with $c = 40$, 29.640 triples are generated.

   In [1] we applied this principle to predict null values in the numerical case. Analyzing the results obtained in each step of the process, we noticed two facts: (i) The number of chosen triples (step 2 and 3) containing attributs whose **ad** — when compared to d — is 2 is minimal, and what is more revealing, (ii) the accuracy of similarity proportions $(a : a :: a : a)$ is considerably higher than that of the pairwise identity $(a : a :: b : b)$ and identity of change $(a : b :: a : b)$. The average accuracy rates for the similarity, pairwise identity and identity of change proportion over the four datasets tested are 88.3 ± 2.21 , 77 ± 7.74, and 77.8 ± 7.34, respectively.

   We used this information for i) reducing the size of the training set, ii) giving priority to the similarity type of analogical proportions when classifying a new item. The corresponding algorithm is described hereafter.

3.2   An Optimized Algorithm

Our algorithm takes as input a training set S, a set D of items to be completed as they contain *null* attribute values (this corresponds to a generalization of the classification problem), and two integers k and r. The steps are:

1. discard from S^3 all the triples (a, b, c) such that $(a_i \neq b_i \wedge b_i = c_i)$ is true for at least one feature;
2. let $s(t)$ be the number of features on which $t = (a, b, c)$ agrees, i.e, $(a_i = b_i = c_i)$; discard from S^3 all the triples t such that $s(t) \leqslant r$;
3. for every object d involving at least one missing attribute value, do:
   (a) for every triple (a, b, c) of S^3, compute $\mathbf{ad}(a, b, c, d)$;

    (b) sort these n triples by increasing value of their ad;

    (c) if the k-th triple has the integer value p for ad, let E be the set of triples (a, b, c) such that $\mathrm{ad}(a, b, c, d) \leq p$

    (d) for each attribute A_j such that $d_j = null$, do:

        i. let $E_j^s \subseteq E$ be the set of triples for which a similarity[1] proportion holds for d_j, and $E_j^n \subseteq E$ the set of triples for which one of the other two types of proportion holds; if $|E_j^s| > 0$, then use E_j^s to solve the analogical equations for d_j; use E_j^n otherwise.

        ii. if A_i is a numerical attribute, compute v as the average of the $|E_j^s|$ (resp. $|E_j^n|$) votes; if A_i is Boolean, compute v as the winner of the $|E_j^s|$ (resp. $|E_j^n|$) votes.

        iii. choose v as the value of d_j.

Step 1 means that we eliminate all the triples containing attribute values for which no analogical answer exists: the patterns $(0, 1, 1, x)$ and $(1, 0, 0, x)$ have no analogical solution no matter what the value of x is. Step 2 discards the triples where the proportion of attributs validating an equality relation is too low. Step 3.d.i means that for each missing attribute of each incomplete object, only the triples satisfying a similarity type of proportion are used in the case where there exists at least one such triple. The other triples are used otherwise.

4   Experimental Results

The main objective of our experimentation is to assess the extent to which a lazy analogical classification method can be optimized by giving priority to the similarity type of analogical proportion. We thus compare our results with those obtained using a classical FADANA implementation [7]. The comparison is both in terms of *accuracy* and *processing time*, the latter being strongly related to the size of the training set.

    In order to evaluate the approach, four datasets from the UCI machine learning repository[2], namely *adult, blood, cancer,* and *energy* have been used. Each dataset is composed of both categorical and numerical attributes, the latter being treated with the *approximate equality relation* — cf. section 2. For each dataset, a sample E has been extracted by randomly choosing 1000 tuples, and a sample M of 50 tuples has been modified (40% of the attribute values of its tuples have been replaced by *null*). Then, FADANA, kNN, and our algorithm (named oF for "optimized FADANA" hereafter) have been run so as to predict the missing values: for each tuple d involving at least one missing value, a random sample D of $E - M$ (thus made of complete tuples) has been chosen. This sample D (training set) was used for running the three algorithms. The size of the training set has

---

[1] For Boolean attributes, a similarity proportion is satisfied by a quadruple (x, y, z, t) if all four elements are equal. For numerical attribute, we use the tolerant view mentioned at the end of Section 2 and we only impose that the absolute pairwise difference between the four elements is at most equal to $\lambda = 0.05$.

[2] http://http://archive.ics.uci.edu/ml/datasets.html

Table 1. Results obtained

		Adult	*Blood*	*Cancer*	*Energy*
FADANA	*accuracy*	72.17	**88.21**	85.09	**89.12**
	nbtriples	29640	29640	29640	29640
OF ($r=0$)	*accuracy*	**75**	88	**87**	87.7
	nbtriples	4432	11136	10578	4387
OF ($r=1$)	*accuracy*	74.18	88	86	86.2
	nbtriples	4138	5923	7059	1963
OF ($r=2$)	*accuracy*	74.18	87.68	85.94	82.69
	nbtriples	4093	3464	3177	901
KNN	*accuracy*	72.5	87.8	86.9	84.9

been set to 40, and the value of k to 10 (complementary experimentations that cannot be presented here due to lack of space showed that these are the parameter values yielding the best classification results in general). The numbers in the table are average values (10 runs have been performed on each dataset). For each method, the first line gives the accuracy (percentage of correct predictions) and the second line indicates the number of triples generated by the algorithm for each value to predict.

As FADANA does not preprocess the training set, its size remains constant. A remarkable result is that, even though OF generates much less triples than FADANA, its accuracy is quite similar. The best performances were obtained by OF (75%), FADANA (88.21%), OF (87%) and FADANA (89.12%) for the datasets *adult, blood, cancer*, and *energy* respectively.

5   What Makes Analogy Work ?

The experimental results described above seem to indicate that the strength of the analogical approach reside mainly in the similarity case. However, some questions remain open: i) what is the role of the other two cases ? ii) what are the main differences with the kNN approach ?

Suppose we want to classify an object x. In order to use the "pairwise identity" or the "identity of change" cases, one needs an object b such that $b \simeq x$, and two elements a and a' such that $a \simeq a'$. One can then build the quadruples $(a : b :: a' : x)$ and $(a : a' :: b : x)$. How valuable is the information extracted from it ? The preliminary experimental results reported above tend to show that such quadruples are not as useful as the similarity ones for a classification purpose, but it is not clear why. Notice that the elements a and a' do not need to be similar to x, so it is reasonable to suppose that they would not be used by a kNN approach (which only exploits the elements that are the closest to x).

In the similarity case, one generates quadruples of the form $(a : b :: c : x)$ such that all four elements are pairwise similar on every attribute. On the other hand, the kNN approach looks for those elements d that are the closest to x as possible, which does not imply that they are close enough in the sense of the relation $\simeq$ used by the analogical approach. Notice also that kNN uses a global

(Euclidean) distance whereas analogy — as defined in our approach — imposes a local proximity on every attribute (for ad to be equal to 0).

Let us emphasize that these questions concern the view of analogical proportion based on approximate equality (which handles numerical values in a way similar to Boolean ones, only relaxing the comparison operator). On the other hand, if one uses arithmetic analogical proportions as defined in [4], i.e., $(a : b :: c : d) \Leftrightarrow (a - b) = (c - d)$, one just needs three elements a, b, and c such that $|a - b| \simeq |c - x|$ in order to form a valid analogical quadruple. However, we have shown in [1] that this view does not always perform better than the approximate-equality-based one. The respective performances of the two approaches depend on the datasets, and further experiments and analyzes are necessary to understand what properties of the data makes one approach more effective than the other.

6   Conclusion

In this paper, we have shown how a classical "classification by analogy" algorithm from the literature could be improved by focusing first on a certain type of analogical proportion. A preliminary experimentation reported here, carried out on four datasets from the UCI machine learning repository, shows that the improvement in terms of processing time is between 62% and 97%, without any loss in terms of accuracy (we even observed an improvement in three cases).

Perspectives concern, among other things, an indepth study, completing the discussion that could only be started here, of the relationship between analogical proportions and kNN in a classification context.

References

1. Correa Beltran, W., Jaudoin, H., Pivert, O.: Analogical prediction of null values: The numerical attribute case. In: Proc. of the 18th East-European Conference on Advances in Databases and Information Systems (ADBIS 2014) (2014)
2. Correa Beltran, W., Jaudoin, H., Pivert, O.: Estimating null values in relational databases using analogical proportions. In: Laurent, A., Strauss, O., Bouchon-Meunier, B., Yager, R.R. (eds.) IPMU 2014, Part III. CCIS, vol. 444, pp. 110–119. Springer, Heidelberg (2014)
3. Miclet, L., Prade, H.: Handling analogical proportions in classical logic and fuzzy logics settings. In: Sossai, C., Chemello, G. (eds.) ECSQARU 2009. LNCS, vol. 5590, pp. 638–650. Springer, Heidelberg (2009)
4. Prade, H., Richard, G., Yao, B.: Enforcing regularity by means of analogy-related proportions — a new approach to classification. IJCISIM 4, 648–658 (2012)
5. Prade, H., Richard, G.: Reasoning with logical proportions. In: Lin, F., Sattler, U., Truszczynski, M. (eds.) KR. AAAI Press (2010)
6. Prade, H., Richard, G.: Analogical proportions and multiple-valued logics. In: van der Gaag, L.C. (ed.) ECSQARU 2013. LNCS, vol. 7958, pp. 497–509. Springer, Heidelberg (2013)
7. Bayoudh, S., Miclet, L., Delhay, A.: Learning by analogy: A classification rule for binary and nominal data. In: IJCAI, pp. 678–683 (2007)

Polyhedral Labellings for Argumentation Frameworks

Cosmina Croitoru

MPII, Saarbrücken, Germany

Abstract. In this paper we introduce *polyhedral labellings* associated to an argumentation framework. The name suggests the use of ideas from *Polyhedral Combinatorics*, an important topic in *Combinatorial Optimization*, mainly concerned with encoding combinatorial problems by means of systems of linear equations and inequalities, making these problems accessible to linear programming techniques. A *polyhedral labelling* for an argumentation framework $AF = (A, D)$ is a polytope P_{AF}, that is, a bounded set of solutions $x \in \mathbb{R}^A$ (x_a is the label of the argument $a \in A$), to a system of linear constraints, such that the set of integral vectors in P_{AF} are exactly the incidence vectors of some specific type of Dung's extensions. The linear constraints vary from the obvious $x_a = 1$ for each non attacked argument a, or $x_a + x_b \leq 1$ for each attack $(a, b) \in D$ (in order to assure Dung's conflict-free condition), to more deep inequalities of the form *"the sum of the label of an argument and the labels of all its attackers is at least 1"* or if (b, a) is an attack then *"the label of a is not greater than the sum of the labels of all attackers of b"*.

1 Introduction

The graph-theoretic model of argumentation frameworks introduced by Dung [5] focuses on the manner in which a specified set A of abstract arguments interact via an attack binary relation D on A. If $(a, b) \in D$ (argument a *attacks* argument b) we have a conflict. A *conflict-free* set of arguments is a set $T \subseteq A$ such that there are no $a, b \in T$ with $(a, b) \in D$. An *admissible* set of arguments is a conflict-free set $T \subseteq A$ such that the arguments in T defend "collectively" against any attack: for each $(a, b) \in D$ with $b \in T$, there is $c \in T$ such that $(c, a) \in D$. Let us denote $\mathscr{A} \subseteq 2^A$ the family of all admissible sets in AF.

In this model, the main aim of argumentation is in deciding the status of some argument by presenting a justification for this. More precisely, the acceptability of an argument a is defined based on its membership in an admissible set of arguments satisfying certain properties. A family $\mathscr{S} \subseteq \mathscr{A}$ of sets of arguments is defined (the predicate $S \in \mathscr{S}$ is called *semantics* in this context) and a is considered *acceptable* if there is $S \in \mathscr{S}$ such that $a \in S$ – *credulous acceptance* – or if $a \in S$ for all $S \in \mathscr{S}$ – *skeptical acceptance*. This kind of rationality, based on the possibility of extending the analyzed argument to a set of "collectively acceptable" arguments, is called *extension based semantics*. A justification for an argument a is an admissible set containing a and satisfying additional properties. In this case the argument a is considered *acceptable*. The *grounded, preferred* and *stable semantics* defined by Dung (see Section 2) formalizes different intuitions about which arguments to accept on the basis of a given framework.

U. Straccia and A. Calì (Eds.): SUM 2014, LNAI 8720, pp. 86–99, 2014.

The popularity of the Dung's graph based model of argumentation is explained by its simplicity and generality, due to the abstract nature of the arguments, and the relationship with non-monotonic reasoning formalisms such as Default Logic and Logic Programming (see, for example, [15]).

An important intuitive way to express Dung's extension-based semantics is using *argument labellings*, as proposed by Caminada [4] (originally introduced in [14]). The idea is to consider symbolic vectors[1] $\lambda \in \{I, O, U\}^A$, such that if $a \in A$ is an argument, then λ_a is the label of a, with the intuitive meaning: $\lambda_a = I$ (i.e. In) if and only if a is accepted, $\lambda_a = O$ (i.e. Out) if and only if a is rejected, and $\lambda_a = U$ (i.e. Undecided) if and only if one abstains from an opinion on whether the argument a is accepted or rejected. Constraining the labels of arguments with respect to the attack relation D of the argumentation framework, Caminada characterized the subsets of $\{I, O, U\}^A$ corresponding to Dung's semantics (see Section 2).

Another interesting approach to the semantics of argumentation frameworks was introduced by Gabbay in [6, 7, 8] under the name *equational approach*, and independently by Gratie and Florea in [9] under the name *fuzzy labellings*. The idea is to consider solutions $x \in [0,1]^A$ of some non-linear systems of equations associated to the argumentation framework and to relate them to Caminada labellings. Using this approach, Gabbay proposes an interesting method to avoid the semantics problems caused by the odd circuits in argumentation frameworks.

In this paper we introduce *polyhedral labellings* associated to an argumentation framework. The name suggests the use of ideas from *Polyhedral Combinatorics*, an important topic in *Combinatorial Optimization*, mainly concerned with encoding combinatorial problems by means of systems of linear equations and inequalities. The interest in such representation is that it makes the corresponding combinatorial optimization problems accessible to linear programming techniques (see, for example [16]). More precisely, if $\chi^S \in \{0,1\}^A$ is the incidence vector of a set $S \subseteq A$ of arguments (that is, $\chi_a^S = 1$ if $a \in S$ and $\chi_a^S = 0$ if $a \notin S$), and $\mathscr{S}$ is a collection of sets of arguments, then the convex hull of their incidence vectors, $\mathrm{conv}\{\chi^S | S \in \mathscr{S}\}$, is a polytope in $\mathbb{R}^A$, therefore there exist a matrix $C \in \mathbb{R}^{m \times |A|}$ and a vector $b \in \mathbb{R}^m$ such that

$$\mathrm{conv}\{\chi^S | S \in \mathscr{S}\} = \{x \in \mathbb{R}^A | Cx \leq b\}.$$

If $w : A \to \mathbb{R}$ is a weight function on A, and we are interested in finding a member S^* of $\mathscr{S}$ of maximum weight (where the weight of S is $w(S) = \sum_{a \in S} w(a)$), since $\mathscr{S}$ is finite and the weight function can be viewed as a linear function on $\mathbb{R}^A$, we could maximize over the convex hull $\mathrm{conv}\{\chi^S | S \in \mathscr{S}\}$, that is, by the above representation, finding $\max\{w^T x | x \in \mathbb{R}^A, Cx \leq b\}$. This is computationally worthwhile when $\mathscr{S}$ is too large to evaluate the weight of each member S in $\mathscr{S}$, but the description $Cx \leq b$ of the above polytope has polynomial size. Then, we can solve in polynomial time the equivalent linear programming problem obtained (for example, using the ellipsoid method [10]). An illustration of this approach is discussed in Section 3, where we describe an

---

[1] Throughout this paper, if A is a finite set, we make no distinction between the set B^A of all functions from A to B and the set $B^{|A|}$ of all vectors with components from B and indexed by the elements of A. Supposing a fixed ordering of A, there is an obvious one to one correspondence between them, and we use the notation B^A for both sets.

interesting polytope encoding the *non-attacked sets* of arguments in an argumentation framework.

In several cases, the desired system of inequalities, $Cx \leq b$, turns out not to be a complete description, but just gives an approximation of the polytope $\text{conv}\{\chi^S | S \in \mathscr{S}\}$. This can still be useful, since in that case the linear programming problem gives a (hopefully good) upper bound for the combinatorial maximum. These bounds are used in designing branch-and-bound algorithms for the combinatorial problem, which are implemented in state-of-the-art integer programming solvers (e.g. CPLEX [13]). In fact, exploiting integer programming encodings of the problems is an usual modeling method in areas related to argumentation, such as non monotonic reasoning ([2]), satisfiability ([11]), or answer set programming ([12]).

Summarizing, a *polyhedral labelling* for an argumentation frameork $AF = (A, D)$ is a polytope P_{AF}, that is, a bounded set of solutions $x \in \mathbb{R}^A$ (x_a is the label of the argument $a \in A$), of a system of linear inequalities (and equations), such that the set of integral vectors in P_{AF} are exactly the incidence vectors of various Dung's admissibility based extensions. The linear constraints varies from the obvious $x_a = 1$ for each non attacked argument a, or $x_a + x_b \leq 1$ for each attack $(a, b) \in D$ (in order to assures the Dung's conflict-free condition), to more sophisticated inequalities of the form *"the label of an argument plus the labels of all its attackers is at least 1"* or if (b, a) is an attack then *"the label of a is not greater than the sum of the labels of all attackers of b"*.

The remainder of this paper is organized as follows. In Section 2, we discuss basic notions of Dung's theory of argumentation. In Section 3, we illustrate the polyhedral approach by a simple but interesting problem in an argumentation framework. In Section 4, we discuss elements of the *Gabbay's equational approach*, which is the starting point of our *labellings polytopes*, introduced and studied in Section 5. Finally, Section 6 concludes the paper.

2   Dung's Theory of Argumentation

In this section we present the basic concepts used for defining classical semantics in abstract argumentation frameworks introduced by Dung in 1995, [5]. All notions and results, if not otherwise cited, are from this paper (even some of them are not literally the same).

Definition 1. An *Argumentation Framework* is a digraph $AF = (A, D)$, where A is a finite and nonempty set; the vertices in A are called *arguments*, and if $(a, b) \in D$ is a directed edge, then *argument a defeats (attacks) argument b*.

Let $AF = (A, D)$ be an argumentation framework. For each $a \in A$ we denote $a^+ = \{b \in A | (a, b) \in D\}$ the set of all arguments *attacked* by a, and $a^- = \{b \in A | (b, a) \in D\}$ the set of all arguments *attacking* a. These notations can be extended to sets of arguments. The set of all arguments *attacked by* (the arguments in) $S \subseteq A$ is $S^+ = \bigcup_{a \in S} a^+$, and the set of all arguments *attacking* (the arguments in) S is $S^- = \bigcup_{a \in S} a^-$. We also have $\emptyset^+ = \emptyset^- = \emptyset$.

The set S of arguments *defends* an argument $a \in A$ if $a^- \subseteq S^+$ (i.e. any a's attacker is attacked by an argument in S). The set of *all arguments defended by* a set S of arguments is denoted by $F(S)$.

For $\mathbb{M} \subseteq 2^A$, **max**$(\mathbb{M})$ denotes the set of maximal (w.r.t. inclusion) members of $\mathbb{M}$ and **min**$(\mathbb{M})$ denotes the set of its minimal (w.r.t. inclusion) members.

The main admissibility extension-based acceptability semantics are defined below.

Definition 2. Let $AF = (A, D)$ be an argumentation framework.

- A *conflict-free set* in AF is a set $S \subseteq A$ with property $S \cap S^+ = \emptyset$ (i.e. there are no attacking arguments in S). We will denote **cf**$(AF) = \{S \subseteq A | S$ is conflict-free set $\}$.
- An *admissible set* in AF is a set $S \in \mathbf{cf}(AF)$ with property $S^- \subseteq S^+$ (i.e. defends its elements). We will denote **adm**$(AF) = \{S \subseteq A | S$ is admissible set $\}$.
- A *complete extension* in AF is a set $S \in \mathbf{cf}(AF)$ with property $S = F(S)$. We will denote **comp**$(AF) = \{S \subseteq A | S$ is complete extension $\}$.
- A *preferred extension* in AF is a set $S \in \mathbf{max}(\mathbf{comp}(AF))$. We will denote **pref**$(AF) :=$ **max**$(\mathbf{comp}(AF))$.
- A *grounded extension* in AF is a set $S \in \mathbf{min}(\mathbf{comp}(AF))$. We will denote **gr**$(AF) :=$ **min**$(\mathbf{comp}(AF))$.
- A *stable extension* in AF is a set $S \in \mathbf{cf}(AF)$ with the property $S^+ = A - S$. We will denote **stab**$(AF) = \{S \subseteq A | S$ is stable extension $\}$.

An equivalent way to express Dung's extension-based semantics is using argument labellings as proposed by Caminada [4] (originally introduced in [14]). The idea underlying the labellings-based approach is to assign to each argument a label from the set $\{I, O, U\}$. The label I (i.e. In) means the argument is accepted, the label O (i.e. Out) means the argument is rejected, and the label U (i.e. Undecided) means one abstains from an opinion on whether the argument is accepted or rejected.

Definition 3. [4] Let $AF = (A, D)$ be an argumentation framework. A *complete labelling* of AF is a function $Lab : A \to \{I, O, U\}$ such that $\forall a \in A$:
- $Lab(a) = I$ if and only if $a^- \subseteq Lab^{-1}(O)$,
- $Lab(a) = O$ if and only if $a^- \cap Lab^{-1}(I) \neq \emptyset$,
- $Lab(a) = U$ if and only if $a^- \cap Lab^{-1}(I) = \emptyset$ and $a^- \cap Lab^{-1}(U) \neq \emptyset$.

A *grounded labelling* of AF is a complete labelling Lab such that there is no complete labelling Lab_1 with $Lab_1^{-1}(I) \subset Lab^{-1}(I)$. A *preferred labelling* of AF is a complete labelling Lab such that there is no complete labelling Lab_1 with $Lab^{-1}(I) \subset Lab_1^{-1}(I)$. A *stable labelling* of AF is a complete labelling Lab such that $Lab^{-1}(U) = \emptyset$.

In [4] it was proved that, for any argumentation framework $AF = (A, D)$ and any semantics $\sigma \in \{\mathbf{comp}, \mathbf{gr}, \mathbf{pref}, \mathbf{stab}\}$, a set $S \subseteq A$ satisfies $S \in \sigma(AF)$ if and only if there is a σ-labelling Lab of AF such that $S = Lab^{-1}(I)$.

3   The Non-attacked Sets Polytope

Definition 4. *Let $AF = (A, D)$ be an argumentation framework. A **non-attacked set** of arguments is a set $N \subseteq A$ such that $N^- \subseteq N$. Let $\mathcal{N}_{AF} := \{N | N \subseteq A, N^- \subseteq N\}$.*

Trivial non-attacked sets are $\emptyset, A \in \mathcal{N}_{AF}$ for any argumentation framework $AF = (A, D)$. The interest in such sets of arguments is given by the following proposition (see also the "directionality principle" in [1]).

Proposition 5. *Let $a \in A$ be an argument in the argumentation framework $AF = (A, D)$ and $X \in \mathcal{N}_{AF}$ be a non-attacked set containing a. There is an admissible set S in AF such that $a \in S$ if and only if there is an admissible set S' in AF' such that $a \in S'$, where AF' is the argumentation framework induced by X in AF.*

Proof. Let S an admissible set in AF such that $a \in S$. Then S is conflict-free and $S^- \subseteq S^+$. Then, $S_X = S \cap X$ is a conflict-free set in AF', and $a \in S_X$. If S_X is not an admissible set in AF', then there is $b \in S_X^- \cap X - S_X^+$. Since S is an admissible set in AF, $b \in S^+$. It follows that there is $c \in S - X$ such that $(c, b) \in D$, that is $c \in X^- - X$, contradicting the hypothesis that X is a non-attacking set in AF.

Conversely, if $S' \subseteq X$ is an admissible set in AF' such that $a \in S'$, then it is a conflict-free set in AF. Since in AF we have $S'^- \subseteq X^- \subseteq X$, it follows that in AF we have $S'^- \subseteq S'^+$, that is, S' is an admissible set in AF.   □

We show now that $\mathcal{N}_{AF}$ has an interesting polyhedral characterization. For each $X \in \mathcal{N}_{AF}$ we consider its incidence vector $\chi^X \in \{0, 1\}^A$, with $\chi_a^X = 1$ if and only if $a \in X$.

Let $\mathbf{N}_{AF} = \{x \in \mathbb{R}^A | x \text{ satisfies } (*)\}$ be the polyhedron defined by

$$(*) \quad \begin{cases} 0 \le x_a \le 1 & \forall a \in A, \\ x_a - x_b \ge 0 & \forall (a, b) \in D. \end{cases}$$

Hence, if $x \in \mathbf{N}_{AF}$ then each argument $a \in A$ is labeled with the real number $x_a \in [0, 1]$ such that $x_a \ge x_b$ whenever the argument a attacks the argument b. This type of constraints are used to model preferences (see, for example, *"value-based argumentation frameworks"*, [3]).

Theorem 6. *Let $AF = (A, D)$ be an argumentation framework. Then,*

$$\mathbf{N}_{AF} = \text{conv}\{\chi^X | X \in \mathcal{N}_{AF}\}.$$

Proof. For $X \in \mathcal{N}_{AF}$, let $y = \chi^X$. Since $y_a \in \{0, 1\}$, the first group of inequalities in $(*)$ is satisfied. Let $(a, b) \in D$. If $|\{a, b\} \cap X| \ne 1$, then $y_a = y_b$ and the second constraint in $(*)$ for (a, b), is satisfied with equality. If $a \in X$ and $b \notin X$ then $1 = y_a > y_b = 0$. Since $X \in \mathcal{N}_{AF}$, we can not have $a \notin X$ and $b \in X$. Hence $\chi^X \in \mathbf{N}_{AF}$ for each $X \in \mathcal{N}_{AF}$. It follows that

$$\text{conv}\{\chi^X | X \in \mathcal{N}_{AF}\} \subseteq \mathbf{N}_{AF}.$$

To prove the converse inclusion, we observe that the integer vectors in $\mathbf{N}_{AF}$ are exactly the incidence vectors of non-attacked sets. Hence it is sufficient to prove that the vertices of $\mathbf{N}_{AF}$ are integral. Let x be a vertex of $\mathbf{N}_{AF}$.

Suppose that $Frac(x) = \{a \in A | 0 < x_a < 1\} \ne \emptyset$, and let $\alpha = \min\{x_a | a \in Frac(x)\}$. Take $\varepsilon > 0$ such that $\alpha - \varepsilon > 0$ and $\alpha + \varepsilon < x_a$ for all $a \in A$ such that $x_a > \alpha$. Then, let $x', x'' \in \mathbb{R}^A$ be such that:

$$x_a' = \begin{cases} \alpha - \varepsilon & \text{if } x_a = \alpha, \\ x_a & \text{if } x_a \ne \alpha \end{cases} \quad \text{and} \quad x_a'' = \begin{cases} \alpha + \varepsilon & \text{if } x_a = \alpha, \\ x_a & \text{if } x_a \ne \alpha. \end{cases}$$

By the choosing of ε, x' and x'' satisfy the first group of inequalities in $(*)$. Since the order of the components in x' and x'' is the same as in x, and since $x \in \mathbf{N}_{AF}$, it follows that the second group of inequalities are satisfied by x' and x''. Hence, $x',x'' \in \mathbf{N}_{AF}$, $x' \neq x''$, and $x = \frac{1}{2}x' + \frac{1}{2}x''$, contradicting the hypothesis that x is a vertex in $\mathbf{N}_{AF}$. $\square$

From Proposition 5, it follows that in order to decide the σ-acceptability of a given argument a in an argumentation framework, for $\sigma \in \{\mathbf{comp}, \mathbf{gr}, \mathbf{pref}\}$, is worthwhile to find a minimum cardinality non-attacked set containing a. This can be obtained with a simple polynomial algorithm (similar to one used for obtaining the grounded extension), but also using a linear programming solver, as a consequence of the Theorem 6.

Indeed, if we solve the linear program $\min\{c^T x | x \in \mathbf{N}_{AF}, x_{a_0} = 1\}$, for $c \in \mathbb{R}^A$ we obtain $\min\{\sum_{a \in X} c_a | X \in \mathscr{N}_{AF}, a_0 \in X\}$. In particular, for $c = \mathbf{1}$ (the vector with all components 1), the minimum value obtained is the minimum cardinality of a non-attacked set of arguments containing a_0. If we solve (in polynomial time) the above linear program and x^0 is the optimal solution, then the set $X = \{a | a \in A, x_a^0 > 0\}$ is the minimum cardinality non-attacking set containing a_0.

4 Gabbay's Equational Approach

An interesting approach to the semantics of argumentation frameworks was introduced in [6, 7, 8] called the **equational approach**. Let $AF = (A, D)$ be an argumentation framework, and let A_0 be the set of arguments not attacked in AF: $A_0 = \{a \in A | a^- = \emptyset\}$. We consider for each argument $a \in A$ a real variable $x_a \in [0, 1]$ and we are searching for real solutions of the following system of non-linear equations:

$$Eq_{\max}(AF) \quad \begin{cases} x_a = 1, \text{ if } a \in A_0 \\ x_a = 1 - \max_{b \in a^-} x_b, \text{ otherwise.} \end{cases}$$

The following theorem holds.

Theorem 7. ([7]) *If* $\lambda : A \to \{I, O, U\}$ *is a Caminada complete labellings of* AF, *then taking* $x_a = 0$ *if* $\lambda(a) = O$, $x_a = \frac{1}{2}$ *if* $\lambda(a) = U$, *and* $x_a = 1$ *if* $\lambda(a) = I$, *we obtain a solution to the system of equations* $Eq_{\max}(AF)$. *If* x *is a solution to the system* $Eq_{\max}(AF)$, *then taking* $\lambda(a) = O$ *if* $x_a = 0$, $\lambda(a) = U$ *if* $0 < x_a < 1$, *and* $\lambda(a) = I$ *if* $x_a = 1$ *we obtain a Caminada complete labelling* $\lambda : A \to \{I, O, U\}$.

Example. For argumentation framework AF in Figure 1 below,

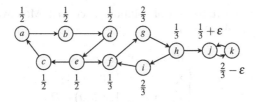

Fig. 1. $Eq_{\max}(AF)$ labellings. ε is a small positive number

the Eq_{max} system is

$$Eq_{max}(AF) \quad \begin{cases} x_a = 1 - x_c, \ x_b = 1 - x_a, \ x_c = 1 - x_b, \ x_d = 1 - x_b, \\ x_e = 1 - \max(x_b, x_f), \ x_f = 1 - \max(x_d, x_e, x_i), \\ x_g = 1 - x_f, \ x_h = 1 - x_g, \ x_i = 1 - x_h, \end{cases}$$

and a set of solutions to $Eq_{max}(AF)$ are suggested.

We can observe that if we translate them to Caminada labellings as in Theorem 7, the result is not so good since, in this case $\mathbf{gr}(AF) = \{\emptyset\}$, and there are odd loops[2]. In order to avoid this, Gabbay [7] proposes the so called *perturbation method* to solve a system of the form $Eq(AF)$. Essentially, by this method, some variables x_a are forced to be 0 by extending the system $Eq(AF)$ with these new equations. The idea for choosing this forcing is to destroy the loops of AF. This gives rise to interesting semantics (which are, in general, not admissibility based), called **LB-semantics** in [8].

If we make the convention that for $X = \emptyset$ then $\max_{x \in X} = 0$ and $\min_{x \in X} = 1$, then the following proposition holds.

Proposition 8. *Let $AF = (A, D)$ an argumentation framework. If x is a solution in $[0, 1]$ of the system $Eq_{max}(AF)$ then $x_a = \min_{b \in a^-} \max_{c \in b^-} x_c$, for each $a \in A$.*

Proof. Let x be a solution in $[0, 1]$ of the system $Eq_{max}(AF)$. With our convention, it follows that the second group of equations in $Eq_{max}(AF)$ is satisfied for each $a \in A$, that is $x_a = 1 - \max_{b \in a^-} x_b$. Then, we have successively,

$$x_a = 1 - \max_{b \in a^-} x_b = 1 - \max_{b \in a^-}(1 - \max_{c \in b^-} x_c) = -\max_{b \in a^-}(-\max_{c \in b^-} x_c) = \min_{b \in a^-} \max_{c \in b^-} x_c. \qquad \square$$

We note that the converse of the above proposition is not true. More precisely, if for a given argumentation framework $AF = (A, D)$, the labelling $x \in [0, 1]^A$ satisfies $x_a = \min_{b \in a^-} \max_{c \in b^-} x_c$ for each $a \in A$, then x is not necessary a solution in $[0, 1]$ of the system $Eq_{max}(AF)$. For example, if $AF = (A, D)$ is a circuit, taking $x_a = 1$ for each $a \in A$, then x satisfies $x_a = \min_{b \in a^-} \max_{c \in b^-} x_c$, but it is not a solution of $Eq_{max}(AF)$.

If, instead of searching the solutions of the system $Eq_{max}(AF)$, we are searching for real solutions in $[0, 1]$ of the following system of non-linear inequalities,

$$Ineq_{max}(AF) \quad \begin{cases} x_a = 1, \ \text{if } a \in A_0 \\ x_a \leq 1 - \max_{b \in a^-} x_b, \ \text{if } a \in A - A_0, \end{cases}$$

then, is not difficult to prove this set of solutions is convex. Moreover, we can easily translate it as the set of solutions to

$$P_{AF} \quad \begin{cases} 0 \leq x_a \leq 1, \forall a \in A \\ x_a = 1, \ \text{if } a \in A_0 \\ x_a + x_b \leq 1, \forall (b, a) \in D, \end{cases}$$

---

[2] In Gabbay's terminology, a loop is a circuit (see the discussion after Theorem 17).

that is, this set is a polytope. It is not difficult to show that the integral vectors in this polytope are exactly the incidence vectors of the conflict-free sets in AF containing the set A_0 of non-attacked arguments, therefore loosing the nice semantics property in Theorem 7.

In the next section, using linear constraints suggested by Proposition 8, we restrict the above polytope in order to obtain argumentation significance of the corresponding set of integral vectors.

5 Labellings Polytopes

Throughout this section we consider only argumentation frameworks $AF = (A,D)$ without isolated arguments, that is without arguments $a \in A$ such that $a^- \cup a^+ = \emptyset$. Clearly, adding or deleting an isolated argument does not influence the acceptability status of the other arguments.

Definition 9. *Let $AF = (A,D)$ be an argumentation framework. The **admissible sets polytope** of AF is the set $P_{\text{adm}}(AF)$ of all vectors in $\mathbb{R}^A$ satisfying:*

$$
\begin{array}{lll}
(1) & x_a \geq 0 & \forall a \in A, \\
(2) & x_a + x_b \leq 1 & \forall (b,a) \in D, \\
(3) & x_a - \sum_{c \in b^-} x_c \leq 0 & \forall (b,a) \in D.
\end{array}
$$

We make the convention that if $b^- = \emptyset$, then the sum $\sum_{c \in b^-} x_c$ in constraints (3) is 0.

Example 1. In Figure 2 we consider a simple argumentation framework $AF = (A,D)$ with $A = \{a,b\}$. We have two constraints of type (2), corresponding to the two attacks in $D = \{(a,a),(a,b)\}$. Note that in any argumentation framework if a is a self-attacking argument, then $x_a \in [0,1/2]$ due to the type (2) constraint $x_a + x_a \leq 1$. The are two type (3) constraints, but the type (3) constraint for a gives $x_a \leq x_a$ which is trivially satisfied. It follows that all constraints giving $P_{\text{adm}}(AF)$ are those given in the middle of Figure 2 and a graphic illustration of the admissible sets polytope is at right. Its vertices are $(0,0),(\frac{1}{2},0)$, and $(\frac{1}{2},\frac{1}{2})$. The only integer point in $P_{\text{adm}}(AF)$ is $(0,0)$.

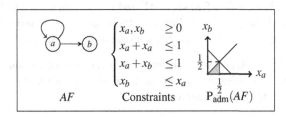

$$
\begin{cases}
x_a, x_b & \geq 0 \\
x_a + x_a & \leq 1 \\
x_a + x_b & \leq 1 \\
x_b & \leq x_a
\end{cases}
$$

AF   Constraints $P_{\text{adm}}(AF)$

Fig. 2. An argumentation framework and its admissible sets polytope

94   C. Croitoru

Example 2. Let $AF = (A,D)$ illustrated in Figure 3 with $A = \{a,b,c\}$ and $D = \{(a,b),$ $(b,a),(b,c)\}$. The two mutual attacks (a,b) and (b,a) generate a single type (2) constraint $x_a + x_b \leq 1$. The only non-trivial type (3) constraint is $x_c \leq x_a$. It follows that all constraints giving $P_{adm}(AF)$ are those given in the middle of Figure 3 and its graphic representation appears at the right. Its vertices are the integral vectors $(0,0,0)^T, (0,1,0)^T,$ $(1,0,0)^T$, and $(1,0,1)^T$ (where, x^T denotes the transpose of the row vector x).

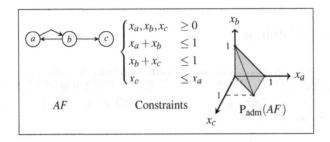

Fig. 3. An admissible sets polytope with integer vertices

    In both above examples the integral vectors in $P_{adm}(AF)$ are exactly the incidence vectors of the admissible sets of the argumentation framework. We will prove next that this holds in general and this justifies the name of the considered polytope.

Lemma 10. *Let $AF = (A,D)$ be an argumentation framework without isolated arguments. If $x \in \mathbb{Z}^A$ satisfies the constraints (1) and (2) then x is a 0-1 vector.*

Proof. For each $a \in A$ there is $b \in A$ such that $(a,b) \in D$ or $(b,a) \in D$. By constraint (2), $x_a + x_b \leq 1$ and, by constraint (1), $x_b \geq 0$. It follows that $x_a \leq 1$, and, since $x_a \in \mathbb{Z}$, we have $x_a \in \{0,1\}$. $\qquad\square$

Lemma 11. *Let $AF = (A,D)$ be an argumentation framework without isolated arguments. An integral vector $x \in \mathbb{Z}^A$ satisfies the constraints (1) and (2) if and only if x is the incidence vector of a conflict-free set in AF.*

Proof. Let $x \in \mathbb{Z}^A$ be an integral vector satisfying the constraints (1) and (2). By Lemma 10, x is a 0-1 vector, and there is $X \subseteq A$ such that $\chi^X = x$. There is no $(a,b) \in D$ with $a,b \in X$, since then $x_a = x_b = 1$ and the constraint (2) for (a,b) is not satisfied. Hence X is a conflict-free set in AF.

    Conversely, if X is a conflict-free set in AF and $x = \chi^X$, then x is a 0-1 vector, hence constraints (1) are trivially satisfied. For any attack $(a,b) \in D$ at most one of a and b are in X, therefore $x_a + x_b \in \{0,1\}$ and the constraints (2) are satisfied. $\qquad\square$

Lemma 12. *Let $AF = (A,D)$ be an argumentation framework without isolated arguments. A 0-1 vector x satisfies the constraints (3) if and only if x is the incidence vector of a set $X \subseteq A$ with the property that $X^- \subseteq X^+$.*

Proof. Let x be a 0-1 vector satisfying the constraints (3) and let $X \subseteq A$ such that $\chi^X = x$. Let $b \in X^-$, that is there is $a \in X$ such that $(b,a) \in D$. By constraint (3), we have

$1 = x_a \leq \sum_{c \in b^-} x_c$. It follows that $b^- \cap X \neq \emptyset$ and there is $a' \in A$ such that $a' \in b^-$ and $x_{a'} = 1$. Hence for each $b \in X^-$ there is $a' \in X$ such that $(a', b) \in D$, that is $X^- \subseteq X^+$.

Conversely, let $X \subseteq A$ with the property that $X^- \subseteq X^+$ and let $x = \chi^X$. Let $(b,a) \in D$. If $a \notin X$, then $x_a = 0$, and the constraint (3) trivially holds since x is a 0-1 vector. If $a \in X$ then $b \in X^-$, and by the hypothesis there is $a' \in X$ such that $(a', b) \in D$. It follows that $\sum_{c \in b^-} x_c \geq x_{a'} = 1 = x_a$, that is the corresponding constraint (3) is satisfied.   □

By Lemmas 10, 11 and 12, the following theorem holds.

Theorem 13. *Let* $AF = (A, D)$ *an argumentation framework without isolated arguments. The integral vectors of* $P_{\mathrm{adm}}(AF)$ *are exactly the incidence vectors of the admissible sets of* AF.

We introduce a set of linear constraints in order to enforce the "directionality principle": the non-attacked arguments must receive label 1, all their attackers must receive label 0, all arguments attacked only by 0 labeled arguments must receive label 1, and so on.

Definition 14. *Let* $AF = (A, D)$ *be an argumentation framework. The* **stable extensions polytope** *of* AF *is the set* $P_{\mathrm{stab}}(AF)$ *of all vectors in* $\mathbb{R}^A$ *satisfying:*

$$
\begin{aligned}
&(1) & x_a &\geq 0 & \forall a \in A, \\
&(2) & x_a + x_b &\leq 1 & \forall (b,a) \in D, \\
&(4) & x_a + \sum_{b \in a^-} x_b &\geq 1 & \forall a \in A.
\end{aligned}
$$

Example 3. In Figure 4 the $P_{\mathrm{stab}}(AF)$ for the argumentation framework in Example 2 is illustrated. Note that in this particular argumentation framework each argument is attacked by exactly one argument, and constraints (2) and (4) gives the equality constraints in the middle of the Figure. The polytope $P_{\mathrm{stab}}(AF)$ is in this case the line segment $\{\lambda x^1 + (1 - \lambda)x^2 | 0 \leq \lambda \leq 1\} = \{(1 - \lambda, \lambda, 1 - \lambda)^T | 0 \leq \lambda \leq 1\}$, where $x^1 = (0,1,0)^T$ and $x_2 = (1,0,1)^T$ are its vertices.

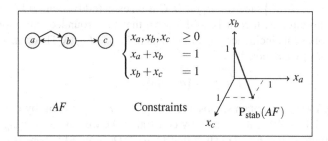

Fig. 4. A stable extensions polytope with integer vertices

In order to relate the vectors in $P_{stab}(AF)$ to complete extensions, we can use Theorem 7 and the following observation.

Proposition 15. *Let* $AF = (A,D)$ *an argumentation framework without isolated arguments. The set of solutions in* $[0,1]$ *to the system of equations* $Eq_{max}(AF)$ *is contained in* $P_{stab}(AF)$.

Proof. Let $x \in [0,1]^A$ be a solution to the system of equations $Eq_{max}(AF)$. Constraints (1) are clearly satisfied. Since $x_a = 1 - \max_{c \in a^-} x_c$ for all $a \in A$, it follows that if $(b,a) \in D$ we have $x_a + x_b \leq x_a + \max_{c \in a^-} x_c = 1 - \max_{c \in a^-} x_c + \max_{c \in a^-} x_c = 1$, that is constraints (2) are satisfied by x. Also constraints (4) are satisfied, since for any $a \in A$ we have $x_a + \sum_{b \in a^-} x_b \geq x_a + \max_{c \in a^-} x_c = 1$. $\square$

The integral vectors of $P_{stab}(AF)$ are interesting as the following Lemma shows.

Lemma 16. *Let* $AF = (A,D)$ *be an argumentation framework without isolated arguments. A* 0-1 *vector* x *satisfies the constraints* (4) *if and only if* x *is the incidence vector of a set* $X \subseteq A$ *with the property that* $A - X \subseteq X^+$.

Proof. Let x be a 0-1 vector satisfying the constraints (4) and let $X \subseteq A$ such that $\chi^X = x$. Let $a \in A - X$, that is $x_a = 0$. By constraint (4), we have $x_a + \sum_{b \in a^-} x_b \geq 1$, and since $x_a = 0$, we have $\sum_{b \in a^-} x_b \geq 1$. It follows that there is $b \in a^-$ such that $x_b = 1$, that is there is $b \in X$ such that $(b,a) \in D$. Therefore $A - X \subseteq X^+$.

Conversely, let $X \subseteq A$ with the property that $A - X \subseteq X^+$ and let $x = \chi^X$. We prove that the constraint (4) holds for every $a \in A$. If $a \in X$, then $x_a = 1$ and, since x is a 0-1 vector, the constraint (4) holds trivially for a. If $a \notin X$, then $x_a = 0$ and since $A - X \subseteq X^+$, there is $b_0 \in a^- \cap X$. Since $x_{b_0} = 1$ it follows that $x_a + \sum_{b \in a^-} x_b \geq 0 + x_{b_0} = 1$, that is the constraint (4) holds for a. $\square$

By Lemmas 10, 11 and 16, the following theorem holds.

Theorem 17. *Let* $AF = (A,D)$ *an argumentation framework without isolated arguments. The integral vectors of* $P_{stab}(AF)$ *are exactly the incidence vectors of the stable extensions of* AF.

The structure of $P_{stab}(AF)$ is strongly dependent on the combinatorial structure of AF, more precisely on its family of circuits. We represent here a circuit in $AF = (A,D)$ as a sequence $C = (a_1, a_2, \ldots, a_k)$ of distinct arguments $a_i \in A$ such that $(a_i, a_{i+1}) \in D$, for each $i \in \{1, \ldots, k-1\}$, and $(a_k, a_1) \in D$. C is an even (odd) circuit if k is even (odd).

If AF has no circuits, then it is well known that the grounded extension is also a stable extension, so its incidence vector belongs to $P_{stab}(AF)$, by Theorem 17. For $x \in P_{stab}(AF)$ we denote by

$$Frac(x) = \{a \in A | 0 < x_a < 1\}$$

its set of fractional components. Observe that if $a \in Frac(x)$ then, by constraint (2), we have $x_b < 1$ for all $b \in a^-$. Since, by constraint (4), we have $x_a + \sum_{b \in a^-} x_b \geq 1$, it follows that there is $b \in Frac(x)$ such that $(b,a) \in D$. Similarly, there is $c \in Frac(x)$ such that $(c,b) \in D$. Since $Frac(x)$ is a finite set, continuing the above argument we find a circuit $C = (a_1, a_2, \ldots, a_k)$ with all $a_i \in Frac(x)$. Hence, if AF has no circuits, then all vectors in $P_{stab}(AF)$ are integral. Since $P_{stab}(AF)$ is non-empty (containing the incidence vector of the grounded extension), it follows by convexity that it has exactly one point. Hence, the following Corollary of the Theorem 17 holds.

Corollary 18. *Let $AF = (A,D)$ an argumentation framework without circuits. Then $P_{stab}(AF)$ has exactly one point, the incidence vector of the grounding extension of AF.*

In order to characterize the incidence vectors of complete extensions of an argumentation framework AF, let us note that, in general, these vectors are not members of $P_{stab}(AF)$ as Lemma 16 shows. On the other hand, in the discussion preceding Corollary 18, we have argued that if $x \in P_{stab}(AF)$ is such that $Frac(x) \neq \emptyset$ then each argument in $Frac(x)$ has an attacker in the same set. Furthermore, if $x_a = 1$ then $a^+ \cap Frac(x) = \emptyset$ and $a^- \cap Frac(x) = \emptyset$ (by constraints (2)). These facts justify the following definition.

Definition 19. *Let $AF = (A,D)$ be an argumentation framework and $P_{adm}(AF)$ be its stable extensions polytope. A vector $x \in P_{stab}(AF)$ is called a **complete vector** if for each $a \in A$ such that $x_a = 0$ there is $b \in A$ such that $(b,a) \in D$ and $x_b = 1$.*

Note that if $x \in P_{stab}(AF)$ is such that $Frac(x) = A$, then x is a complete vector. The following characterization of **comp**(AF) holds.

Theorem 20. *Let $AF = (A,D)$ an argumentation framework without isolated arguments. The integral vectors obtained from complete vectors of $P_{stab}(AF)$ by replacing the fractional components with 0 are exactly the incidence vectors of the complete extensions of AF.*

Proof. Let S be a complete extension of AF and $y = \chi^S$. Let $x \in \mathbb{R}^A$ defined by

$$x_a = \begin{cases} y_a & \text{if } y_a = 1, \\ 0 & \text{if } y_a = 0 \text{ and } \exists b \in S \text{ s.t. } (b,a) \in D, \\ \frac{1}{2} & \text{otherwise.} \end{cases}$$

Clearly, y is obtained from x by replacing its fractional components with 0. We show that x is a complete vector of $P_{stab}(AF)$. x trivially satisfies the constraints (1), and constraints (2) hold since S is a conflict-free set. Let $a \in A$. If $x_a = 1$ then constraint (4) holds trivially. If $x_a = 0$ then $a \notin S$ and there is $b \in S$ such that $(b,a) \in D$. It follows that $y_b = x_b = 1$ and $\sum_{c \in a^-} x_c \geq x_b = 1$, hence constraint (4) holds. If $x_a = \frac{1}{2}$ it follows that $a \notin S$ and there is no $b \in S$ such that $(b,a) \in D$. Since S is a complete extension, we have $F(S) = S$. Since $a \notin S$, there is $b \in A - S \cup S^+$ such that $(b,a) \in D$. It follows that $x_b = \frac{1}{2}$. Hence $x_a + \sum_{c \in a^-} x_c \geq x_a + x_b = \frac{1}{2} + \frac{1}{2} = 1$, that is, the constraint (4) holds.

Conversely, let $x \in P_{stab}(AF)$ be a complete vector, and let $y \in \{0,1\}^A$ be the vector obtained from x by replacing its fractional components with 0. Let $S \subseteq A$ such that $\chi^S = y$. We show that S is a complete extension in AF. Clearly, S is a conflict-free set because x satisfies the constraints (2). Let $a \in S^-$, that is, there is $b \in S$ such that $(a,b) \in D$. Then $a \notin S$, hence $y_a = 0$, and moreover $x_a = 0$, because $x_a + x_b \leq 1$ and $x_b = 1$. Since x is a complete vector, it follows that there is $c \in S$ with $x_c = 1$ such that $(c,a) \in D$. Hence, we proved that $S^- \subseteq S^+$, that is S is an admissible set. Suppose that there is $a \notin S \cup S^+$ such that $a^- - (S \cup S^+) = \emptyset$. Since $a \notin S^+$, we must have $a \in Frac(x)$ and from the constraint (4) for a we have $x_a + \sum_{b \in a^-} x_b = x_a + 0 = x_a \geq 1$, contradiction. Hence we have obtained that S is an admissible set and $F(S) = S$, that is S is a complete extension. $\qquad\square$

6 Discussion

In this paper, two polytopes associated with an argumentation framework AF are introduced in order to represent Dung's classical extensions of AF. The first one, $P_{adm}(AF)$, has as integral points exactly the set of incidence vectors of the admissible sets of AF. That is, a 0-1 vector is the incidence vector of a conflict-free set of arguments defending collectively against any attack if and only if it belongs to $P_{adm}(AF)$. The second, $P_{stab}(AF)$, characterizes the stable extensions: a 0-1 vector is the incidence vector of a conflict-free set of arguments attacking all others arguments if and only if is a member of $P_{stab}(AF)$. Furthermore, it is proved that $P_{stab}(AF)$ encodes also the complete extensions: a set of arguments is a complete extension if and only if it is the set of 1-coordinates of a point x in $P_{stab}(AF)$ with the property that each 0-coordinate of x is attacked by an 1-coordinate of x.

These characterizations open the way of using techniques from (integer) linear programming to select such extensions after considering appropriate weights for arguments. As future work we intend to study the vertices of the above polytopes. For example, we believe that $P_{stab}(AF)$ has integer vertices for argumentation frameworks AF with no odd circuits. Also, consequences of linear duality theorems may be of interest. Beside the computational motivation, the approach introduced in this paper can be viewed as a new attempt to relate abstract argumentation to game theory, where the different solution concepts are studied using linear programming techniques.

References

[1] Baroni, P., Giacomin, M.: On principle-based evaluation of extension-based argumentation semantics. Artificial Intelligence 171, 675–700 (2007)
[2] Bell, C., Nerode, A., Ng, R.T., Subrahmanian, V.S.: Mixed integer programming methods for computing nonmonotonic deductive databases. Journal of ACM 41, 1178–1215 (1994)
[3] Bench-Capon, T.: Persuasion in practical argument using value-based argumentation frameworks. Journal of Logic and Computation 13, 429–448 (2003)
[4] Caminada, M.: On the issue of reinstatement in argumentation. In: Fisher, M., van der Hoek, W., Konev, B., Lisitsa, A. (eds.) JELIA 2006. LNCS (LNAI), vol. 4160, pp. 111–123. Springer, Heidelberg (2006)
[5] Dung, P.M.: On the acceptability of arguments and its fundamental role in nonmonotonic reasoning, logic programming and n-person games. Artificial Intelligence 77, 321–357 (1995)
[6] Gabbay, D.: Introducing equational semantics for argumentation networks. In: Liu, W. (ed.) ECSQARU 2011. LNCS, vol. 6717, pp. 19–35. Springer, Heidelberg (2011)
[7] Gabbay, D.: An equational approach to argumentation networks. Argumentation and Computation 3(2-3) (2012)
[8] Gabbay, D.: The equational approach to cf2 semantics. In: Proc. of COMMA 2012, pp. 141–152 (2012)
[9] Gratie, C., Florea, A.M.: Fuzzy labeling for argumentation frameworks. In: McBurney, P., Parsons, S., Rahwan, I. (eds.) ArgMAS 2011. LNCS, vol. 7543, pp. 1–8. Springer, Heidelberg (2012)
[10] Khachiyan, L.G.: A polynomial algorithm in linear programmming. Soviet Mathematics Doklady 20, 191–194 (1979)

[11] Li, R., Zhou, D., Du, D.: Satisfiability and integer programming as complementary tools. In: Proc. of ASPDAC 2004 (2004)
[12] Liu, G., Janhunen, T., Niemelä, I.: Answer set programming via mixed integer programming. In: Proc. of KR 2012 (2012)
[13] Lowe, J.: CPLEX 10 Solver Manual. GAMS Development Corporation (2012)
[14] Pollock, J.L.: Cognitive Carpentry. A Blueprint for How to Build a Person. MIT Press, Cambridge (1995)
[15] Rahwan, I., Simari, G. (eds.): Argumentation in Artificial Intelligence. Springer (2009)
[16] Schrijver, A.: Combinatorial Optimization - Polyhedra and Efficiency. Springer (2003)

Update Operators for Inconsistent Query Answering: A New Point of View

Madalina Croitoru[1] and Ricardo Oscar Rodriguez[2]

[1] University of Montpellier 2, France
croitoru@lirmm.fr
[2] University of Buenos Aires, Argentina
ricardo@dc.uba.ar

Abstract. This paper investigates the definition of belief revision operators that correspond to the IAR and ICR inconsistency tolerant semantics in the Ontology Based Data Access (OBDA) setting. By doing this equivalence our long term goal is to provide an axiomatic characterisation of the above mentioned semantics.

1 Introduction and Motivation

We position ourselves in the Ontology Based Data Access (OBDA) setting where a query is being asked over a set of knowledge bases defined over a common ontology. When the union of knowledge bases along with the ontology is inconsistent, several semantics have been defined [5,11] which are tolerant to inconsistency. They all rely on computing repairs, i.e. maximal (in terms of set inclusion) subsets of the knowledge bases. The inconsistency tolerant semantics (Intersection of All Repairs: IAR, All Repairs: AR, Intersection of Closed Repairs: ICR) have been studied [5,11], from a productivity point of view and a complexity point of view.

In this paper we take a new point of view. Our long term aim is to define axiomatic characterisations of two such semantics (IAR and ICR). We argue that such characterisation can provide an alternative way of comparing the semantics and can provide new insights into their properties. Furthermore such axiomatisation can be used when proposing a generalisation of inconsistency tolerant semantics.

In order to provide the axiomatic characterisation we define belief revision operators that correspond to IAR and ICR. The belief revision operator corresponding to the AR semantics is left for future work. Please note that while a lot of work has been done in belief revision and OBDA, none of the approaches deal with the final goal of their axiomatic characterisations. The paper is structures as follows: Section 2 introduces the rule based OBDA language used in the paper, Section 3 defines the revision operators that correspond to ICR and IAR. Based on the axiomatic characterisation of the belief operators we can provide the axiomatic characterisation of IAR and ICR. Finally, Section 4 concludes the paper.

2 Rule Based Knowledge Representation

There are two major approaches in the literature used to represent an ontology for the OBDA problem and namely Description Logics (such as $\mathcal{EL}$([3]) and DL-Lite [7] families) and rule based languages. The most notable rule based language is the Datalog$^+$ [6]

U. Straccia and A. Calì (Eds.): SUM 2014, LNAI 8720, pp. 100–105, 2014.

language, a generalization of Datalog that allows for existentially quantified variables in the head of the rules. Despite Datalog$^+$ undecidability when answering conjunctive queries, there exist decidable fragments of Datalog$^+$ are studied in the literature [4]. These fragments generalize the above mentioned Description Logics families.

In this paper we represent the ontology via rules. We consider a (potentially inconsistent) knowledge base composed of a set $\mathcal{F}$ of facts corresponding to existentially closed conjunctions of atoms, which can contain n-ary predicates; a set of negative constraints $\mathcal{N}$ which represent the negation of a fact and an ontology composed of a set of rules $\mathcal{R}$ that represent general implicit knowledge that can introduce new variables in their head (conclusion).

A rule is *applicable* to set of facts $\mathcal{F}$ if and only if the set entails the hypothesis of the rule. If rule R is applicable to the set $\mathcal{F}$, the application of R on $\mathcal{F}$ produces a new set of facts obtained from the initial set with additional information from the rule conclusion. We then say that the new set is an *immediate derivation* of $\mathcal{F}$ by R denoted by $R(\mathcal{F})$. Let $\mathcal{F}$ be a set of facts and let $\mathcal{R}$ be a set of rules. A set $\mathcal{F}_n$ is called an $\mathcal{R}$-*derivation* of $\mathcal{F}$ if there is a sequence of sets (*derivation sequence*) $(\mathcal{F}_0, \mathcal{F}_1, \dots, \mathcal{F}_n)$ such that: (i) $\mathcal{F}_0 \subseteq \mathcal{F}$, (ii) $\mathcal{F}_0$ is $\mathcal{R}$-consistent, (iii) for every $i \in \{1, \dots, n-1\}$, it holds that $\mathcal{F}_i$ is an immediate derivation of $\mathcal{F}_{i-1}$.

Given a set $\{F_0, \dots, F_k\}$ and a set of rules $\mathcal{R}$, the closure of $\{F_0, \dots, F_k\}$ with respect to $\mathcal{R}$, denoted $\mathtt{Cl}_{\mathcal{R}}(\{F_0, \dots, F_k\})$, is defined as the smallest set (with respect to $\subseteq$) which contains $\{F_0, \dots, F_k\}$, and is closed for $\mathcal{R}$-derivation (that is, for every $\mathcal{R}$-derivation F_n of $\{F_0, \dots, F_k\}$, we have $F_n \subseteq \mathtt{Cl}_{\mathcal{R}}(\{F_0, \dots, F_k\})$). Finally, we say that a set $\mathcal{F}$ and a set of rules $\mathcal{R}$ *entail* a fact G (and we write $\mathcal{F}, \mathcal{R} \models G$) iff the closure of the facts by all the rules entails G (i.e. if $\mathtt{Cl}_{\mathcal{R}}(\mathcal{F}) \models G$). Given a set of facts $\{F_1, \dots, F_k\}$, and a set of rules $\mathcal{R}$, the set of facts is called $\mathcal{R}$-*inconsistent* if and only if there exists a constraint $N = \neg F$ such that $\mathtt{Cl}_{\mathcal{R}}(\{F_1, \dots, F_k\}) \models F$. A set of facts is said to be $\mathcal{R}$-*consistent* iff it is not $\mathcal{R}$-inconsistent. A knowledge base $\mathcal{K} = (\mathcal{F}, \mathcal{R}, \mathcal{N})$ is said to be *consistent* if and only if $\mathcal{F}$ is $\mathcal{R}$-consistent. A knowledge base is *inconsistent* if and only if it is not consistent.

Several semantics have been proposed to handle consistency based in the concept of data *repairs* [5,11]. Once the repairs are computed, various strategies can be adapted to answer a query. We can consider querying *all repairs* (AR-semantics), the *intersection of all repairs* (IAR-semantics) or the *intersection of closed repairs* (ICR-semantics).

Definition 1. *[5,11] Let $\mathcal{K} = (\mathcal{F}, \mathcal{R}, \mathcal{N})$ be a knowledge base and let α be a query. Then α is **AR-entailed** from $\mathcal{K}$, written $\mathcal{K} \models_{AR} \alpha$ iff for every repair $A' \in \mathcal{R}epair(\mathcal{K})$, it holds that $\mathtt{Cl}_{\mathcal{R}}(A') \models \alpha$.*

Definition 2. *[5,11] Let $\mathcal{K} = (\mathcal{F}, \mathcal{R}, \mathcal{N})$ be a knowledge base and let α be a query. Then α is **IAR-entailed** from $\mathcal{K}$, written $\mathcal{K} \models_{IAR} \alpha$ iff $\mathtt{Cl}_{\mathcal{R}}(\bigcap_{A' \in \mathcal{R}epair(\mathcal{K})}) \models \alpha$.*

Definition 3. *[5,11] Let $\mathcal{K} = (\mathcal{F}, \mathcal{R}, \mathcal{N})$ be a knowledge base and let α be a query. Then α is **ICR-entailed** from $\mathcal{K}$, written $\mathcal{K} \models_{ICR} \alpha$ iff $\bigcap_{A' \in \mathcal{R}epair(\mathcal{K})} \mathtt{Cl}_{\mathcal{R}}(A') \models \alpha$.*

3   Belief Revision Operators

In the context of Belief Revision, the problem of making an inconsistent belief base consistent was solved by Hanson in [10] who proposed a new operation called *Consolidation*. An inconsistent belief base can be consolidated by removing some of its elements. The consolidation of a belief base B is denoted $B!$. A plausible way to perform consolidation is to contract by falsum (contradiction), i.e. $B! = B \div \perp$. In order to express this notion more precisely, we need to introduce the concept of *remainder set* proposed by Alchourrón and Makinson in [2]. For any belief base B and any formula α, the remainder set of B by α, $B \perp \alpha$, is the set of maximal subsets of B that do not imply α. In other words, in our language:

Definition 4. *(Updated AM81 [2]) Let $(\mathcal{F}, \mathcal{R}, \mathcal{N})$ a knowledge base and α a fact in $\mathcal{F}$. The set $\mathcal{F} \perp \alpha$ (read "$\mathcal{F}$ less α") is the set such that A belongs to $\mathcal{F} \perp \alpha$ if and only if $A \subseteq \mathcal{F}$, $\mathtt{Cl}_\mathcal{R}(A) \not\models \alpha$ and there is no set A' such that $A \subset A' \subseteq \mathcal{F}$ and $\mathtt{Cl}_\mathcal{R}(A') \not\models \alpha$.*

Definition 5. *A selection function is a function γ such that for every set $\mathcal{F}$ of formulae and any fact α it holds: $\gamma(\mathcal{F}\perp\alpha)$ is a non-empty subset of $\mathcal{F}\perp\alpha$ if this set is non-empty, and, on the contrary, $\gamma(\mathcal{F}\perp\alpha) = \{\mathcal{F}\}$.*

Definition 6. *(Updated AGM85 [1]) Let $\mathcal{F}$ be a set of facts in a knowledge base $\mathcal{K}$. Let $\mathcal{F}\perp\alpha$ and γ be the set of all maximal subsets of $\mathcal{F}$ that do not imply α and a selection function, respectively. The* partial meet contraction *on $\mathcal{F}$ that is generated by γ is the operation $\sim_\gamma$ such that for all facts α:*

$$\mathcal{F} \sim_\gamma \alpha = \cap\gamma(\mathcal{F}\perp\alpha)$$

Two limiting cases have been very well studied: when γ gives back either only one element of $\mathcal{F}\perp\alpha$ or all members of $\mathcal{F}\perp\alpha$. In the first case, we are talking about *Maxichoice contraction* and in the second it is called *Full meet contraction*.

There are other special and interesting cases when the selection function is based on a relation (that may be thought of as a preference relation).

Definition 7. *A selection function γ for a belief base $\mathcal{F}$ in a knowledge base $\mathcal{K}$, and the contraction operator based on it, are*

1. *relational if and only if there is a binary relation $\sqsubseteq$ such that for all fact α, if $\mathcal{F}\perp\alpha$ is non-empty, then*

$$\gamma(\mathcal{F}\perp\alpha) = \{A \in \mathcal{F}\perp\alpha \mid C \sqsubseteq A \text{ for all } C \in \mathcal{F}\perp\alpha\}$$

2. *transitively relational if and only if there is such a relation that is transitive.*

Based on partial meet contraction, one can define a partial meet consolidation as $\mathcal{F} \sim_\gamma \perp$ which is the intersection of the "most preferred" maximal consistent subsets of $\mathcal{F}$, i.e. $\mathcal{F}! = \mathcal{F} \sim_\gamma \perp = \cap\gamma(\mathcal{F}\perp\perp)$ where $\perp$ denotes logical contradiction.

Partial meet consolidation has been axiomatically characterized as follows:

Theorem 1. *(Updated [9]) An operation is a partial meet consolidation if and only if for all sets $\mathcal{F}$ of facts the following are satisfied:*

Consistency: $\mathcal{F}!$ *is $\mathcal{R}$-consistent.*
Inclusion: $\mathcal{F}! \subseteq \mathcal{F}$.
Relevance: *If $\alpha \in \mathcal{F} \setminus \mathcal{F}!$, then there is some $\mathcal{F}'$ with $\mathcal{F}! \subseteq \mathcal{F}' \subseteq \mathcal{F}$, such that $\mathcal{F}'$ is $\mathcal{R}$-consistent and $\mathcal{F}' \cup \{\alpha\}$ is $\mathcal{R}$-inconsistent.*

*In addition, it is a **full meet consolidation** if and only if it also satisfies:*

Core identity: *$\beta \in \mathcal{F}!$ if and only if $\beta \in \mathcal{F}$ and there is no $\mathcal{F}' \subseteq \mathcal{F}$ such that $\mathcal{F}'$ is $\mathcal{R}$-consistent but $\mathcal{F}' \cup \{\beta\}$ is $\mathcal{R}$-inconsistent.*

*On the other hand, an operator is a **maxi-choice consolidation** if and only if it satisfies the postulates consistency, inclusion, and:*

Fullness: *If $\beta \in \mathcal{F}$ and $\beta \notin \mathcal{F}!$ then $\mathcal{F}! \cup \{\beta\}$ is $\mathcal{R}$-inconsistent.*

There are five ways to characterise a belief revision function: axiomatic, using remainder sets, using kernel sets, epistemic entrenchment and spheres system. In the following we will define the operators corresponding to the inconsistency tolerant semantics above using its axiomatic characterisation.

3.1   Consolidation Operators for Existing Semantics

We want to define an operator of consolidation that given a inconsistent knowledge base $\mathcal{K} = (\mathcal{F}, \mathcal{R}, \mathcal{N})$, returns a new consistent knowledge base $\mathcal{K}! = (\mathcal{F}!, \mathcal{R}, \mathcal{N})$.

As a first step, we consider the maximal consistent subsets (repairs or remainders) of $\mathcal{F}$ denoted $\mathcal{K} \perp \perp$. More precisely:

$$\mathcal{K} \perp \perp = \{\mathcal{M} | \mathcal{M} \subseteq \mathcal{F}, (\mathcal{M}, \mathcal{R}, \mathcal{N}) \text{ maximal consistent set } \}$$

We are now able to define an operator of consolidation as $\mathcal{K}! = Op(Ch(\mathcal{K} \perp \perp))$ where Ch is a *choice function* and Op an intersection operator (or an any operator defined over the results of choice). It is easy to check the following:

1. **If Ch returns all elements and Op is the intersection of sets then we get the IAR semantics.**

Note that in this case we are talking about a full meet consolidation and it allows us to give immediately an axiomatic characterization of IAR semantics.

In order to obtain the equivalent of ICR semantics we need to work on sets closed by $Cl_{\mathcal{R}}$. First, we have to introduce some properties.

Observation 1 *If $\mathcal{M} \in \mathcal{K} \perp \perp$ then $Cl_{\mathcal{R}}(\mathcal{M}) \cap \mathcal{F} \subseteq \mathcal{M}$. Hence, if $\mathcal{F}$ is closed by $\mathcal{R}$-derivation (i.e. $\mathcal{F} = Cl_{\mathcal{R}}(\mathcal{F})$) then $\mathcal{M}$ is as well.*

We now define the remainders or repairs sets closed by $Cl_{\mathcal{R}}$ in the following way:

$$\mathcal{K} \perp_{Cl_{\mathcal{R}}} \perp = \{\mathcal{M} | \mathcal{M} \subseteq Cl_{\mathcal{R}}(\mathcal{F}), (\mathcal{M}, \mathcal{R}, \mathcal{N}) \text{ maximal consistent set } \}$$

We are now able to define a new family of consolidation operators using these remainders. This kind of operators defined over closed sets by $\mathcal{R}$-derivations will be called closure partial meet consolidation operators and denoted by $\dagger$. We also cover the ICR semantics as follows:

2. If Ch returns all elements and Op is the intersection of sets over $\{\mathcal{M}|\mathcal{M} \in \mathcal{K} \perp_{\mathtt{Cl}_{\mathcal{R}}} \perp\}$ then we get the ICR semantics.

Note that we describe a full meet consolidation but over remainders closed by $\mathcal{R}$-derivations. In order to give an axiomatization of this kind of consolidations, we need to included an axiom which mirrors this fact. This will be done in the next section.

3.2 Axiom Compliance

In this section we go one step further in the definition of consolidation operators in relationship with ICR semantics through a set of postulates. In order to give logical properties of that kind of consolidation operators, we first rephrase Hansson's postulates within our framework. Let $\mathcal{K} = (\mathcal{F}, \mathcal{R}, \mathcal{N})$ be a knowledge base, the original postulates can be rewritten in following way:

Closure: $\mathcal{F}\dagger = \mathtt{Cl}_{\mathcal{R}}(\mathcal{F}\dagger)$.
Consistency: $\mathcal{K}\dagger = (\mathcal{F}\dagger, \mathcal{R}, \mathcal{N})$ is consistent.
Inclusion: $\mathcal{K}\dagger = (\mathcal{F}\dagger, \mathcal{R}, \mathcal{N}) \sqsubseteq \mathcal{K} = (\mathcal{F}, \mathcal{R}, \mathcal{N})$.
Core identity: If $f \in \mathcal{F}\dagger$ if and only if $f \in \mathtt{Cl}_{\mathcal{R}}(\mathcal{F})$ and $\nexists \mathcal{X} \subseteq \mathtt{Cl}_{\mathcal{R}}(\mathcal{F})$, such that $\mathcal{X}$ is $\mathcal{R}$-consistent and $\mathcal{X} \cup f$ does not.

Theorem 2. *An operator $\dagger$ is an operation of closure full meet consolidation if and only if it satisfies Closure, Consistency, Inclusion and Core identity.*

Proof *Checking that operations of closure full meet consolidation satisfy postulates: Closure, Consistency and Inclusion follow directly from definition and Observation 1. To see that closure full meet consolidation satisfies Core identity, let $f \in \mathtt{Cl}_{\mathcal{R}}(\mathcal{F})$. Then $f \notin \mathcal{F}\dagger$ if and only if there is some $\mathcal{X} \in \mathcal{K} \perp_{\mathtt{Cl}_{\mathcal{R}}} \perp$ such that $f \notin \mathcal{X}$. Since $\mathcal{X}$ is a maximal $\mathcal{R}$-consistent set it follows that $\mathcal{X} \cup f$ does not. Hence, $\mathcal{X}$ satisfies the required properties by Core Identity.*

On the contrary direction, let $\dagger$ be an operator that satisfies the four postulates mentioned in the theorem. We need to show that $\mathcal{F}\dagger = \bigcap \mathcal{K} \perp_{\mathtt{Cl}_{\mathcal{R}}} \perp$. We assume again, that $f \in \mathtt{Cl}_{\mathcal{R}}(\mathcal{F})$. Suppose that $f \notin \bigcap \mathcal{K} \perp_{\mathtt{Cl}_{\mathcal{R}}} \perp$ then there is $\mathcal{X} \in \mathcal{K} \perp_{\mathtt{Cl}_{\mathcal{R}}} \perp$ such that $f \notin \mathcal{X}$. Since $\mathcal{X}$ is maximal $\mathcal{R}$-consistent then $\mathcal{X} \cup \{f\}$ does not. From that and Core Identity we conclude that $f \notin \mathcal{F}\dagger$. On the other hand, since $\dagger$ satisfies Inclusion we can conclude that $\mathcal{F}\dagger \subseteq \mathtt{Cl}_{\mathcal{R}}(\mathcal{F})$. In addition, because $\dagger$ satisfies Closure and Consistency, we may assume that there exists $f \notin \mathcal{F}\dagger$. It follows by Core Identity that for each $f \notin \mathcal{F}\dagger$ there exists $\mathcal{X} \subset \mathtt{Cl}_{\mathcal{R}}(\mathcal{F})$ such that $\mathcal{X}$ is $\mathcal{R}$-consistent and $\mathcal{X} \cup f$ does not. In the usual way we are able to extend $\mathcal{X}$ to a maximal $\mathcal{R}$-consistent set $\mathcal{X}'$ such that $f \notin \mathcal{X}'$. Since $\mathcal{X} \subset \mathtt{Cl}_{\mathcal{R}}(\mathcal{F})$ then $\mathcal{X}' \subset \mathtt{Cl}_{\mathcal{R}}(\mathcal{F})$ and $\mathcal{X}' \in \mathcal{K} \perp_{\mathtt{Cl}_{\mathcal{R}}} \perp$. From the last, we may conclude that $f \notin \mathcal{K} \perp_{\mathtt{Cl}_{\mathcal{R}}} \perp$. This finishes the proof.

4   Conclusion and Future Work

In this paper we showed how to get first results towards an axiomatic characterisation of the IAR and ICR semantics in the OBDA setting. We did this by covering the semantics using known consolidation operators from belief revision. The AR semantics poses some problems. Indeed, to cover the AR semantics let $\{Ch_i\}$ with $1 \leq i \leq \#(\mathcal{K} \perp \perp)$ be a family of choice functions such that each Ch_i returns only one different element of $\mathcal{K} \perp_{Cl_\mathcal{R}} \perp$, i.e. we enumerate the elements of $\mathcal{K} \perp_{Cl_\mathcal{R}} \perp$ and each Ch_i gives back only one of it. If we consider the family of maxichoice consolidation defined by $\{Ch_i\}$ and we take the intersection among the classical consequence closure of each $\mathcal{K}_i!$ then we get the AR semantics. In this case, we are considering a family of maxichoice consolidations which have also a well known axiomatic characterization. But it is not clear how to axiomatize the combination proposed here. This will be left for future research. Note that by using the equivalence showed in [8] we also obtained here an axiomatic characterisation of some argumentation semantics.

Acknowledgements. The first author was funded by the ANR ASPIQ project.

References

1. Alchourrón, C., Gärdenfors, P., Makinson, D.: On the logic of theory change: partial meet contraction and revision function. Journal Symbolic Logic 50, 510–530 (1985)
2. Alchourrón, C., Makinson, D.: Hierarchies of regulations and their logic. In: Hilpinen, R. (ed.) Deontic Logic:Introductory and Systematic Readings, pp. 125–148. Reidel Publishing Company, Dordrecht (1981)
3. Baader, F., Brandt, S., Lutz, C.: Pushing the el envelope. In: Proc. of IJCAI 2005 (2005)
4. Baget, J.F., Mugnier, M.L., Rudolph, S., Thomazo, M.: Walking the complexity lines for generalized guarded existential rules. In: Proceedings of the 22nd International Joint Conference on Artificial Intelligence (IJCAI 2011), pp. 712–717 (2011)
5. Bienvenu, M.: On the complexity of consistent query answering in the presence of simple ontologies. In: Proc of AAAI (2012)
6. Calì, A., Gottlob, G., Lukasiewicz, T.: A general datalog-based framework for tractable query answering over ontologies. In: Proceedings of the Twenty-Eigth ACM SIGMOD-SIGACT-SIGART Symposium on Principles of Database Systems, pp. 77–86. ACM (2009)
7. Calvanese, D., De Giacomo, G., Lembo, D., Lenzerini, M., Rosati, R.: Tractable reasoning and efficient query answering in description logics: The dl-lite family. J. Autom. Reasoning 39(3), 385–429 (2007)
8. Croitoru, M., Vesic, S.: What can argumentation do for inconsistent ontology query answering? In: Liu, W., Subrahmanian, V.S., Wijsen, J. (eds.) SUM 2013. LNCS, vol. 8078, pp. 15–29. Springer, Heidelberg (2013)
9. Hansson, S.: Belief Base Dynamics. Ph.D. thesis, Uppsala (1991)
10. Hansson, S.: Semi-revision. Journal of Applied Non-Classical Logics 7(1-2), 151–175 (1997)
11. Lembo, D., Lenzerini, M., Rosati, R., Ruzzi, M., Savo, D.F.: Inconsistency-tolerant semantics for description logics. In: Hitzler, P., Lukasiewicz, T. (eds.) RR 2010. LNCS, vol. 6333, pp. 103–117. Springer, Heidelberg (2010)

Conflicts of Belief Functions: Continuity and Frame Resizement

Milan Daniel[1] and Jianbing Ma[2]

[1] Institute of Computer Science, Academy of Sciences of the Czech Republic
Pod Vodárenskou věží 2, CZ – 182 07 Prague 8, Czech Republic
milan.daniel@cs.cas.cz
[2] Faculty of Science and Technology, Bournemouth University
Bournemouth, UK, BH12, 5BB
jma@bournemouth.ac.uk

Abstract. Plausibility and pignistic conflict of belief functions are briefly recalled in this study. These measures of conflict are based on two different probability transformations of belief functions, normalised plausibility of singletons Pl_C and Smets' pignistic probability $BetP$.

Continuity properties and relationship of these conflict measures to extension and refinement of a frame of discernment are investigated here. A new continuous improvement of both the measures which is preserved by a frame extension is introduced. A relation of the new conflict measures to refinement of a frame of discernment is also discussed. Finally a comparison between the new measure and the two original measures as well as W. Liu's degree of conflict cf is presented.

Keywords: Belief functions, Dempster-Shafer theory, uncertainty, plausibility conflict, pignistic conflict, degree of conflict, continuity, extension of a frame of discernment, refinement of a frame of discernment.

1 Introduction

When combining belief functions (BFs) by the conjunctive rules of combination, conflicts often appear (which are assigned to $\emptyset$ by non-normalised conjunctive rule $\odot$ or normalised by Dempster's rule of combination $\oplus$). Combination of conflicting BFs and interpretation of conflicts are often questionable in real applications. Thus a series of papers were published on alternative combination rules, conflicting belief functions, e.g. [2,4,13,15,16,22], and measures of conflicts, e.g. [12,17,18].

A new interpretation of conflicts of BFs was introduced in [6]. Important distinction of conflicts between BFs due to internal conflict of a single BF, and due to the difference between BFs was introduced there. The most elaborated perspective of the three approaches initiated in [6] — plausibility conflict of BFs — was analysed in [9] and improved in [10]. An alternative pignistic conflict based on Smets' pignistic probability $BetP$ was introduced there as well.

The presented study investigates plausibility and pignistic conflicts from the point of view of continuity and resizement of a frame of discernment: extension

U. Straccia and A. Calì (Eds.): SUM 2014, LNAI 8720, pp. 106–119, 2014.

and refinement of the frame. New improvements of both the measures of conflicts between BFs with respect to these properties are presented.

Similarly to [6,9,10] we use W. Liu's assumption that conflict between BFs appears when the BFs strongly support mutually non-compatible hypotheses [16], and also the assumptions from [6] that there is no conflict between BFs when the BFs (strongly) support same or compatible hypotheses. Moreover, starting from Section 4, we assume continuity of conflict measures defined in Section 3; starting from Section 5, we assume keeping of conflictness/non-conflictness when extending a frame of discernment; and further, starting from Section 6, we assume also keeping of conflictness/non-conflictness when refining the frame. Section 7 compares and summarizes the presented results, several ideas for a future research are stated in Section 8.

2   State of the Art

We assume classic definitions of basic notions from theory of belief functions [19] on finite frames of discernment $\Omega_n = \{\omega_1, \omega_2, ..., \omega_n\}$. Due to a limited space we do not repeat all the notions used in [6,9,10], but only the important of those, which were introduced there.

A *basic belief assignment (bba)* $m : \mathcal{P}(\Omega) \longrightarrow [0,1]$, $\sum_{A \subseteq \Omega} m(A) = 1$, its values are called *basic belief masses (bbms)*; a *belief function (BF)* $Bel : \mathcal{P}(\Omega) \longrightarrow [0,1]$, $Bel(A) = \sum_{\emptyset \neq X \subseteq A} m(X)$. A *plausibility function* $Pl(A) = \sum_{\emptyset \neq A \cap X} m(X)$. There is a unique correspondence between m and the corresponding Bel and Pl; thus we often speak about m as a belief function. A *focal element* is a subset X of the frame of discernment such that $m(X) > 0$. *Normalised plausibility of singletons* corresponding to Bel: $Pl\_P(\omega) = \frac{Pl(\{\omega\})}{\sum_{\omega' \in \Omega} Pl(\{\omega'\})}$ [3,5] (this is normalised contour function); *pignistic probability*: $BetP(\omega) = \sum_{\omega \in X \subseteq \Omega} \frac{1}{|X|} \frac{m(X)}{1 - m(\emptyset)}$ [20].

We say that $\omega \in \Omega$ is *supported* or *preferred by* a *belief function* Bel defined on Ω_n when $Pl\_P(\omega) > \frac{1}{n}$, ω is *opposed* by Bel if $Pl\_P(\omega) < \frac{1}{n}$. Analogously for $BetP(\omega)$ if Smets pignistic probability is used. U_n is a BF on Ω_n such that $m(\omega) = \frac{1}{n}$ for all $\omega \in \Omega_n$.

Conflict between BFs is distinguished from internal conflict in [6,9,10], where internal conflict of a BF is included inside the individual BF. Total conflict of two BFs Bel_1, Bel_2, which is equal to sum of all conflicting belief mases: $m_{\bigcirc}(\emptyset) = \sum_{X \cap Y = \emptyset} m_1(X) m_2(Y)$, includes internal conflicts of individual BFs Bel_1, Bel_2 and a conflict between them. Thus two definitions were introduced in [6]; we are interested in conflict between belief BFs in this study.

Definition 1. *The* internal plausibility conflict *Pl-IntC of BF Bel is defined as*

$$Pl\text{-}IntC(Bel) = 1 - max_{\omega \in \Omega} Pl(\{\omega\}),$$

where Pl is the plausibility corresponding to Bel.

Definition 2. *Let Bel_1, Bel_2 be two belief functions on Ω_n given by bbms m_1 and m_2 which have normalised plausibility of singletons Pl_P_1 and Pl_P_2. The* conflicting set $\Omega_{PlC}(Bel_1, Bel_2)$ *is defined to be the set of* elements $\omega \in \Omega_n$ *with conflicting* Pl_P *masses it is conditionally extended with union of sets* max Pl_P_i *value elements under condition that they are disjoint. Formally we have* $\Omega_{PlC}(Bel_1, Bel_2) = \Omega_{PlC_0}(Bel_1, Bel_2) \cup \Omega_{smPlC}(Bel_1, Bel_2)$, *where* $\Omega_{PlC_0}(Bel_1, Bel_2) = \{\omega \in \Omega_n \mid (Pl\_P_1(\omega) - \frac{1}{n})(Pl\_P_2(\omega) - \frac{1}{n}) < 0\}$, $\Omega_{smPlC}(Bel_1, Bel_2) = \{\omega \in \Omega_n \mid \omega \in \{max_{\omega \in \Omega_n} Pl\_P_1(\omega)\} \cup \{max_{\omega \in \Omega_n} Pl\_P_2(\omega)\}$ $\& \{max_{\omega \in \Omega_n} Pl\_P_1(\omega)\} \cap \{max_{\omega \in \Omega_n} Pl\_P_2(\omega)\} \neq \emptyset\}$.

Plausibility conflict between BFs Bel_1 *and* Bel_2 *is then defined by the formula*

$$Pl\text{-}C(Bel_1, Bel_2) = min(\, Pl\text{-}C_0(Bel_1, Bel_2), (m_1 \ominus m_2)(\emptyset)\,),$$

where[1]

$$Pl\text{-}C_0(Bel_1, Bel_2) = \sum_{\omega \in \Omega_{PlC}(Bel_1, Bel_2)} \frac{1}{2} |Pl\_P(Bel_1)(\omega) - Pl\_P(Bel_2)(\omega)|.$$

There are two reasons for minimising with $(m_1 \ominus m_2)(\emptyset)$ (briefly with $m_\ominus(\emptyset)$ if Bel_1 and Bel_2 are clear from a context): at first the original from [6], see Example 1, where two obviously non-conflicting BFs have non-empty conflicting set and positive $Pl\text{-}C_0$, whereas $m_\ominus(\emptyset) = 0$; the second is that $m_\ominus(\emptyset)$ was found to be an upper bound for conflict between BFs [10].

Example 1. Let us suppose two categorical BFs on Ω_6 given by $m_1(\{\omega_1\}) = 1$ and $m_2(\{\omega_1, \omega_2, \omega_3, \omega_4\}) = 1$, thus $Pl\_P_1 = (1,0,0,0,0,0)$ and $Pl\_P_2 = (0.25, 0.25, 0.25, 0.25, 0, 0)$ thus $\Omega_{PlC} = \{\omega_2, \omega_3, \omega_4\}$ as $Pl\_P_1(\omega_i) = 0 < \frac{1}{6}$ and $Pl\_P_2(\omega_i) = 0.25 > \frac{1}{6}$ for $i = 1, 2, 3, 4$ (other elements are non-conflicting), hence $Pl\text{-}C_0 = 0.375$ and this should be minimised with $m_\ominus(\emptyset) = 0$.

Four variants of $\Omega_{PlC}(Bel_1, Bel_2)$ are defined and analysed in [10]: Ω_{smPlC}, $\Omega_{spPlC} = \Omega_{smPlC}(Bel_1, Bel_2) \cup \Omega_{PlC_0}(Bel_1, Bel_2)$ (as above), Ω_{cpPlC} which includes ω with different order of $Pl\_P_i(\omega)$ values, and $\Omega_{cbPlC} = \Omega_{cpPlC}(Bel_1, Bel_2)$ $\cup \Omega_{PlC_0}(Bel_1, Bel_2)$. I.e., Ω_{PlC} is constructed using either max Pl_P_i values, ordering Pl_P_i values, support/opposition of elements of the frame of discernment (+ max Pl_P_i values), or combination of these options; for detail see [10]. (All of these variants coincide on Ω_2).

Example 2. Four variants of conflicting sets. Let us suppose Bel_1, Bel_2 on Ω_5.

X :	$\{\omega_1\}$	$\{\omega_2\}$	$\{\omega_3\}$	$\{\omega_4\}$	$\{\omega_5\}$	$\{\omega_1,\omega_2\}$	$\{\omega_2,\omega_4\}$	$\{\omega_1-\omega_3\}$	$\{\omega_1..\omega_4\}$	Ω_5
$m_1(X)$:	0.225	0.195	0.19	0.19	0.01	0.02	0.01	0.03	0.11	0.02
$m_2(X)$:	0.110	0.410	0.16	0.00	0.01	0.03	0.04	0.02	0.09	0.05

We obtain $Pl\_P_1 = (0.27, 0.25, 0.24, 0.22, 0.02)$, $Pl\_P_2 = (0.20, 0.40, 0.24, 0.12, 0.04)$, and $\Omega_{smPlC} = \{\omega_1, \omega_2\}$, $\Omega_{PlC_0} = \{\omega_4\}$, $\Omega_{spPlC} = \{\omega_1, \omega_2, \omega_4\}$, $\Omega_{cpPlC} = \{\omega_1, \omega_2, \omega_3\}$, and $\Omega_{cbPlC} = \{\omega_1, \omega_2, \omega_3, \omega_4\}$, hence Bel_1 and Bel_2 are considered to be mutually conflicting by all variants of $Pl\text{-}C$, but values of conflict are different in this case: $smPl\text{-}C(Bel_1, Bel_2) = 0.11$, $spPl\text{-}C = 0.16$, $cpPl\text{-}C = 0.11$, $cbPl\text{-}C = 0.16$. (ω_3 has different order of Pl_P_i values thus it is included in both Ω_{cpPlC} and Ω_{cbPlC} but both its Pl_P_i values are the same: $Pl\_P_1(\omega_3) = Pl\_P_2(\omega_3) = 0.24$, thus $smPl\text{-}C = cpPl\text{-}C$ and $spPl\text{-}C = cbPl\text{-}C$ in this special case.

---

[1] $Pl\text{-}C_0$ is not a separate measure of conflict in general; it is just a component of $Pl\text{-}C$.

Definition 3. *Let Bel_1, Bel_2 be two belief functions on Ω_n given by bbms m_1 and m_2 which have pignistic probabilities $BetP_1$ and $BetP_2$. The pignistic conflicting set $\Omega_{BetC}(Bel_1, Bel_2)$ is defined analogously to plausibility conflicting set $\Omega_{PlC}(Bel_1, Bel_2)$, having analogously four variants Ω_{smBetC}, Ω_{spBetC}, Ω_{cpPlC} and Ω_{cbPlC}, see [10].*

Pignistic conflict between BFs Bel_1 and Bel_2 is then defined by the formula

$$Bet\text{-}C(Bel_1, Bel_2) = min(\, Bet\text{-}C_0(Bel_1, Bel_2), (m_1 \varominus m_2)(\emptyset)\,),$$

where[2]

$$Bet\text{-}C_0(Bel_1, Bel_2) = \sum_{\omega \in \Omega_{BetC}(Bel_1, Bel_2)} \frac{1}{2} |BetP_1(\omega) - BetP_2(\omega)|.$$

Quantitative aspect of conflict — conflictness/non-conflictness is classified by emptyness/non-emptyness of related conflicting set by both $Pl\text{-}C$ and $Bet\text{-}C$. Quantitative conflict is then computed only for mutually conflicting BFs.

Whereas qualitative values are computed for any pair of BFs by Liu's two-component degree of conflict $cf = (dif BetP_{m_i}^{m_j}, m_\cap(\emptyset))$ [10,16], where $dif BetP_{m_i}^{m_j} = max_{A \subseteq \Omega}(|BetP_{m_i}(A) - BetP_{m_j}(A)|)$, $m_\cap(\emptyset) = (m_1 \varominus m_2)(\emptyset)$. Qualitative question of conflictness/non-conflictness is not addressed there, in fact; and 'high conflictness' / 'not high conflictness' is determined from the qualitative values using empirically/heuristically given threshold of conflict tolerance ε.

Unfortunately, jumps of $Pl\text{-}C$ and $Bet\text{-}C$ values were observed, see the following examples. Such a jump in conflict values is counter-intuitive, moreover neither $m_\cap(\emptyset)$ nor the other component $dif BetP_{m_i}^{m_j}$ of Liu's degree of conflict cf have similar jumps. Hence, we are interested in how to remove the jumps from conflict measures in this study, i.e., how to modify measures of conflict $Pl\text{-}C$ and $Bet\text{-}C$ to be continuous, or jump-free.

Example 3. Let us suppose two BFs on Ω_2: $Bel_1 = (m_1(\{\omega_1\}), m_1(\{\omega_2\}) = (0.8, 0.1)$ $(m_1(\{\omega_1, \omega_2\}) = 1 - m_1(\{\omega_1\}) - m_1(\{\omega_2\}) = 0.1)$, $Bel_2 = (0.3, 0.3)$, thus we obtain $Pl\_P_1 = (0.888, 0.111)$, $Pl\_P_2 = (0.5, 0.5)$, hence these two BFs are non-conflicting. Let us suppose a very small change of Bel_2, thus we expect zero conflict again or a very small conflict value corresponding to the very small change. Let $Bel_2' = (0.3, 0.31)$, thus $Pl\_P_2' = (0.4964, 0.5036)$, hence $\Omega_{PlC}(Bel_1, Bel_2') = \Omega_2$ and $Pl\text{-}C(Bel_1, Bel_2') = 0.3925$, which is significantly higher than the slight changes made on $m_2(\{\omega_2\})$ and $m_2(\{\omega_1, \omega_2\})$ by 0.01. Analogously, we have $BetP_1 = (0.85, 0.15)$, $BetP_2 = (0.5, 0.5)$, and $BetP_2' = (0.495, 0.505)$, which leads to $BetP\text{-}C(Bel_1, Bel_2) = 0$, $BetP\text{-}C(Bel_1, Bel_2') = 0.355$.

Let us suppose a free BF $Bel_1 = (a_1, b_1)$ and a fixed BF $Bel_2 = (a_2, b_2)$ on Ω_2, such that $Pl\_P_2 = (u, 1 - u)$ where $u \geq \frac{1}{2}$. We can show how the value $Pl\text{-}C(Bel_1, Bel_2)$ depends on the value $Pl\_P_1(\omega_1)$ (i.e., $Pl\_P_1(\omega_1) = \frac{1-b_1}{2-a_1-b_1}$, as $Pl\_P_1(\omega_2) = 1 - Pl\_P_1(\omega_1)$ and u is fixed). A jump is obvious at $Pl\_P_1(\omega_1) = \frac{1}{2}$, see Fig. 1. For another example of jumps of conflict see Example 4.

---

[2] $Bet\text{-}C_0$ is again not a separate measure of conflict in general; it is just a component of $Bet\text{-}C$.

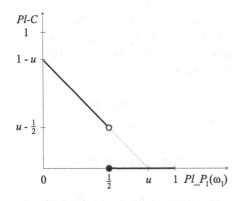

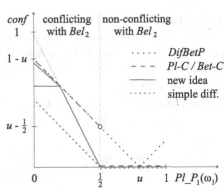

Fig. 1. Jump of *Pl-C*   **Fig. 2.** A comparison of approaches

Example 4. Let us suppose BFs on Ω_4 given by the following table now:

X :	$\{\omega_1\}$	$\{\omega_2\}$	$\{\omega_3\}$	$\{\omega_4\}$	$\{\omega_1,\omega_2\}$	$\{\omega_4,\omega_3\}$	Ω_4
$m_1(X)$:	0.15	0.15	0.15	0.15	0.01		0.39
$m_2(X)$:	0.15	0.15	0.15	0.15		0.01	0.39
$m_3(X)$:	0.20	0.15	0.10	0.05	0.15	0.05	0.30
$m_4(X)$:	0.90	0.03	0.02		0.01	0.01	0.03

$Pl\_C_1 = (0.2523, 0.2523, 0.2477, 0.2477)$, $Pl\_C_2 = (0.2477, 0.2477, 0.2523, 0.2523)$, $Pl\_C_3 = (0.3095, 0.2857, 0.2143, 0.1905)$, $Pl\_C_4 = (0.8468, 0.0631, 0.0541, 0.0360)$, $Pl\text{-}C(Bel_1, Bel_3) = 0$, $Pl\text{-}C(Bel_2, Bel_3) = 0.0998$, which is about ten times larger than the changes on $\{\omega_1, \omega_2\}$ and $\{\omega_3, \omega_4\}$ by 0.01. $smPl\text{-}C(Bel_1, Bel_4) = cpPl\text{-}C(Bel_1, Bel_4) = 0$, $spPl\text{-}C(Bel_1, Bel_4) = cbPl\text{-}C(Bel_1, Bel_4) = 0.2027$, $smPl\text{-}C(Bel_2, Bel_4) = 0.5068$, $cpPl\text{-}C(Bel_2, Bel_4) = 0.5991$, $spPl\text{-}C(Bel_2, Bel_4) = 0.5068$, $cbPl\text{-}C(Bel_2, Bel_4) = 0.5991$, all the changes on conflict values are significantly greater than the difference between m_1 values and m_2 values (i.e., 0.01).

3   Continuity of Measures of Conflict between Belief Functions

Let us define continuity of a measure of conflict using the conventional $\epsilon - \delta$ way. That is, we first define a δ-surrounding for any BF, and we then use δ-surrounding to define continuity of conflict measures.

Formally, we have the following definitions.

Definition 4. *We say that a belief function Bel' is in δ surrounding of a belief function Bel (briefly $Bel' \in \delta(Bel)$) if $|m(X) - m'(X)| \leq \delta$.*

Definition 5. *We say that a measure of conflict of belief functions conf is continuous if for any $\varepsilon > 0$ and any BFs Bel_1, Bel_2, there exits a δ surrounding of Bel_1, such that for any $Bel' \in \delta(Bel_1)$, $|conf(Bel', Bel_2) - conf(Bel_1, Bel_2)| \leq \varepsilon$.*

Lemma 1. *(i)* $(m_1 \odot m_2)(\emptyset)$ *is continuous measure of conflict.*
(ii) $min(conf(Bel_1, Bel_2), (m_1 \odot m_2)(\emptyset))$ *is continuous for any continuous measure of conflict conf and any pair of BFs* Bel_1, Bel_2 *given by* m_1 *and* m_2.
(iii) $min(difBet_{Bel_1}^{Bel_2}, (m_1 \odot m_2)(\emptyset))$ *is a continuous measure of conflict of BFs.*

Proof. Proofs are verifications of the statements, for detail see [11].

4 Continuous Improvement of Plausibility and Pignistic Conflicts

There is a non-conflicting area around any BF (a half of the belief triangle in the case of BFs on Ω_2; there is a possibility of different variants of such areas using different conflicting sets Ω_{PlC} (or Ω_{BetC}) for BFs on Ω_n, $n > 2$). The idea is that conflict is zero on the border of conflicting area and it should continually increase without any jump behind the border. In the case of Ω_2, we compute a difference of Pl_P (or $BetP$) from U_2; for obtaining continuity we use minimal difference. Its value should be doubled to obtain normalised conflict, i.e. to obtain conflict between $(1, 0)$ and $(0, 1)$ equal to 1; see green line in Fig. 2, for simple (non doubled) difference see blue line. This is equal to the sum of minimal differences over Ω_{PlC} (it is $\emptyset$ or entire Ω_2 in the case of Ω_2). Thus we obtain the following modification of $Pl\text{-}C_0$:

$$Pl\text{-}C_1(Bel_1, Bel_2) = Pl\text{-}C_0(Bel_1, Bel_2) = 0,$$

if $(Bel_1(\{\omega_1\}) - Bel_1(\{\omega_2\}))(Bel_2(\{\omega_1\}) - Bel_2(\{\omega_2\})) \geq 0$.

$$Pl\text{-}C_1(Bel_1, Bel_2) = 2 \, min \, (\, |Pl\_P(Bel_1)(\omega_1) - \frac{1}{2}|, \, |Pl\_P(Bel_2)(\omega_1) - \frac{1}{2}| \,)$$

$$= \sum_{i=1,2} min \, (\, |Pl\_P(Bel_1)(\omega_i) - \frac{1}{2}|, \, |Pl\_P(Bel_2)(\omega_i) - \frac{1}{2}| \,);$$

as $Pl\_P(Bel_i)(\omega_2) = 1 - Pl\_P(Bel_i)(\omega_1)$ for $i = 1, 2$.

The situation is more complicated on $\Omega_n = \{\omega_1, \omega_2, ... \omega_n\}$: We have to distinguish two parts of $\Omega_{PlC}(Bel_1, Bel_2)$: $\Omega_{PlC}^{opp}(Bel_1, Bel_2)$ ... ω's which are opposed by Bel_1 and Bel_2 (i.e. where $(Bel_1(\omega) - \frac{1}{n})(Bel_2(\omega) - \frac{1}{n}) < 0$) and $\Omega_{PlC}^{ord}(Bel_1, Bel_2) = \Omega_{PlC}(Bel_1, Bel_2) \setminus \Omega_{PlC}^{opp}(Bel_1, Bel_2)$ ... ω's from corresponding Ω_{cpPlC} and Ω_{cbPlC} which have different order of $Pl\_P(Bel_i)(\omega)$ values, but they are not opposed by Bel_1 and Bel_2, thus we have to handle them separately.

Let us assume that all ω's from $\Omega_{PlC}(Bel_1, Bel_2)$ are opposed by Bel_1 and Bel_2, i.e. $\Omega_{PlC}^{ord}(Bel_1, Bel_2) = \emptyset$, now. We can compute $Pl\text{-}C_1(Bel_1, Bel_2)$ 'per elements' directly in the same way as it is computed on Ω_2:

$$Pl\text{-}C_1(Bel_1, Bel_2) = \sum_{\omega \in \Omega_{PlC}(Bel_1, Bel_2)} min \, (\, |Pl\_P(Bel_1)(\omega) - \frac{1}{n}|, \, |Pl\_P(Bel_2)(\omega) - \frac{1}{n}| \,).$$

Let us look at the following example of belief functions Bel_1, ..., Bel_4 on $\Omega_{10} = \{\omega_1, \omega_2, ..., \omega_{10}\}$ such that, $Pl\_P_1(\omega_1) = 1$, $Pl\_P_2(\omega_{10}) = 1$. In this case,

we obtain $\Omega_{PIC}(Bel_1, Bel_2) = \Omega_{PIC}^{opp}(Bel_1, Bel_2) = \{\omega_1, \omega_{10}\}$, $Pl\text{-}C_1(Bel_1, Bel_2)$
$= \sum_{\omega \in \Omega_{PIC}} min(|Pl\_P_1(\omega) - \frac{1}{10}|, |Pl\_P_2(\omega) - \frac{1}{10}|) = \frac{1}{10} + \frac{1}{10} = \frac{2}{10}$.
$Pl\_P_3(\omega_1) = Pl\_P_3(\omega_2) = \frac{1}{2}$, $Pl\_P_4(\omega_9) = Pl\_P_4(\omega_{10}) = \frac{1}{2}$, $\Omega_{PIC}(Bel_3, Bel_4)$
$= \Omega_{PIC}^{opp}(Bel_3, Bel_4) = \{\omega_1, \omega_2, \omega_9, \omega_{10}\}$, thus $Pl\text{-}C_1(Bel_3, Bel_4) = 4 \cdot \frac{1}{10} = \frac{4}{10}$.
The conflict between two different categorical singletons $Pl\text{-}C_1(Bel_1, Bel_2)$ should
be maximal/greatest (as different elements (disjoint hypotheses) are fully (cat-
egorically) supported). More precisely, it should be equal to 1 for normalised
conflict. Moreover conflict $Pl\text{-}C_1(Bel_1, Bel_2)$ should be the same or greater than
conflict $Pl\text{-}C_1(Bel_3, Bel_4)$, definitely not a half of it.

Considering the above example, we have to proportionalise comparison of
$Pl\_P(Bel_j)(\omega_i)$ with $\frac{1}{n}$; i.e., to multiply $|Pl\_P(Bel_j)(\omega_i) - \frac{1}{n}|$ by appropriate
coefficient(s):
- a coefficient $\frac{n}{2}$ determined by the size of frame of discernment;
  this factor is equal to $\frac{2}{2} = 1$ for $n = 2$;
- a coefficient $\frac{1}{2}(Pl\_P(Bel_1)(\omega_i) + Pl\_P(Bel_2)(\omega_i))$,
  i.e., by the relative size of sum of relative plausibilities of corresponding ω_i.
Thus we obtain:

$$Pl\text{-}C_{11}(Bel_1, Bel_2) = \sum_{\omega \in \Omega_{PIC}(Bel_1, Bel_2)} \frac{n}{2} \frac{Pl\_P_1(\omega) + Pl\_P_2(\omega)}{2} min_{i=1,2}(|Pl\_P_i(\omega) - \frac{1}{n}|).$$

For proving of continuity of $Pl\text{-}C_{11}$ we will use the following technical lemma,
for proofs see [11].

Lemma 2. (i) *For any BFs Bel and Bel' on Ω_n such that $Bel' \in \delta(Bel)$ for
$\delta = \frac{\varepsilon}{2^{n-1}}$ it holds that $|Pl\_P(\omega) - Pl\_P'(\omega)| \leq \varepsilon$ for any $\omega \in \Omega_n$.*
(ii) *For any BFs Bel_1, Bel_2 and Bel' on Ω_n such that $Bel' \in \delta(Bel_1)$ for $\delta \leq
min_{\omega_i \in \Omega_n}|Pl_P_1(\omega_i) - \frac{1}{n}|$ it holds that $\Omega_{PIC}(Bel', Bel_2) = \Omega_{PIC}(Bel_1, Bel_2)$.*

Analogously to the original version $Pl\text{-}C_0$ we need to minimize $m_{\bigodot}(\emptyset)$ (also
for $Pl\text{-}C_{11}$), see BFs and Ω_{PIC} from Example 1 again. From the previous proof
and Lemma 1 we obtain also continuity of $min(Pl\text{-}C_{11}, m_{\bigodot}(\emptyset))$.

For $\omega \in \Omega_{PIC}^{ord}$ we cannot use min of $(Pl\_P(\omega) - \frac{1}{n})$ as both the BFs are in
accordance with respect to ω, and both of them support (or oppose) ω (thus
min may be relatively high for BFs with same or similar $Pl\_P(\omega)$ and, on the
other hand, it is very small for $Pl\_P_1(\omega)$ close to $\frac{1}{n}$ and $Pl\_P_2(\omega)$ close to 0 or
1.) Hence we have to use difference of differences, i.e., we have $||Pl_P_1(\omega) - \frac{1}{n}| -
|Pl_P_2(\omega) - \frac{1}{n}|| = |Pl_P_1(\omega) - Pl_P_2(\omega)|$ as it is in $Pl\text{-}C$ (see [10]).

Thus we obtain the following formula:

$$Pl\text{-}C_{12}(Bel_1, Bel_2) = \sum_{\omega \in \Omega_{PIC}^{opp}(Bel_1, Bel_2)} \frac{n}{2} \frac{Pl\_P_1(\omega) + Pl\_P_2(\omega)}{2} min_{i=1,2}(|Pl\_P_i(\omega) - \frac{1}{n}|)$$

$$+ \sum_{\omega \in \Omega_{PIC}^{ord}(Bel_1, Bel_2)} (|Pl\_P_1(\omega) - Pl\_P_2(\omega)|).$$

A difference of Pl_P values is continuous, thus continuity is not lost upgrading $Pl\text{-}C_{11}$ to $Pl\text{-}C_{12}$.

The situation is analogous for $Bet\text{-}C$. For proving of continuity of $Bet\text{-}C_{12}$, where $BetP$'s are used instead of Pl_P's; we use the analogy of Lemma 2:

Lemma 3. *(i) For any BFs Bel and Bel' on Ω_n such that $Bel' \in \delta(Bel)$ for $\delta = \frac{\varepsilon}{2^{n-1}}$ it holds that $|BetP(\omega) - BetP'(\omega)| \leq \varepsilon$ for any $\omega \in \Omega_n$.*
(ii) For any BFs Bel_1, Bel_2 and Bel' on Ω_n such that $Bel' \in \delta(Bel_1)$ for $\delta \leq min_{\omega_i \in \Omega_n}|BetP_1(\omega_i) - \frac{1}{n}|$ it holds that $\Omega_{BetC}(Bel', Bel_2) = \Omega_{BetC}(Bel_1, Bel_2)$.

For proofs see [11].

5   Extension of a Frame of Discernment

We have to note a relationship of $Pl\text{-}C_{12}(Bel_1, Bel_2)$ to resizement of a frame of discernment. An extension of a frame of discernment is to add one or more elements into the frame of discernment but keeping the BFs not changed. More precisely, let us suppose a frame $\Omega_m = \{\omega_1, \omega_2, ..., \omega_m\}$ and BFs Bel_i's given by bbms m_i's. Let us extend the frame with $\{\omega_{m+1}, ..., \omega_{m+k}\}$ for $m \geq 2$, $k \geq 1$. Let Bel_i''s be given by m_i''s such that $m_i'(X) = m_i(X)$ for $X \subseteq \Omega_m$, $m_i'(X) = 0$ for $X \cap \{\omega_{m+1}, ..., \omega_{m+k}\} \neq \emptyset$. Thus we have $Pl\_P_i'(\omega) = Pl\_P_i(\omega)$ for $\omega \in \Omega_m$ and $Pl\_P_i'(\omega) = 0$ for $\omega \in \Omega_{m+k} \setminus \Omega_m$. Comparing $Pl\_P_i'(\omega)$ with $\frac{1}{m+k} < \frac{1}{m}$ some ω's may be opposed by one of both Bel_i's but supported by both Bel_i''s. If such ω is opposed by just one of Bel_i's and supported by both Bel_i''s, or it is opposed by both Bel_i's and supported just by one of Bel_i''s, there may be $\Omega_{PlC}'^{opp} \subsetneq \Omega_{PlC}^{opp}$ or $\Omega_{PlC}'^{opp} \supsetneq \Omega_{PlC}^{opp}$ (or $\Omega_{PlC}'^{opp} = \Omega_{PlC}^{opp}$ of course). Hence conflict may be increased or decreased with greater or less conflicting set on the extended frame of discernment. $\Omega_{PlC}'^{ord} = \Omega_{PlC}^{ord}$ as $Pl\_P_i'(\omega) = Pl\_P_i(\omega)$ for $\omega \in \Omega_m$ and $Pl\_P_i'(\omega) = 0$ out of Ω_m. See the following example:

Let $\Omega_m = \Omega_3$, $\Omega_{m+k} = \Omega_4$ and Bel_i's are given by m_i's as follows:

X :	$\{\omega_1\}$	$\{\omega_2\}$	$\{\omega_3\}$	$\{\omega_1, \omega_2\}$	$\{\omega_1, \omega_3\}$	$\{\omega_2, \omega_3\}$	$\{\omega_1, \omega_2, \omega_3\}$
$m_1(X)$:	0.60	0.10	0.1				0.2
$m_2(X)$:	0.20	0.05	0.35			0.3	0.1
$m_3(X)$:	0.45	0.15	0.10	0.1			0.2
$m_4(X)$:		0.25	0.40			0.2	0.15

Thus we obtain the following normalised plausibilities:

$Pl\_P_1 = (0.6, 0.2, 0.2)$ ... $Pl\_P_1' = (0.6, 0.2, 0.2, 0.0)$,
$Pl\_P_2 = (0.2, 0.3, 0.5)$ ... $Pl\_P_2' = (0.2, 0.3, 0.5, 0.0)$.

Only ω_1 is supported on both the frames by Bel_1 and Bel_1'; ω_2, ω_3 (and ω_4) are opposed. Only ω_3 is supported by Bel_2 on Ω_3, but ω_2 and ω_3 are supported by Bel_2' on Ω_4, thus we obtain different conflicting sets on the frames, namely, $\Omega_{PlC}(Bel_1, Bel_2) = \{\omega_1, \omega_3\}$ on Ω_3, whereas $\Omega_{PlC}(Bel_1', Bel_2') = \{\omega_1, \omega_2, \omega_3\}$ on Ω_4, as $\frac{1}{4} < Pl\_P_2(\omega_2) = 0.3 = Pl\_P_2'(\omega_2) < \frac{1}{3}$ and $Pl\_P_1(\omega_2) = 0.2 < \frac{1}{4} < \frac{1}{3}$. Hence conflicting set was increased with the extension of the frame.

Analogously $\Omega_{PlC}(Bel_3, Bel_4) = \{\omega_1, \omega_2, \omega_3\} \supsetneq \Omega_{PlC}(Bel_3', Bel_4') = \{\omega_1, \omega_3\}$.

Behaviour of conflict $Pl\text{-}C_{12}$ may be even more different on the original and extended frames in some cases. Let us look at the following example:

X :	$\{\omega_1\}$	$\{\omega_2\}$	$\{\omega_3\}$	$\{\omega_1,\omega_2\}$	$\{\omega_1,\omega_3\}$	$\{\omega_2,\omega_3\}$	$\{\omega_1,\omega_2,\omega_3\}$
$m_5(X)$:	0.15	0.05	0.15	0.2		0.1	0.35
$m_6(X)$:	0.20			0.6			0.2
$m_7(X)$:	0.50	0.05	0.05		0.3		0.1
$m_8(X)$:	0.20		0.20		0.2		0.4

Thus we obtain the following normalised plausibilities:
$Pl\_P_5 = (0.35, 0.35, 0.3) \dots Pl\_P_5' = (0.35, 0.35, 0.3, 0.0)$,
$Pl\_P_6 = (0.5, 0.4, 0.1) \dots Pl\_P_6' = (0.5, 0.4, 0.1, 0.0)$.
ω_1 and ω_2 are supported by both Bel_5 and Bel_6 on both the frames, ω_3 is opposed by both BFs on Ω_3 but only by Bel_6' on Ω_4 whereas supported by Bel_5', thus conflicting set $\{\omega_3\}$ appears when extending the frame. Hence the extension makes conflict between two originally $Pl\text{-}C_{12}$ non-conflicting BFs Bel_5 and Bel_6. Analogously $\Omega_{PlC}(Bel_7, Bel_8) = \{\omega_3\}$ becomes empty when extending the frame.

A very simple example is $Pl\text{-}C_{12}$ from BFs $(0.4, 0.6)$ and $(0.65, 0.35)$ on Ω_2 which becomes non-conflicting by any extension of the frame.

The above problem of $Pl\text{-}C_{12}$ with a change of conflicting sets when extending the frame of discernment is related only to Ω_{PlC}^{opp} not to Ω_{PlC}^{ord} as orderings of $Pl\_P(\omega_i)$ values are the same on the original Ω_m and on corresponding extended Ω_{m+k}. Thus the problem is related only to $spPl\text{-}C$, $cbPl\text{-}C$, not to $smPl\text{-}C$, $cpPl\text{-}C$. Hence we have obtained a new argument for using the latter versions of $Pl\text{-}C$.

We have either to accept a strange behaviour of $Pl\text{-}C_{12}$ when extending the frame of discernment, or to change the definition of supporting/opposing elements by BFs to be independent of extension of the frame of discernment or to concentrate ourselves to $smPl\text{-}C$, $cpPl\text{-}C$ versions, as it is in the following.

Unfortunately, we are again at the beginning of the continuity problems. As $smPl\text{-}C((0.5, 0.5), (0.9, 0.1)$ is zero for BFs on Ω_2 as ω_1 has maximal Pl_P for both of them, but $smPl\text{-}C((0.49, 0.51), (0.9, 0.1)) = \frac{1}{2}(0.41 + 0.41) = 0.41$. Moreover we have the same results and same problem regardless computing $\frac{1}{2}(|Pl\_P_1(\omega_1) - Pl\_P_2(\omega_1)| + |Pl\_P_1(\omega_2) - Pl\_P_2(\omega_2)|)$ or $\frac{1}{2}(|Pl\_P_1(\omega_1) - Pl\_P_1(\omega_2)| + |Pl\_P_2(\omega_1) - Pl\_P_2(\omega_2)|)$ on Ω_2; thus regardless whether we use differences of Pl_P_i per elements or differences between max and max but one value of the same Pl_P per BFs.

Nevertheless, we can apply the 'min idea' from $Pl\text{-}C_1$ to differences of the max Pl_P's values from max but one values of Pl_P's (Pl_P_1 and Pl_P_2) instead of differences of these values from $\frac{1}{n}$. Thus we obtain

$$Pl\text{-}C_{13}(Bel_1, Bel_2) = min\left(|Pl\_P_1(\omega_1) - Pl\_P_1(\omega_2)|, |Pl\_P_2(\omega_1) - Pl\_P_2(\omega_2)|\right)$$

for Bel_1, Bel_2 on Ω_2 such that $\Omega_{smPlC}(Bel_1, Bel_2) = \Omega_2$ on Ω_2; and generally

$$Pl\text{-}C_{13}(Bel_1, Bel_2) = min\left(|Pl\_P_1(\omega_i) - Pl\_P_1(\omega_j)|, |Pl\_P_2(\omega_k) - Pl\_P_2(\omega_l)|\right)$$

for Bel_1, Bel_2 on general finite frame Ω_n, where max $Pl\_P(\omega)$ values appear for ω_i and ω_k, $i \neq k$, and $Pl\_P_1(\omega_j)$, $Pl\_P_2(\omega_l)$ are max but one values of Pl_P's;

$$Pl\text{-}C_{13}(Bel_1, Bel_2) = 0$$

if sets of max values of Pl_P_1 and Pl_P_2 are not disjoint.

A proof of continuity. Let suppose a pair of BFs Bel_1, Bel_2 and a BFs Bel' in δ surrounding of Bel_1, such that $|m_1(X) - m'(X)| \leq \delta = \frac{\varepsilon}{2^n}$, thus $|Pl_1(X) - Pl'(X)| \leq 2^{n-1}\frac{\varepsilon}{2^n} = \frac{\varepsilon}{2}$ for any $X \subseteq \Omega$ and sequently also $|Pl\_P_1(\omega) - Pl\_P'(\omega)| \leq \frac{\varepsilon}{2}$ for any $\omega \in \Omega$, hence we obtain $|Pl\_P_1(\omega_i) - Pl\_P_1(\omega_j)| - |Pl\_P'(\omega_i) - Pl\_P'(\omega_j)| \leq |Pl\_P_1(\omega_i) - Pl\_P'(\omega_i)| + |Pl\_P_1(\omega_j) - Pl\_P'(\omega_j)| \leq \varepsilon$ and sequently $|Pl\text{-}C_{13}(Bel_1, Bel_2) - Pl\text{-}C_{13}(Bel', Bel_2)| = |min(|Pl\_P_1(\omega_i) - Pl\_P_1(\omega_j)|, |Pl\_P_2(\omega_k) - Pl\_P_2(\omega_l)|) - min(|Pl\_P'(\omega_i) - Pl\_P'(\omega_j)|, |Pl\_P_2(\omega_k) - Pl\_P_2(\omega_l)|)| \leq \varepsilon$ (for detail see [11]). Hence $Pl\text{-}C_{13}$ is continuous.

Values of m, Bel, Pl and of Pl_P are kept with an extension of the frame of discernment, thus also conflictness/non-conflictness and the size of $Pl\text{-}C_{13}$ are kept with a frame extension. Thus using $Pl\text{-}C_{13}$ instead of $Pl\text{-}C_0$ we obtain a continuous improvement $min(Pl\text{-}C_{13}, m_\cap(\emptyset))$ of $Pl\text{-}C$ which is preserved when extending the frame of discernment.

$Pl\text{-}C_{13}$ is a modification or analogy of $smPl\text{-}C$ in fact: if max but one value of Pl_P_1 appears for element(s) which has/have the max value of Pl_P_2 and vice versa then $Pl\text{-}C_{13}$ coincides with sm version of $Pl\text{-}C$ in some cases, but not in general. Thus, it seems neither easy nor useful useful to try to define a similar continuous improvement which is a modification of $cpPl\text{-}C$.

The above problems of Ω_{PlC}^{opp} and $Pl\text{-}C_{12}$ are in the same way relevant also to Ω_{BetC}^{opp} (not to Ω_{BetC}^{ord}) and $Bet\text{-}C_{12}$. Thus completely analogously to $Pl\text{-}C_{13}$, just using $BetP$'s instead of Pl_P's we can define $Bet\text{-}C_{13}$. Having Lemma 3, we can use also the above proof of continuity substituting Pl_P's with $BetP$'s. Values of $BetP$ are also kept with an extension of the frame of discernment, thus conflictness/non-conflictness and the size of $Bet\text{-}C_{13}$ are kept with a frame extension as well. Thus using $Bet\text{-}C_{13}$ instead of $Bet\text{-}C_0$ we obtain a continuous improvement $min(Bet\text{-}C_{13}, m_\cap(\emptyset))$ of $Bet\text{-}C$ which is preserved when extending the frame of discernment.

As in the case of $Pl\text{-}C_{13}$, $Bet\text{-}C_{13}$ is a modification of sm version conflict measure and it does not seems to be useful to try to define similar modification of $cpBet\text{-}C$.

6   Refinement of a Frame of Discernment

There is a completely different case of resizement of a frame of discernment, or the refinement of a frame. In this case, there are no new elements added but some of the original is/are split into one or more new one(s), thus bbm(s) of the split singleton(s) is/are transferred to the corresponding resulting set(s) and bbms of sets containing split element(s) are transferred to the corresponding larger sets. Pl_P and $BetP$ have different behaviour in this case, hence Pl-C and Bet-C as well.

We can easily show that using neither Bet-C_{12} or Bet-C_{13} conflictness or non-conflictness of a pair of BFs is kept when refining the corresponding frame of discernment. It is enough to show the simple examples of BFs on Ω_2 and its refinement to $\{\omega_{11}; \omega_{12}; \omega_2\}$. Let us suppose a non-conflicting pair $(\mathbf{0.6}, 0.4)$ and $(\mathbf{0.8}, 0.2)$ where both the BFs support ω_1 and oppose ω_2. Refining the frame, we obtain $m'_1(\{\omega_{11}, \omega_{12}\}) = 0.6$, $m'_2(\{\omega_{11}, \omega_{12}\}) = 0.8$ and $BetP'_1 = (0.3, 0.3, \mathbf{0.4})$, $BetP'_2 = (\mathbf{0.4}, \mathbf{0.4}, 0.2)$, where ω_{11}, ω_{12} are supported by Bel'_2 but opposed by Bel'_1 and ω_2 is supported by Bel'_1 but opposed by Bel'_2. Thus Bet-C_{12} conflict has appeared when refining the frame. ω_2 has max $BetP'_1$ value, but max $BetP'_2$ value appears at ω_{11} and ω_{12}, hence also Bet-C_{13} conflict has appeared.

Let us further suppose Bel_3 given by $(0.2, \mathbf{0.8})$ on Ω_2. Refining the frame we obtain $m'_3(\{\omega_{11}, \omega_{12}\}) = 0.2$ and $BetP'_3 = (0.1, 0.1, \mathbf{0.8})$. Thus ω_{11}, ω_{12} are opposed by Bel'_3 and ω_2 is supported by Bel'_3, as by Bel'_1; moreover max $BetP'_3$ value appears at ω_2 as in the case of $BetP'_1$. Hence two conflicting BFs Bel_1 and Bel_3 became both Bet-C_{12} and Bet-C_{13} non-conflicting when the frame was refined.

Note that we can use the same examples to show the same property for Bet-C and Bet-C_{11}.

On the other hand the Pl-C_{13} conflictness/non-conflicteness is preserved by refinement of the frame (see Corollary 2; for proof of the lemma see [11]):

Lemma 4. *Ordering of Pl_P values is not changed with a refinement of a frame of discernment.*[3]

Corollary 1. *The sets of elements with the maximal (minimal) value of Pl_P are the same (up to refinement) for a belief functions Bel and Bel' on an extended frame of discernment.*

Corollary 2. *Measure of conflict Pl-C_{13} keeps conflictness/non-conflictness of a pair of belief functions when the frame of discernment is refined.*

The situation is more complicated for Pl-C, Pl-C_{11} and Pl-C_{12}: Orderings of the Pl_P values (and max/min values) are kept when refining a frame; thus also sm and cp versions of conflictness/non-conflictness. But there is possibility of change of support/opposition of other elements; thus change of $sp\Omega_{PlC}$ and $cb\Omega_{PlC}$ and also of sp and cb versions of Pl-C, Pl-C_{11} and Pl-C_{12} conflictness/non-conflictness.

7   A Comparison of the Presented Measures of Conflict

Comparing the series of Pl-C using Pl-C_0, Pl-C_{11}, Pl-C_{12}, Pl-C_{13} (and analogously Bet-C using Bet-C_0, Bet-C_{11}, Bet-C_{12}, Bet-C_{13}) we see step-wise improvement from Pl-C_0 to Pl-C_{13} (and from Bet-C_0 to Bet-C_{13}) from the point of view of the investigated properties, see Table 1.

---

[3] Unfortunately, after completion of this text, we have realized, that Lemma 4 and its corollaries hold true only under a special condition; for correction see [11].

Table 1. A comparison of properties of conflict measures and their components

property → measure ↓	Cont. distinguish.	Conf/NonC distinguish.	SmallVal	BigVal	Extens. equal.	Extension Conf/NonC	Refinement Conf/NonC	Refin. equal.
$\emptyset$ (Diff Pl_P)	+	-	-	+/-	=	=	+/- / N.A.	≠
cf	+	-	-	+/-	=	=	C↔N / N.A.	≠
Pl-C	-	+	-	+/-	≠	C↔N	(*)	≠
Bet-C	-	+	-	+/-	≠	C↔N	C↔N	≠
Pl-C_{11}	+	+	+/-	+/-	≠	C↔N	(*)	≠
Bet-C_{11}	+	+	+/-	+/-	≠	C↔N	C↔N	≠
Pl-C_{12}	+	+	+/-	+/-	≠	C↔N	(*)	≠
Bet-C_{12}	+	+	+/-	+/-	≠	C↔N	C↔N	≠
Pl-C_{13}	+	+	+/-	+/-	=	=	+	≠
Bet-C_{13}	+	+	+/-	+/-	=	=	C↔N	≠

Explanation:

+   property is satisfied,

-    property is not satisfied / values are not acceptable,

+/-   we can accept the values as an approximation of values of conflict,

C↔N   conflicting pair of BFs may become a non-conflicting (and vice versa) when resizing the frame of discernment,

(*)   elements with maximal preprerence / opposition are the same, nevertheless the property is not satisfied in general (cp and cb conflicts).

Further we have to note that:

"$\emptyset$ (Diff Pl_P)" is a Pl_P version of cf (is has not been mentioned anywhere, it is here just for a comparison of the properties);

each of Pl-C, Bet-C, Pl-C_{12}, Bet-C_{12} have four variants (sm, sp, cp, cb according to 4 variants of conflicting sets Ω_{PlC} or Ω_{BetC});

Pl-C_{11}, Bet-C_{11} suppose $\Omega_{PlC}^{ord} = \emptyset$, $\Omega_{BetC}^{ord} = \emptyset$, thus there are two variants of each of them (sm and sp);

Pl-C_{13}, Bet-C_{13} classify conflictness/non-conflictness according to max and max but one values of Pl_P_i, $BetP_i$ (there is the only variant analogous to sm but not the same as 2–4 elements play their role here).

The conflict measures using Pl-C_{13} and Bet-C_{13} are also simpler in comparison with previous both theoretically and from the computational point of view. Only a modified version of simple conflicting set is used there, hence there are not four variants (sm, sp, cp, cb) and thus the computation is also simpler or equal in comparison with the previous measures.

In the case of the series of the measures based on $BetP$, Bet-C using Bet-C_{13} is also improvement of Liu's degree of conflict cf from the point of view all these properties, whereas original version using Bet-C_0 was improvement only from the point of view better and clearer distinguishing of conflictness/non-conflictness and using $m_\odot$ as upper bound, on the other hand continuity and robustness with respect to an extension of a frame of discernment was lost.

When comparing Pl-C with Bet-C we have obtained a new argument in favour of Pl-C, that is its keeping of conflictness/non-conflictness when a frame of discernment is refined. The original arguments mentioned already in [10] are better interpretation of Pl-C and its compatibility with Dempster's rule based on commutativity of Pl_P with Dempster's rule [3,5]. It is also strengthened by keeping zero/non-zero values by $difPl\_P_{m_i}^{m_j}$ (a Pl_P version of $difBetP_{m_i}^{m_j}$ when a frame of discernment is refined, see value "+/- / N.A." in "$\emptyset$" row of Table 1.

8   Open Problems and Ideas for a Future Research

Investigating and improving measures of conflicts of BFs we have met the following open problems:

- $Pl\text{-}C_{13}$ does not use conflicting sets, there is no problem with BFs from Example 1, thus there is a question whether it holds $Pl\text{-}C_{13}(Bel_1, Bel_2) \leq (m_1 \ominus m_2)(\emptyset)$ or not.
- Analogously whether it holds $Bet\text{-}C_{13}(Bel_1, Bel_2) \leq (m_1 \ominus m_2)(\emptyset)$ or not.
- To look for an alternative support/opposition of ω by a BF not depending from resizement of a frame of discernment.
- Investigation of an idea to use Pl_P or $BetP$ for classification of conflictness/non-conflictness only, and look for an appropriate distance of BFs (not transformed to probabilities) to use it for determination of conflict of BFs which were already classified as conflicting, i.e., which are in some positive conflict. (This partial "step back" may be either useful or a dead end procedure).

9   Conclusion

A series of gradual improvements of two measures of conflict between belief functions, plausibility conflict $Pl\text{-}C$ and pignistic conflict $Bet\text{-}C$, are presented in this theoretical contribution. The measures are improved from the point of view of their continuity and robustness with respect to resizing of a frame of discernment: its extension and refinement. $Bet\text{-}C$ is now a real improvement of Liu's degree of conflict cf.

Higher robustness of $Pl\text{-}C$ with respect to frame refinement is a new argument in favour of the measure based on normalised plausibility of singletons against the measure based on Smets' pignistic probability.

Improved conflict measures both increase our general understanding of the nature of belief functions and can be applied in better combination of conflicting belief functions in numerous applications of the real world.

Acknowledgments. The authors are grateful to Weiru Liu for her useful comments and for establishing of their cooperation.

This research is supported by the grant P202/10/1826 of the Czech Science Foundation (GAČR). The support by the EU INFER project under grant: 251617 and the partial institutional support of RVO 67985807 from the Institute of Computer Science is also acknowledged.

References

1. Almond, R.G.: Graphical Belief Modeling. Chapman & Hall, London (1995)
2. Ayoun, A., Smets, P.: Data association in multi-target detection using the transferable belief model. Int. Journal of Intelligent Systems 16(10), 1167–1182 (2001)

3. Cobb, B.R., Shenoy, P.P.: A Comparison of Methods for Transforming Belief Function Models to Probability Models. In: Nielsen, T.D., Zhang, N.L. (eds.) ECSQARU 2003. LNCS (LNAI), vol. 2711, pp. 255–266. Springer, Heidelberg (2003)
4. Daniel, M.: Distribution of Contradictive Belief Masses in Combination of Belief Functions. In: Bouchon-Meunier, B., Yager, R.R., Zadeh, L.A. (eds.) Information, Uncertainty and Fusion, pp. 431–446. Kluwer Acad. Publ., Boston (2000)
5. Daniel, M.: Probabilistic Transformations of Belief Functions. In: Godo, L. (ed.) ECSQARU 2005. LNCS (LNAI), vol. 3571, pp. 539–551. Springer, Heidelberg (2005)
6. Daniel, M.: Conflicts within and between Belief Functions. In: Hüllermeier, E., Kruse, R., Hoffmann, F. (eds.) IPMU 2010. LNCS, vol. 6178, pp. 696–705. Springer, Heidelberg (2010)
7. Daniel, M.: Non-conflicting and Conflicting Parts of Belief Functions. In: Coolen, F., et al. (eds.) ISIPTA 2011, pp. 149–158. Studia Universitätsverlag, Innsbruck (2011)
8. Daniel, M.: Introduction to an Algebra of Belief Functions on Three-Element Frame of Discernment — A Quasi Bayesian Case. In: Greco, S., Bouchon-Meunier, B., Coletti, G., Fedrizzi, M., Matarazzo, B., Yager, R.R., et al. (eds.) IPMU 2012, Part III. CCIS, vol. 299, pp. 532–542. Springer, Heidelberg (2012)
9. Daniel, M.: Properties of Plausibility Conflict of Belief Functions. In: Rutkowski, L., Korytkowski, M., Scherer, R., Tadeusiewicz, R., Zadeh, L.A., Zurada, J.M. (eds.) ICAISC 2013, Part I. LNCS, vol. 7894, pp. 235–246. Springer, Heidelberg (2013)
10. Daniel, M.: Belief Functions: a Revision of Plausibility Conflict and Pignistic Conflict. In: Liu, W., Subrahmanian, V.S., Wijsen, J. (eds.) SUM 2013. LNCS, vol. 8078, pp. 190–203. Springer, Heidelberg (2013)
11. Daniel, M., Ma, J.: Plausibility and Pignistic Conflicts of Belief Functions: Continuity and Frame Resizement. Technical report V-1207, ICS AS CR, Prague (2014)
12. Destercke, S., Burger, T.: Toward an axiomatic definition of conflict between belief functions. IEEE Transactions on Cybernetics 43(2), 585–596 (2013)
13. Dubois, D., Liu, W., Ma, J., Prade, H.: Toward a general framework for information fusion. In: Torra, V., Narukawa, Y., Navarro-Arribas, G., Megías, D. (eds.) MDAI 2013. LNCS, vol. 8234, pp. 37–48. Springer, Heidelberg (2013)
14. Hájek, P., Havránek, T., Jiroušek, R.: Uncertain Information Processing in Expert Systems. CRC Press, Boca Raton (1992)
15. Lefèvre, É., Elouedi, Z., Mercier, D.: Towards an Alarm for Opposition Conflict in a Conjunctive Combination of Belief Functions. In: Liu, W. (ed.) ECSQARU 2011. LNCS, vol. 6717, pp. 314–325. Springer, Heidelberg (2011)
16. Liu, W.: Analysing the degree of conflict among belief functions. Artificial Intelligence 170, 909–924 (2006)
17. Martin, A., Jousselme, A.-L., Osswald, C.: Conflict measure for the discounting operation on belief functions. In: Fusion 2008, Proceedings of 11th International Conference on Information Fusion, Cologne, Germany (2008)
18. Martin, A.: About Conflict in the Theory of Belief Functions. In: Denœux, T., Masson, M.-H. (eds.) Belief Functions: Theory & Appl. AISC, vol. 164, pp. 161–168. Springer, Heidelberg (2012)
19. Shafer, G.: A Mathematical Theory of Evidence. Princeton University Press, Princeton (1976)
20. Smets, P.: The combination of evidence in the transferable belief model. IEEE-Pattern Analysis and Machine Intelligence 12, 447–458 (1990)
21. Smets, P.: Decision Making in the TBM: the Necessity of the Pignistic Transformation. Int. Journal of Approximate Reasoning 38, 133–147 (2005)
22. Smets, P.: Analyzing the combination of conflicting belief functions. Information Fusion 8, 387–412 (2007)

Improving Inconsistency Resolution by Considering Global Conflicts

Cristhian Ariel David Deagustini[1,3], Maria Vanina Martiínez[2],
Marcelo A. Falappa[1,3], and Guillermo Ricardo Simari[1]

[1] Artificial Intelligence Research and Development Laboratory
Department of Computer Science and Engineering
Universidad Nacional del Sur - Alem 1253, (8000) Bahía Blanca, Buenos Aires
[2] Department of Computer Science
University of Oxford - Wolfson Building, Parks Road, Oxford OX13QD, UK
[3] Consejo Nacional de Investigaciones Científicas y Técnicas, Av. Rivadavia 1917,
Ciudad Autónoma de Buenos Aires - República Argentina

Abstract. Over the years, inconsistency management has caught the attention of researchers of different areas. Inconsistency is a problem that arises in many different scenarios, for instance, ontology development or knowledge integration. In such settings, it is important to have adequate automatic tools for handling potential conflicts. Here we propose a novel approach to belief base consolidation based on a refinement of kernel contraction that accounts for the relation among kernels using clusters. We define cluster contraction based consolidation operators as the contraction by *falsum* on a belief base using cluster incision functions, a refinement of (smooth) kernel incision functions. A cluster contraction-based approach to belief bases consolidation can successfully obtain a belief base satisfying the expected consistency requirement. Also, we show that the application of cluster contraction-based consolidation operators satisfy minimality regarding loss of information and are equivalent to operators based on maxichoice contraction.

Keywords: Inconsistency Management, Belief Consolidation, Minimal Loss of Information.

1 Introduction

Inconsistency management is admittedly an important problem that has to be faced, *e. g.,* when knowledge provided by different users is expected to be exploited by a reasoning process. Although the integrated knowledge may be inconsistent, it is obvious there is still value in that information even in the presence of (potential) conflicts, and it is highly possible the existence of information that is not related and/or affected by those conflicts. Consider the following simple example that we use in the rest of the paper as the running example.

Suppose that we are gathering information about sports activities of early alumnus of a college and some official records have been lost. We are particularly interested in several remarkable students for which we wish to compile their

U. Straccia and A. Calì (Eds.): SUM 2014, LNAI 8720, pp. 120–133, 2014.

doings and achievements in the college. As the first step for this activity we ask for help from staff and faculty members; for a particular alumni, called Martin, we obtain the following information from three different people that were in the college at the same time as Martin:

- Staff member S_1 tells us Martin used to play soccer, and he thinks he remembers he also coached the school's basketball team; let's denote the first proposition with p and the second with q .
- P.E. professor S_2, who used to be one of Martin's college mates, states that he thinks Martin used to play in the basketball team; we will refer to this proposition as r.
- An old class mate of Martin is not sure but she remembers the soccer team used to be very proud and demanding at that time, so definitely if Martin played soccer he did not play basketball; let this be proposition s.

We have not yet provided a formal definition of consistency, however, it is rather intuitive that it is not possible for all these statements to hold together. Several important approaches used to address the handling of inconsistency had been proposed in Artificial Intelligence (AI), specially in the areas of *belief revision* and *argumentation*. In particular, belief revision deals with the general problem of the dynamics of knowledge, *i.e.,* how belief states change and evolve through time, solving possible inconsistencies in the process. One particular way to deal with the above situation is to try to modify the information contained in the knowledge base as little as possible in order to make it consistent; this is known as knowledge consolidation in the belief revision community. In this work, we define consolidation operators that takes an inconsistent belief base and apply special functions, called incisions functions, so that inconsistencies are resolved. The main contributions of this paper are:

- We first analyze a class of consolidation operators based on kernel contraction [11,12]; these operators make *incisions* on the minimal conflictive subsets of the inconsistent belief base. In Section 3 we demonstrate the operators' behavior and show there are cases in which such operators may not yield minimal loss of information; we also show that this problem arises from treating inconsistency in a localized way, isolating minimal conflicting sets in the consistency restoration process.
- In order to prevent unnecessary loss of information, we develop an alternative and novel class of consolidation operators, called *cluster contraction-based consolidation operators*; these operators aim to address conflicts globally by means of the use of *clusters* [18] instead of minimal conflictive sets.
- In Section 4, we show that cluster incision functions are refinements of smooth kernel incision functions [11], therefore the application of cluster contraction-based consolidation operators produces in general, the deletion of a smaller number of formulæ from the original knowledge base than any smooth kernel contraction-based operators would produce.
- Finally, in Section 5 we show that cluster incision functions satisfy the minimality requirement regarding loss of information and we conclude that a

consolidation operator is a cluster contraction-based consolidation operator if and only if it is a maxichoice contraction-based consolidation operator, completing in this way the spectrum of possibilities arising from the treatment of inconsistency by means of minimal conflicts.

2   Preliminaries

We begin by introducing the notation necessary for our presentation and the required concepts that will be used throughout the paper. Also, we present the research context of belief change theory from which revision operators had arisen.

We assume a propositional language $\mathcal{L}$ built from a set of propositional symbols $\mathcal{P}$. This language is closed under the classical propositional logic symbols $\neg$ (negation), $\wedge$ (conjunction), $\vee$ (disjunction), $\rightarrow$ (implication), and $\leftrightarrow$ (equivalence). We denote propositional letters using lower-case Latin letters, possibly using subscripts (e. g., a, b, c, a_1, a_2) and propositional formulæ using lower-case Greek letters, possibly using subscripts (e. g., $\alpha, \beta, \gamma, \alpha_1, \alpha_2$); but, we reserve ρ and ϱ to represent incision functions.

An interpretation is a total function from $\mathcal{P}$ to $\{0, 1\}$, and the set of all interpretations is denoted with $\mathcal{W}$. An interpretation $\omega \in \mathcal{W}$ is a model of a formula α iff it makes α true in the classical way, denoted with $\omega \models \alpha$. The set of all models of a formula α is denoted with $\text{mods}(\alpha)$, i.e., $\text{mods}(\alpha) = \{\omega \in \mathcal{W} \mid \omega \models \alpha\}$. Finally, $\vdash$ stands for the usual deduction relation on propositional logic, and $\perp$ stands for an arbitrary contradiction.

We assume finite sets of propositional formulæ $\{\alpha_1, \alpha_2, \ldots, \alpha_n\}$, which are called belief bases and are denoted with upper-case Latin letters, usually K. We extend the notion of models of a formula to sets of formulæ in the natural way, i.e., $\text{mods}(K) = \{\omega \in \mathcal{W} \mid \omega \models \alpha \text{ for all } \alpha \in K\}$. Additionally, $K_{\mathcal{L}}$ denotes the set of every belief base K containing formulæ in $\mathcal{L}$. Finally, a consistent belief base K must have at least one model; formally, we say that K is consistent iff $\text{mods}(K) \neq \emptyset$. Also, K is inconsistent iff K is not consistent.

The work of Alchourrón, Gärdenfors and Makinson where the *AGM* model is presented [1], is currently considered the cornerstone from which belief change theory has evolved (see [19]). In the AGM model, three basic change operators are defined; these can be defined over a knowledge base K as follows: the result of *expanding* K by a sentence α is a possibly larger set that infers α, the result of *contracting* K by α is a possibly smaller set that does not infer α, and finally, the result of *revising* K by α is a set K' that infers α and possibly neither extends nor is part of the set K. In particular, if K infers $\neg\alpha$ then the result of the revision of K by α is a consistent set K' that infers α. AGM provides an axiomatic characterizations of contraction and revision in terms of rationality postulates. AGM contractions can be realized by *partial meet contractions*, which are based on a selection among (maximal) subsets of K that do not imply α (the input sentence). Particular cases of partial meet contractions are *full meet contractions* and *maxichoice contractions*. The former stands for an approach that is as cautious as possible (i.e., only retaining formulæ that belong to every maximal consistent subset), while the latter has the desirable property that it minimizes the loss of information, in the sense that it preserves as most formulæ as possible, since basically it selects one among all maximal consistent subsets. Another possible approach for contraction is based on a selection among the (minimal) subsets of K that contribute to make K imply α; *kernel contraction* [11] is one of such approaches and it is known to be more general than partial meet contraction, and hence to the AGM approach to contraction [11,12]. Finally, Hansson presents a refinement of kernel contraction, known as *smooth kernel contraction*, that aims to solve a problem attached to the generality of the former, as sometimes kernel contraction may produce unnecessary deletions.

In this work we focus on a different belief change operation called *consolidation*; this operation is inherently different from contraction and revision as the ultimate goal of consolidation is to obtain a consistent belief base rather than revising the knowledge base by a specific formula or removing a particular formula from it. A natural way of achieving this is to take an inconsistent belief base and restore its consistency by attending every conflict in it, a process that is known in the belief revision literature as contraction by *falsum* [10].

3   Kernel and Cluster Contraction-based Belief Base Consolidation

The work of Hansson in [11] describes how a contraction operation on belief bases can be modeled by defining *incision functions*. These functions contract a belief base, by a formula α by taking minimal sets that entail α (called α-*kernels*) and producing "incisions" on those sets so they no longer entail α. The resulting belief base is formed by the union of all formulæ that are not removed by the function. This approach is known as *kernel contraction*.

Here, we define the consolidation process as the application of incision functions over the minimal inconsistent subsets of a belief base. Following the terminology proposed by Hansson [11] we will call such sets $\perp$-kernels, or kernels for short; in the following we recall the formal definition from [11].

Definition 1 (Kernels). *Let K be a belief base. The set of kernels of K, denoted $K^{\perp\!\!\!\perp}\perp$, is the set of all $X \subseteq K$ such that $\mathrm{mods}(X) = \emptyset$ and for every $X' \subsetneq X$ it holds that $\mathrm{mods}(X') \neq \emptyset$.*

Example 1. Consider the inconsistent belief base $K = \{a, b \rightarrow \neg a, b, c, \neg c, d\}$. For K we have two kernels: $K^{\perp\!\!\!\perp}\perp = \{\kappa_1, \kappa_2\}$, with $\kappa_1 = \{a, b \rightarrow \neg a, b\}$ and $\kappa_2 = \{c, \neg c\}$. As expected by the definition of kernels, if we remove at least one formula from them, the result is consistent.

Once the set of kernels is identified, we need to establish how the inconsistencies are to be resolved. A *kernel incision function* takes a set of kernels and selects formulæ in them to be deleted from K [11].

Definition 2 ((Smooth) Kernel Incision Function). *Let K be a belief base and $K^{\perp\!\!\!\perp}\perp$ be the set of kernels for K. A* kernel incision function *is a function $\rho : 2^{K_{\mathcal{L}}} \mapsto K_{\mathcal{L}}$ such that the following conditions hold:*

- *$\rho(K^{\perp\!\!\!\perp}\perp) \subseteq \bigcup(K^{\perp\!\!\!\perp}\perp)$, and*
- *for all $X \in K^{\perp\!\!\!\perp}\perp$, if $X \neq \emptyset$ then $(X \cap \rho(K^{\perp\!\!\!\perp}\perp)) \neq \emptyset$.*

A kernel incision function ρ is said to be smooth *if and only if for all $X \subset K$ such that $X \vdash \beta$ and $\beta \in \rho(K^{\perp\!\!\!\perp}\perp)$, we have then $X \cap \rho(K^{\perp\!\!\!\perp}\perp) \neq \emptyset$.*

The second condition on Definition 2 requires from the incision function to select *at least* one formula to be deleted from every kernel. An incision function may remove several formulæ from a kernel; however, note that given the minimality of kernels, removing only one formula from each kernel suffices to restore its consistency. The last condition ensures that a kernel incision function is smooth [11]. Smoothness is characterized by the relative closure postulate [11] that aims to retain as much from the original knowledge base as possible; it states that the result of contracting a knowledge base K must contain those of its own logical consequences that are also elements of K. Intuitively, smoothness captures the set of incisions that yield contractions that can be obtained by performing the contraction by any incision function and then adding back the elements from K that were unnecessarily dropped by the incision function.

Based on (smooth) kernel incision functions we define kernel contraction-based belief consolidation operators as follows.

Definition 3 (Kernel Contraction-based Consolidation Operator). *Given belief base K, let $K^{\perp\!\!\!\perp}\perp$ be the set of kernels for K and ρ a kernel incision function. A* kernel contraction-based consolidation operator Υ_ρ *for K is defined as:*

$$\Upsilon_\rho(K) = K \setminus \rho(K^{\perp\!\!\!\perp}\perp)$$

Furthermore, if ρ is a smooth kernel incision function then Υ_ρ is a smooth kernel contraction-based consolidation operator.

Note that operator $\Upsilon_\rho(\cdot)$ is parameterized by the incision function ρ; the result of applying such operator will be a consistent belief base since every conflict is attended to by the kernel incision function. However, if we strive for minimal loss of information (as it is usually assumed in the management of inconsistent information), then this operator defined as it is, has the important drawback of solving conflicts locally to every kernel; even if the function only removes one formula from each kernel, the incisions may be too drastic from a global point of view and the operator might end up giving up more formulæ than the ones that are absolutely necessary. To see this problem, consider the following example:

Example 2. Consider $K = \{p, q, r, p \to \neg r, \neg(q \wedge r)\}$. This KB comes from our running example, regarding Martin's sports activities. The fourth proposition corresponds to proposition s. Furthermore, we have added one more proposition, namely $\neg(q \wedge r)$; it is common sense to assume that it is not possible for the same person to be both a player and the coach of a basketball team. Clearly, K

is inconsistent. As we want to obtain a consistent belief base, we will apply a kernel contraction-based consolidation operator. For belief base K we have that $K^{\perp\!\!\!\perp}\bot = \{\kappa_1, \kappa_2\}$, where $\kappa_1 = \{p, r, p \to \neg r\}$ and $\kappa_2 = \{q, r, \neg(q \wedge r)\}$.

The following table shows all possible incision functions that delete exactly one formula for each kernel (other incisions are possible deleting more than one formula from each kernel):

Possible Kernel Incision Functions	
$\rho(\kappa_1) = \{p\}$ and $\rho(\kappa_2) = \{q\}$	$\rho(\kappa_1) = \{p\}$ and $\rho(\kappa_2) = \{r\}$
$\rho(\kappa_1) = \{p\}$ and $\rho(\kappa_2) = \{\neg(q \wedge r)\}$	$\rho(\kappa_1) = \{r\}$ and $\rho(\kappa_2) = \{q\}$
$\rho(\kappa_1) = \{r\}$ and $\rho(\kappa_2) = \{r\}$	$\rho(\kappa_1) = \{r\}$ and $\rho(\kappa_2) = \{\neg(q \wedge r)\}$
$\rho(\kappa_1) = \{p \to \neg r\}$ and $\rho(\kappa_2) = \{q\}$	$\rho(\kappa_1) = \{p \to \neg r\}$ and $\rho(\kappa_2) = \{r\}$
$\rho(\kappa_1) = \{p \to \neg r\}$ and $\rho(\kappa_2) = \{\neg(q \wedge r)\}$	

All the above possibilities restore consistency in K, but clearly there are some choices that are better with respect to the amount of information lost in the process. For instance, suppose we choose the functions that perform the following incisions $\rho(\kappa_1) = \{r\}$ and $\rho(\kappa_2) = \{q\}$; we then have that: $\Upsilon_\rho(K) = K \setminus \rho(K^{\perp\!\!\!\perp}\bot) = K \setminus \{q, r\}$. As we can see, for κ_2 we have deleted q from K in order to solve the conflict. However, this is not actually necessary, as r (*i.e.*, the proposition that says that Martin played at the school's basketball team) will not be in the final belief base anyway, since it is deleted to solve the conflict in κ_1, and thus the conflict in kernel κ_2 is already resolved, that is there is no need to further remove propositions from κ_2. The reason behind this choice is that a kernel contraction-based operator solves conflicts *locally* to the kernels and there is no mechanism in its definition to consider any interaction among them.

Clearly, it is possible to address the problem described above by analyzing all possible incisions and computing the combination that makes the best choice globally. However, this would involve traversing the (possibly) enormous search space of all possible incision functions; in the following we present an approach that avoids this by contemplating only incisions that are globally optimal with respect to the amount of information loss. The proposal is based on the use of *clusters*, first introduced in [18] and further analyzed as a foundation for inconsistency management in [16,17]. This construction will allow us to have a more global vision of conflicts, and, as we shall see latter, will also have a direct impact on the consolidation process. Clusters are obtained by defining an overlapping relation among kernels.

Definition 4 (Overlapping Kernels, Equivalence). *Let K be a belief base, and $K^{\perp\!\!\!\perp}\bot$ be the set of kernels for K. Given kernels $\kappa_1, \kappa_2 \in K^{\perp\!\!\!\perp}\bot$ we say they overlap, denoted $\kappa_1 \theta \kappa_2$, iff for some $\alpha \in \kappa_1$ and $\beta \in \kappa_2$ it holds that $\alpha \models \beta$. Furthermore, we denote as θ^* the equivalence relation obtained over $K^{\perp\!\!\!\perp}\bot$ through the reflexive and transitive closure of θ.*

Example 3. Consider a belief base K such that $K^{\perp\!\!\!\perp}\bot = \{\kappa_1, \kappa_2\}$ where $\kappa_1 = \{a, \neg a \wedge \neg b\}$ and $\kappa_2 = \{b \vee a, \neg a \wedge \neg b\}$. Clearly: $\kappa_1 \theta \kappa_2$, as $\neg a \wedge \neg b \models \neg a \wedge \neg b$.

As another example of overlapping, consider belief base K' such that $K'^{\perp\!\!\!\perp}\perp = \{\kappa_1, \kappa_2, \kappa_3\}$ where $\kappa_1 = \{a \wedge b, \neg b\}$, $\kappa_2 = \{a, \neg a\}$ and $\kappa_3 = \{a \wedge b, \neg a\}$. Then, it holds that for instance $\kappa_1 \theta \kappa_2$ because $a \wedge b \models a$.

Below, we recall the notion of *clusters*, that formalizes the way in which conflicts will be structured; intuitively, a cluster groups together kernels that stand for related conflicts, in a transitive way.

Definition 5 (Clusters [18]). *Let K be a belief base, $K^{\perp\!\!\!\perp}\perp$ be the set of kernels for K, and θ the overlapping relation. A cluster of K is a set $\varsigma = \bigcup_{\kappa \in [\kappa]} \kappa$, where $[\kappa] \in K^{\perp\!\!\!\perp}\perp/\theta^*$. We use $K^{\perp\!\!\!\perp\!\!\!\perp}\perp$ to denote the set of all clusters for K.*

Example 4. Consider a belief base K such that $K^{\perp\!\!\!\perp}\perp = \{\kappa_1, \kappa_2, \kappa_3\}$, with $\kappa_1 = \{\alpha, \beta\}$, $\kappa_2 = \{\beta, \gamma\}$, and $\kappa_3 = \{\delta, \epsilon\}$. Then, we have the following set of clusters $K^{\perp\!\!\!\perp\!\!\!\perp}\perp = \{\varsigma_1, \varsigma_2\}$, where $\varsigma_1 = \{\alpha, \beta, \gamma\}$ and $\varsigma_2 = \{\delta, \epsilon\}$. Note that, κ_3 does not overlap with any other kernel in $K^{\perp\!\!\!\perp}\perp$, but $[\kappa_3] \in K^{\perp\!\!\!\perp}\perp/\theta^*$ is such that $[\kappa_3] = \{\kappa_3\}$, then it constitutes a cluster in itself (*i.e.*, $\varsigma_2 = \{\delta, \epsilon\}$).

The use of clusters instead of kernels can help in preventing situations like the one in Example 2 since the cluster structure allows us to identify kernels that overlap; thus, we can contemplate incisions that make global considerations of optimality. Moreover, the proposed notion of overlapping helps to identify only useful clusters. To see this consider belief base K'' such that $K''^{\perp\!\!\!\perp}\perp = \{\kappa_1, \kappa_2\}$ where $\kappa_1 = \{a \wedge b, \neg b \wedge \neg c\}$ and $\kappa_2 = \{b \wedge c, \neg b \wedge \neg d\}$. Formulæ $a \wedge b$ and $b \wedge c$ share models, however they do not overlap under Definition 4. Considering these two kernels together does not help in improving the consistency restoration process as, for instance, the removal of $a \wedge b$ does not resolve the conflict in κ_2. We have chosen to not consider these cases as overlaps in this work, but clearly this decision depends directly on the way conflicts are allowed to be resolved; for a consistency restoration technique not based on deleting entire formulæ from the clusters a different notion of overlapping could prove more useful.

Remember that by design the simple removal of any single formula within a kernel makes the set no longer inconsistent; however, this is not necessarily the case for clusters [16]. Therefore, in order to define incision functions over clusters, we cannot simply reuse Definition 2, as the following example shows.

Example 5. Continuing with Example 4, consider a kernel incision function ρ and the cluster $\varsigma_1 \in K^{\perp\!\!\!\perp\!\!\!\perp}\perp$. We could have, for instance, that $\rho(\varsigma_1) = \{\alpha\}$. Then, the intersection between the cluster and the result of the incision function is non empty and the selected formula belongs to the union of clusters, fulfilling the conditions on the definition of kernel incision functions, but the inconsistency remains as $\kappa_2 = \{\beta, \gamma\}$ still is an inconsistent set.

We now introduce cluster incision functions; these functions are refinements of the ones introduced earlier in the paper.

Definition 6 (Cluster Incision Function). *Let K be a belief base and $K^{\perp\!\!\!\perp}\perp$ and $K^{\perp\!\!\!\perp\!\!\!\perp}\perp$ be the set of kernels and clusters for K, respectively. A cluster incision function is a function $\varrho : 2^{K_{\mathcal{L}}} \mapsto K_{\mathcal{L}}$ such that:*

- $\varrho(K^{\perp\!\!\!\perp\!\!\!\perp}\bot) \subseteq \bigcup(K^{\perp\!\!\!\perp\!\!\!\perp}\bot)$, and
- for all $X \in K^{\perp\!\!\!\perp\!\!\!\perp}\bot$ and $Y \in K^{\perp\!\!\perp}\bot$ such that $Y \subseteq X$ it holds that for some $\alpha \in Y$, $Y \cap \varrho(K^{\perp\!\!\!\perp\!\!\!\perp}\bot) = \{\alpha\}$.

From Definition 6, we have that for any kernel Y included in some cluster X in a belief base a cluster incision function selects exactly one formula to remove, (i.e., $\{Y \cap \varrho(K^{\perp\!\!\!\perp\!\!\!\perp}\bot)\}$ is a singleton). Now, based on cluster incision functions we define a new operator, namely cluster contraction-based consolidation operator.

Definition 7 (Cluster Contraction-based Consolidation Operator). *Given a belief base K, let $K^{\perp\!\!\perp}\bot$ and $K^{\perp\!\!\!\perp\!\!\!\perp}\bot$ be the set of kernels and clusters for K, respectively, and ϱ be a cluster incision function. A cluster contraction-based operator Ψ_ϱ for K is defined as follows:*

$$\Psi_\varrho(K) = K \setminus \varrho(K^{\perp\!\!\!\perp\!\!\!\perp}\bot)$$

The last condition in Definition 6 ensures that all conflicts are resolved once we delete the selected formulæ. Example 6 shows the behavior of a cluster contraction-based operator Ψ_ϱ over the belief base K from Example 2.

Example 6. Consider once again belief base K from Example 2; we have the following set of clusters $K^{\perp\!\!\!\perp\!\!\!\perp}\bot = \{\varsigma_1\}$, with $\varsigma_1 = \{p, q, r, p \rightarrow \neg r, \neg(q \wedge r)\}$, because r belongs to both kernels in $K^{\perp\!\!\perp}\bot$; thus, $r \models r$. For a cluster contraction-based operator Ψ_ϱ based on a cluster incision function ϱ, we have the following possible incisions, narrowing the previous ones shown in Example 2:

Possible Cluster Incision Functions	
$\varrho(\varsigma_1) = \{p, q\}$	$\varrho(\varsigma_1) = \{p, \neg(q \wedge r)\}$
$\varrho(\varsigma_1) = \{r\}$	$\varrho(\varsigma_1) = \{p \rightarrow \neg r, q\}$
$\varrho(\varsigma_1) = \{p \rightarrow \neg r, \neg(q \wedge r)\}$	

Consider option $\varrho(\varsigma_1) = \{p, q\}$. Thus, $\Psi_\varrho(K) = K \setminus \varrho(K^{\perp\!\!\!\perp\!\!\!\perp}\bot) = K \setminus \{p, q\}$. Note that, even if we prefer proposition r over proposition q, the minimal loss of information principle is still fulfilled, as the non-minimal options in Example 2 are not even considered by any cluster incision function. For example, if we were to choose r for deletion then to also choose q (i.e., the option considered in Example 2) is no longer a viable option for a cluster incision function, as if we choose both formulæ then the set $\varrho(K^{\perp\!\!\!\perp\!\!\!\perp}\bot) \cap \kappa_2$ will no longer be a singleton set, violating the second condition from Definition 6.

Proposition 1 shows that the consistency restoration process based on cluster contraction fulfils the consistency requirement.

Proposition 1. *Let K be a belief base, and Ψ_ϱ be a cluster contraction-based consolidation operator. Then, $\mathrm{mods}(\Psi_\varrho(K)) \neq \emptyset$.*

For space reasons we do not include the proof of results. The formal proof for Proposition 1 relies on the fact that, by definition, cluster incision function select one formula for every kernel that composes every clusters, effectively resolving the inconsistency for each one since kernels are minimal inconsistent sets.

4   Relationship with Kernel Contraction-Based Consolidation

In the previous section we introduced an approach for belief base consolidation that works as a refinement of the approach based on kernel contraction. In this section we focus on the relationship between (smooth) kernel contraction-based consolidation and cluster contraction-based consolidation. Specifically, we seek to establish this relationship from the point of view of loss of information.

We first show that cluster incision functions are refinements of kernel incision functions, that is, we have that every cluster contraction-based consolidation operators is also a kernel contraction-based one.

Proposition 2. *Let Ψ_ϱ be a cluster contraction-based consolidation operator. Then, Ψ_ϱ is a kernel contraction-based consolidation operator.*

To prove that Ψ_ϱ is a kernel contraction-based merging operator it is enough to show that cluster incision functions are also kernel incision functions; if we consider the tables in Examples 5 and 6 showing all possible kernel and cluster incisions, respectively, we can see that every possible cluster incision is indeed a kernel incision. The converse from Proposition 2 does not hold, as kernel incision functions are not necessarily cluster incision functions, since the former not always satisfy the last condition from Definition 6 as we illustrate below.

Example 7. Consider belief base K from Example 2 and suppose we have $\Upsilon_\rho(K) = K \setminus \rho(K^{\perp\!\!\!\perp}\perp) = K \setminus \{q, r\}$. The kernel incision function that gives rise to this operator performs the following incisions: $\rho(\kappa_1) = \{r\}$ and $\rho(\kappa_2) = \{q\}$. Note that ρ is not a valid cluster incision function since $|\kappa_2 \cap \rho(K^{\perp\!\!\!\perp\!\!\!\perp}\perp)| = \{q, r\}$, i.e., $|\kappa_2 \cap \rho(K^{\perp\!\!\!\perp\!\!\!\perp}\perp)| > 1$. Furthermore, it is not possible for any valid cluster incision function to yield this result.

As hinted by Examples 2 and 6, a benefit of using cluster-based consolidation operators over (smooth) kernel-based ones is that unnecessary deletions can be avoided by clustering conflicts. The characteristics of the choices made by cluster incision functions have an important impact on the number of formulæ deleted: we can show that for any kernel contraction-based consolidation operator there is a cluster contraction-based one that removes at most the same number of formulæ than the former; the following proposition formalizes the result.

Proposition 3. *Let K be a belief base. Then, for any kernel contraction-based consolidation operator Υ_ρ over K there exists a cluster contraction-based consolidation operator Ψ_ϱ over K such that $\Upsilon_\rho(K) \subseteq \Psi_\varrho(K)$.*

Above, we have shown that cluster-based operators are refinements of "pure" kernel contraction-based ones. We can shown that cluster incision functions refines smooth incision functions as well, and hence the operators based on them.

Proposition 4. *If Ψ_ϱ is a cluster contraction-based consolidation operator, then Ψ_ϱ is a smooth kernel contraction-based consolidation operator.*

The converse of Proposition 4 does not hold; consider the following example.

Example 8. Consider a belief base $K = \{p, q, r, p \to \neg q, p \to \neg r\}$. We have that $K^{\perp\!\!\!\perp} \perp = \{\kappa_1, \kappa_2\}$ where $\kappa_1 = \{p, q, p \to \neg q\}$ and $\kappa_2 = \{p, r, p \to \neg r\}\}$, and thus $\perp\!\!\!\perp\!\!\!\perp(K) = \{\varsigma_1\}$ where $\varsigma_1 = \{p, q, r, p \to \neg q, p \to \neg r\}$. Now, consider an incision function $\rho(K^{\perp\!\!\!\perp} \perp) = \{p, r\}$. We can see that ρ satisfies smoothness (cf. Def. 2). However, we have that $\rho(K^{\perp\!\!\!\perp} \perp) \cap \kappa_2$ is not a singleton. Then, ρ is not a cluster incision function. Note that, once we choose to remove p from K, it is unnecessary to remove r, and any valid cluster incision function will avoid that.

From the previous example we can conclude that smooth incision functions, although a proper refinement of kernel incision functions, can still produce unnecessary loss of information. In the next section we characterize a notion of optimality of incision functions and position our proposal with respect to AGM-based approaches to consolidation.

5   Connection with Maxichoice Contraction-Based Consolidation

In this section we further analyze cluster incision functions and the consolidation operators based on them; particularly, we focus in the relationship with maxichoice contraction-based consolidation, as maxichoice contractions [1] are as conservative as possible. Maxichoice contraction is based on the use of selection functions that select one among all possible maximal consistent subsets of the knowledge base. Maxichoice contraction-based consolidation operators are those based on maxichoice contraction in the same manner as the operators defined previously. To formally characterize optimality of incision functions, we recall the notion of *minimality* from [9] and adapt it for cluster incision functions.

Definition 8 (Minimality). *An incision function ϱ for a belief base K is minimal if no proper subset of $\varrho(K^{\perp\!\!\!\perp} \perp)$ defines an incision function.*

Next we show that cluster incision functions are minimal.

Proposition 5. *Let ϱ be an incision function. Then, ϱ is a cluster incision function iff it is a minimal incision function.*

The proof for Proposition 5 is based on the fact that for every cluster $X \in K^{\perp\!\!\!\perp} \perp$ such that $X \cap \varrho(K^{\perp\!\!\!\perp} \perp) = A$ no proper subset of A in itself restores consistency, and hence no subset of it gives raise to a proper incision function.

The relationship between selection and incision functions was previously analyzed in [9]. As noted there, minimality of incision functions corresponds to maximality of contraction; a direct result of this is the following proposition.

Proposition 6 (adapted from [9]). *Let ϱ be an incision function, Ψ_ϱ be its associated consolidation operator and K a belief base. Then, ϱ is a minimal incision function iff there exists $H \subseteq K$ such that $H = K \setminus \Psi_\varrho(K)$ where (1) H is consistent, and (2) there is no $H \subsetneq H' \subsetneq K$ such that H' is consistent.*

Proposition 6 states that there is a one-to-one correlation between maximal consistent subsets of a K and minimal incision functions. More specifically, the result in [9] shows that any kernel contraction-based on minimal incision functions is a maxichoice contraction. The proof in [9] can be slightly modified for our setting; intuitively the validity of this result relies on the fact that if there is a subset H of K that is maximally consistent, then adding any further formulæ will make it inconsistent, thus, any incision ϱ such that $\varrho \cap K = H$ must be minimal; if this were not the case then there would exist a subset $\varrho' \subsetneq \varrho$ such that $K \setminus \varrho'$ is consistent, and therefore H would not be maximally consistent since $K \setminus H \subsetneq K \setminus \varrho'$. Conversely, if an incision ϱ is minimal then it generates a maximally consistent subset because no proper subset of ϱ is an incision function, *i.e.*, for any $A \subsetneq \varrho$ it holds that $K \setminus A$ is inconsistent.

Although Falappa *et al.* elaborate on the relationship between kernel and maxichoice contractions further, no class of incision functions satisfying minimality (thus corresponding to maxichoice contractions) is identified. As a corollary of the previous results we can conclude that our approach is equivalent to consolidation through maxichoice contraction, but arising from minimal incision functions, which means that operators retain as much information as possible.

Corollary 1. Ψ_ϱ *is a cluster contraction-based consolidation operator iff* Ψ_ϱ *is a maxichoice contraction-based consolidation operator.*

Discussion. Corollary 1 completes the spectrum of possibilities arising from the treatment of minimal conflicts. Nevertheless, it is important to note that, although our approach is equivalent to consolidation through maxichoice contraction in terms of the final belief base obtained, there is still importance in the difference in how this belief base is obtained. While maxichoice operators have to deal with maximal consistent subsets, ours deal with minimal inconsistent ones. There is an interesting ongoing discussion about which approach is better. In [20] examples are shown that indicate that for some instances it is faster to use kernel contraction while for others it is faster to use partial meet (or maxichoice) contraction. As noticed in [20], whether it is possible to detect when it is better to use one or the other method is still an open problem. It can be argued that the final choice will depend on the application environment and the language selected; clearly, different needs from the point of users prompt choosing one approach over the other. As an example, consider the setting from [17], where knowledge bases are relational databases considered together with functional dependencies, there it is possible to efficiently (polynomial time in the number of tuples in the database, assuming a fixed schema) compute and maintain the set of clusters by means of indexes; alternatively, in such setting an inconsistency management approach based on the manipulation of maximal consistent subsets (other than simply computing one of them) would require higher computational effort and possibly the utilization of tools outside the DBMS.

6   Related Work

The problem of inconsistency handling has been addressed differently over the years in diverse environments, *e. g.,* relational databases, propositional knowledge bases or fragments of first order logics such as logic programming.

As stated in the introduction, the area of Belief Revision has undoubtedly produced great advances in the handling of inconsistencies. Particularly, the work by Hansson [11] is the foundation and inspiration for this paper. More recently, the work in [13] presents an approach for merging belief bases, stemming from inconsistency minimization, which removes exactly one formula in each minimal inconsistent subset of formulas. As shown, to remove one formula from minimal inconsistent sets may not be sufficient to ensure that nothing is given up without reason: it is still important *how* such formula is chosen, considering other minimal inconsistent sets as well (cf. Examples 2 and 6). As shown in the paper, the structure of cluster helps in such choice by only considering optimal incisions, minimizing loss of information. In [8] an approach for revising a propositional knowledge base by a set of sentences is presented, where every sentence in the input set can be independently accepted but there may exist inconsistencies when considering the whole set. The main difference between this work and ours is that they first solve inconsistencies in the set of sentences, in this manner they can decide which subset of it will characterize the revision. Furthermore, in our approach no preponderance to particular formulæ is given in the process as our proposal is based on consolidation instead of revision.

Also within Artificial Intelligence, the works by Baral *et al.* in knowledge bases combination [3], Brewka's preferred subtheories [7], and several other works on entailment from inconsistent knowledge bases such as [4,5], are based on the idea of selecting maximal subsets of the knowledge base (or the combination of several ones) that are consistent *w.r.t.* a set the integrity constraints. All of these approaches can be defined in the AGM framework as specific partial meet contraction functions by adequately specifying the selection function. As shown in the previous section, our operators are equivalent to operators based on maxichoice contraction, a specific class of partial meet contraction.

In the area of Databases one of the most influential works is the one by Arenas *et al.* [2] on *Consistent Query Answering.* Their treatment of inconsistencies does not attempt to obtain a consistent database, instead, the consistent answers to a query correspond to the set of (classical) answers to the query in *every repair* of the inconsistent database, which are the consistent subsets (or supersets, depending on the type of integrity constraints) of the original database that differs minimally from it. Similar in spirit are some of the syntactic approaches analyzed in [6]. Unlike our approach, these approaches can be seen as an "on the fly" consistency restoration, guided for particular queries, targeting the subset of the knowledge that matters for that query and not the whole knowledge base.

Finally, regarding the use of clusters the most closely related research to the work presented here is the one by Lukasiewicz *et al.* [16]. There, the authors define a general framework for inconsistency-tolerant query answering in Datalog+/− ontologies based on the notion of incision functions. Besides the obvious difference in the language, the aims of their work and ours are clearly different; their work follows the same idea of Arenas *et al.* [2], focusing on enforcing consistency at query time obtaining (lazy) consistent answers. Clearly, this process must be carried on for every query posed to the system, while our approach allows to obtain a new knowledge base that can be queried without considering inconsistency issues. As usual the choice of one approach over the other heavily depend on the application environment.

7   Conclusions and Future Work

In this paper we focus on an approach to consistency restoration (consolidation) of belief bases defined on terms of belief base contractions [1]. We developed a new class of belief consolidation operators, called cluster contraction-based operators, based on incision functions that aims for a globally efficient conflict resolution. The results show that a cluster contraction-based consolidation operator do not only yields a consistent belief base, as expected, but also does it satisfying minimality requirements regarding loss of information.

This family of operators are defined based on *cluster incision functions*. We have shown that cluster incision functions are refinements of smooth kernel incision functions, which implies that cluster contraction-based operators are at least as efficient as (smooth) kernel contraction-based operators from the point of view of minimal loss of information in the consolidation process. Furthermore, we show that our operators are equivalent to consolidation operators based on maxichoice contraction, completing the spectrum of possibilities for approaches arising from the treatment of minimal conflicts. As recent findings indicates [20], in some cases it is better to use kernel-based approaches, while in others the contrary holds. Clearly, the choice of one approach over the other depend on particular aspects of the application environment.

For future work, we plan to implement the different operators and perform empirical trials over different scenarios. Also, in this first step towards the formalization of this new class of consolidation operators we have not considered any form of ranking in the definition of the incision functions. In the future, we plan to define entrenchment relations in terms of orderings among formulæ based on generic measures, and to study the merits of extensions of cluster incision function to account for such orderings. Furthermore, we intend to analyze the behavior of the operators for particular measures, for instance measures of amount of information in a knowledge base in the presence of inconsistency (*e. g.*, [15]) and measures of the degree of inconsistency (as considered in [14]).

Acknowledgements. This work was partially supported by CONICET Argentina, SeGCyT Universidad Nacional del Sur in Bahia Blanca, UK Engineering and Physical Sciences Research Council (EPSRC) grant EP/J008346/1 (PrOQAW).

References

1. Alchourrón, C., Gärdenfors, P., Makinson, D.: On the logic of theory change: Partial meet contraction and revision functions. J. Symb. Log. 50(2), 510–530 (1985)
2. Arenas, M., Bertossi, L., Chomicki, J.: Consistent query answers in inconsistent databases. In: PODS, pp. 68–79 (1999)
3. Baral, C., Kraus, S., Minker, J., Subrahmanian, V.S.: Combining knowledge bases consisting of first-order analysis. Comput. Intell. 8, 45–71 (1992)
4. Benferhat, S., Cayrol, C., Dubois, D., Lang, J., Prade, H.: Inconsistency management and prioritized syntax-based entailment. In: IJCAI, pp. 640–645 (1993)
5. Benferhat, S., Dubois, D., Lang, J., Prade, H., Saffiotti, A., Smets, P.: A general approach for inconsistency handling and merging information in prioritized knowledge bases. In: KR, pp. 466–477 (1998)
6. Benferhat, S., Dubois, D., Prade, H.: Some syntactic approaches to the handling of inconsistent knowledge bases: A comparative study part 1: The flat case. Studia Logica 58(1), 17–45 (1997)
7. Brewka, G.: Preferred subtheories: An extended logical framework for default reasoning. In: IJCAI, pp. 1043–1048 (1989)
8. Delgrande, J., Jin, Y.: Parallel belief revision: Revising by sets of formulas. Artif. Intell. 176(1), 2223–2245 (2012)
9. Falappa, M., Fermé, E., Kern-Isberner, G.: On the logic of theory change: Relations between incision and selection functions. In: ECAI, pp. 402–406 (2006)
10. Hansson, S.: Belief Base Dynamics. Ph.D. thesis, Uppsala University, Department of Philosophy, Uppsala, Sweden (1991)
11. Hansson, S.: Kernel contraction. J. Symb. Log. 59(3), 845–859 (1994)
12. Hansson, S.: A Textbook of Belief Dynamics: Solutions to Exercises. Kluwer Academic Publishers, Norwell (2001)
13. Hué, J., Würbel, E., Papini, O.: Removed sets fusion: Performing off the shelf. In: ECAI, vol. 178, pp. 94–98 (2008)
14. Hunter, A., Konieczny, S.: On the measure of conflicts: Shapley inconsistency values. Artif. Intell. 174(14), 1007–1026 (2010)
15. Lozinskii, E.: Information and evidence in logic systems. JETAI 6(2), 163–193 (1994)
16. Lukasiewicz, T., Martinez, M.V., Simari, G.I.: Inconsistency handling in Datalog+/- ontologies. In: ECAI, pp. 558–563 (2012)
17. Martinez, M.V., Parisi, F., Pugliese, A., Simari, G.I., Subrahmanian, V.S.: Policy-based inconsistency management in relational databases. IJAR 55(2), 501–528 (2014)
18. Martinez, M.V., Pugliese, A., Simari, G.I., Subrahmanian, V.S., Prade, H.: How dirty is your relational database? An axiomatic approach. In: Mellouli, K. (ed.) ECSQARU 2007. LNCS (LNAI), vol. 4724, pp. 103–114. Springer, Heidelberg (2007)
19. Peppas, P.: Belief revision. In: Handbook of Knowledge Representation, ch. 8, pp. 317–359. Foundations of AI, Elsevier (2008)
20. Ribeiro, M.M.: Belief Revision in Non-Classical Logics. Springer Briefs in Computer Science. Springer (2013)

Probabilistic Argumentation Frameworks –
A Logical Approach

Dragan Doder[1] and Stefan Woltran[2]

[1] Computer Science and Communications
University of Luxembourg
Rue Richard Coudenhove-Kalergi 6, L-1359 Luxembourg
dragan.doder@uni.lu
[2] Institute of Information Systems
Vienna University of Technology
Favoritenstrasse 9–11, 1040 Vienna, Austria
woltran@dbai.tuwien.ac.at

Abstract. Abstract argumentation is nowadays a vivid field within artificial intelligence and has seen different developments recently. In particular, enrichments of the standard Dung frameworks have been proposed in order to model scenarios where probabilities or uncertain information have to be expressed. As for standard approaches of abstract argumentation, a uniform logical formalization for such frameworks is of great help in order to understand and compare different approaches. In this paper, we take a first step in this direction and characterize different semantics from the approach of Li et al in terms of probabilistic logic. This not only provides a uniform logical formalization but also might pave the way for future implementations.

1 Introduction

Within the last decade, abstract argumentation has emerged as a central field in Artificial Intelligence [2]. Hereby, it is only the relation between arguments which is taken into account when evaluating a certain scenario; the actual contents of the arguments do not play a role. The most simple objects used in abstract argumentation are Dung's argumentation frameworks (AFs) [6]. AFs are just directed graphs where vertices represent the arguments and edges indicate a certain conflict between the two connected arguments. The goal is to identify jointly acceptable sets of arguments for which a large selection of different semantics is available. One particular line of research in abstract argumentation concerns the formalization of such argumentation semantics in terms of logics, see e.g. [1,3,12,11,15]. Such formalizations not only provide a uniform definition of the different semantics, they also can lay the foundations for efficient systems (see e.g. [4,9,25] where SAT-solvers are used to evaluate formulas of propositional logic which express certain argumentation problems). Another benefit of logically characterizing argumentation problems is the potential of direct derivation of important properties (for example, complexity results based on characterizations in Monadic Second-Order Logic are given in [8,10]).

For many applications, Dung's AFs appear too simple in order to conveniently model all aspects of an argumentation problem. One such shortcoming is the lack of handling

U. Straccia and A. Calì (Eds.): SUM 2014, LNAI 8720, pp. 134–147, 2014.
© Springer International Publishing Switzerland 2014

levels of uncertainty, an aspect which typically occurs in domains, where diverging opinions are raised. This calls for augmenting simple AFs with probabilities. Several proposals have been recently made [7,17,20,24]. The most detailed overview is probably the article by Hunter [18].

As for standard abstract argumentation, a uniform logical formalization for AFs with probabilities is of great help in order to understand and compare these different approaches. Not surprisingly, the most suitable logic for this purpose is probabilistic logic [13,22]. In this paper, we want to take a first step towards such a unified view. Actually, we focus here on the approach of Li, Oren and Norman [20]. Hereby, an argumentation framework is enriched by probabilities assigned to both arguments and conflicts. These probabilities are then used to calculate probabilities of subgraphs of the given AF via the "independency" assumption. The probability for a given set S to be an extension (with respect to a particular semantics) is obtained from the sum of the probabilities of the subgraphs for which S is such an extension in the standard way. Our main goal is to express this entire process by logical means.

More specifically, our main contributions are as follows:

- We start with propositional formulae which allow us to handle subgraphs of AFs in a direct way. These encodings follow a different approach than the standard ones from [3] and will be the basis for our characterizations in terms of probabilistic logic.
- We build a probabilistic logic suitable for modelling probabilistic argumentation frameworks. In order to define semantics for the logic, we characterize the class of probability measures induced by the "independence" assumption. We apply the probabilistic operators to the constructed propositional formulas in order to formally represent the probabilities of different extensions.
- In particular, for a given set of arguments, we associate to an argumentation framework a formula whose models correspond to probabilistic extensions of the framework, under a given semantics. We also check if a set of arguments is acceptable with a given probability (w.r.t. a given semantics) by checking satisfiability of the corresponding formula.

2   Background

Dung's Abstract Argumentation Frameworks. We first introduce (abstract) argumentation frameworks [6] and recall the semantics we study in this paper.

Definition 1. *An* argumentation framework (AF) *is a pair* $F = \langle A, R \rangle$ *where* A *is a set of arguments and* $R \subseteq A \times A$ *is the attack relation. The pair* $(a, b) \in R$ *means that* a *attacks* b. *We say that an argument* $a \in A$ *is* defended *(in* F*) by a set* $S \subseteq A$ *if, for each* $b \in A$ *such that* $(b, a) \in R$, *there exists a* $c \in S$ *such that* $(c, b) \in R$.

Semantics for argumentation frameworks are given via a function σ which assigns to each AF $F = \langle A, R \rangle$ a set $\sigma(F) \subseteq 2^A$ of extensions. We consider for σ the functions stb, adm, com, grd and prf, which stand for stable, admissible, complete, grounded and preferred extensions, respectively.

Definition 2. *Let* $F = \langle A, R \rangle$ *be an AF. A set* $S \subseteq A$ *is* conflict-free (in F), *if there are no* $a, b \in S$, *such that* $(a, b) \in R$. $cf(F)$ *denotes the collection of conflict-free sets of* F. *For a conflict-free set* $S \in cf(F)$, *it holds that*

- $S \in stb(F)$, *if, for all* $a \in A \setminus S$, *there exists a* $b \in S$, *such that* $(b, a) \in R$;
- $S \in adm(F)$, *if each* $a \in S$ *is defended (in* F) *by* S;
- $S \in com(F)$, *if* $S \in adm(F)$ *and for each* a *defended (in* F) *by* S, $a \in S$;
- $S \in grd(F)$, *if* $S \in com(F)$ *and there is no* $T \in com(F)$ *with* $T \subset S$;
- $S \in prf(F)$, *if* $S \in adm(F)$ *and there is no* $T \in adm(F)$ *with* $S \subset T$.

We recall that for each AF F, the grounded semantics yields a unique extension, while all other semantics might yield multiple extensions. The only semantics where an empty set of extensions is possible, is the stable semantics.

Definition 3. *A set* S *of arguments is* consistent *in an AF* F *w.r.t. a semantics* σ *if it is a subset of an extension* $\Gamma \in \sigma(F)$.

Example 1. Let $F = \langle A, R \rangle$, where $A = \{a, b, c, d\}$ and $R = \{(a, b), (b, a), (b, c), (d, c)\}$. Then:

- $cf(F) = \{\emptyset, \{a\}, \{b\}, \{c\}, \{d\}, \{a, c\}, \{a, d\}, \{b, d\}\}$;
- $stb(F) = \{\{a, d\}, \{b, d\}\}$;
- $adm(F) = \{\emptyset, \{a\}, \{b\}, \{d\}, \{a, d\}, \{b, d\}\}$;
- $com(F) = \{\{d\}, \{a, d\}, \{b, d\}\}$;
- $grd(F) = \{\{d\}\}$;
- $prf(F) = \{\{a, d\}, \{b, d\}\}$.

Thus, the set $\{b\}$ is consistent in F w.r.t. to the stable, admissible, complete and preferred semantics, but it is not consistent w.r.t. grounded semantics; $\{a, b\}$ is not consistent w.r.t. any semantics, and $\{d\}$ is consistent with respect to all the semantics.

Subgraphs of Argumentation Frameworks. Next, we provide a few definitions concerned with subgraphs of AFs. The intuition of these subgraphs is to express possible interpretations of the original AF F, whereby it is not sure that all arguments from F (or all its attacks) actually belong to the framework.

Before making this concept formal, we need one definition. For an AF $F = \langle A, R \rangle$ and a set of arguments $A' \subseteq A$, we denote by $R_{A'}$ the restriction of the attack relation R to $A' \times A'$, i.e. $R_{A'} = \{(a, b) \in R \mid a, b \in A'\}$.

Definition 4. *Let* $F = \langle A, R \rangle$ *be an AF. A* subgraph G *of* F *(in symbols* $G \sqsubseteq F$) *is any pair* $\langle A', R' \rangle$ *such that* $A' \subseteq A$ *and* $R' \subseteq R_{A'}$. *We denote by* $s(F)$ *the set of all subgraphs of* F, *i.e.* $s(F) = \{G \mid G \sqsubseteq F\}$. *A subgraph* $G = \langle A', R' \rangle$ *of* F *is a* full subgraph *of* G *(in symbols* $G \sqsubseteq_f F$) *if* $R' = R_{A'}$. *The set of full subgraphs of* F *is denoted by* $fs(F)$.

Remark 1. Every $A' \subseteq A$ induces a unique full subgraph.

Example 2. If F is the AF from Example 1, then $fs(F) = \{F, \langle\{a, b, c\}, \{(a, b), (b, a),$
$(b, c)\}\rangle, \langle\{a, b, d\}, \{(a, b), (b, a)\}\rangle, \langle\{a, c, d\}, \{(d, c)\}\rangle, \langle\{b, c, d\}, \{(b, c), (d, c)\}\rangle,$
$\langle\{a, b\}, \{(a, b), (b, a)\}\rangle, \langle\{a, c\}, \emptyset\rangle, \langle\{a, d\}, \emptyset\rangle, \langle\{b, c\}, \{(b, c)\}\rangle, \langle\{b, d\}, \emptyset\rangle, \langle\{c, d\},$
$\{(d, c)\}\rangle, \langle\{a\}, \emptyset\rangle, \langle\{b\}, \emptyset\rangle, \langle\{c\}, \emptyset\rangle, \langle\{d\}, \emptyset\rangle, \langle\emptyset, \emptyset\rangle\}.$

Subgraphs are obtained by the full subgraphs by deleting any set of attacks. For example, $\langle\{a, b, c\}, \{(a, b), (b, c)\}\rangle \in s(F)$, and $\langle\{a, b, c\}, \{(a, b), (b, c), (c, b)\}\rangle \notin s(F)$.

Given a potential extension Γ we are now able to use the concept of subgraphs in order to determine which subframeworks of a given AF are having Γ as its extension. This will serve as a basis for probabilistic AFs which we introduce later in Section 3.

Definition 5. *Let* $F = \langle A, R \rangle$ *be an AF,* $\Gamma \subseteq A$, *and* σ *a semantics. The set of subgraphs of F for which* Γ *is a* σ-*extension is denoted by*

$$Q_F^\sigma(\Gamma) = \{G \in s(F) \mid \Gamma \in \sigma(G)\}.$$

Likewise, we define $Q_{F,f}^\sigma(\Gamma) = Q_F^\sigma(\Gamma) \cap fs(F)$ *as the set of full subgraphs of G for which* Γ *is a* σ-*extension.*

Example 3. Let F be the AF from Example 1, and let $\Gamma = \{a, c\}$. Considering the preferred semantics, we have:

$$Q_F^{prf}(\Gamma) = \{ \langle\{a, b, c\}, \{(a, b), (b, a), (b, c)\}\rangle, \langle\{a, b, c\}, \{(a, b), (b, c)\}\rangle,$$
$$\langle\{a, b, c\}, \{(a, b), (b, a)\}\rangle, \langle\{a, c\}, \emptyset\rangle\};$$
$$Q_{F,f}^{prf}(\Gamma) = \{ \langle\{a, b, c\}, \{(a, b), (b, a), (b, c)\}\rangle, \langle\{a, c\}, \emptyset\rangle\}.$$

Finally, analogous concepts can be defined for the weaker notion of consistency.

Definition 6. *Let* $F = \langle A, R \rangle$ *be an AF,* $\Gamma \subseteq A$, *and* σ *a semantics. The set of subgraphs of F for which* Γ *is consistent w.r.t.* σ *is denoted by*

$$C_F^\sigma(\Gamma) = \{G \in s(F) \mid \Gamma \text{ is consistent in } G \text{ w.r.t. } \sigma\}.$$

Likewise, we define $C_{F,f}^\sigma(\Gamma) = C_F^\sigma(\Gamma) \cap fs(F)$.

Probability measures and distributions. Let S be a nonempty set, and let A be an algebra of subsets of S, i.e., a set of subsets of S such that (i) $S \in A$, (ii) if $H_1, H_2 \in A$, then $H_1^c \in A$ and $H_1 \cup H_2 \in A$.

A finitely additive probability measure is a function $Pr : A \longrightarrow [0, 1]$, such that the following equations hold:

1. $Pr(S) = 1$,
2. $Pr(H_1 \cup H_2) = Pr(H_1) + Pr(H_2)$, whenever $H_1 \cap H_2 = \emptyset$.

The triple (S, A, Pr) is called a probability space, and elements of A are called measurable sets. For any probability measure Pr and $H, H' \in A$ such that $Pr(H) > 0$, the conditional probability $Pr(H'|H)$ is defined in the usual way, i.e.

$$Pr(H'|H) = \frac{Pr(H' \cap H)}{Pr(H)}. \tag{1}$$

It is known that the function $Pr(\cdot|H)$ is also a probability measure. In this paper, we will denote the induced measure $Pr(\cdot|H)$ by Pr^H. For a finite set S, a probability distribution is any function $p : S \longrightarrow [0, 1]$ such that

$$\sum_{s \in S} p(s) = 1.$$

Each probability distribution p on a finite set S induces a function Pr on the set of all subsets of S, $Pr(H) = \sum_{s \in H} p(s)$, which is a (finitely additive) probability measure. Also, any probability measure Pr on the power set of S induces a distribution $p(s) = Pr(\{s\})$ (i.e. on finite sets "measures=distributions").

Probabilistic logic. Probabilistic logic can be understood as a tool for reasoning with unreliable, incomplete or imprecise knowledge, where the uncertainty of the premises is expressed by qualitative or quantitative probabilistic statements. The typical form of the qualitative statement is "α is more probable than β", and the typical form of the quantitative probabilistic statements is "the probability of α is at least r". In this paper, we suppose that (non-probabilistic) knowledge is represented by propositional formulas. We will use the probabilistic operators $P_{\geq r}$ from the papers [22,23].

Suppose that that $\mathcal{P}$ is a set of propositional letters and $For_\mathcal{P}$ the corresponding set of propositional formulas. The set of probabilistic formulas is built over the set of propositional formulas $For_\mathcal{P}$.

Definition 7. *If $\alpha \in For_\mathcal{P}$ and $r \in [0, 1]$, then $P_{\geq r}\alpha$ is a basic probabilistic formula. The set of probabilistic formulas is the smallest set containing the basic probabilistic formulas, closed under boolean connectives.*

We can also formally introduce other types of inequalities from the basic formulas, as abbreviations: $P_{<r}\alpha$ is $\neg P_{\geq r}\alpha$, $P_{\leq r}\alpha$ is $P_{\geq 1-r}\neg\alpha$, $P_{>r}\alpha$ is $\neg P_{\leq r}\alpha$ and $P_{=r}\alpha$ is $P_{\geq r}\alpha \wedge P_{\leq r}\alpha$.

For instance, $P_{=0}(\alpha \wedge \beta) \rightarrow P_{\leq \frac{1}{2}}\beta$, is read as "If the probability of $\alpha \wedge \beta$ is zero, then the probability that β holds is at most one half." There are also various more complex logical languages, which can formalize higher order probabilities, conditional probabilities or qualitative probabilities.

The notion of satisfiability of a propositional formula $\alpha \in For_\mathcal{P}$ under a valuation $v : \mathcal{P} \longrightarrow \{\top, \bot\}$ ($v \models \alpha$) is introduced in the standard way. Semantics of probabilistic logic consists on probability measures μ on sets of valuations. The class of measurable sets is (or, alternatively, includes) the set $\{[\alpha] \mid \alpha \in For_\mathcal{P}\}$, where $[\alpha] = \{v \mid v \models \alpha\}$. In the following definition, we take conjunction and negation as primitive connectives.

Definition 8. *The satisfiability of a probabilistic formulas is defined recursively, as follows:*

- $\mu \models P_{\geq r}\alpha$ *iff* $\mu([\alpha]) \geqslant r$,
- $\mu \models \neg\alpha$ *iff* $\mu \not\models \alpha$,
- $\mu \models \alpha \wedge \beta$ *iff* $\mu \models \alpha$ *and* $M \models \beta$.

3   Probabilistic Argumentation Frameworks

We start with an overview of [20] and [18].

Definition 9. *A probabilistic argumentation framework (PAF) F_{pr} is a tuple $\langle A, R, P_A, P_R \rangle$, where $\langle A, R \rangle$ is an argumentation framework, $P_A : A \longrightarrow (0, 1]$ and $P_R : R \longrightarrow (0, 1]$.*

The intuition hereby is as follows: for $a \in A$, $P_A(a)$ is the probability that a belongs to an arbitrary subgraph of $\langle A, R \rangle$ (while $1 - P_A(a)$ is the probability that a does not belong to an arbitrary subgraph); for $r = (a, b) \in R_A$, whenever $a, b \in A' \subseteq A$, $P_R(r)$ is the probability that r belongs to an arbitrary subgraph of $\langle A', R_{A'} \rangle$. Using the independency assumption, those probabilities allow calculation of the probability of a subgraph.

Definition 10 ([20], slightly modified). *Let $F_{pr} = \langle A, R, P_A, P_R \rangle$ be a probabilistic argumentation framework and $G = \langle A', R' \rangle \sqsubseteq \langle A, R \rangle$, then $p_{F_{pr}}(G)$, the probability of the subgraph G, is*

$$(\prod_{a \in A'} P(a))(\prod_{a \in A \backslash A'} (1 - P(a)))(\prod_{r \in R'} P_R(r))(\prod_{r \in R_A \backslash R'} (1 - P_R(r))).$$

Remark 2. In [18], Hunter only considers probabilistic argumentation frameworks of the form $\langle A, R, P_A \rangle$, $P_A : A \longrightarrow [0, 1]$, and corresponding full subgraphs $G = \langle A', R' \rangle$. Thus, he defines the probability of G as

$$(\prod_{a \in A'} P(a))(\prod_{a \in A \backslash A'} (1 - P(a))).$$

This value is equal to $p_{F_{pr}}(G)$, where $F_{pr} = \langle A, R, P_A, P_R \rangle$ extends $\langle A, R, P_A \rangle$ such that $P_R(r) = 1$, for every $r \in R$.

Example 4. Let $F\langle A, R \rangle$ be the AF from Example 1, and let $F_{pr} = \langle A, R, P_A, P_R \rangle$ be the PAF, where P_A and P_R are defined in the following way:

- $P_A(a) = P_A(b) = P_A(c) = 1$, $P_A(d) = 0.7$;
- $P_R((a, b)) = P_R((b, a)) = P_R((d, c)) = 1$, $P_R((b, c)) = 0.4$.

Then there are only four subgraphs with nonzero probability: $G_1 = F$, $G_2 = \langle \{a, b, c, d\}, \{(a, b), (b, a), (d, c)\} \rangle$, $G_3 = \langle \{a, b, c\}, \{(a, b), (b, a), (b, c)\} \rangle$ and $G_4 = \langle \{a, b, c\}, \{(a, b), (b, a)\} \rangle$. Their probability values are $p_{F_{pr}}(G_1) = 0.28$, $p_{F_{pr}}(G_2) = 0.42$, $p_{F_{pr}}(G_3) = 0.12$ and $p_{F_{pr}}(G_4) = 0.18$.

Theorem 1 ([20]). *For any PAF $F_{pr} = \langle A, R, P_A, P_R \rangle$, the function $p_{F_{pr}}$ is a probability distribution on the set $s(\langle A, R \rangle)$, i.e., a nonnegative function such that $\sum_{G \sqsubseteq \langle A, R \rangle} p_{F_{pr}}(G) = 1$.*

Now the probability that a given set of arguments is an extension, or a consistent subset, of the framework can be defined as follows.

Definition 11. *Let $F = \langle A, R \rangle$ be an AF and let $F_{pr} = \langle A, R, P_A, P_R \rangle$ be a PAF. The probability that $\Gamma \subseteq A$ is a σ extension (where $\sigma = \{cf, ad, co, pr, gr, st\}$) is defined as*

$$P^{\sigma}_{F_{pr}}(\Gamma) = \sum_{G \in Q^{\sigma}_F(\Gamma)} p_{F_{pr}}(G),$$

and the probability of $S \subseteq A$ being consistent according to σ semantics is defined as

$$Cn^{\sigma}_{F_{pr}}(S) = \sum_{G \in C^{\sigma}_F(S)} p_{F_{pr}}(G).$$

$Cn^{\sigma}_{F_{pr}}$ is defined in [20], while $P^{\sigma}_{F_{pr}}$ is defined in [18] (for the full subgraphs only).

Example 5. Continuing Example 4, let $\Gamma = \{a, c\}$. Let us calculate the probability that Γ is a preferred extension of F_{pr}. Since $P^{prf}_{F_{pr}}(\Gamma) = \sum_{G \in Q^{prf}_F(\Gamma)} p_{F_{pr}}(G)$, using $Q^{prf}_F(\Gamma)$ from Example 3 and the fact that only G_1, G_2, G_3 and G_4 have nonzero probability, we obtain

$$P^{prf}_{F_{pr}}(\Gamma) = p_{F_{pr}}(G_3) + p_{F_{pr}}(G_4) = 0.12 + 0.18 = 0.3.$$

We now relate these concepts to certain probability measures. By Theorem 1, for any probabilistic argumentation framework $F_{pr} = \langle A, R, P_A, P_R \rangle$, $p_{F_{pr}}$ is a probability distribution on the set $s(F)$, where $F = \langle A, R \rangle$. Thus, the function $Pr_{F_{pr}}$ on subsets of $s(F)$, defined by $Pr_{F_{pr}}(H) = \sum_{G \in H} p_{F_{pr}}(G)$ is a probability measure. We can reformulate Definition 11 as follows.

Lemma 1. *Let F_{pr} be a PAF, Γ a set of arguments. Then, $P^{\sigma}_{F_{pr}}(\Gamma) = Pr_{F_{pr}}(Q^{\sigma}_G(\Gamma))$, $Cn^{\sigma}_{F_{pr}}(S) = Pr_{F_{pr}}(C^{\sigma}_G(S))$.*

For $a \in A$, we define H_a as the set of all subgraphs of $\langle A, R \rangle$ whose sets of arguments contain a:

$$H_a = \{G = \langle A', R' \rangle \in s(F) \mid a \in A'\}.$$

Also, for $B \subseteq A$, let H_B denotes he set of all subgraphs whose sets of arguments contain B as subset. Then

$$H_B = \{G = \langle A', R' \rangle \in s(F) \mid B \subseteq A'\} = \bigcap_{a \in B} H_a.$$

Finally, by $\overline{H}_B$ we denote the set $(\bigcap_{a \in B} H_a) \cap \bigcap_{a \notin B} (H_a)^c$ (the superscript c denotes the set operator "complement").

Lemma 2. *$Pr_{F_{pr}}(H_B) = \prod_{a \in B} P_A(a)$. In particular, $Pr_{F_{pr}}(H_a) = P_A(a)$.*

As in the case of arguments, we define sets H_r (the set of all subgraphs whose attacks contain r) as

$$H_r = \{G = \langle A', R' \rangle \in s(F) \mid r \in R'\},$$

and $H_{R''}$ for $R'' \subseteq R$ as

$$H_{R''} = \{G = \langle A', R' \rangle \in s(F) \mid R'' \subseteq R'\} = \bigcap_{r \in R''} H_r.$$

Lemma 3. *For $A' \subseteq A$ and $R'' \subseteq R_{A'}$, $Pr_{F_{pr}}(H_{R''}|\overline{H}_{A'}) = \prod_{r \in R''} P_R(r)$. In particular, for $r \in R_{A'}$, $Pr_{F_{pr}}(H_r|\overline{H}_{A'}) = P_R(r)$.*

But it is not the case that all probability measures are defined by probabilistic argumentation frameworks. Thus, we introduce the following definition that describe all probability measures that correspond to probabilities of arguments and attacks in the sense of Definition 10.

Definition 12. *Let $F = \langle A, R \rangle$ be an argumentation framework. A probability measure Pr on subsets of $s(F)$ is F-independent, iff the following conditions hold:*

1. *For every $a \in A$ and $B \subseteq A \setminus \{a\}$, $Pr(H_a|H_B) = Pr(H_a)$.*
2. *For every $A' \subseteq A$, $r' \in R_{A'}$ and $R'' \subseteq R_{A'} \setminus \{r'\}$ such that $Pr(H_{R''}) > 0$, $Pr^{\overline{H}_{A'}}(H_{r'}|H_{R''}) = Pr^{\overline{H}_{A'}}(H_{r'})$.*

Thus, F-independency means that arguments are independent, and attacks are independent within every set of arguments.

Theorem 2 (Representation theorem for PAFs). *For a given argumentation framework $F = \langle A, R \rangle$, a probability measure on subsets of $s(F)$ is induced by some probabilistic argumentation framework $\langle A, R, P_A, P_R \rangle$, iff it is F-independent.*

4 Expressing Subgraphs of AFs in Propositional Logic

In this section, we provide propositional formulae which allow us to reason about subgraphs of AFs. In particular, we want to characterize the sets $Q^\sigma_F(\Gamma)$ (resp. $Q^\sigma_{F,f}(\Gamma)$, $C^\sigma_F(\Gamma)$, and $C^\sigma_{F,f}(\Gamma)$) as models of propositional formulas for a given extension candidate Γ. Note that in contrast to the encodings from [3] (where the framework remains fixed and the extensions are characterized by models), we fix here the extension and want to characterize the appropriate subgraphs.

We start with some basic notation for propositional logic which are required for our purpose.

Definition 13. *For a given set of arguments A, the set of propositional letters $\mathcal{P}_A$ is $A \cup \{r_{ab}|\, a, b \in A\}$, and the set of formulas For_A is the set of all Boolean combinations of $\mathcal{P}_A$.*

Definition 14. *Let A be a set of arguments and $v : \mathcal{P}_A \longrightarrow \{\top, \bot\}$ a valuation (over $\mathcal{P}_A$). We define the argumentation framework characterized by v as $F_v = \langle A_v, R_v \rangle$, where $A_v = \{a \in A \mid v(a) = \top\}$ and $R_v = \{(a,b) \in A^2_v \mid v(r_{ab}) = \top\}$.*

Our goal here in formal terms is as follows: given a semantics $\sigma \in \{cf, stb, adm, com, grd, prf\}$, an AF $F = \langle A, R \rangle$ and a set $\Gamma \subseteq A$, we want to construct a formula $\psi^\sigma_F(\Gamma) \in For_A$ which corresponds to the set $Q^\sigma_F(\Gamma)$, in the sense that, for all valuations v over $\mathcal{P}_A$,[1]

$$v \models \psi^\sigma_F(\Gamma) \quad \text{iff} \quad F_v \in Q^\sigma_F(\Gamma).$$

For this purpose, we first need to express a restriction to $\langle A', R' \rangle$ of an AF F such that $\Gamma \subseteq A'$.

---

[1] As usual, we write $v \models \phi$ to denote that a valuation v is a classical model of ϕ.

Proposition 1. *For $F = \langle A, R \rangle$ and $\Gamma \subseteq A$, let*

$$\varphi_F(\Gamma) = \bigwedge_{a \in \Gamma} a \wedge \bigwedge_{(a,b) \in A^2 \setminus R} (r_{ab} \to (\neg a \vee \neg b)).$$

Then, for all valuations v over $\mathcal{P}_A$, $v \models \varphi_F(\Gamma)$ iff $F_v \in \{\langle A', R' \rangle \in s(F) \mid \Gamma \subseteq A'\}$.

The next result collects the required formulas for characterizing subgraphs which – seen as AF — possess a given extension.

Theorem 3. *For an $F = \langle A, R \rangle$ and $\Gamma \subseteq A$, let*

1. $\psi_F^{cf}(\Gamma) = \varphi_F(\Gamma) \wedge \bigwedge_{a,b \in \Gamma} \neg r_{ab}$
2. $\psi_F^{stb}(\Gamma) = \psi_F^{cf}(\Gamma) \wedge \bigwedge_{b \in A \setminus \Gamma} (b \to \bigvee_{a \in \Gamma} r_{ab})$
3. $\psi_F^{adm}(\Gamma) = \psi_F^{cf}(\Gamma) \wedge \bigwedge_{c \in A \setminus \Gamma} (c \to \bigwedge_{a \in \Gamma} (\neg r_{ca} \vee \bigvee_{b \in \Gamma} r_{bc}))$
4. $\psi_F^{com}(\Gamma) = \psi_F^{adm}(\Gamma) \wedge \bigwedge_{a,b \in A \setminus \Gamma} ((a \wedge b) \to (r_{ab} \to ((\bigvee_{c \in \Gamma} r_{cb}) \bigwedge_{c \in \Gamma} \neg r_{ca})))$
5. $\psi_F^{grd}(\Gamma) = \psi_F^{com}(\Gamma) \wedge \bigwedge_{\Gamma' \subset \Gamma} \neg \psi_F^{com}(\Gamma')$
6. $\psi_F^{prf}(\Gamma) = \psi_F^{adm}(\Gamma) \wedge \bigwedge_{A \supseteq \Gamma' \supset \Gamma} \neg \psi_F^{adm}(\Gamma')$

Then, for all valuations v over $\mathcal{P}_A$, $v \models \psi_F^\sigma(\Gamma)$ iff $F_v \in Q_F^\sigma(\Gamma)$ ($\sigma \in \{cf, stb, adm, com, grd, prf\}$).

Note that the formulas for cf, stb, adm, and com can be constructed in linear time in the size of F. This does not hold for the formulas for grd and prf. In fact, one can efficiently construct QBFs for these two semantics, but for the sake of simplicity and uniformity, we decided to stay in propositional logic.

We now provide an adaptation of the encodings for full subgraphs, i.e. we define formulae $\psi_{F,f}^\sigma(\Gamma)$ which correspond to the sets $Q_{F,f}^\sigma(\Gamma)$ in such a way that $v \models \psi_{F,f}^\sigma(\Gamma)$ holds iff $F_v \in Q_{F,f}^\sigma(\Gamma)$.

Theorem 4. *For an $F = \langle A, R \rangle$, $\Gamma \subseteq A$, and $\sigma \in \{cf, stb, adm, com, grd, prf\}$, let*

$$\psi_{F,f}^\sigma(\Gamma) = \psi_F^\sigma(\Gamma) \wedge \bigwedge_{(a,b) \in R} ((a \wedge b) \to r_{ab}).$$

Then, for all valuations v over $\mathcal{P}_A$ $v \models \psi_{F,f}^\sigma(\Gamma)$ iff $F_v \in Q_{F,f}^\sigma(\Gamma)$.

Consequently, we can state for given $F = \langle A, R \rangle$ and $\Gamma \subseteq A$,

$$Q_F^\sigma(\Gamma) = \{F_v \mid v : \mathcal{P}_A \longrightarrow \{\top, \bot\}, v \models \psi_F^\sigma(\Gamma)\}, \tag{2}$$

and respectively,

$$Q_{F,f}^\sigma(\Gamma) = \{F_v \mid v : \mathcal{P}_A \longrightarrow \{\top, \bot\}, v \models \psi_{F,f}^\sigma(\Gamma)\}.$$

For a subgraph $G = \langle A', R' \rangle$ of the argumentation framework $F = \langle A, R \rangle$, we define the formula α_G that specifies G:

$$\alpha_G = \bigwedge_{a \in A'} a \wedge \bigwedge_{a \in A \setminus A'} \neg a \wedge \bigwedge_{(a,b) \in R'} r_{ab} \wedge \bigwedge_{(a,b) \in R_{A'} \setminus R'} \neg r_{ab}.$$

It is easy to check that $v \models \alpha_G$ iff $G = F_v$. The following result follows in straightforward way.

Corollary 1. *Let $F = \langle A, R \rangle, G \in s(F)$ ($G \in fs(F)$), $\Gamma \subseteq A$ and $\sigma \in \{cf, stb, adm, com, grd, prf\}$. Then Γ is a σ extension of G iff the formula $\psi_F^\sigma(\Gamma) \wedge \alpha_G$ ($\psi_{F,f}^\sigma(\Gamma) \wedge \alpha_G$, respectively) is satisfiable.*

Our final results of this section are concerned with the notion of consistency. For a set of arguments $S \subseteq A$, and AF $F = \langle A, R \rangle$ and $\sigma \in \{cf, stb, adm, com, grd, prf\}$, we define the formulas $\theta_F^\sigma(S) = \bigvee_{\Gamma:S \subseteq \Gamma} \psi_F^\sigma(\Gamma)$ and $\theta_{F,f}^\sigma(\Gamma) = \bigvee_{\Gamma:S \subseteq \Gamma} \psi_{F,f}^\sigma(\Gamma)$.

Proposition 2. *Let $\sigma \in \{cf, stb, adm, com, grd, prf\}$, $F = \langle A, R \rangle$ and $S \subseteq A$, then*

$$C_F^\sigma(S) = \{F_v \mid v : \mathcal{P}_A \longrightarrow \{\top, \bot\}, v \models \theta_F^\sigma(S)\},$$

$$C_{F,f}^\sigma(S) = \{F_v \mid v : \mathcal{P}_A \longrightarrow \{\top, \bot\}, v \models \theta_{F,f}^\sigma(S)\}.$$

Corollary 2. *Let $\sigma \in \{cf, stb, adm, com, grd, prf\}$, $F = \langle A, R \rangle$, $G \in s(F)$ ($G \in fs(F)$) and $S \subseteq A$, then S in consistent w.r.t. σ in G iff the formula $\theta_F^\sigma(S) \wedge \alpha_G$ ($\theta_{F,f}^\sigma(S) \wedge \alpha_G$, respectively) is satisfiable.*

5   A Logic for Probabilistic Argumentation

In this section, we use probabilistic logic to express whether a set of arguments is acceptable with some given probability value.

To this end, we require the following logical framework. For given argumentation framework $F = \langle A, R \rangle$, we define the language and class of models for the logic $LP_F^\mathbb{R}$. We choose the set of propositional formulas $For_A = A \cup \{r_{ab} \mid a, b \in A\}$ from Definition 13. The set of probabilistic formulas is built over For_A, using the probabilistic operators $P_{\geq r}$, as in Definition 7.

The corresponding semantics for probabilistic logics generally consist of probability measures on sets of valuations of For_A. In order to interpret probabilities on subgraphs of F, we restrict the class of models of the logic $LP_F^\mathbb{R}$ to those probability measures μ such that

$$\mu(([a] \cap [b])^c \cap [r_{ab}]) = 0, \tag{3}$$

for all $a, b \in A$, and

$$\mu([r_{ab}]) = 0, \quad \text{whenever } (a, b) \notin R. \tag{4}$$

The relation $\models$ is defined as in Definition 8. The notions of satisfiability and validity are as usual. Also, for a probability measure μ and a set of valuations V, we denote by μ^V the measure $\mu(\cdot | V)$.

Now we introduce the logic $LP_{F,ind}^\mathbb{R}$ which has the same syntax as $LP_F^\mathbb{R}$, but restricted semantics. In order to capture the properties of Definition 12, we restrict the set of models of the logic $LP_{F,ind}^\mathbb{R}$ to consist of those probability measures μ from semantics of $LP_F^\mathbb{R}$ such that

$$\mu([a] | [\bigwedge_{b \in B} b]) = \mu([a]), \tag{5}$$

whenever $B \subseteq A \setminus \{a\}$, and

$$\mu^{[\bigwedge_{a \in A'} a \wedge \bigwedge_{a \notin A'} \neg a]}([r_{ab}]|[\bigwedge_{(c,d) \in R''} r_{cd}]) = \mu([r_{ab}]), \tag{6}$$

whenever $A' \subseteq A$, $(a, b) \in R_{A'}$ and $R'' \subseteq R_{A'} \setminus \{(a, b)\}$ such that $\mu([\bigwedge_{(c,d) \in R''} r_{cd}]) > 0$. By Theorem 2, the introduced class of models corresponds to the measures induced by probabilistic argumentation frameworks.

Remark 3. If we want to work with full subgraphs only, as in [18], it is enough to restrict to the set of models μ such that $\mu([r_{ab}]) = 1$, whenever $(a, b) \in R$.

Definition 15. *If $F - \langle A, R \rangle$ is an argumentation framework and μ a probability measure on sets of valuations of For_A, we define the probabilistic argumentation framework $F_\mu = \langle A, R, P_A^\mu, P_R^\mu \rangle$, where:*

- *$P_A^\mu(a) = \mu([a])$ for all $a \in A$,*
- *$P_R^\mu((a, b)) = \mu([r_{ab}])$, for all $(a, b) \in R$.*

Similarly as in Section 4, for a given argumentation framework $F = \langle A, R \rangle$, a set of arguments $\Gamma \subseteq A$, $\sigma \in \{cf, stb, adm, com, grd, prf\}$ and $r \in [0, 1]$, we want to construct the probabilistic formula $\rho_F^\sigma(\Gamma, r)$ such that the models μ of $\rho_F^\sigma(\Gamma, r)$ correspond to probabilistic argumentation frameworks F_μ for which Γ is σ extension with probability at least r, i.e.,

$$\mu \models \rho_F^\sigma(\Gamma, r) \quad \text{iff} \quad P_{F_\mu}^\sigma(\Gamma) \geq r.$$

where $\models$ denotes the satisfiability relation of the logic $LP_{F,ind}^{\mathbb{R}}$.

Theorem 5. *Let $F = \langle A, R \rangle$ be a given argumentation framework, $\Gamma \subseteq A$, $\sigma \in \{cf, stb, adm, com, grd, prf\}$ and $r \in [0, 1]$. Then*

$$\rho_F^\sigma(\Gamma, r) = P_{\geq r} \psi_F^\sigma(\Gamma).$$

Remark 4. In the previous theorem, as well as in the propositions below, we can replace $\geq r$ with $= r$, $> r$, $\leq r$ and $< r$.

Definition 16. *For $F_{pr} = \langle A, R, P_A, P_R \rangle$ a PAF, define $\mu_{F_{pr}}$ as a probabilistic measure on For_A, such that*

- *$\mu_{F_{pr}}(\{v\}) = p_{F_{pr}}(F_v)$, if $F_v \in s(\langle A, R \rangle)$,*
- *$\mu_{F_{pr}}(\{v\}) = 0$, otherwise;*

where $F_v = \langle \{a \in A \mid v \models a\}, \{(a, b) \in A^2 \mid v \models r_{ab}\} \rangle$.

Lemma 4. *$\mu_{F_{pr}}(V) = Pr_{F_{pr}}(\{F_v \mid v \in V\})$.*

Corollary 3. *For a given AF $\langle A, R \rangle$ and $\sigma \in \{cf, stb, adm, com, grd, prf\}$, the set of probabilistic argumentation frameworks with the set of arguments A and the set of attacks R such that the probability that $\Gamma \subseteq A$ is a σ extension is at least r is given by $\{F_\mu \mid \mu \models P_{\geq r} \psi_F^\sigma(\Gamma)\}$.*

For a probabilistic argumentation framework $F_{pr} = \langle A, R, P_A, P_R \rangle$, we define

$$\chi_{F_{pr}} = \bigwedge_{a \in A} P_{=Pr_{F_{pr}}(H_a)} a \wedge \bigwedge_{r \in R} P_{=Pr_{F_{pr}}(H_r)} r.$$

Corollary 4. *For a PAF F_{pr} and $\sigma \in \{cf, stb, adm, com, grd, prf\}$, the set of arguments Γ is a σ extension of F_{pr} with probability at least r iff the formula $P_{\geq r} \psi_F^\sigma(\Gamma) \wedge \chi_{F_{pr}}$ is satisfiable.*

Remark 5. All the previous results about probability of extensions can be transformed to probability of consistency of S w.r.t. σ, by replacing the formulas $\psi_F^\sigma(S)$ with $\theta_F^\sigma(S)$.

6 Discussion

In this work we extended ideas of Besnard and Doutre [3] (who addressed the problem of acceptability of sets of arguments under Dung's semantics via propositional logic) to probabilistic argumentation. We have focused on the approaches from [20,18], where argumentation frameworks $\langle A, R \rangle$ are enriched with probabilities on A and R. Those probabilities determine certain probabilities of subgraphs of $\langle A, R \rangle$, using the "independency" assumption. We have provided a logical formalization in terms of propositional probabilistic logic using set of formulae depending on a given PAF and class of models that correspond to the probability measures induced by PAFs.

Although we follow the initial paper on probabilistic argumentation frameworks [20], it is easy to modify our results to cover the more general case, where the independency assumptions are not used (e.g. [19]). In that case, the probability of a subgraph cannot be calculated from the probabilities of the arguments and attacks. Instead it has to be taken as a constituent of the framework. In other words, instead of using P_A and P_R, the frameworks are of the form $\langle A, R, p \rangle$, where p is a probability distribution on the set of subgraphs of $\langle A, R \rangle$. Using p, the probability of an extension is then defined as in Definition 11. Consequently, one can now omit the independency constraints (5) and (6) from the semantics of probabilistic logic. All results then carry over, with the difference that in Corollary 4 we cannot use $\chi_{F_{pr}}$ as syntactic description of framework, but the formula $\bigwedge_{G \sqsubseteq \langle A,R \rangle} P_{=p(G)} \alpha_G$.

We propose three avenues for further research on this topic. First, we plan to extend our results constructing formulas which can be used to formally compare the probabilities of different extensions. For example, we wish to express statements like "Γ_1 is more probably a stable extension than Γ_2", or "the probability that Γ is preferred extension is less then the probability that it is grounded extension" and to check truthfulness of the statements by checking satisfiability of associated probabilistic formulas. The natural choice would be to enrich the logic from this paper with the qualitative probability operator [21,5].

Second, we plan to investigate how our logical representation of probabilistic argumentation can be exploited for implementing systems. Recent complexity results from the area [14] show that for stable and admissible semantics, the problem of computing the probability that a set of argument is an extension is tractable, while for the other semantics the problem is hard. This suggests to study how these insights can be used to improve our encodings.

Third, we want to extend our approach to other realizations of probabilistic argumentation. For example, Grossi and van der Hoek [16] give a probabilistic account on Dung's game for grounded semantics. They consider several graphs over a given set of arguments, and a probability distribution on the set of graphs. Another example is the work by Thimm [24] who has a different starting point. For a given framework $\langle A, R \rangle$, he does not consider uncertainty that elements of the graph belong to the graph, but uses probability distributions on subsets of A. Although our approach is not directly applicable to these works, we plan to study logical characterizations for them as well.

Acknowledgment. This work was supported by the National Research Fund (FNR) of Luxembourg through project PRIMAT, and by the Austrian Science Fund (FWF): project I1102.

References

1. Arieli, O., Caminada, M.W.A.: A QBF-based formalization of abstract argumentation semantics. J. Applied Logic 11(2), 229–252 (2013)
2. Bench-Capon, T.J.M., Dunne, P.E.: Argumentation in artificial intelligence. Artif. Intell. 171(10-15), 619–641 (2007)
3. Besnard, P., Doutre, S.: Checking the acceptability of a set of arguments. In: Proc. NMR 2004, pp. 59–64 (2004)
4. Cerutti, F., Dunne, P.E., Giacomin, M., Vallati, M.: A SAT-based Approach for Computing Extensions on Abstract Argumentation. In: Proc. TAFA (2013)
5. Doder, D.: A logic with big-stepped probabilities that can model nonmonotonic reasoning of system P. Publications de l'Institut Mathématique 90(104), 13–22 (2011)
6. Dung, P.M.: On the acceptability of arguments and its fundamental role in nonmonotonic reasoning, logic programming and n-person games. Artif. Intell. 77(2), 321–358 (1995)
7. Dung, P.M., Thang, P.M.: Towards (probabilistic) argumentation for jury-based dispute resolution. In: Proc. COMMA. FAIA, vol. 216, pp. 171–182. IOS Press (2010)
8. Dunne, P.E.: Computational properties of argument systems satisfying graph-theoretic constraints. Artif. Intell. 171(10-15), 701–729 (2007)
9. Dvořák, W., Järvisalo, M., Wallner, J.P., Woltran, S.: Complexity-sensitive decision procedures for abstract argumentation. Artif. Intell. 206, 53–78 (2014)
10. Dvořák, W., Szeider, S., Woltran, S.: Abstract argumentation via monadic second order logic. In: Hüllermeier, E., Link, S., Fober, T., Seeger, B. (eds.) SUM 2012. LNCS, vol. 7520, pp. 85–98. Springer, Heidelberg (2012)
11. Dyrkolbotn, S.K.: The same, similar, or just completely different? equivalence for argumentation in light of logic. In: Libkin, L., Kohlenbach, U., de Queiroz, R. (eds.) WoLLIC 2013. LNCS, vol. 8071, pp. 96–110. Springer, Heidelberg (2013)
12. Egly, U., Woltran, S.: Reasoning in argumentation frameworks using quantified boolean formulas. In: Proc. COMMA. FAIA, vol. 144, pp. 133–144. IOS Press (2006)
13. Fagin, R., Halpern, J.Y., Megiddo, N.: A logic for reasoning about probabilities. Inf. Comput. 87(1/2), 78–128 (1990)
14. Fazzinga, B., Flesca, S., Parisi, F.: On the complexity of probabilistic abstract argumentation. In: Rossi, F. (ed.) Proc. IJCAI. IJCAI/AAAI (2013)

15. Grossi, D.: Argumentation in the view of modal logic. In: McBurney, P., Rahwan, I., Parsons, S. (eds.) ArgMAS 2010. LNCS, vol. 6614, pp. 190–208. Springer, Heidelberg (2011)
16. Grossi, D., van der Hoek, W.: Audience-based uncertainty in abstract argument games. In: Rossi, F. (ed.) Proc. IJCAI. IJCAI/AAAI (2013)
17. Hunter, A.: Some foundations for probabilistic abstract argumentation. In: Verheij, B., Szeider, S., Woltran, S. (eds.) Proc. COMMA. Frontiers in Artificial Intelligence and Applications, vol. 245, pp. 117–128. IOS Press (2012)
18. Hunter, A.: A probabilistic approach to modelling uncertain logical arguments. Int. J. Approx. Reasoning 54(1), 47–81 (2013)
19. Hunter, A.: Probabilistic qualification of attack in abstract argumentation. Int. J. Approx. Reasoning 55(2), 607–638 (2014)
20. Li, H., Oren, N., Norman, T.J.: Probabilistic argumentation frameworks. In: Modgil, S., Oren, N., Toni, F. (eds.) TAFA 2011. LNCS, vol. 7132, pp. 1–16. Springer, Heidelberg (2012)
21. Ognjanović, Z., Perović, A., Rašković, M.: Logics with the qualitative probability operator. Logic Journal of the IGPL 16(2), 105–120 (2008)
22. Ognjanović, Z., Rašković, M.: Some probability logics with new types of probability operators. J. Log. Comput. 9(2), 181–195 (1999)
23. Ognjanović, Z., Rašković, M.: Some first-order probability logics. Theor. Comput. Sci. 247(1-2), 191–212 (2000)
24. Thimm, M.: A probabilistic semantics for abstract argumentation. In: Proc. ECAI. FAIA, vol. 242, pp. 750–755. IOS Press (2012)
25. Wallner, J.P., Weissenbacher, G., Woltran, S.: Advanced SAT Techniques for Abstract Argumentation. In: Leite, J., Son, T.C., Torroni, P., van der Torre, L., Woltran, S. (eds.) CLIMA XIV 2013. LNCS, vol. 8143, pp. 138–154. Springer, Heidelberg (2013)

Computing Skyline from Evidential Data

Sayda Elmi[1], Karim Benouaret[2], Allel Hadjali[3],
Mohamed Anis Bach Tobji[4], and Boutheina Ben Yaghlane[5]

[1] LARODEC, Tunis University, Tunisia
elmi.sayda@gmail.com
[2] LT2C, Jean Monnet University, France
karim.benouaret@univ-st-etienne.fr
[3] LIAS/ENSMA, Poitiers, France
allel.hadjali@ensma.fr
[4] LARODEC, Tunis University, ISG, Tunisia
anis.bach@isg.rnu.tn
[5] LARODEC - Carthage University, IHEC, Tunisia
boutheina.yaghlane@ihec.rnu.tn

Abstract. The skyline operator is a powerful means in multi-criteria decision-making since it retrieves the most interesting objects according to a set of attributes. On the other hand, uncertainty is inherent in many real applications. One of the most powerful approaches used to model uncertainty is the evidence theory. Databases that manage such type of data are called evidential databases. In this paper, we tackle the problem of skyline analysis on evidential databases. We first introduce a skyline model that is appropriate to the evidential data nature. We then develop an efficient algorithm to compute this kind of skyline. Finally, we present a thorough experimental evaluation of our approach.

1 Introduction

Skyline analysis has been shown to be a powerful means in multi-criteria decision-making. Given a set of database objects, defined on a set of attributes, an object o_i is said to dominate (in the Pareto sense) another object o_j if and only if o_i is better than or equal to o_j in all attributes and better in at least one attribute. The skyline comprises those objects that are not dominated by any other object.

On the other hand, due to the exploding number of information stored and shared over Internet, and the introduction of new technologies to capture and transit data, uncertain data analysis is an important issue in many applications such as decision-making and data integration, to name just a few. To deal with uncertain values of database attributes, several models were proposed. The most studied and known models are: probabilistic databases [11,12,1], possibilistic databases [7,8] and evidential databases (based on Dempster-shafer theory) [19,22,23,4,3]. The advantage of the evidential databases model is twofold: (i) it allows modeling both uncertainty and imprecision (due to the lack of information) in data; and (ii) it represents a generalization of both probabilistic and

U. Straccia and A. Calì (Eds.): SUM 2014, LNAI 8720, pp. 148–161, 2014.

possibilistic models. Considering the example borrowed from [19] about the location of some conferences for the next year, which should be either in Europe or U.S.A. Assume that we know the following probability distribution ⟨Europe, 0.5⟩ and ⟨U.S.A, 0.5⟩. Now, if it is in Europe, it will be in Paris or London. If it is in US.A., it will be in Phoenix, Iowa City or Kansas City. But we don't know any probability distribution for these locations. It is then natural to represent this information as ⟨{Paris, London}, 0.5⟩ and ⟨{Phoenix, Iowa City, Kansas City}, 0.5⟩. One can observe that data are pervaded both by uncertainty and imprecision. Imprecision is due to the lack of information of probability distribution between cities. Modeling this kind of lack of information is one of motivations behind Dempster-Shafer theory of evidence. Probabilistic and possibilistic models cannot support the presence of imperfection in data.

Substantial research work has addressed the problem of skyline analysis on uncertain data from different perspectives and within various communities, including, databases; e.g., [24,18,20,31,6], Web services; e.g., [28,5], and so on. These works are important and useful, but they focus on either probabilistic data or possibilistic data. However, as mentioned above, these models (i.e., probabilistic and possibilistic) have some limitations.

In sharp contrast with these approaches, in this paper, we tackle the problem of skyline analysis on evidential data. To the best of our knowledge, this is the first attempt to introduce skyline queries on uncertain data where uncertainty is modeled thanks to evidential theory. Specifically, we address two main challenges. The first is about modeling skyline on evidential data: how can we capture the dominance relationship between the objects of an evidential database? And what should be the skyline on those objects? The second is about computing this kind of skyline: can we provide techniques for computing the skyline on evidential data efficiently?

Contributions: We tackle the above-mentioned challenges with the following major contributions.

- Given two objects of an evidential database, we calculate the belief that each object dominates the other. Based on this dominance relationship, we propose the notion of b-dominant skyline, which comprises the objects that are not dominated with some belief threshold b.
- We develop a suitable algorithm based on an efficient comparison method between two objects for computing efficiently the b-dominant skyline.
- We perform an extensive experimental evaluation to demonstrate the scalability of the algorithm proposed for the evidential skyline .

The rest of the paper is organized as follows. Section 2 contains a reminder about skyline on certain data and provides the basic notions of evidential theory and evidential databases. In Section 3, we formally define the dominance relationship and the skyline on evidential data, while in Section 4 we present our algorithm. Our experimental experimental evaluation is reported in Section 5. Related work is presented in Section 6. Finally, Section 7 concludes the paper.

2   Background

In this section, we first present a reminder about skyline on certain data. Then, we provide the basic notions of evidence theory and evidential databases.

2.1   Skyline on Certain Data

Consider a set of objects $\mathcal{O} = \{o_1, o_2, \ldots, o_n\}$ defined on a set of attributes $\mathcal{A} = \{a_1, a_2, \ldots, a_d\}$. We use $o_i.a_k$ to denote the kth attribute of object o_i. For simplicity, we assume throughout this paper that the higher the value, the more preferable.

Definition 1. *(Dominance)*
Given two objects $o_i, o_j \in \mathcal{O}$, o_i dominates o_j, denoted as $o_i \succ o_j$, if and only if o_i is as good or better than o_j in all attributes and better in at least one attribute, i.e., $\forall a_k \in \mathcal{A} : o_i.a_k \geq o_j.a_k \wedge \exists a_\ell \in \mathcal{A} : o_i.a_\ell > o_j.a_\ell$.

Definition 2. *(Skyline)*
The skyline of $\mathcal{O}$, denoted by $Sky_\mathcal{O}$, comprises those objects in $\mathcal{O}$ that are not dominated by any other object, i.e., $Sky_\mathcal{O} = \{o_i \in \mathcal{O} \mid \nexists\, o_j \in \mathcal{O}, o_j \succ o_i\}$.

Example 1. Consider a set of weight-loss products $\mathcal{O} = \{o_1(18, 70), o_2(15, 60), o_3(13, 80), o_4(17, 20)\}$ and assume that the values of each product denote the amount of weight loss per month (kilograms) and the repayment (%) if the user is not satisfied, respectively. We have, product o_1 dominates product o_3 and o_4; and products o_1 and o_2 are not dominated, they thus form the skyline of $\mathcal{O}$.

2.2   Basic Notions about Evidence Theory

In order to access to the most accurate and reliable information, we need to represent and reason with uncertain, imprecise and incomplete information. In this context, Shafer [25] introduced in 1976 the mathematical theory of evidence which is a subjective evaluation used to characterize the truth of a proposition. The theory of evidence, also known in the literature as the "theory of belief functions" and "theory of Dempster-Shafer", is a generalization of the Bayesian theory of subjective probability [14]. This theory represents a set of propositional hypotheses by a frame of discernment.

Definition 3. *(Frames of Discernment)*
A frame of discernment, usually denoted as Θ where $\Theta = \{h_1, h_2, \ldots, h_m\}$ contains mutually exclusive and exhaustive propositional hypotheses, one and only one of which is true.

Definition 4. *(Mass Function)*
A function, $m : 2^\Theta \longrightarrow [0, 1]$, is called a basic probability assignment on a frame Θ if it satisfies the following two conditions: $m(\emptyset) = 0$ and $\sum_{A \subseteq \Theta} m(A) = 1$.

Definition 5. *(Belief function)*
For every subset A of Θ, the belief of A, denoted as $bel(A)$, is defined as the sum of the masses assigned to every subset B of A, i.e., $bel(A) = \sum\limits_{B \subseteq A} m(B)$.

Definition 6. *(Plausibility function)*
For every subset A of Θ, the plausibility of A, denoted as $pl(A)$, is defined as the sum of the masses assigned to every subset B of Θ that intersects A, i.e., $pl(A) = \sum\limits_{B \cap A \neq \emptyset} m(B)$.

2.3   Evidential Databases

As mentioned above, many real-world applications deal with imperfect data. Evidential databases allow representing missing, uncertain or imprecise information thanks to the evidence theory. An evidential database, is a collection of objects $\mathcal{O} = \{o_1, o_2, \ldots, o_n\}$ defined on a set of attributes $\mathcal{A} = \{a_1, a_2, \ldots, a_d\}$ where each attribute a_k has a domain $Dom(a_k)$, and each attribute a_k of an object o_i, denoted by $o_i.a_k$ contains a normalized basic probability assignment called mass function. That is: $o_i.a_k = \{\langle A, m_{ik}(A)\rangle \mid A \subseteq Dom(a_k), m_{ik}(A) > 0\}$, where $m_{ik} : 2^{Dom(a_k)} \longrightarrow [0, 1]$, with $m_{ik}(\emptyset) = 0$ and $\sum_{A \subseteq Dom(a_k)} m_{ik}(A) = 1$. Each set $A \subseteq Dom(a_k) : m_{ik}(A) > 0$ is called an item set; and each tuple $\langle A, m_{ik}(A)\rangle$, i.e., an item set associated with its mass function, is called a focal element.

We obtain such a database, by collecting different experts opinions. Experts are first grouped into schools of thought, then opinions are aggregated [17]. The following example depicts an evidential database.

Table 1. Evidential data example

Product	Weight loss per month (kilograms)	Repayment (%)
o_1	$\langle\{15, 16, 18\}, 0.1\rangle, \langle\{19, 20\}, 0.9\rangle$	$\langle 90, 0.3\rangle, \langle\{90, 100\}, 0.7\rangle$
o_2	$\langle 7, 0.7\rangle \; \langle\{8, 9\}, 0.3\rangle$	$\langle\{70, 80\}, 0.8\rangle, \langle 80, 0.2\rangle$
o_3	$\langle\{1, 4\}, 0.1\rangle, \langle 5, 0.9\rangle$	$\langle\{70, 80\}, 0.7\rangle, \langle 100, 0.3\rangle$
o_4	$\langle 10, 0.2\rangle, \langle 12, 0.2\rangle, \langle\{13, 14\}, 0.6\rangle$	$\langle 100, 1\rangle$
o_5	$\langle\{12, 13, 14\}, 0.2\rangle, \langle 17, 0.4\rangle, \langle 19, 0.4\rangle$	$\langle\{20, 30\}, 0.6\rangle, \langle 30, 0.4\rangle$

Example 2. Consider in Table 1 a set of weight-loss products, defined over two attributes; weight loss per month and repayment (if the user is not satisfied). Each product may have one or more focal elements w.r.t. each attribute. For example, the weight loss per month of product o_1 comprises two focal elements $\langle\{15, 16, 18\}, 0.1\rangle$ and $\langle\{19, 20\}, 0.9\rangle$. That is, we believe that the attribute value is either 15, 16 or 18 with mass function 0.1 or one of the values 19 or 20 with mass 0.9. However, we do not know how credible each single element is. We use this example throughout the rest of the paper.

3   Skyline on Evidential Data

In this section, we extend the dominance relationship to evidential data, then, we introduce the notion of evidential skyline.

Given a set of objects $\mathcal{O} = \{o_1, o_2, \ldots, o_n\}$ defined on a set of attributes $\mathcal{A} = \{a_1, a_2, \ldots, a_d\}$, with $o_i.a_k$ denoting the set of focal elements of object o_i w.r.t. attribute a_k; for example[1], $o_1.wl = \{\langle\{15, 16, 18\}, 0.1\rangle\}, \langle\{19, 20\}, 0.9\rangle$ and $o_1.r = \{\langle 90, 0.3\rangle, \langle\{90, 100\}, 0.7\rangle\}$. The belief that an object o_i is better than or equal to another object o_j w.r.t. an attribute a_k is given by [4]:

$$bel(o_i.a_k \geq o_j.a_k) = \sum_{A \subseteq Dom(a_k)} (m_{jk}(A) \sum_{B \subseteq Dom(a_k),\, A \leq^\forall B} m_{ik}(B)) \quad (1)$$

Where $A \geq^\forall B$ stands for $a \geq b$ for all $(a, b) \in A \times B$. For instances, in our example, we have, $bel(o_1.wl \geq o_3.wl) = 0.3 \cdot (0.1 + 0.9) + 0.7 \cdot (0.1 + 0.9) = 1$, and $bel(o_1.r \geq o_3.r) = 0.7 \cdot 0.7 + 0.3 \cdot 0.7 = 0.7$.

Let us now extend the dominance relationship to evidential data. Given two objects o_i and o_j in $\mathcal{O} : o_i \neq o_j$, the belief that o_i dominates o_j is given by:

$$bel(o_i \succ o_j) = \prod_{a_k \in \mathcal{A}} bel(o_i.a_k \geq o_j.a_k) \quad (2)$$

For example, since $bel(o_1.wl \geq o_3.wl) = 1$ and $bel(o_1.r \geq o_3.r) = 0.7$, the belief that o_1 dominates o_3 is $bel(o_1 \succ o_3) = 1 \cdot 0.7 = 0.7$. Table 2 shows the dominance belief that each object in lines dominates another object in columns.

Table 2. Dominance beliefs

Objects	o_1	o_2	o_3	o_4	o_5
o_1	1	1	0.7	0	0.92
o_2	0	1	0.7	0	0
o_3	0	0	1	0	0
o_4	0	1	1	1	0
o_5	0	0	0	0	1

Observe that this definition (equation 2) boils down to the usual dominance relationship when the descriptions of the objects are certain, i.e., the belief is 1 if $o_i \succ o_j$ and 0 otherwise. This is because the condition $\geq$ is sufficient for an object o_i to dominate another object $o_j \neq o_i$.

Based on this relationship, we define the notion of b-dominance as follows.

Definition 7. *(b-dominance)*
Given two objects $o_i, o_j \in \mathcal{O}$ and a belief threshold b, o_i b-dominates o_j, denoted by $o_i \succ_b o_j$ if and only if $bel(o_i \succ o_j) \geq b$.

---

[1] For short, we use wl and r to denote the weight loss per month and the repayment attributes, respectively

For instance, o_1 0.9-dominates both o_2 and o_5. However, it does not 0.9-dominate o_4 since $bel(o_1 \succ o_4) = 0 < 0.9$.

We can now use this b-dominance relationship to define the notion of evidential skyline. Intuitively, an object is in the evidential skyline if it is not dominated with some threshold. Thus, we define the notion of b-dominant skyline as follows.

Definition 8. *(b-dominant skyline)*
The b-dominant skyline of $\mathcal{O}$, denoted by b-$Sky_{\mathcal{O}}$, comprises those objects in $\mathcal{O}$ that are not b-dominated by any other object, i.e., b-$Sky_{\mathcal{O}} = \{o_i \in \mathcal{O} \mid \nexists\, o_j \in \mathcal{O}, o_j \succ_b o_i\}$.

For example, the 0.4-dominant skyline comprises objects o_1 and o_4, since they are not 0.4-dominated by any other object, while the 0.2 contains only o_1 as o_4 is 0.2 dominated by o_1. From, this observation, we illustrate a key property of the b-dominant skyline.

Theorem 1. *Given two belief thresholds b and b', if $b < b'$ then the b-dominant skyline is a subset of the b'-dominant skyline, i.e., $b < b' \Rightarrow b$-$Sky_{\mathcal{O}} \subseteq b'$-$Sky_{\mathcal{O}}$.*

Proof. Assume that there exists an object o_i such that $o_i \in b$-$Sky_{\mathcal{O}}$ and $o_i \notin b'$-$Sky_{\mathcal{O}}$. Since $o_i \notin b'$-$Sky_{\mathcal{O}}$, there must exists another object, say o_j, that b'-dominates o_i. Thus, $bel(o_j \succ o_i) > b'$. But, $b < b'$. Therefore, $bel(o_j \succ o_i) > b$. Hence, $o_j \succ_b o_i$, which leads to a contradiction as $o_i \in b$-$Sky_{\mathcal{O}}$.

Theorem 1 indicates that the size of the b-dominant skyline is smaller than that of the b'-dominant skyline if $b < b'$. Roughly speaking, from Theorem 1, we can see that users have the flexibility to control the size of the retrieved evidential skyline by varying the belief threshold.

4   Computing the b-Dominant Skyline

In this section, we first propose an appropriate algorithm to compute the evidential skyline, minimizing the dominance checks. We then devise an efficient method for optimizing the dominance checks.

A straightforward algorithm to compute the b-dominant skyline is to compare each object o_i against the other objects. If o_i is not b-dominated, then it belongs to the evidential skyline. However, this approach results in a high computational cost (see Section 5) as it needs to compare each object with every others.

Also, consider the objects depicted in Table 3. We have, $bel(o_x \succ o_y) = 0.3$, $bel(o_y \succ o_z) = 0.4$ and $bel(o_x \succ o_z) = 0.035$. Observe that, o_x 0.3-dominates o_y and o_y 0.3-dominates o_z, but o_x does not 0.3-dominate o_z. Thus, the b-dominance relationship is not transitive. Therefore, an object cannot be eliminated from the comparison even if it is b-dominated since it will be useful for eliminating other objects. For this reason, we propose a two phase algorithm (see Algorithm 1) that follows the principle of the two scan algorithm [9].

Our proposed algorithm, named TPA, computes the evidential skyline through two phases. In the first phase (lines 2–13), a set of candidate objects b-$Sky_{\mathcal{O}}$ is

Table 3. Evidential data example

Product	Weight loss per month (kilograms)	Repayment (%)
o_x	$\langle\{10,11\},0.5\rangle, \langle\{19,20\},0.5\rangle$	$\langle 70,0.3\rangle, \langle\{80,90\},0.7\rangle$
o_y	$\langle\{17,18\},1\rangle$	$\langle\{60,70\},0.6\rangle, \langle 100,0.4\rangle$
o_z	$\langle\{9,12\},0.7\rangle, \langle\{15,16\},0.3\rangle$	$\langle 80,0.1\rangle, \langle\{90,100\},0.9\rangle$

selected by comparing each object o_i in $\mathcal{O}$ with those selected in $b\text{-}Sky_\mathcal{O}$. If an object o_j in $b\text{-}Sky_\mathcal{O}$ is b-dominated by o_i, then o_j is removed from the set of candidate objects since it is not part of the evidential skyline. At the end of the comparison of o_i with objects of $b\text{-}Sky_\mathcal{O}$, if o_i is not b-dominated by any object then, it is added to $b\text{-}Sky_\mathcal{O}$ as a candidate object. After the first phase, $b\text{-}Sky_\mathcal{O}$ comprises a set of objects that may be part of the b-dominated skyline.

To avoid the situation illustrated by the example in Table 3, a second phase is needed (lines 14–17). To determine if an object o_i in $b\text{-}Sky_\mathcal{O}$ is indeed in the b-dominant skyline it is sufficient to compare o_i with those in $\mathcal{O} \setminus \{b\text{-}Sky_\mathcal{O} \cup undom(o_i) \cup \{o_i\}\}$ that occurs earlier than o_i since the other ones have been already compared against o_i, where $undom(o_i)$ is the set of objects that occurs before o_i and that do not b-dominate o_j. This set is computed in the first phase in order to reduce the dominance checks in the second phase.

Applying TPA to our example (see Table 2), with $b = 0.4$, both objects o_1 and o_4 will be inserted into $b\text{-}Sky_\mathcal{O}$ in the first phase, then o_1 and o_4 are not 0.4-dominated, thus after the second phase o_1 and o_4 will be returned. However, applying TPA to the objects in Table 3, with $b = 0.3$, both o_x and o_z will be inserted in $b\text{-}Sky_\mathcal{O}$ in the first phase, but o_z will be eliminated by o_y in the second phase. Thus the algorithm returns only o_x.

Even if, TPA minimizes the number of dominance checks. It also may result in a high computational cost. In particular, when the average number of focal elements per object is large. Thus, it is crucial to optimize the dominance checks to improve the performance of TPA. In the following, we devise an efficient method that overcomes this problem using the minimum and the maximum values of each object w.r.t. each attribute. Given an object o_i, we denote by $o_i.a_k^-$ and by $o_i.a_k^+$ respectively the minimum value and the maximum value of o_i on attribute a_k. For example, $o_1.wl^- = 15$ and $o_1.wl^+ = 20$. Next, we delve into some useful properties that help us improve the dominance checks. Given two objects $o_i, o_j \in \mathcal{O}$, and an attribute $a_k \in \mathcal{A}$ it is easy to check the following properties.

Property 1. if $o_i.a_k^+ < o_j.a_k^-$ then $bel(o_i \succ o_j) = 0$ since $bel(o_i.a_k \geq o_j.a_k) = 0$.

Property 2. if $o_i.a_k^- > o_j.a_k^+$ then $bel(o_i \succ o_j) = \prod_{a_\ell \in \mathcal{A} \setminus \{a_k\}} bel(o_i.a_\ell \geq o_j.a_\ell)$ since $bel(o_i.a_k \geq o_j.a_k) = 1$.

To determine if an object o_i b-dominates another object o_j Property 1 and Property 2 show that it is not necessary to iterate the focal elements of all at-

---

Algorithm 1. TPA

Input: Objects $\mathcal{O}$; belief threshold b;
Output: Evidential skyline $b\text{-}Sky_{\mathcal{O}}$;

```
 1  begin
 2  |   foreach oi in O do
 3  |   |   isSkyline ← true;
 4  |   |   foreach oj in b-SkyO do
 5  |   |   |   if isSkyline then
 6  |   |   |   |   if oj ≻b oi then
 7  |   |   |   |   |   isSkyline ← false;
 8  |   |   |   |   else
 9  |   |   |   |   |   undom(oi) ← undom(oi) ∪ {oj};
10  |   |   |   if oi ≻b oj then
11  |   |   |   |   remove oj from b-SkyO;
12  |   |   if isSkyline then
13  |   |   |   insert oi in b-SkyO;
14  |   foreach oi in b-SkyO do
15  |   |   foreach oj in O \ (b-SkyO ∪ undom(oi) ∪ {oi}), pos(oj) < pos(oi) do
16  |   |   |   if oj ≻b oi then
17  |   |   |   |   remove oi from b-SkyO;
18  |   return b-SkyO;
19  end
```

---

tributes. Based on these properties, we propose an efficient method (Algorithm 2) that checks if a given objet o_i b-dominates or not another object o_j.

The details of the b-dominates function are as follows. For each attribute $a_k \in \mathcal{A}$, $o_i.a_k^+$ is compared against $o_j.a_k^-$. If there is any attribute a_k for which $o_i.a_k^+ < o_j.a_k^-$ holds then return false (loop in line 1); since o_i cannot b-dominate o_j according to Property 1. Otherwise, the belief that o_i dominates o_j, bel, is computed by considering only the attributes where $o_i.a_k^- \leq o_j.a_k+$ (loop in line 5); since if $o_i.a_k^- > o_j.a_k^+$ then $bel(o_i.a_k \geq o_j.a_k) = 1$ and thus it does not affect the result; see Property 2. The method returns "false" as soon as bel is less than the threshold b; since $bel \times bel(o_i.a_k \geq o_j.a_k)$ (line 7) will be less than b for the rest of iterations. Finally, if bel is greater than b, the function returns true.

For example, comparing object o_1 against object o_2 in our example. The method can quickly identify that $bel(o_1 \succ o_2) = 1$ and returns true, without iterating on the focal element of each attribute. Similarly, to compare o_2 against o_1, it directly returns false.

5   Experimental Evaluation

In this section, we present an extensive experimental evaluation of our approach. More specifically, we focus on two issues: (i) the size of the evidential skyline;

Algorithm 2. b-dominates(o_i, o_j, b)

---

1 **foreach** a_k *in* $\mathcal{A}$ **do**
2   **if** $o_i.a_k^+ < o_j.a_k^-$ **then**
3        $\lfloor$ **return** false;

4 $bel \leftarrow 1$;
5 **foreach** a_k *in* $\mathcal{A}$ **do**
6   **if** $o_i.a_k^- \leq o_j.a_k^+$ **then**
7        $bel \leftarrow bel \times bel(o_i.a_k \geq o_j.a_k)$;
8   **if** $bel < b$ **then**
9           $\lfloor$ **return** false;

10 **return** true;

---

and (ii) the scalability of our proposed techniques for computing the evidential skyline. For comparison purposes, we also implemented a baseline algorithm referred to as BA (baseline algorithm). In addition, to show the benefits resulting from the use of the dominance check function, we also implemented TPA without this function; we refer to this algorithm as BTPA (basic TPA).

5.1 Experimental Setup

Due to the limited availability of real evidential databases, existing approaches use synthetically generated datasets for their evaluation; we also follow this direction. The generation of the sets of evidential data is controlled by the parameters in Table 4, which lists the parameters under investigation, their examined and default values. In each experimental setup, we investigate the effect of one parameter, while we set the remaining ones to their default values. The data generator and the algorithms, i.e., BA, BTPA and TPA were, implemented in Java, and all experiments were conducted on a 2.3 GHz Intel Core i5 processor, with 8GB of RAM, running Mac OS X.

Table 4. Parameters and Examined Values

Parameter	Symbol	Values	Default
Number of objects	n	1K, 5K, 10K, 50K, 100K	10K
Number of attributes	d	2, 3, 4, 5, 6	4
belief threshold	b	0.001, 0.002, 0.003, 0.004, 0.005	0.003
Number of focal elements/attribute	f	8, 9, 10, 11, 12	10

5.2 Size of the Evidential Skyline

Fig. 1 shows the size (i.e., the number of objects returned) of the b-dominant skyline w.r.t. n, d, b and f. Fig. 1a shows that the size on the evidential skyline

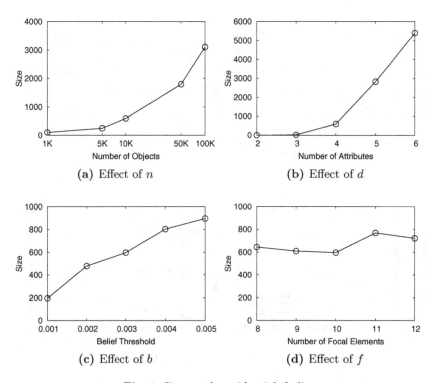

Fig. 1. Size on the evidential skyline

increases with higher n since when n increases more objects have chances not to be dominated. As shown in Fig. 1b the cardinality of the evidential skyline increases significantly with the increase of d. This is because with the increase of d an object has better opportunity to be not dominated in all attributes. Fig. 1c shows that the size of the evidential skyline increases with the increase of the b since the b-dominant skyline contains the b'-dominant skyline if $b > b'$; see Theorem 1 – recall that from this property users have the flexibility to control the size of the returned objects. In contrast to n, d, and b, f has no apparent effect on the size of the evidential skyline as shown in Fig. 1d varying f some objects have better chances not to be dominated, while other have better chances to be dominated.

5.3 Performance and Scalability

Fig. 2 depicts the execution time of the implemented algorithms w.r.t. n, d, b and f. Overall, TPA outperforms BA and BTPA. More specifically, TPA is faster than BTPA, which in turn is faster than BA. As expected, Fig. 2a shows that the performance of the algorithms deteriorates with the increase of n. Observe that TPA is one order of magnitude faster BA and BTPA since it can quickly identifies if an object is dominated or not. As shown in Fig. 2b BTPA does not

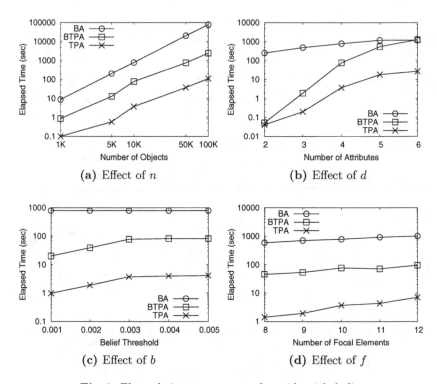

Fig. 2. Elapsed time to compute the evidential skyline

scale with d. This is because when d increases the size of the evidential skyline becomes larger, thus a large number of objects will be selected to the second phase. Hence, BTPA performs a large number of dominance checks with a basic function. Even if TPA performs the same number of dominance checks than BTPA, TPA is efficient than BTPA since it can detect immediately whether an object dominates or not another. As shown in Fig. 2c, BA is not affected by b as it computes the belief dominance between all objects. However, TPA and BTPA increases slightly because the size of the evidential skyline increases with the increase of b, thus, less objects will be eliminated. Fig. 2d shows that the execution time of the algorithm slightly with the increase of f as the size of the skyline is not very affected by f, each algorithm performs practically the same number of dominance check. Still, observe that, TPA is more than one order of magnitude faster than BA and BTPA.

6   Related Work

Our proposal can be related to the previous work on modeling uncertainty and computing skyline from uncertain data. Below we provide a brief overview on works done in both of these two topics.

Several models were proposed to represent uncertainty in Databases. The most studied and known models are : probability theory, possibility theory [15] and evidential theory [26]. Probability theory is surely the oldest theory allowing to model uncertainty. It is a quantitative model and generally used to model uncertainty due the variability of observed natural phenomena (i.e., randomness). Possibility theory which is rather a qualitative model, is used to represent uncertainty due to the lack of knowledge or missing information (i.e., incompleteness). It is less demanding than single probability distributions. Dempster-Shafer theory of evidence is used to model both imprecision and uncertainty in data. This model makes sense in the applications where the nature of information is uncertain and the complete information is often not available (for instance, in the example about conference location given in the Introduction, imprecision is present in the probability distribution between the cities). Note also that the evidential model can generalize both probabilistic and possibilistic models. For more details about uncertainty theories, the reader can refer to [16].

Since its proposal, the skyline queries have been recognized as a useful and practical technique to help users make intelligent decisions over complex data. In the years that followed the emergence of the concept of skyline queries, computing the skyline was the major concern, most of the works were about designing efficient evaluation algorithms under different conditions and in different contexts, see for instance [9] [10] [29] [13]. In the last decade, skyline computation over uncertain data has also attracted the interest of many researchers. Probabilistic skyline on uncertain data is first tackled by Pei et al [24] where skyline objects are retrieved based on skyline probabilities. This idea is also developed and improved in [18]. Efficient techniques are proposed following the bounding-pruning-refining framework. There are other studies which have adopted the probabilistic skyline model. Lian et al. [21] combine reverse skyline with uncertain semantics and study the probabilistic reverse skyline problem in both monochromatic and bichromatic fashion. Atallah and Qi [2] develop sub-quadratic algorithms to compute skyline probabilities for every object. Zhang et al. [31] tackle the problem of efficiently on-line computing probabilistic skyline over sliding windows. Yong et al. [27] studied the problem of supporting skyline queries for uncertain data with maybe confidence. This paper is the first skyline work considering maybe confidence semantics, where each uncertain tuple is associated with the probability of existence. In [30], a novel variant of skyline, called stochastic skyline, is proposed. It captures the preference of users and guarantees to provide the minimum set of candidates for the optimal solutions over all possible monotonic multiplicative utility functions.

As for uncertain data modeled by possibilistic distributions, the only work proposed in the literature is the one by Bosc et al [6]. It aims at computing the possibility that a given tuple is not dominated by any other tuple. While in this work, we addressed the problem of computing skyline from evidential databases.

7   Conclusion

In this paper, we have addressed the problem of skyline analysis on evidential data. We introduced a novel kind of skyline for this problem, and we developed an efficient algorithm for computing this skyline. Our experimental evaluation demonstrated the flexibility of the proposed evidential skyline and the scalability of our algorithm. An interesting future direction is to develop techniques for ranking the objects retrieved by the evidential skyline.

References

1. Aggarwal, C.C., Yu, P.S.: A survey of uncertain data algorithms and applications. IEEE Trans. Knowl. Data Eng. 21(5), 609–623 (2009)
2. Atallah, M.J., Qi, Y.: Computing all skyline probabilities for uncertain data. In: PODS, pp. 279–287 (2009)
3. Bach-Tobji, M.A., Ben-Yaghlane, B., Mellouli, K.: A new algorithm for mining frequent itemsets from evidential databases. In: IPMU, pp. 1535–1542 (2008)
4. Bell, D.A., Guan, J.W., Lee, S.K.: Generalized union and project operations for pooling uncertain and imprecise information. Data Knowl. Eng. 18(2), 89–117 (1996)
5. Benouaret, K., Benslimane, D., HadjAli, A.: Selecting skyline web services from uncertain qos. In: IEEE SCC, pp. 523–530 (2012)
6. Bosc, P., Hadjali, A., Pivert, O.: On possibilistic skyline queries. In: Christiansen, H., De Tré, G., Yazici, A., Zadrozny, S., Andreasen, T., Larsen, H.L. (eds.) FQAS 2011. LNCS, vol. 7022, pp. 412–423. Springer, Heidelberg (2011)
7. Bosc, P., Pivert, O.: About projection-selection-join queries addressed to possibilistic relational databases. IEEE T. Fuzzy Systems 13(1), 124–139 (2005)
8. Bosc, P., Pivert, O.: Modeling and querying uncertain relational databases: a survey of approaches based on the possible worlds semantics. International Journal of Uncertainty, Fuzziness and Knowledge-Based Systems 18(5), 565–603 (2010)
9. Chan, C.Y., Jagadish, H.V., Tan, K.L., Tung, A.K.H., Zhang, Z.: Finding k-dominant skylines in high dimensional space. In: SIGMOD Conference, pp. 503–514 (2006)
10. Chomicki, J., Godfrey, P., Gryz, J., Liang, D.: Skyline with presorting. In: ICDE, pp. 717–719 (2003)
11. Dalvi, N.N., Suciu, D.: Efficient query evaluation on probabilistic databases. VLDB J. 16(4), 523–544 (2007)
12. Dalvi, N.N., Suciu, D.: Management of probabilistic data: foundations and challenges. In: PODS, pp. 1–12 (2007)
13. Das Sarma, A., Lall, A., Nanongkai, D., Lipton, R.J., Xu, J.: Representative skylines using threshold-based preference distributions. In: Proceedings of the 2011 IEEE 27th International Conference on Data Engineering, ICDE 2011, pp. 387–398. IEEE Computer Society, Washington, DC (2011)
14. Dempster, A.P.: A generalization of bayesian inference. Journal of the Royal Statistical Society 30(B), 205–247 (1968)
15. Dubois, D., Prade, H.: Possibility Theory: An Approach to Computerized Processing of Uncertainty. Plenum Press (1988)

16. Dubois, D., Prade, H.: Formal representations of uncertainty. In: Bouyssou, D., Dubois, D., Pirlot, M., Prade, H. (eds.) Decision-Making - Concepts and Methods, ch. 3, pp. 85–156. Wiley (2009)
17. Ha-Duong, M.: Hierarchical fusion of expert opinions in the transferable belief model, application to climate sensitivity. Int. J. Approx. Reasoning 49(3), 555–574 (2008)
18. Jiang, B., Pei, J., Lin, X., Yuan, Y.: Probabilistic skylines on uncertain data: model and bounding-pruning-refining methods. J. Intell. Inf. Syst. 38(1), 1–39 (2012)
19. Lee, S.K.: An extended relational database model for uncertain and imprecise information. In: VLDB, pp. 211–220 (1992)
20. Lian, X., Chen, L.: Monochromatic and bichromatic reverse skyline search over uncertain databases. In: SIGMOD Conference, pp. 213–226 (2008)
21. Lian, X., Chen, L.: Probabilistic inverse ranking queries over uncertain data. In: Zhou, X., Yokota, H., Deng, K., Liu, Q. (eds.) DASFAA 2009. LNCS, vol. 5463, pp. 35–50. Springer, Heidelberg (2009)
22. Lim, E.P., Srivastava, J., Shekhar, S.: Resolving attribute incompatibility in database integration: An evidential reasoning approach. In: ICDE, pp. 154–163 (1994)
23. Lim, E.P., Srivastava, J., Shekhar, S.: An evidential reasoning approach to attribute value conflict resolution in database integration. IEEE Trans. Knowl. Data Eng. 8(5), 707–723 (1996)
24. Pei, J., Jiang, B., Lin, X., Yuan, Y.: Probabilistic skylines on uncertain data. In: VLDB, pp. 15–26 (2007)
25. Shafer, G.: A Mathematical Theory of Evidence. Princeton University Press, Princeton (1976)
26. Yager, R.R., Kacprzyk, J., Fedrizzi, M.: Advances in the dempster-shafer theory of evidence. John Wiley & Sons, Inc., New York (1994)
27. Yong, H., Lee, J., Kim, J., won Hwang, S.: Skyline ranking for uncertain databases. Information Systems (2014)
28. Yu, Q., Bouguettaya, A.: Computing service skyline from uncertain qows. IEEE T. Services Computing 3(1), 16–29 (2010)
29. Zhang, M., Alhajj, R.: Skyline queries with constraints: Integrating skyline and traditional query operators. Data Knowl. Eng. 69(1), 153–168 (2010)
30. Zhang, W., Lin, X., Zhang, Y., Cheema, M.A., Zhang, Q.: Stochastic skylines. ACM Trans. Database Syst. 37(2) (2012)
31. Zhang, W., Lin, X., Zhang, Y., Wang, W., Yu, J.X.: Probabilistic skyline operator over sliding windows. In: ICDE, pp. 1060–1071 (2009)

A Two-Level Approach to Maximum Entropy Model Computation for Relational Probabilistic Logic Based on Weighted Conditional Impacts

Marc Finthammer and Christoph Beierle

Dept. of Computer Science, University of Hagen, Germany

Abstract. The principle of maximum entropy allows to define the semantics of a knowledge base consisting of a set of probabilistic relational conditionals by a unique model having maximum entropy. Using the concept of a conditional structure of a world, we define the notion of weighted conditional impacts and present a two-level approach for maximum entropy model computation based on them. Once the weighted conditional impact of a knowledge base has been determined, a generalized iterative scaling algorithm is used that fully abstracts from concrete worlds. The weighted conditional impact may be reused when only the quantitative aspects of the knowledge base are changed. As a further extension of previous work, also deterministic conditionals may be present in the knowledge base, and a special treatment of such conditionals reduces the problem size.

1 Introduction

When enriching propositional logic with probabilities for modeling uncertainty (e.g. [15,18,4,9]), can play a vital role. Relational probabilistic conditionals are useful for modeling uncertain knowledge in scenarios where relations among individual objects are important. For instance, given a set of connected personal computers, stating that the probability that a malware infected PC sends a message to another PC is 0.7 while for a non-infected PC it is only 0.1, could be formally denoted by the conditionals $(sendsMail(X,Y)|infected(X))[0.7]$ and $(sendsMail(X,Y)|\neg infected(X))[0.1]$. Having a knowledge base $\mathcal{R}$ consisting of a set of such conditionals, there may be many different probability distributions satisfying them. The idea of the *principle of maximum entropy (ME)* [20,17,10,11] is to select among all models the model adding as little information as possible and thus being the most unbiased one. Recently, different approaches to applying the ME principle not only to the propositional case, but also in a relational first-order setting have been proposed [13,2]. In these approaches, ME reasoning amounts to compute the probability of a formula F under the ME model of $\mathcal{R}$, and determining the ME model of a knowledge base is the most crucial step for reasoning under ME semantics.

In this paper, we present AGGME, a system that implements the ME model computation for probabilistic relational conditionals under *aggregating ME semantics* [13] which requires solving a complex optimization problem. In [5], a

U. Straccia and A. Calì (Eds.): SUM 2014, LNAI 8720, pp. 162–175, 2014.

generalized iterative scaling (GIS) algorithm is proposed for this task. The approach implemented in AGGME refines and extends the proposal of [5] in several directions. While in [5] only non-deterministic conditionals are allowed, AGGME also treats deterministic conditionals having probability 0 or 1 which is required in many application scenarios. While [5] uses conditional structures introduced by Kern-Isberner [10] for defining equivalences of worlds, AGGME extends the use of conditional structures and introduces a two-phase ME computation. For the first phase, an algorithm WCI is developed for computing what we call the *weighted conditional impact of* $\mathcal{R}$; this algorithm is based solely on the qualitative parts of the conditionals in $\mathcal{R}$. The second phase employs a GIS algorithm $\text{GIS}_{\odot}^{\gamma\mathcal{R}}$ that fully abstracts from worlds by just using the weighted conditional impact and the probabilities given in the conditionals in $\mathcal{R}$. The modular design of AGGME allows for an easy exchange of alternative computation methods for both phases. It also supports the reuse of the weighted conditional impact of $\mathcal{R}$ for a modified knowledge base $\mathcal{R}'$ obtained from $\mathcal{R}$ by just changing the probabilities of the conditionals, a situation that is quite common when developing a knowledge base. AGGME is implemented in Java and is available as a plugin for KREATOR[1] [6], an integrated development environment for relational probabilistic logic.

After briefly recalling the basics of aggregating semantics (Sec. 2), Sec. 3 addresses the treatment of deterministic conditionals under ME semantics. In Sec. 4, weighted conditional impacts are defined and illustrated, leading to an alternative formulation of the ME optimization problem solved by the AGGME algorithms presented in Sec. 5. Some examples and first evaluation results are given in Sec. 6, and in Section 7 we conclude and point out further work.

2   Background

We consider a quantifier-free first-order language $\mathcal{L}$ over a set of predicates *Pred* and a finite set of constants *Const*. For formulas $A, B \in \mathcal{L}$, AB abbreviates the conjunction $A \wedge B$, and $\text{gnd}(A)$ denotes the set of ground instances of A. By introducing the operator $|$, we obtain the language $(\mathcal{L}|\mathcal{L})^{prob}$ of *probabilistic conditionals* of the form $(B(\boldsymbol{X})|A(\boldsymbol{X}))[d]$ with $\boldsymbol{X}$ containing the variables of the formulas A and B, and where $d \in [0,1]$ is a probability; $(B(\boldsymbol{X})|\top)[d]$ is a *probabilistic fact*. The conditional is *deterministic* iff $d = 0$ or $d = 1$; otherwise, it is *non-deterministic*. A finite set $\mathcal{R} \subseteq (\mathcal{L}|\mathcal{L})^{prob}$ is called a *knowledge base*. We always implicitly consider $\mathcal{R}$ together with the respective sets *Pred* and *Const*.

$\mathcal{H}$ denotes the *Herbrand base*, i.e. the set containing all ground atoms over *Pred* and *Const*, and $\Omega = \mathfrak{P}(\mathcal{H})$ is the set of all possible worlds (i.e. *Herbrand interpretations*), where $\mathfrak{P}$ is the power set operator. The *probabilistic interpretations* for $(\mathcal{L}|\mathcal{L})^{prob}$ are given by the set $\mathcal{P}_{\Omega}$ of all probability distributions $P : \Omega \to [0,1]$ over possible worlds. P is extended to ground formulas $A(\boldsymbol{a})$, with $A(\boldsymbol{a}) \in \text{gnd}(A(\boldsymbol{X}))$, by defining $P(A(\boldsymbol{a})) := \sum_{\omega \models A(\boldsymbol{a})} P(\omega)$. The *aggregation semantics* [13] extends P to conditionals and resembles the definition of a

---

[1] KREATOR and AGGME can be found at `http://kreator-ide.sourceforge.net/`

conditional probability by summing up the probabilities of all respective ground formulas; it defines the satisfaction relation $\models_{\odot}$ for $r = (B(\boldsymbol{X})|A(\boldsymbol{X}))[d]$ by

$$P \models_{\odot} r \quad \text{iff} \quad \frac{\displaystyle\sum_{(B(\boldsymbol{a})|A(\boldsymbol{a}))\in \text{gnd}(B(\boldsymbol{X})|A(\boldsymbol{X}))} P\left(A(\boldsymbol{a})B(\boldsymbol{a})\right)}{\displaystyle\sum_{(B(\boldsymbol{a})|A(\boldsymbol{a}))\in \text{gnd}(B(\boldsymbol{X})|A(\boldsymbol{X}))} P\left(A(\boldsymbol{a})\right)} = d \qquad (1)$$

Where $\sum_{(B(\boldsymbol{a})|A(\boldsymbol{a}))\in \text{gnd}(B(\boldsymbol{X})|A(\boldsymbol{X}))} P\left(A(\boldsymbol{a})\right) > 0$. If $P \models_{\odot} r$ holds, we say that P *satisfies* r or P *is a model of* r. P *satisfies* a set of conditionals $\mathcal{R}$ if it satisfies every element of $\mathcal{R}$, and $Mod(\mathcal{R}) := \{P \in \mathcal{P}_{\Omega} \mid P \models_{\odot} \mathcal{R}\}$. $\mathcal{R}$ is *consistent* iff $Mod(\mathcal{R}) \neq \emptyset$. The *entropy* $H(P) := -\sum_{\omega \in \Omega} P(\omega) \log P(\omega)$ of a probability distribution P measures the indifference within P. The principle of *maximum entropy* (*ME*) chooses the distribution P where $H(P)$ is maximal among all distributions satisfying $\mathcal{R}$ [17,10]. The ME model $P_{\mathcal{R}}^{*}$ for $\mathcal{R}$ based on aggregation semantics is uniquely defined [13] by the solution of the convex optimization problem

$$P_{\mathcal{R}}^{*} := \arg \max_{P \in \mathcal{P}_{\Omega} : P \models_{\odot} \mathcal{R}} H(P) \qquad (2)$$

3   Null-Worlds and Maximum Entropy

For illustrating knowledge bases with relational probabilistic conditionals and as a running example, we consider the following scenario:

Example 1 (Antivirus, $\mathcal{R}_{\text{vir}}$). Suppose we want to model some knowledge about virus infected computers (cf. Sec. 1): If an infected computer sends mail to another computer without antivirus protection, the other computer is likely to get infected (with probability 0.9). Computers with antivirus on very rarely get infected (probability 0.01). Infected computers are likely to send email to any computer (0.7), while uninfected computers do this only with probability 0.1. Moreover, we know that in our scenario to be modeled, computers do not send email to themselves. The following knowledge base $\mathcal{R}_{\text{vir}}$ represents this:

$r_1 : (infected(Y)|sendsMail(X,Y) \wedge infected(X) \wedge \neg antiVirOn(Y))[0.9]$

$r_2 : (infected(X)|antiVirOn(X))[0.01]$

$r_3 : (sendsMail(X,Y)|infected(X))[0.7]$

$r_4 : (sendsMail(X,Y)|\neg infected(X))[0.1]$

$r_5 : (sendsMail(X,X)|\top)[0.0]$

Note that r_5 is a deterministic conditional, and the presence of the deterministic conditionals prohibits applying the GIS algorithm approach of [5] directly to $\mathcal{R}_{\text{vir}}$. In the following, we will show how the restriction to nondeterministic conditionals required in [5] can be removed. For the rest of this paper, we assume

$$\mathcal{R} := \mathcal{R}^{\approx} \cup \mathcal{R}^{=}, \quad \mathcal{R}^{\approx} := \underbrace{\{r_1, \ldots, r_m\}}_{m \text{ non-deterministic}}, \quad \mathcal{R}^{=} := \underbrace{\{r_{m+1}, \ldots, r_{m+M}\}}_{M \text{ deterministic}} \qquad (3)$$

Where $\mathcal{R}$ is a consistent set consisting of m non-deterministic and M deterministic conditionals. Furthermore, let $\mathcal{R}^{=0} := \{r_i \in \mathcal{R}^= \mid d_i = 0\}$ and $\mathcal{R}^{=1} := \{r_i \in \mathcal{R}^= \mid d_i = 1\}$ denote the set of deterministic conditionals with probability 0 and 1, respectively.

For a relational conditional $r_i = (B_i(\boldsymbol{X})|A_i(\boldsymbol{X}))[d_i]$, the *counting functions* (cf. [12] and also [5]) $ver_i, fal_i : \Omega \to \mathbb{N}_0$ are given by:

$$ver_i(\omega) := \left| \left\{ (B_i(\boldsymbol{a})|A_i(\boldsymbol{a})) \in \text{gnd}(B_i(\boldsymbol{X})|A_i(\boldsymbol{X})) \mid \omega \models A_i(\boldsymbol{a})B_i(\boldsymbol{a}) \right\} \right| \quad (4)$$

$$fal_i(\omega) := \left| \left\{ (B_i(\boldsymbol{a})|A_i(\boldsymbol{a})) \in \text{gnd}(B_i(\boldsymbol{X})|A_i(\boldsymbol{X})) \mid \omega \models A_i(\boldsymbol{a})\neg B_i(\boldsymbol{a}) \right\} \right| \quad (5)$$

For a world $\omega \in \Omega$, $ver_i(\omega)$ yields the number of ground instances of the qualitative part of r_i which are *verified* by ω; and analogously, $fal_i(\omega)$ yields the number of ground instances of the qualitative part of r_i which are *falsified* by ω. In the following, when talking about a conditional, we will not distinguish explicitly the qualitative part of a conditional and the conditional and we may just drop the probability if the context is clear.

Example 2. Consider the five conditionals of $\mathcal{R}_{\text{vir}}$ from Example 1 together with the set of constants $Const = \{a, b, c\}$. Then each of the conditionals r_1, r_3, and r_4 has nine ground instances and both r_2 and r_5 have three ground instances. When abbreviating *infected* by *in*, *antiVirOn* by *an*, and *sendsMail* by *se*, these ground instances are:

$r_{1,1}: (in(a)|se(a,a) \wedge in(a) \wedge \neg an(a))$	$r_{3,1}: (se(a,a)|in(a))$	$r_{4,1}: (se(a,a)|\neg in(a))$
$r_{1,2}: (in(a)|se(b,a) \wedge in(b) \wedge \neg an(a))$	$r_{3,2}: (se(a,b)|in(a))$	$r_{4,2}: (se(a,b)|\neg in(a))$
$r_{1,3}: (in(a)|se(c,a) \wedge in(c) \wedge \neg an(a))$	$r_{3,3}: (se(a,c)|in(a))$	$r_{4,3}: (se(a,c)|\neg in(a))$
$r_{1,4}: (in(b)|se(a,b) \wedge in(a) \wedge \neg an(b))$	$r_{3,4}: (se(b,a)|in(b))$	$r_{4,4}: (se(b,a)|\neg in(b))$
$r_{1,5}: (in(b)|se(b,b) \wedge in(b) \wedge \neg an(b))$	$r_{3,5}: (se(b,b)|in(b))$	$r_{4,5}: (se(b,b)|\neg in(b))$
$r_{1,6}: (in(b)|se(c,b) \wedge in(c) \wedge \neg an(b))$	$r_{3,6}: (se(b,c)|in(b))$	$r_{4,6}: (se(b,c)|\neg in(b))$
$r_{1,7}: (in(c)|se(a,c) \wedge in(a) \wedge \neg an(c))$	$r_{3,7}: (se(c,a)|in(c))$	$r_{4,7}: (se(c,a)|\neg in(c))$
$r_{1,8}: (in(c)|se(b,c) \wedge in(b) \wedge \neg an(c))$	$r_{3,8}: (se(c,b)|in(c))$	$r_{4,8}: (se(c,b)|\neg in(c))$
$r_{1,9}: (in(c)|se(c,c) \wedge in(c) \wedge \neg an(c))$	$r_{3,9}: (se(c,c)|in(c))$	$r_{4,9}: (se(c,c)|\neg in(c))$

$r_{2,1}: (in(a)|an(a))$	$r_{5,1}: (se(a,a)|\top)$
$r_{2,2}: (in(b)|an(b))$	$r_{5,2}: (se(b,b)|\top)$
$r_{2,3}: (in(c)|an(c))$	$r_{5,3}: (se(c,c)|\top)$

For the world $\omega' = \{in(a), in(b), an(c), se(a,c), se(b,c)\}$, we have

$$\omega' \models in(a) \wedge se(a,c) \quad \text{and} \quad \omega' \models in(a) \wedge \neg se(a,b)$$

Since $in(a) \in \omega'$, $se(a,c) \in \omega'$, and $se(a,b) \notin \omega'$. Thus, ω' verifies the ground instance $r_{3,3}: (se(a,c)|in(a))$, and ω' falsifies the ground instance $r_{3,2}: (se(a,b)|in(a))$. Overall, ω' verifies two and falsifies four ground instances of r_3, i.e. $ver_3(\omega') = 2$ and $fal_3(\omega') = 4$ holds.

For characterizing the behavior of $P_{\mathcal{R}}^*$ on $\mathcal{R}^{\approx}$ and $\mathcal{R}^=$, we employ the counting functions (4) and (5).

Definition 1 (Null-Worlds and Potentially Positive Worlds). *The set*

$$\Omega_{\text{null}(\mathcal{R})} := \left\{ \omega \in \Omega \mid (\exists r_i \in \mathcal{R}^{=0} : ver_i(\omega) > 0) \vee (\exists r_i \in \mathcal{R}^{=1} : fal_i(\omega) > 0) \right\}$$

is called the set of null-worlds with respect to $\mathcal{R}$. *The set of potentially positive worlds with respect to* $\mathcal{R}$ *is given by* $\Omega_{\text{pos}(\mathcal{R})} := \Omega \setminus \Omega_{\text{null}(\mathcal{R})}$.

Extending Paris' *open-mindedness principle* [17], we can show:

Proposition 1 (Null-Worlds and ME). *If* $P \in Mod(\mathcal{R})$, *then* $P(\omega) = 0$ *for all* $\omega \in \Omega_{\text{null}(\mathcal{R})}$, *and* $P_{\mathcal{R}}^*(\omega) > 0$ *for all* $\omega \in \Omega_{\text{pos}(\mathcal{R})}$.

Thus, for the ME model $P_{\mathcal{R}}^*$ and for any null-world ω, $P_{\mathcal{R}}^*(\omega) = 0$ holds, but for every potentially positive world, $P_{\mathcal{R}}^*$ yields a non-zero probability, i.e. the worlds in $\Omega_{\text{pos}(\mathcal{R})}$ are indeed positive under $P_{\mathcal{R}}^*$.

Example 3. For $\mathcal{R}_{\text{vir}}$ from Example 2 the world

$$\omega_0' = \{an(b), in(c), se(a,a), se(b,c)\}$$

Is a null-world, because ω_0' verifies the ground instance $r_{5,1}: (se(a,a)|\top)$ of the deterministic conditional r_5, i.e. $ver_5(\omega_0') > 0$ holds. So every world containing a ground atom $se(a,a)$, $se(b,b)$, or $se(c,c)$ is a null-world due to r_5. In fact, $28,672 (= 7 \cdot 2^{12})$ of the $32,768 (= 2^{15})$ worlds contained in Ω are null-worlds, i.e. there are merely $4,096 (= 2^{12})$ potentially positive worlds.

4   Weighted Conditional Impact

For propositional conditionals, the satisfaction relation can be expressed by using feature functions (e.g. [7]). For a relational conditional $r_i = (B_i(\boldsymbol{X})|A_i(\boldsymbol{X}))[d_i]$, the *feature function* $f_i : \Omega \to \mathbb{R}$ with

$$f_i(\omega) := ver_i(\omega)(1 - d_i) - fal_i(\omega)d_i \qquad (6)$$

given in [5] uses the counting functions (4), (5). While the satisfaction relation (1) can be expressed using these feature functions by observing

$$P \models_{\odot} r_i \text{ iff } \sum_{\omega \in \Omega} P(\omega)f_i(\omega) = 0, \qquad (7)$$

Here we will transform the f_i so that they do not have to consider worlds any more.

In [10], Kern-Isberner investigates the behavior of worlds with respect to propositional conditionals and introduces the concept of *conditional structure*, formalized as a product in a free Abelian group, of a world with respect to a set of propositional conditionals. Kern-Isberner's idea of a conditional structure carries over to the relational case by employing the functions ver_i, fal_i counting the number of verified and falsified ground instances [12]. In the following, we will employ the conditional structure of a world in a relational setting and extend it to the case where also deterministic conditionals may be present. Instead of a free Abelian group notation as in [12], we will use a concrete representation using ordered tuples of pairs of natural numbers as in [5] and call these tuples *conditional impact*.

Definition 2 (Conditional Impact). *Let $\mathcal{R} = \mathcal{R}^{\approx} \cup \mathcal{R}^{=}$ be as in (3). The conditional impact caused by a world ω is given by the function*

$$\gamma_{\mathcal{R}} : \Omega_{\text{pos}(\mathcal{R})} \to (\mathbb{N}_0 \times \mathbb{N}_0)^m \quad with$$

$$\gamma_{\mathcal{R}}(\omega) := \Big((ver_1(\omega), fal_1(\omega)), \dots, (ver_m(\omega), fal_m(\omega)) \Big) \tag{8}$$

Note that the conditional impact caused by a world does neither take any deterministic conditionals nor any probabilities into account, i.e. it just considers the logical part of non-deterministic conditionals in $\mathcal{R}$.

Example 4. As in Example 2, consider again the world

$$\omega' = \{in(a), in(b), an(c), se(a, c), se(b, c)\}. \tag{9}$$

Then $\gamma_{\mathcal{R}}(\omega') = ((0, 0), (0, 1), (2, 4), (0, 3))$ holds, because ω'

- neither verifies nor falsifies any ground instances of r_1,
  i.e. $ver_1(\omega') = 0$, $fal_1(\omega') = 0$, and
- verifies none and falsifies one ground instance of r_2,
  i.e. $ver_2(\omega') = 0$, $fal_2(\omega') = 1$, and
- verifies two and falsifies four ground instances of r_3,
  i.e. $ver_3(\omega') = 2$, $fal_3(\omega') = 4$, and
- verifies none and falsifies three ground instances of r_4,
  i.e. $ver_4(\omega') = 0$, $fal_4(\omega') = 3$, .

So one can say that $\gamma_{\mathcal{R}}(\omega')$ indicates the conditional impact on ground instances caused by the world ω'. Now consider the worlds

$$\omega'' = \{in(b), in(c), an(a), se(b, a), se(c, a)\} \tag{10}$$
$$\omega''' = \{in(a), in(c), an(b), se(a, b), se(c, b)\} \tag{11}$$

Then determining the values of ver_i and fal_i for ω'' and ω''' reveals that all three worlds have the same conditional impact

$$\gamma' := ((0, 0), (0, 1), (2, 4), (0, 3)) \tag{12}$$
$$= \gamma_{\mathcal{R}}(\omega') = \gamma_{\mathcal{R}}(\omega'') = \gamma_{\mathcal{R}}(\omega''')$$

Since each of these worlds verifies and falsifies, respectively, the same number of ground instances with respect to each conditional. Note that it does not matter which concrete ground instances of a conditional are verified and falsified, but just the numbers of verified and of falsified ground instances of a conditional are relevant. For instance, each of the three worlds falsifies exactly one ground instance of r_2, i.e., ω' falsifies $(in(c)|an(c))$, ω'' falsifies $(in(a)|an(a))$, and ω''' falsifies $(in(b)|an(b))$.

We now fully abstract from worlds by introducing weighted conditional impacts obtained from the images of $\gamma_{\mathcal{R}}$ and their preimage cardinalities.

Definition 3 (Weighted Conditional Impact). *Let* $\mathcal{R} = \mathcal{R}^{\approx} \cup \mathcal{R}^{=}$ *as in (3).*

- *A tuple* $\gamma \in (\mathbb{N}_0 \times \mathbb{N}_0)^m$ *is a* conditional impact *of* $\mathcal{R}$ *iff there is a world* ω *with* $\gamma_{\mathcal{R}}(\omega) = \gamma$.
- *For such a* γ, $\mathrm{wgt}(\gamma) := |\gamma_{\mathcal{R}}^{-1}(\gamma)|$ *is the* weight *of* γ, *and* wgt *is called the* weighting function *of* $\mathcal{R}$.
- $\Gamma_{\mathcal{R}}$ *denotes the set of all conditional impacts of* $\mathcal{R}$.
- $(\Gamma_{\mathcal{R}}, \mathrm{wgt})$ *is called the* weighted conditional impact *of* $\mathcal{R}$.

Example 5. As in Example 4, consider again $\gamma' = ((0,0), (0,1), (2,4), (0,3))$ as given in (12). This tuple γ' is a conditional impact of $\mathcal{R}_{\mathrm{vir}}$, i.e. $\gamma' \in \Gamma_{\mathcal{R}}$ holds, because for instance $\gamma_{\mathcal{R}}(\omega') = \gamma'$ where ω' is as in (9).

Overall, there exist 328 different conditional impacts of $\mathcal{R}_{\mathrm{vir}}$, i.e. the set $\Gamma_{\mathcal{R}}$ has 328 elements. However, the tuple

$$((0,0), (\mathbf{3}, \mathbf{2}), (2,4), (0,3)) \in (\mathbb{N}_0 \times \mathbb{N}_0)^m \tag{13}$$

Is not in $\Gamma_{\mathcal{R}}$. In fact, (13) cannot be a conditional impact of $\mathcal{R}_{\mathrm{vir}}$, because r_2 has only 3 ground instances and therefore it is not possible that both $ver_2(\omega) = 3$ and $fal_2(w) = 2$ holds for any world $\omega \in \Omega_{\mathrm{pos}(\mathcal{R})}$. Furthermore, also the tuple

$$((0,0), (0,1), (\mathbf{1}, \mathbf{4}), (0,3)) \in (\mathbb{N}_0 \times \mathbb{N}_0)^m \tag{14}$$

Is not a conditional impact of $\mathcal{R}_{\mathrm{vir}}$ either. Here, the reason is not as obvious as it is for (13). A closer lock at the ground instances of r_3 reveals that $ver_3(\omega) + fal_3(w) \in \{0, 3, 6, 9\}$ must hold for every world $\omega \in \Omega_{\mathrm{pos}(\mathcal{R})}$, since always three ground instances of r_3 share the same antecedence, implying that $ver_3(\omega) + fal_3(w)$ must be a multiple of 3. Thus, (14) cannot be a conditional impact of $\mathcal{R}_{\mathrm{vir}}$, since it cannot be caused by any world.

When determining the conditional impacts of $\mathcal{R}_{\mathrm{vir}}$ caused by each world, it becomes evident that apart from the three worlds ω', ω'', and ω''' given in (9), (10), and (11), there is no other world $\omega \in \Omega_{\mathrm{pos}(\mathcal{R})}$ with $\gamma_{\mathcal{R}}(\omega) = \gamma'$. Therefore, $\mathrm{wgt}(\gamma') = 3$ holds, i.e. the weight of γ is 3 since there are exactly three worlds which cause the conditional impact γ'.

Figure 1 shows some conditional impacts of $\mathcal{R}_{\mathrm{vir}}$ and their weights, together with the worlds causing these impacts.

For $\gamma = ((ver_1, fal_1), \ldots, (ver_m, fal_m)) \in \Gamma_{\mathcal{R}}$ let $\gamma_{|i,v}$ denote the value ver_i and let $\gamma_{|i,f}$ denote the value fal_i. Then for $r_i \in \mathcal{R}^{\approx}$, the *feature function* $f_i^{\Gamma} : \Gamma_{\mathcal{R}} \to \mathbb{R}$ *on conditional impacts* is given by:

$$f_i^{\Gamma}(\gamma) := \gamma_{|i,v} \cdot (1 - d_i) - \gamma_{|i,f} \cdot d_i \tag{15}$$

As in [5], we get *normalized feature functions* $\hat{f}_i^{\Gamma} : \Gamma_{\mathcal{R}} \to [0,1]$ and an additional *correctional feature function* $\hat{f}_{\hat{m}}^{\Gamma} : \Gamma_{\mathcal{R}} \to [0,1]$ with $\hat{m} = m + 1$ by

$$\hat{f}_i^{\Gamma}(\gamma) := \frac{f_i^{\Gamma}(\gamma) + d_i g_i}{G} \quad \text{and} \quad \hat{f}_{\hat{m}}^{\Gamma}(\gamma) := 1 - \sum_{i=1}^{m} \hat{f}_i^{\Gamma}(\gamma) \tag{16}$$

$\gamma \in \Gamma_{\mathcal{R}}$	$\mathrm{wgt}(\gamma)$	$\omega \in \Omega_{\mathrm{pos}(\mathcal{R})}$ with $\gamma_{\mathcal{R}}(\omega) = \gamma$
$((0,0),(0,1),(2,4),(0,3))$	3	$\{in(a),\ in(b),\ an(c),\ se(a,\ c),\ se(b,\ c)\}$,
		$\{in(b),\ in(c),\ an(a),\ se(b,\ a),\ se(c,\ a)\}$,
		$\{in(a),\ in(c),\ an(b),\ se(a,\ b),\ se(c,\ b)\}$
$((0,0),(0,0),(0,9),(0,0))$	1	$\{in(a),\ in(b),\ in(c)\}$
$((0,0),(0,2),(0,3),(0,6))$	3	$\{in(a),\ an(b),\ an(c)\}$,
		$\{in(b),\ an(a),\ an(c)\}$,
		$\{in(c),\ an(a),\ an(b)\}$
$\dots$	$\dots$	$\dots$

Fig. 1. Some conditional impacts of $\mathcal{R}_{\mathrm{vir}}$ and their weights (Example 5)

where g_i denotes the number of ground instances of $r_i \in \mathcal{R}^{\approx}$ and $G := \sum_{r_i \in \mathcal{R}^{\approx}} g_i$. The expected values of these functions remain as in [5]:

$$\hat{\varepsilon}_i = \frac{d_i g_i}{G} \qquad \text{and} \qquad \hat{\varepsilon}_{\hat{m}} = 1 - \sum_{i=1}^{m} \hat{\varepsilon}_i \qquad (17)$$

Example 6. As in Example 4, consider again the conditional impact $\gamma' = ((0,0),(0,1),(2,4),(0,3))$ of $\mathcal{R}_{\mathrm{vir}}$ as given in (12). The four feature functions on conditional impacts f_1^{Γ} to f_4^{Γ} corresponding to the four probabilistic conditionals r_1 to r_4 have the following values on γ':

$$f_1^{\Gamma}(\gamma') = 0 \cdot (1 - 0.9)\ \ -0 \cdot 0.9\ =\ \ 0 \qquad f_3^{\Gamma}(\gamma') = 2 \cdot (1 - 0.7)\ -4 \cdot 0.7 = -\ 2.2$$
$$f_2^{\Gamma}(\gamma') = 0 \cdot (1 - 0.01)\ -1 \cdot 0.01 = -\ 0.01 \qquad f_4^{\Gamma}(\gamma') = 0 \cdot (1 - 0.1)\ -3 \cdot 0.1 = -\ 0.3$$

Since the conditionals have $g_1 = g_3 = g_4 = 9$ and $g_2 = 3$ ground instances, respectively, there is an overall number of $G = 30$ ground instances (cf. Example 2). Therefore, the corresponding normalized feature functions $\hat{f}_i^{\Gamma}$ have the following values on γ' and the following expected values $\hat{\varepsilon}_i$:

$$\hat{f}_1^{\Gamma}(\gamma') = (\quad 0 \quad + 0.9 \ \cdot 9) / 30 = 0.27 \qquad \text{and} \qquad \hat{\varepsilon}_1 = \frac{0.9 \cdot 9}{30} = 0.27$$
$$\hat{f}_2^{\Gamma}(\gamma') = (- 0.01 + 0.01 \ \cdot 3) / 30 = 0.000\overline{6} \qquad \text{and} \qquad \hat{\varepsilon}_2 = \frac{0.01 \cdot 3}{30} = 0.001$$
$$\hat{f}_3^{\Gamma}(\gamma') = (- 2.2 \quad + 0.7 \ \cdot 9) / 30 = 0.13\overline{6} \qquad \text{and} \qquad \hat{\varepsilon}_3 = \frac{0.7 \cdot 9}{30} = 0.21$$
$$\hat{f}_4^{\Gamma}(\gamma') = (- 0.3 \quad + 0.1 \ \cdot 9) / 30 = 0.02 \qquad \text{and} \qquad \hat{\varepsilon}_4 = \frac{0.1 \cdot 9}{30} = 0.03$$

So for the correctional feature function $\hat{f}_{\hat{m}}^{\Gamma}$ and the expected value $\hat{\varepsilon}_{\hat{m}}$ we get:

$$\hat{f}_{\hat{m}}^{\Gamma}(\gamma') = 1 - (0.27 + 0.000\overline{6} + 0.13\overline{6} + 0.02) = 0.572\overline{6}$$
$$\hat{\varepsilon}_{\hat{m}} = 1 - (0.27 + 0.001 + 0.21 + 0.03) = 0.489$$

For every $\omega', \omega'' \in \gamma_{\mathcal{R}}^{-1}(\gamma)$, we have $P_{\mathcal{R}}^*(\omega') = P_{\mathcal{R}}^*(\omega'')$ (cf. Corollary 1 in [5]). Thus, setting $P_{\mathcal{R}}^*(\gamma) := P_{\mathcal{R}}^*(\omega')$ for an arbitrary $\omega' \in \gamma_{\mathcal{R}}^{-1}(\gamma)$ yields a well-defined function $P_{\mathcal{R}}^* : \Gamma_{\mathcal{R}} \to [0,1]$. Using this function, we can express the satisfaction relation (7) with respect to the ME model $P_{\mathcal{R}}^*$ as follows:

Proposition 2 (Satisfaction for $P_\mathcal{R}^*$). *Let $\mathcal{R} = \mathcal{R}^\approx \cup \mathcal{R}^=$ be as in (3). Then for any probabilistic conditional $r_i \in \mathcal{R}^\approx$, we have*

$$P_\mathcal{R}^* \models_\odot r_i \quad \text{iff} \quad \sum_{\gamma \in \Gamma_\mathcal{R}} \hat{f}_i^\Gamma(\gamma) \cdot \mathrm{wgt}(\gamma) \cdot P_\mathcal{R}^*(\gamma) = \hat{\varepsilon}_i \qquad (18)$$

Compared to (1) and (7), the satisfaction relation (18) employs feature functions on conditional impacts. Thereby it allows us to solve the ME optimization problem induced by (2) by a two-level algorithmic approach: First, the weighted conditional impact $(\Gamma_\mathcal{R}, \mathrm{wgt})$ is determined, then a generalized iterating scaling algorithm working on $(\Gamma_\mathcal{R}, \mathrm{wgt})$ is used to determine the ME distribution $P_\mathcal{R}^*$.

In the following, we omit the index Γ of f_i^Γ and $\hat{f}_i^\Gamma$ in order to ease our notation as it will be clear from the context when we use feature functions operating on the set of conditional impacts $\Gamma_\mathcal{R}$ rather than on worlds.

5   Computing ME Models using Weighted Conditional Impacts

The algorithm WCI implemented in AGGME for computing the weighted conditional impact $(\Gamma_\mathcal{R}, \mathrm{wgt})$ of any $\mathcal{R}$ is given in Fig. 2. The algorithm starts with an empty set $\Gamma_\mathcal{R}$ in step (1). Then the elements of the set $\Gamma_\mathcal{R}$ and the values for the weighting function wgt are successively determined by performing the following steps once for each world $\omega \in \Omega$: In step (2a), the deterministic conditionals are exploited to check if ω is a null-world. If ω is a null-world, no further steps are performed on ω. Otherwise ω is a positive world and in step (2b), the conditional impact $\gamma_\mathcal{R}(\omega)$ is determined. In step (2c), $\gamma_\mathcal{R}(\omega)$ is appended to $\Gamma_\mathcal{R}$ if its not already there, and its weight $\mathrm{wgt}(\gamma_\mathcal{R}(\omega))$ is adjusted.

Having determined $(\Gamma_\mathcal{R}, \mathrm{wgt})$ for a knowledge base $\mathcal{R}$, the algorithm $\mathrm{GIS}_\odot^{\gamma_\mathcal{R}}$ given in Fig. 3 is used for the second phase of computing the ME model $P_\mathcal{R}^*$. As in [5], a generalized iterative scaling approach is used, but $\mathrm{GIS}_\odot^{\gamma_\mathcal{R}}$ fully abstracts from worlds. Instead of referring to worlds as the algorithms $\mathrm{GIS}_\odot^\alpha$ and $\mathrm{GIS}_\odot^{\equiv_\mathcal{R}}$ in [5], $\mathrm{GIS}_\odot^{\gamma_\mathcal{R}}$ performs all steps on $(\Gamma_\mathcal{R}, \mathrm{wgt})$. That way, $\mathrm{GIS}_\odot^{\gamma_\mathcal{R}}$ can also cope with deterministic conditionals which are not allowed in [5].

For any consistent set $\mathcal{R}$ of probabilistic conditionals as in Fig. 3, $\mathrm{GIS}_\odot^{\gamma_\mathcal{R}}$ computes values $\alpha_0, \alpha_1, \ldots, \alpha_{m+M}$. Based on the method of Lagrange multipliers [3], these alpha-values determine the ME model as a Gibbs distribution [8] by

$$P_\mathcal{R}^*(\omega) = \alpha_0 \prod_{i=1}^{m+M} \alpha_i^{f_i(\omega)} \qquad (19)$$

With the feature functions f_i as given in (6).

6   Examples and First Evaluation Results

We apply the two-phase ME model computation implemented in AGGME to different knowledge bases; the results are shown in Fig. 4.

Input: a set $\mathcal{R} = \{r_1, \ldots, r_m\} \cup \{r_{m+1}, \ldots, r_{m+M}\}$
of m non-deterministic and M deterministic probabilistic conditionals

Output: the weighted conditional impact $(\Gamma_{\mathcal{R}}, \mathrm{wgt})$ of $\mathcal{R}$

1. $\Gamma_{\mathcal{R}} := \emptyset$ // *initialize value*

2. for each $\omega \in \Omega$:

   (a) // *check if ω is a null-world by evaluating deterministic conditionals*

       for each $r_j = (B_j(\boldsymbol{X})|A_j(\boldsymbol{X})) \in \mathcal{R}^=$: // *for determ. cond. r_j, $m+1 \leq j \leq M$*

                                       // *consider all ground instances of r_j*

          for each $(B_j(\boldsymbol{a})|A_j(\boldsymbol{a})) \in \mathrm{gnd}(B_j(\boldsymbol{X})|A_j(\boldsymbol{X}))$:

             if $(r_j \in \mathcal{R}^{=0})$

                if $(\omega \models A_j(\boldsymbol{a})B_j(\boldsymbol{a}))$ // *if ω verifies this ground conditional*

                then **break** to step 2 // *ω is a null-world, so check is finished*

             else // *$r_j \in \mathcal{R}^{=1}$ holds*

                if $(\omega \models A_j(\boldsymbol{a})\overline{B_j(\boldsymbol{a})})$ // *if ω falsifies this ground conditional*

                then **break** to step 2 // *ω is a null-world, so check is finished*

         end loop

      end loop

      // *ω is a positive world, since no determ. cond. proved ω to be a null-world*

   (b) // *determine $\gamma_{\mathcal{R}}(\omega)$*

      $\gamma_{\mathcal{R}}(\omega) := ((0,0), \ldots, (0,0))$ // *initialize all vf-pairs with $(0,0)$*

      for each $r_i = (B_i(\boldsymbol{X})|A_i(\boldsymbol{X})) \in \mathcal{R}^{\approx}$: // *for non-determ. cond. r_i, $1 \leq i \leq m$*

                                       // *consider all ground instances of r_i*

         for each $(B_i(\boldsymbol{a})|A_i(\boldsymbol{a})) \in \mathrm{gnd}(B_i(\boldsymbol{X})|A_i(\boldsymbol{X}))$:

           if $(\omega \models A_i(\boldsymbol{a})B_i(\boldsymbol{a}))$ // *if ω verifies this ground conditional*

           then $\gamma_{\mathcal{R}}(\omega)_{|i,\mathrm{v}} := \gamma_{\mathcal{R}}(\omega)_{|i,\mathrm{v}} + 1$ // *then increment verify count*

           if $(\omega \models A_i(\boldsymbol{a})\overline{B_i(\boldsymbol{a})})$ // *if ω falsifies this ground conditional*

           then $\gamma_{\mathcal{R}}(\omega)_{|i,\mathrm{f}} := \gamma_{\mathcal{R}}(\omega)_{|i,\mathrm{f}} + 1$ // *then increment falsify count*

        end loop

      end loop

   (c) // *check if value $\gamma_{\mathcal{R}}(\omega)$ is already contained in $\Gamma_{\mathcal{R}}$*

      if $\gamma_{\mathcal{R}}(\omega) \in \Gamma_{\mathcal{R}}$

      then $\mathrm{wgt}(\gamma_{\mathcal{R}}(\omega)) := \mathrm{wgt}(\gamma_{\mathcal{R}}(\omega)) + 1$ // *increment cardinality of $\gamma_{\mathcal{R}}(\omega)$*

      else

          $\Gamma_{\mathcal{R}} := \Gamma_{\mathcal{R}} \cup \{\gamma_{\mathcal{R}}(\omega)\}$ // *add new value $\gamma_{\mathcal{R}}(\omega)$ to $\Gamma_{\mathcal{R}}$*

         $\mathrm{wgt}(\gamma_{\mathcal{R}}(\omega)) := 1$ // *initialize cardinality of $\gamma_{\mathcal{R}}(\omega)$ with 1*

  end loop

Fig. 2. Algorithm WCI computing the weighted conditional impact of $\mathcal{R}$

Input: - a consistent set $\mathcal{R} = \{r_1, \ldots, r_m\} \cup \{r_{m+1}, \ldots, r_{m+M}\}$ of
m non-deterministic and M deterministic probabilistic conditionals
- the weighted conditional impact $(\Gamma_{\mathcal{R}}, \text{wgt})$ of $\mathcal{R}$

Output: - alpha-values $\alpha_0, \alpha_1, \ldots, \alpha_{m+M}$ determining the ME-distribution $P_{\mathcal{R}}^*$

1. for each $1 \leq i \leq \hat{m}$: $\hat{\alpha}_{(0),i} := 1$ // *initialize normalized $\hat{\alpha}$-values*

2. for each $\gamma \in \Gamma_{\mathcal{R}}$: $P_{(0)}(\gamma) := \dfrac{1}{\sum_{\gamma' \in \Gamma_{\mathcal{R}}} \text{wgt}(\gamma')}$ // *initial. to uniform probabilities*

3. $k := 0$ // *initialize iteration counter*

4. repeat until an abortion condition holds:

   (a) $k := k + 1$ // *increment iteration counter k*

   (b) for each $1 \leq i \leq \hat{m}$:   // *determine current scaling factors $\beta_{(k),i}$*
   $$\beta_{(k),i} := \frac{\hat{\varepsilon}_i}{\sum_{\gamma \in \Gamma_{\mathcal{R}}} \text{wgt}(\gamma) P_{(k-1)}(\gamma) \hat{f}_i(\gamma)}$$

   (c) for each $\gamma \in \Gamma_{\mathcal{R}}$:   // *scale all probabilities $P'_{(k)}(\gamma)$*
   $$P'_{(k)}(\gamma) := P_{(k-1)}(\gamma) \prod_{i=1}^{\hat{m}} \left(\beta_{(k),i}\right)^{\hat{f}_i(\gamma)}$$

   (d) for each $\hat{\alpha}_{(k),i}, 1 \leq i \leq \hat{m}$:   // *scale all $\hat{\alpha}$-values $\hat{\alpha}_{(k),i}$*
   $$\hat{\alpha}_{(k),i} := \hat{\alpha}_{(k-1),i} \cdot \beta_{(k),i}$$

   (e) for each $\gamma \in \Gamma_{\mathcal{R}}$:   // *normalize all probabilities $P_{(k)}(\gamma)$*
   $$P_{(k)}(\gamma) := \frac{P'_{(k)}(\gamma)}{\sum_{\gamma' \in \Gamma_{\mathcal{R}}} \text{wgt}(\gamma') P'_{(k)}(\gamma)}$$

   end loop

5. for each $1 \leq i \leq \hat{m}$: $\hat{\alpha}_i := \hat{\alpha}_{(k),i}$ // *define final $\hat{\alpha}$-values*
   $$\hat{\alpha}_0 := \left(\sum_{\gamma \in \Gamma_{\mathcal{R}}} \text{wgt}(\gamma) \prod_{i=1}^{\hat{m}} \hat{\alpha}_i^{\hat{f}_i(\gamma)}\right)^{-1}$$ // *define $\hat{\alpha}_0$-value*

6. for each $1 \leq i \leq m$: $\alpha_i := \left(\frac{\hat{\alpha}_i}{\hat{\alpha}_{\hat{m}}}\right)^{\frac{1}{G}}$ // *define α-values for $\mathcal{R}^{\approx}$*
   for each $1 \leq j \leq M$: $\alpha_{m+j} := 0$ // *define α-values for $\mathcal{R}^=$*
   $\alpha_0 := \hat{\alpha}_0 \hat{\alpha}_{\hat{m}} \prod_{i=1}^{m} \alpha_i^{d_i g_i}$ // *define α_0-value*

Fig. 3. Algorithm $\text{GIS}_{\odot}^{\gamma_{\mathcal{R}}}$ for aggregation semantics operating on $(\Gamma_{\mathcal{R}}, \text{wgt})$

Example 7 ($\mathcal{R}_{\text{vir}}$ (cont.)). When considering $\mathcal{R}_{\text{vir}}$ from Example 1 together with five constants, the size of Ω is 2^{35} and even $\Omega_{\text{pos}(\mathcal{R})}$ still contains 2^{30} worlds. Since the method given in [5] for computing the ME model requires to keep all worlds in memory, it cannot be applied to that example due to memory limitations. However, the algorithms WCI and $\text{GIS}_{\odot}^{\gamma_{\mathcal{R}}}$ can easily cope with the example, since they just require to keep 18,720 conditional impacts in memory. Fig. 4 also illustrates that for increasing sizes of Ω, the computation of the weighted conditional impact becomes the dominating part in the overall computation time.

KB	$\|Const\|$	Ω	Size of $\Omega_{pos(\mathcal{R})}$	$\Gamma_{\mathcal{R}}$	Iteration Steps	Computation Time WCI	$\text{GIS}_{\odot}^{\gamma_{\mathcal{R}}}$
$\mathcal{R}_{mky}$	4	2^{20}	$6,561 \approx 2^{12}$	156	6,721	< 1 sec	< 1 sec
$\mathcal{R}_{mky}$	5	2^{30}	$1,419,857 \approx 2^{20}$	530	7,912	6 min 41 sec	1 sec
$\mathcal{R}_{cty}$	3+4	2^{21}	$1,404,928 \approx 2^{20}$	992	4,228	5 sec	1 sec
$\mathcal{R}_{cty}$	4+4	2^{28}	$157,351,936 \approx 2^{27}$	3,601	4,947	9 min 41 sec	3 sec
$\mathcal{R}_{vir}$	4	2^{24}	$1,048,576 = 2^{20}$	2,742	5,730	13 sec	4 sec
$\mathcal{R}_{vir}$	5	2^{35}	$1,073,741,824 = 2^{30}$	18,720	4,088	4 h 36 min	15 sec

Fig. 4. Results for example knowledge bases (GIS accuracy threshold: $\delta_\beta = 0.001$)

Example 8 (Monkeys, $\mathcal{R}_{mky}$). Suppose we have a zoo with a population of monkeys exhibiting a peculiar feeding behavior. The predicate *feeds*(X, Y) expresses that a monkey X feeds another monkey Y and *hungry*(X) says that a monkey X is hungry. $\mathcal{R}_{mky}$ contains the following conditionals:

$$r_1 : (feeds(X, Y) \mid \neg hungry(X) \land hungry(Y)) \; [0.80]$$
$$r_2 : (\neg feeds(X, Y) \mid hungry(X)) \; [1.0]$$
$$r_3 : (\neg feeds(X, Y) \mid \neg hungry(X) \land \neg hungry(Y)) \; [0.90]$$
$$r_4 : (feeds(X, charly) \mid \neg hungry(X)) \; [0.95]$$
$$r_5 : (feeds(X, X) \mid \top) \; [0.0]$$

r_1 states that is very likely that a not-hungry monkey feeds a hungry monkey. r_2 expresses the certain knowledge that a hungry monkey never feeds another one. r_3 says that it is very probable that a not-hungry monkey is not fed by another one. r_4 makes a statement about an individual monkey: it is most probable that if a monkey is not hungry, he feeds the monkey Charly, i. e. albeit Charly is hungry or not (perhaps because Charly is an underfed baby monkey suffering from an eating disorder). Thus, r_4 describes a special case for Charly, because according to r_3, one would have suspected that the feeding of Charly (by a not-hungry monkey) depends on whether Charly is hungry or not. r_5 expresses that a monkey does not feed itself.

Example 9 (European Cities, $\mathcal{R}_{cty}$). This example makes use of *typed* constants and predicates; there are a certain number of constants of type *Person*, and the constants *london, paris, rome*, and *vienna* are of type *EuropeanCity*. The predicate *visitsEUcity*(P, E) expresses that a person P visits a European city E. The predicates *likesSightseeing*(P), *livesInEurope*(P), and *likesChurches*(P) express that a person P likes sightseeing, lives in Europe, and likes churches, respectively. The set $\mathcal{R}_{cty}$ contains four conditionals:

$$r_1 : (visitsEUcity(P, C) \mid \top) \; [0.1]$$
$$r_2 : (visitsEUcity(P, C) \mid likesSightseeing(P)) \; [0.3]$$
$$r_3 : (visitsEUcity(P, C) \mid livesInEurope(P)) \; [0.6]$$
$$r_4 : (visitsEUcity(P, rome) \mid likesChurches(P) \land likesSightseeing(P)) \; [1.0]$$

Looking at Examples 7–9 and the corresponding numbers in Fig. 4, we would like to point out the following aspects. By allowing deterministic conditionals, there is no more need to approximate probabilities 0 or 1 as a workaround. For instance, if the probabilities of r_2 and r_5 in $\mathcal{R}_{mky}$ were approximated by 0.999 and 0.001, respectively, then $\Omega_{pos(\mathcal{R})} = \Omega$ would hold and, in case of five constants, more than a billion worlds would have to be processed in the expensive step 2b of algorithm WCI. Furthermore, the size of $\Gamma_{\mathcal{R}}$ would increase significantly as well, increasing the runtime of $\text{GIS}_\odot^{\gamma_\mathcal{R}}$.

Using weighted conditional impacts $(\Gamma_{\mathcal{R}}, \text{wgt})$ and pre-computing them reduces the overall computation time. For instance, computing the ME model for $\mathcal{R}_{cty}$ with 3 constants of type *Person* and 4 constants of type *EuropeanCity* by a straightforward implementation of a GIS algorithm on $\Omega_{pos(\mathcal{R})}$, requires over 17 min., compared to just 6 sec. overall for WCI and $\text{GIS}_\odot^{\gamma_\mathcal{R}}$ as shown in Fig. 4.

Another important benefit of pre-computing $(\Gamma_{\mathcal{R}}, \text{wgt})$ is that it can be reused if the probabilities of some conditionals of $\mathcal{R}^\approx$ are modified since $(\Gamma_{\mathcal{R}}, \text{wgt})$ only depends on the logical part of $\mathcal{R}^\approx$. Since $|\Gamma_{\mathcal{R}}|$ is much smaller than $|\Omega_{pos(\mathcal{R})}|$, working with $(\Gamma_{\mathcal{R}}, \text{wgt})$ also reduces the memory requirements for the ME model computation significantly. In fact, while the algorithm from [5] has a memory requirement of $O(|\Omega_{pos(\mathcal{R})}|)$, preventing it to handle some of the examples given in Fig. 4, the memory requirements during all phases of the ME computation in AGGME are limited by $O(|\Gamma_{\mathcal{R}}|)$.

As pointed out in Ex. 7, the numbers in Fig. 4 illustrate that increasing the size of *Const* and thus the size of Ω is the limiting factor for ME model computation in the current AGGME version. An advantage of the two-level approach is that WCI can be replaced by another algorithm computing the weighted conditional impact of $\mathcal{R}$ more efficiently without having to change $\text{GIS}_\odot^{\gamma_\mathcal{R}}$.

7   Conclusions and Further Work

For knowledge bases $\mathcal{R}$ with probabilistic relational conditionals, we presented a two-level approach for computing the ME model $P_\mathcal{R}^*$ under aggregation semantics, thereby improving on previous work. While our approach can handle larger examples and also deterministic conditionals, it is desirable to develop alternative methods for computing the weighted conditional impact of $\mathcal{R}$ without having to enumerate all possible worlds as in step (2.) of the WCI algorithm. Therefore, we are currently working on employing a combinatorial approach to construct $(\Gamma_{\mathcal{R}}, \text{wgt})$ directly, without considering worlds explicitly. That way, the exponential blow-up in Ω could be circumvented when computing $(\Gamma_{\mathcal{R}}, \text{wgt})$, allowing to handle domains with significantly more constants. We are also investigating which alternative algorithms could be employed to solve the ME optimization problem on $(\Gamma_{\mathcal{R}}, \text{wgt})$. For instance, instead of using a generalized iterative scaling approach as in our $\text{GIS}_\odot^{\gamma_\mathcal{R}}$ algorithm, an alternative approach like L-BFGS [21] could be considered.

Furthermore, we will exploit the concept of weighted conditional impacts for actual ME inference, i.e. for determining the probability of an arbitrary conditional under the ME model $P_\mathcal{R}^*$. To accomplish that, a technique should be developed which operates on the impact of an arbitrary conditional, i.e. the actual query, and the already determined weighted conditional impact of $\mathcal{R}$; for this, methods of lifted inference [19,14] might be applicable.

References

1. Adams, E.: The Logic of Conditionals. D. Reidel, Dordrecht (1975)
2. Beierle, C., Finthammer, M., Kern-Isberner, G., Thimm, M.: Evaluation and comparison criteria for approaches to probabilistic relational knowledge representation. In: Bach, J., Edelkamp, S. (eds.) KI 2011. LNCS, vol. 7006, pp. 63–74. Springer, Heidelberg (2011)
3. Boyd, S., Vandenberghe, L.: Convex Optimization. Cambridge University Press, New York (2004)
4. Fagin, R., Halpern, J.Y.: Reasoning about knowledge and probability. J. ACM 41(2), 340–367 (1994)
5. Finthammer, M., Beierle, C.: Using equivalences of worlds for aggregation semantics of relational conditionals. In: Glimm, B., Krüger, A. (eds.) KI 2012. LNCS, vol. 7526, pp. 49–60. Springer, Heidelberg (2012)
6. Finthammer, M., Thimm, M.: An integrated development environment for probabilistic relational reasoning. Logic Journal of the IGPL 20(5), 831–871 (2012)
7. Fisseler, J.: Learning and Modeling with Probabilistic Conditional Logic, Dissertations in Artificial Intelligence, vol. 328. IOS Press, Amsterdam (2010)
8. Geman, S., Geman, D.: Stochastic relaxation, Gibbs distributions, and the Bayesian restoration of images. IEEE Transactions on Pattern Analysis and Machine Intelligence 6, 721–741 (1984)
9. Halpern, J.: Reasoning About Uncertainty. MIT Press (2005)
10. Kern-Isberner, G.: Conditionals in Nonmonotonic Reasoning and Belief Revision. LNCS (LNAI), vol. 2087, p. 27. Springer, Heidelberg (2001)
11. Kern-Isberner, G., Lukasiewicz, T.: Combining probabilistic logic programming with the power of maximum entropy. Artif. Intell. 157(1-2), 139–202 (2004)
12. Kern-Isberner, G., Thimm, M.: A ranking semantics for first-order conditionals. In: ECAI 2012, pp. 456–461. IOS Press (2012)
13. Kern-Isberner, G., Thimm, M.: Novel semantical approaches to relational probabilistic conditionals. In: Proc. of KR 2010, pp. 382–392. AAAI Press (May 2010)
14. Milch, B., Zettlemoyer, L., Kersting, K., Haimes, M., Kaelbling, L.P.: Lifted probabilistic inference with counting formulas. In: AAAI 2008, pp. 1062–1068. AAAI Press (2008)
15. Nilsson, N.: Probabilistic logic. Artificial Intelligence 28, 71–87 (1986)
16. Nute, D., Cross, C.: Conditional logic. In: Gabbay, D., Guenther, F. (eds.) Handbook of Philosophical Logic, 2nd edn., vol. 4, pp. 1–98. Kluwer Academic Publishers (2002)
17. Paris, J.: The uncertain reasoner's companion – A mathematical perspective. Cambridge University Press (1994)
18. Pearl, J.: Probabilistic Reasoning in Intelligent Systems. Morgan Kaufmann, San Mateo (1988)
19. de Salvo Braz, R., Amir, E., Roth, D.: Lifted first-order probabilistic inference. In: IJCAI 2005, pp. 1319–1325. Professional Book Center (2005)
20. Shore, J., Johnson, R.: Axiomatic derivation of the principle of maximum entropy and the principle of minimum cross-entropy. IEEE Transactions on Information Theory IT-26, 26–37 (1980)
21. Zhu, C., Byrd, R.H., Lu, P., Nocedal, J.: Algorithm 778: L-BFGS-B: Fortran subroutines for large-scale bound-constrained optimization. ACM Trans. Math. Softw. 23(4), 550–560 (1997), http://doi.acm.org/10.1145/279232.279236

Solving Hidden-Semi-Markov-Mode Markov Decision Problems

Emmanuel Hadoux, Aurélie Beynier, and Paul Weng

Sorbonne Universités, UPMC Univ Paris 06, UMR 7606, LIP6, Paris, France
firstname.surname@lip6.fr

Abstract. Hidden-Mode Markov Decision Processes (HM-MDPs) were proposed to represent sequential decision-making problems in non-stationary environments that evolve according to a Markov chain. We introduce in this paper Hidden-Semi-Markov-Mode Markov Decision Processes (HS3MDPs), a generalization of HM-MDPs to the more realistic case of non-stationary environments evolving according to a semi-Markov chain. Like HM-MDPs, HS3MDPs form a subclass of Partially Observable Markov Decision Processes. Therefore, large instances of HS3MDPs (and HM-MDPs) can be solved using an online algorithm, the Partially Observable Monte Carlo Planning (POMCP) algorithm, based on Monte Carlo Tree Search exploiting particle filters for belief state approximation. We propose a first adaptation of POMCP to solve HS3MDPs more efficiently by exploiting their structure. Our empirical results show that the first adapted POMCP reaches higher cumulative rewards than the original algorithm. However, in larger instances, POMCP may run out of particles. To solve this issue, we propose a second adaptation of POMCP, replacing particle filters by exact representations of beliefs. Our empirical results indicate that this new version reaches high cumulative rewards faster than the former adapted POMCP and still remains efficient even for large problems.

1 Introduction

Markov Decision Processes (MDPs) provide a general formal framework for sequential decision-making under uncertainty. They have proved to be powerful for solving many planning problems [14]. However, MDPs run under the assumption that the environment is stationary, i.e., the transition function and/or the reward function do not evolve through time. In many real-world applications, this assumption does not hold and the sources of non-stationarity are diverse. For instance, the environment may change due to external events. In finance, when investing on the stock market, a financial crisis changes the dynamics of stock prices. Another example of non-stationary environment concerns multi-agent systems. Indeed, from the viewpoint of one agent, a change of behavior (e.g., due to learning) of another agent may affect the environment of the first agent.

U. Straccia and A. Calì (Eds.): SUM 2014, LNAI 8720, pp. 176–189, 2014.

Planning in a non-stationary environment is a difficult problem to tackle in the general case. We focus instead on a subclass of problems where non-stationary environments evolve according to a small number of non-observable modes, which are modeled as MDPs and represent different possible dynamics and rewards of that environment. An example of problem belonging to this subclass is that of elevator control [6] where the environment can typically be in three modes: morning rush-hour, late-afternoon rush-hour and non-rush-hour. Planning in such non-stationary environments has already been studied in the MDP framework [8] and in the reinforcement learning framework [7,9,15]. In all those works, non-stationary environments are represented with multiple modes. The model of Hidden-Mode Markov Decision Processes (HM-MDPs) proposed by Choi et al.[8] formalizes this idea. HM-MDPs constitute a subclass of Partially Observable MDPs. In HM-MDPs, the environmental changes are described by a Markov chain and thus occur at each decision step. However, we argue that this assumption is not always realistic. Indeed, in the elevator problem for instance, allowing, even with a small probability, the environment to be able to change between different rush modes at every move of the elevator is debatable.

In this paper, we propose a natural extension of HM-MDPs, called Hidden-Semi-Markov-Mode Markov Decision Process (HS3MDP), where the non-stationary environment evolves according to a semi-Markov chain. This new model is to hidden semi-Markov models [17] what HM-MDPs are to hidden Makov models. In HS3MDPs, when the environment stochastically changes to a new mode, it stays in that mode during a stochastically drawn duration. While HM-MDPs assume that environmental changes follow a geometric law, this assumption is relaxed in HS3MDPs.

In order to solve large-sized HS3MDPs, we exploit the Partially Observable Monte Carlo Planning (POMCP) algorithm [16], an online algorithm proposed for approximately solving POMDPs, based on Monte Carlo Tree Search and particle filters for belief state approximation. We present two improvements of POMCP for solving HS3MDPs more efficiently. The first adaptation exploits the special structure of HS3MDPs and the second furthermore represents belief states exactly instead of using particle filters. Finally, we experimentally validate those algorithms showing their effectiveness on a diverse range of domains.

In Sect. 2, we recall the necessary notations and definitions. Then, in Sect. 3, we introduce our new model. In Sect. 4, we present two adapted algorithms for solving HS3MDPs. Experimental results are presented in Sect. 5. Finally, we conclude in Sect. 6.

2   Background

Markov Decision Process. A *Markov Decision Process* (MDP) [14] is defined by $\langle \mathbf{S}, \mathbf{A}, T, R \rangle$ where $\mathbf{S}$ is a finite set of states, $\mathbf{A}$ is a finite set of actions, $T(s, a, s')$ is the probability of reaching state s' from s after executing action a and $R(s, a) \in \mathbb{R}$ is the immediate reward obtained after performing action a in s. A *policy* π is a sequence $(\delta_0, \delta_1, \dots, \delta_t, \dots)$ of *decision rules* such as each

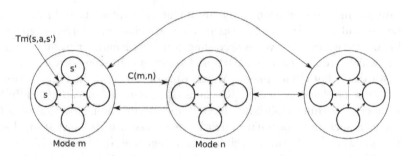

Fig. 1. HM-MDP representation with 3 modes and 4 states

decision rule $\delta_t : \mathbf{S} \rightarrow \mathbf{A}$ dictates which action to take for each state at timestep t. In a state s, a policy π can be valued by the expected discounted total reward it yields:

$$V^\pi(s) = E_\pi \left(\sum_t \gamma^t R(S_t, A_t) \,|\, S_0 = s \right) . \tag{1}$$

where $\gamma \in [0, 1[$ is a discount factor. Function V^π is called the *value function* of π. Solving an MDP consists in finding an *optimal policy*, i.e., a policy that maximizes the expected discounted sum of rewards. One of the main limitations of the standard MDP framework is that it requires the transition and reward functions to be stationary.

Hidden-Mode MDP. *Hidden-Mode MDPs* (HM-MDPs) formalize a subclass of non-stationary problems where environmental changes are limited to a fixed and known number n of modes. Each mode represents a possible stationary environment, formalized as an MDP. Transitions between modes represent environmental changes. Formally, an HM-MDP is defined as follows [8]. For $i \in \{1, \ldots, n\}$, let $m_i = \langle \mathbf{S}, \mathbf{A}, T_i, R_i \rangle$ be a mode, i.e., an MDP. An HM-MDP is characterized by $\langle \mathbf{M}, C \rangle$ where $\mathbf{M} = \{m_1, \ldots, m_n\}$ and $C : \mathbf{M} \times \mathbf{M} \rightarrow [0, 1]$ is a transition function over modes. Notice that $\mathbf{S}$ and $\mathbf{A}$ are shared by all m_i's and that an HM-MDP with $n = 1$ is a standard MDP. In HM-MDPs, the only observable information is the current state $s \in \mathbf{S}$, the current mode $m \in \mathbf{M}$ is not observable. Figure 1, showing a 3-mode, 4-state HM-MDP, depicts how HM-MDPs can be visualized. To illustrate further the definition of an HM-MDP, we present a simple example:

Example 1. The elevator problem consists in controlling e elevators in a f-floor building. At each decision step, a user may call an elevator at any floor and, once inside, select any desired floor to go. The number of states is then $2^{f(e+1)} \times f^e$. The modes are the different types of rush-hour and therefore have an influence on which buttons are pressed, which is described by different transition functions over states. Three actions can be applied to each elevator: go up/down by one floor and open the doors, leading to an action set of size 3^e. Finally, in this problem, the reward function is identical for all modes, the controller receives a penalty for each unsatisfied user.

Considering an office building of 2 floors with 1 elevator: $\mathbf{M} = \{$morning rush-hour, late-afternoon rush-hour, non-rush-hour$\}$, $\mathbf{S} = \{1^{st}$ floor call button states$\} \times \{2^{nd}$ floor call button states$\} \times \{1^{st}$ floor drop-off button states$\} \times \{2^{nd}$ floor drop-off button states$\} \times \{$elevator positions$\}$, $\mathbf{A} = \{$open, up, down$\}$. In this small example, there are 32 states, 3 actions and 3 modes. The transition function in the morning rush-hour mode describes the situation where it is more probable for the elevator to be called at the first floor. In the late-afternoon rush-hour mode, it describes the opposite situation where users tend to leave the office. For the non-rush-hour mode, the transition function models the normal operating situation.

Partially Observable MDP. *Partially Observable MDPs* (POMDPs) extends MDPs to partially observable settings [14] and are defined by $\langle \mathcal{S}, \mathcal{A}, \mathcal{O}, \mathcal{T}, \mathcal{Q}, \mathcal{R} \rangle$ where $\mathcal{S}$ is a set of POMDP states, $\mathcal{A}$ a set of actions, $\mathcal{O}$ a set of observations, $\mathcal{T} : \mathcal{S} \times \mathcal{A} \times \mathcal{S} \to \mathbb{R}$ is a transition function over POMDP states, $\mathcal{Q} : \mathcal{S} \times \mathcal{O} \to \mathbb{R}$ is a probability distribution over observations and $\mathcal{R} : \mathcal{S} \times \mathcal{A} \to \mathbb{R}$ is a reward function. Since the agent does not observe the POMDP state, she has to act based on her only available information (i.e., at step t, the probability distribution over the initial states and her history of observations and actions up to the current step t) which can be represented as a probability distribution over states, called *belief state* [2]. Optimal algorithms have been proposed to solve POMDPs [10,3], but they do not scale to large-sized problems. Indeed, finding an optimal policy for infinite-horizon POMDPs is PSPACE-Complete [13].

Choi et al. have shown that an HM-MDP can be seen as a POMDP $\langle \mathcal{S}, \mathcal{A}, \mathcal{O}, \mathcal{T}, \mathcal{Q}, \mathcal{R} \rangle$ where $\mathcal{S} = \mathbf{M} \times \mathbf{S}$, $\mathcal{A} = \mathbf{A}$, $\mathcal{O} = \mathbf{S}$, $\mathcal{T}(\langle m, s \rangle, a, \langle m', s' \rangle) = T_m(s, a, s') \times C(m, m')$, $\mathcal{Q}(\langle m, s \rangle, a, o) = 1$ if $s = o$ and 0 otherwise, $\mathcal{R}(\langle m, s \rangle, a) = R_m(s, a)$. They have also proposed algorithms to optimally solve HM-MDPs [6,8]. They have adapted exact POMDP solving methods in order to exploit the structure of HM-MDPs. Those adapted methods can solve larger instances of HM-MDPs than the original ones, but they also suffer from the curse of dimensionality. Indeed, solving an HM-MDP is still PSPACE-Complete [5]. Like exact POMDP solving algorithms, exact HM-MDP solving algorithms does not scale. In that case, one has to resort to approximate algorithms like POMCP.

POMCP. The *Partially Observable Monte-Carlo Planning* (POMCP) algorithm [16] is one of the most efficient online algorithms to approximately solve large-sized POMDPs. To choose an action at a given timestep, POMCP (Alg. 1) runs an effective version of Monte-Carlo Tree Search (MCTS), called UCT (Upper Confidence Bounds (UCB) applied to Trees) [11], using a black-box simulator of the environment and a particle filter to approximate a belief state. This search tree is built iteratively. POMCP uses the simulator to run a fixed number of simulations in order to evaluate the actions before performing in the real environment the best action found in the search tree. At one decision step, to choose which action to perform, SEARCH(τ) is invoked with τ the current history, i.e., the sequence of past observations and actions. This history can be expanded with an action a giving τa and an observation o giving τao. The root of the search tree is a node matching the last seen observation. Its children are all possible actions, whose own children are the respective possible observations given an action. A node of the tree is a triplet $\langle N(\tau), V(\tau), B(\tau) \rangle$ associated to τ where the components are respectively the number of times τ has been visited, its mean value and the set of particles (i.e., POMDP states) for this history. During a simulation, the algorithm randomly draws a particle p from the particle set $B(\tau)$ and uses the simulator $\mathcal{G}(p, a)$ to get the new particle p', the observation o and the reward r. Actions are selected (Line 19 of Alg. 1) following the UCB1 procedure guaranteeing a good exploration-exploitation compromise. Once all simulations have been done, a step is performed in the real environment with the action returned by SEARCH, i.e., the best action found in the search tree. The algorithm sets the new root to the node matching this observation and prunes the tree.

At the beginning, POMCP is initialized with an empty history and an initial (e.g., uniform) distribution $\mathcal{I}$ over states. Two important parameters have to be set to guarantee that a good action is selected: the tree depth and the number of simulations. The tree depth d can be deduced from the discount factor γ for a given precision $\epsilon > 0$ as follows: $d = \lfloor \log(\epsilon) / \log(\gamma) \rfloor$. The higher the number of simulations, the better the estimation of the values of the actions but the longer it takes to run. This parameter is generally determined by time constraints. However, as the number of simulations tends to infinity, this algorithm is theoretically guaranteed to choose the optimal action at each step. Finally, notice that the size of the initial particle filter is generally set in function of the number of simulations.

3   HS3MDP

The HM-MDP framework is not always the most suitable model for representing sequential decision-making in non-stationary environments as it assumes that the environment may change at every timestep. For instance, modeling the elevator problem with an HM-MDP is problematic as decisions have to be made every (say) second, while a mode (rush hour or not) can last several hours. In a problem where this assumption does not hold, the usual modeling trick is to set a low probability of transition between modes. However, from a theoretical viewpoint, this is more than questionable when mode transitions are not geometrically distributed. One of the main contributions of this paper is to propose a more natural model for such cases where the environment dynamics evolve according to a semi-Markov chain. More precisely, the new model we propose, called *Hidden-Semi-Markov-Mode MDP*, represents environmental changes with hidden semi-Markov models [17] while in HM-MDPs, they were represented with hidden Makov models.

Definition of HS3MDP. Formally, *Hidden-Semi-Markov-Mode Markov Decision Process* (HS3MDP) is defined by $\langle \mathbf{M}, C, H \rangle$ where $\mathbf{M} = \{m_1, \ldots, m_n\}$ is

---

Algorithm 1. POMCP

---

procedure SEARCH(τ)

1 **foreach** *simulations* **do**
2   **if** $\tau = empty$ **then**
3      $p \sim \mathcal{I}$
4   **else**
5      $p \sim B(\tau)$
6   SIMULATE($p, \tau, 0$)
7 **return** $\arg\max_b V(\tau b)$

procedure ROLLOUT($p, \tau, depth$)

8 **if** $\gamma^{depth} < \epsilon$ **then**
9   **return** 0
10 $a \sim \pi_{rollout}(\tau, \cdot)$
11 $(p', o, r) \sim \mathcal{G}(p, a)$
12 **return** $r + \gamma.$ROLLOUT($p', \tau ao, depth + 1$)

procedure SIMULATE($p, \tau, depth$)

13 **if** $\gamma^{depth} < \epsilon$ **then**
14   **return** 0
15 **if** $\tau \notin Tree$ **then**
16   **forall the** $a \in \mathbf{A}$ **do**
17      $Tree(\tau a) \leftarrow (N_{init}(\tau a), V_{init}(\tau a), \emptyset)$
18   **return** ROLLOUT($p, \tau, depth$)
19 $a \leftarrow \arg\max_b V(\tau b) + c\sqrt{\frac{\log N(\tau)}{N(\tau b)}}$
20 $(p', o, r) \sim \mathcal{G}(p, a)$
21 $R \leftarrow r + \gamma.$SIMULATE($p', \tau ao, depth + 1$)
22 $B(\tau) \leftarrow B(\tau) \cup \{p\}$
23 $N(\tau) \leftarrow N(\tau) + 1$
24 $N(\tau a) \leftarrow N(\tau a) + 1$
25 $V(\tau a) \leftarrow V(\tau a) + \frac{R - V(\tau a)}{N(\tau a)}$
26 **return** R

---

a set of modes, $C : \mathbf{M} \times \mathbf{M} \to [0, 1]$ is a transition function over modes and $H : \mathbf{M} \times \mathbf{M} \times \mathbb{N} \to [0, 1]$ is a *mode duration function*. Transition $C(m, m')$ represents the probability of moving to new mode m' from current mode m knowing that the *duration* in m (i.e., the number of remaining timesteps to stay in m) is null. Value $H(m, m', h)$ represents the probability of staying h timesteps in new mode m' when the current mode is m. Both the mode and the duration are not observable.

At each timestep, after a state transition in current mode m, the next mode m' and its duration h' are determined as follows:

$$\begin{cases} \text{if } h > 0 \ m' = m \text{ and } h' = h - 1 \\ \text{if } h = 0 \ m' \sim C(m, \cdot) \\ \qquad h' = k - 1 \text{ where } k \sim H(m, m', \cdot) \end{cases} \quad (2)$$

where h is the duration of current mode m. If h is positive, the environment does not change. But, if h is null, the environment evolves according to transition function C and the number of steps to stay in the new mode is drawn following conditional probability H.

Like HM-MDPs, HS3MDPs form a subclass of POMDPs. An HS3MDP can be reformulated as a POMDP $\langle S, A, O, T, Q, R \rangle$ whose components are defined by: $S = \mathbf{M} \times \mathbf{S} \times \mathbb{N}$, $A = \mathbf{A}$, $O = \mathbf{S}$, $T(\langle m, s, h \rangle, a, \langle m', s', h' \rangle) = \alpha T_m(s, a, s')$ with

$$\alpha = \begin{cases} C(m, m') \times H(m, m', h') \text{ if } h = 0, \\ 1 \text{ if } h' = h - 1 \text{ and } m' = m, \\ 0 \text{ otherwise} \end{cases} \quad (3)$$

$Q(\langle m, s, h \rangle, a, o) = 1$ if $s = o$ and 0 otherwise, $R(\langle m, s, h \rangle, a) = R_m(s, a)$.

Discussions. When considering non-stationary environments in MDPs, an environmental change may impact each component of the quadruplet $\langle \mathbf{S}, \mathbf{A}, T, R \rangle$. Indeed, some states may become impossible or new states may become reachable, some actions may become infeasible or new actions may appear, the transition function and the reward function can of course also change after the environment evolves. Interestingly, a change in the set of states and/or the set actions may always be modeled by a change in the transition and reward functions by considering the set of all possible states for $\mathbf{S}$ and the set of all possible actions for $\mathbf{A}$ at the beginning.

It is easy to notice that HM-MDPs form a subclass of HS3MDPs. In fact, a problem represented as an HS3MDP can also be exactly represented as an HM-MDP by augmenting the modes. The two models are equivalent in the following sense. A model $\mathcal{M}$ is *equivalent* to a model $\mathcal{M}'$ if and only if a problem that can be represented in model $\mathcal{M}$ can also be exactly represented in model $\mathcal{M}'$ and vice-versa.

Proposition 1. *HM-MDPs are equivalent to HS3MDPs.*

Proof. $\Rightarrow$ Given an HM-MDP, we can define an equivalent HS3MDP by setting a mode duration function H such that $\forall m, m', H(m, m', 1) = 1$ and $H(m, m', h) = 0, \forall h \neq 1$. At each timestep, $h = 0$, thus leading only to the first alternative of 3. This turns out to be the exact formulation of an HM-MDP.

$\Leftarrow$ Given an HS3MDP, we show how to build an equivalent HM-MDP. To that aim, we build a sequence of equivalent HS3MDPs. Denote $\langle \mathbf{M}_1, C_1, H_1 \rangle$ the initial HS3MDP. We repeat the following operation to build the sequence: If, for $\langle \mathbf{M}_i, C_i, H_i \rangle$, there exist $m, m' \in \mathbf{M}_i$ and $h \neq 1$ such that $H_i(m, m', h) > 0$, we define the next HS3MDP $\langle \mathbf{M}_{i+1}, C_{i+1}, H_{i+1} \rangle$ as follows:

$$\mathbf{M}_{i+1} = \mathbf{M}_i \cup \bigcup_{h'\neq 1}\{m'_0,\ldots,m'_{h'-1}|H_i(m,m',h') > 0\}$$
$$C_{i+1}(m,m'_{h'-1}) = C_i(m,m') \times H(m,m',h')$$
$$C_{i+1}(m'_j,m'_{j-1}) = 1, \forall j > 0$$
$$C_{i+1}(m_1,m_2) = C_i(m_1,m_2), \forall (m_1,m_2) \neq (m,m') \tag{4}$$
$$H_{i+1}(m,m'_{h'-1},1) = H_{i+1}(m'_j,m'_{j-1},1) = 1, \forall h' > 0, j > 0$$
$$H_{i+1}(m_1,m_2,h') = H_i(m_1,m_2,h'), \forall (m_1,m_2) \neq (m,m'), \forall h'$$

Where for all j, m'_j is a duplicate of m' and C_{i+1} and H_{i+1} are null for the unspecified cases. When this operation cannot be iterated, in the last HS3MDP, unreachable modes can be removed. Finally, the resulting HS3MDP corresponds to an equivalent HM-MDP.   □

However, representing HS3MDPs in such a way feels unnatural and leads to a higher number of modes, which moreover, would have a negative impact on the solving time. It is also obvious that, if the maximum duration is unbounded, the equivalent HM-MDP would have an infinite number of modes, making it difficult to solve.

As a final note, the models of HM-MDPs and HS3MDPs are particular instances of Mixed-Observable MDPs (MOMDPs) [12,1], a subclass of POMDPs. Therefore, MOMDPs algorithms could be used for solving HS3MDPs. We chose to base our solving method on POMCP, because it tends to be more efficient than specialized algorithms on MOMDPs and more generally on factored POMDPs, even when POMCP is run on the non-factored representations [16].

4   Solving an HS3MDP

As HM-MDPs form a subclass of HS3MDPs, solving exactly an HS3MDPs is a PSPACE-complete problem [5]. In order to be able to tackle large instances of problems, we therefore focus on an approximate solving algorithm. A first naive approach is to apply POMCP (see Sect. 2) to directly solve the POMDP derived from an HS3MDP. In that case, a particle in POMCP represents a mode m, a state s and a duration h of the HS3MDP. We propose in this section two possible improvements to this naive approach. Notice that, as a subclass of HS3MDPs, these solving methods can also be applied to HM-MDPs. In the remaining of the paper, we will therefore focus only on HS3MDPs.

Adaptation to the Structure. In large instances, POMCP can suffer from a lack of particles to approximate the belief state especially if the number of states in the POMDP and/or the horizon are large. To tackle this issue, a particle reinvigoration technique can be used in the original algorithm. However, it is often insufficient. When POMCP runs out of particles, it samples the action set according to a uniform distribution, which obviously leads to suboptimal decisions.

We propose a first adaptation of POMCP that exploits the structure of HS3MDPs to postpone the lack of particles. In fact, in the derived POMDP, as the agent observes a part of the state of the POMDP, a particle needs only to represent non-observable information, that is, the mode m and the number of steps to stay h. This adaptation allows us to initially distribute particles over a set whose cardinality is much smaller. However, the size of the particle set $|B(\tau)|$ still depends on the number of simulations. This modification of POMCP is introduced at line 3 of Alg. 1.

Exact Representation of the Belief State. When solving large-sized problems, the above adaptation of POMCP still suffers from lack of particles. We thus propose a second adaptation where we replace the particle set B by an exact representation of the belief state. This representation consists of a probability distribution μ over $\mathbf{M} \times \mathbb{N}$ (modes and duration in the current mode).

Lines 3 and 5 of Alg. 1 are modified as particles are now drawn according to a probability distribution. Line 22 is not needed anymore. This probability distribution is updated after a new observation using the following equation:

$$\mu'(m', h') = \frac{1}{K} \big(T_{m'}(s, a, s') \times \mu(m', h' + 1) + \tag{5}$$

$$\sum_{m \in \mathbf{M}} C(m, m') \times T_m(s, a, s') \times \mu(m, 0) \times H(m, m', h' + 1)\big) \ .$$

where K is the normalization term and elements s, s', a are respectively the previous observation, the new observation given by the real environment and the action performed and given by the procedure SEARCH. This update is performed after every action executed in the real environment.

In HM-MDPs we can rewrite the above equation knowing $\mu(m', h' + 1) = 0, \forall m', h'$ and $H(m, m', 1) = 1$. We then obtain:

$$\mu'(m') = \frac{1}{K} \big(\sum_{m \in \mathbf{M}} C(m, m') \times T_m(s, a, s') \times \mu(m) \big) \ . \tag{6}$$

We fall back to the HM-MDP update equation described by [8].

Unlike the previous adaptation, the spatial complexity of this one does not depend on the number of simulations. Indeed, μ is a probability distribution over $\mathbf{M} \times \mathbb{N}$. Assuming a finite maximum number $h_{\max}$ of timesteps to stay in a mode, which is often the case in practice, there always exists a number of simulations N for which the size of the particle set is greater than the length of this distribution. In such a case, this second adaptation will be more interesting to consider. The time complexity of the update of the exact representation is $\mathcal{O}(|\mathbf{M}| \times h_{\max})$. It is to be compared to the particle invigoration of the original POMCP and the first adaption which is $\mathcal{O}(N)$ with N being the number of simulations.

5   Experimental Results

We tested POMCP and our two adapted versions on four non-stationary problems. The first three environments are problems of the literature [8]. We solved an extended version of each problem modeled as an HS3MDP. Recall that those adapted versions of the problems cannot be represented as efficiently with HM-MDPs (see Prop. 1). Results for this model are thus not reported.

$H(m, m', \cdot)$ is defined as a truncated Gaussian probability distribution on duration h of the mode m' after a transition from m. The mean of the Gaussian is uniform randomly drawn between 1 and 5 when creating the environment.

We present the results for the original POMCP and for our adaptations of POMCP: the Structure Adapted (SA) and Structure Adapted combined with the Exact Representation (SAER) of belief states. We also show the results of the optimal policy when it could be computed, using *Finite Grid* [4] and *MO-IP* [1]. We also used *MO-SARSOP* [12] with one hour of policy computation time when the model could be generated for offline computing. We present the performances of the algorithms for several numbers of simulations to study how the quality of the solutions evolves. For each number of simulations we averaged the cumulative discounted rewards over 1000 runs. We reported results that could be obtained within one hour on a computer equipped with an Intel XeonX5690 4.47 Ghz core. We chose to present the raw results for the original POMCP and percentages for the others. Reported percentages correspond to the improvement in the average cumulated discounted rewards between our modified versions and the original POMCP.

Traffic Light. In the traffic light problem, the environment is a two-way road where the system has to choose which side to let pass. It has to decide which traffic light to switch on, knowing only the current state of the lights and the presence or not of cars on each side of the road. In this problem, the HS3MDP has two modes: rush on the left or on the right and two actions to choose which light to switch on. The model contains eight states depending on the presence or not of cars on the left, on the right and on the light state. The reward function gives a negative reward when a car waits on a side of the road whose light is shut off. At each timestep, the environment has a probability of 0.9 to stay in the same mode and 0.1 to change. Finally, the transition function over the state depends on the probability of cars arriving on each side, according to the current mode. Exact probabilities for the original problem can be found in [6].

Table 1 describes results for the traffic light problem, using different algorithms: original POMCP (orig.), Structure Adapted (SA), Structure Adapted combined with Exact Representation of belief states (SAER) and Finite Grid, MO-IP and MO-SARSOP. The last three algorithms yield the same results, which are presented in column "Opt." to give an idea of the optimal value. The performances of the original POMCP almost strictly increase with the number of simulations. They therefore get closer to the optimal value, which translates into decreasing percentages in Column "Opt." of Table 1. Since our modified versions of POMCP performs better than the original one (positive percentages for columns "SA" and "SAER"), they also get closer to the optimal. For instance, with 512 simulations, 4.7% of improvement for SAER compared to 9.3% for Column "Opt." means that the performances of SAER are half-way between

Table 1. Results for traffic light   **Table 2.** Results for sailboat (7×7 grid)

Sim.	Orig.	SA	SAER	Opt.
1	-3,42	0.0%	0.0%	38.5%
2	-2,86	3.0%	**4.0%**	26.5%
4	-2,80	8.1%	**8.8%**	25.0%
8	-2,68	6.0%	**9.4%**	21.7%
16	-2,60	**8.0%**	8.0%	19.2%
32	-2,45	5.3%	**6.9%**	14.3%
64	-2,47	**10.0%**	9.1%	14.9%
128	-2,34	**4.3%**	3.4%	10.4%
256	-2,41	8.5%	**10.5%**	12.7%
512	-2,32	**5.6%**	4.7%	9.3%
1024	-2,31	5.1%	**7.0%**	9.3%
2048	-2,38	9.0%	**10.5%**	11.8%

Sim.	Orig.	SA	SAER	MO
1	60	**11.7%**	6.7%	408.3%
2	63	**30.2%**	**30.2%**	384.1%
4	55	38.2%	**54.5%**	454.5%
8	70	8.6%	**27.1%**	335.7%
16	59	13.6%	**88.1%**	416.9%
32	66	28.8%	**92.4%**	362.1%
64	90	21.1%	**45.6%**	238.9%
128	94	53.2%	**71.3%**	224.5%
256	119	48.7%	**76.5%**	156.3%
512	159	**31.4%**	27.0%	91.8%
1024	177	20.9%	**28.8%**	72.3%
2048	206	**13.6%**	10.2%	48.1%
4096	226	12.4%	**16.4%**	35.0%
8192	227	20.7%	**25.6%**	34.4%

those of the original POMCP and the optimal value. Note that a decreasing percentage does not mean a raw decrease in the performances. It means that the increase of the performances of the original POMCP is higher than those of the other algorithms. Nonetheless, the percentages being positive, the later still perform better.

Theoretically, POMCP converges towards the optimal solution while the number of simulations increases. Experimental results (Table 1) show that it is also the case for our adapted versions whose performances are always at least as good as the original POMCP.

In the traffic light problem, both adaptations of POMCP are roughly even. In fact, the size of the problem is quite small so the original POMCP and the structured adapted POMCP do not lack particles. Moreover, there are enough particles to draw a high quality estimation of the belief state. That is why, the exact representation of belief states does not significantly outperform other POMCP versions. Nonetheless, our adaptations of POMCP both outperform the original version since exploiting the structure of the HS3MDP leads to more accurate belief states.

Sailboat. The sailboat problem is about controlling a boat from a corner of a finite grid to the opposite corner. The states are possible positions in the grid and the modes are the different wind directions, limited to North, South, West and East. Two possible actions manage the sail orientation between North-South and East-West. The transition function over states depends on the sail orientation given the wind direction. The environment has a probability of 0.5 to stay in the same mode, 0.2 to go to an adjacent one and 0.1 to go to the opposite one. The reward function gives a reward of 1 when the goal is reached. This problem can be enlarged as needed by increasing the size of the grid. Results for a 7 × 7 grid are reported in Table 2.

Table 3. Results for elevator ($f = 7$, $e = 1$)

Sim.	Orig.	SA	SAER
1	-10.56	0.0%	**1.1%**
2	-10.60	0.0%	0.0%
4	-10.50	2.2%	**3.6%**
8	-10.49	**4.2%**	3.9%
16	-10.44	**5.2%**	5.0%
32	-10.54	**6.2%**	**6.2%**

Table 4. Results for elevator ($f = 4$, $e = 2$)

Sim.	Orig.	SA	SAER
1	-7.41	**1.0%**	0.4%
2	-7.35	**0.3%**	0.0%
4	-7.44	**1.5%**	1.3%
8	-7.35	**0.4%**	0.0%
16	-7.30	**19.1%**	17.2%
32	-7.25	**22.1%**	21.6%
64	-7.17	**24.3%**	**24.3%**
128	-7.22	**27.0%**	**27.0%**

Due to probabilities of transition between modes, the environment can stay several steps in the same mode thus giving the same wind direction. When the boat is on an edge of the grid, it cannot move until the wind changes to a more favorable configuration. This particularity of the environment leads to a big set of runs where the boat cannot reach the goal and gets stuck on an edge until the end of the run. Moreover, the small drops in the original POMCP performances can be explained with the low number of simulations. If this number is not high enough to explore efficiently, the impact of the random can lead to a high variance. Results show that our adaptations always perform better than the original method and that SAER performs almost always better than SA. Column "MO" stands for the results of MO-SARSOP. We can see that SAER converges toward those results as the number of simulations increases.

Elevators. In the elevator problem, the environment can stay in the current mode (see Example 1) with a probability of 0.1 and has a probability 0.45 to change to the other two when the duration is null. Table 3 contains results for an instance with 7 floors and 1 elevator whereas Table 4 shows results for a 4 floors and 2 elevators instance. We were not able to compute the optimal policy for these instances because of their high dimensionality. The results of our adaptations are roughly even since the size of the problem remains quite limited and does not lead to lack particles. However, it is important to note that our methods always outperform the original POMCP whose performances increase with the number of simulations and converge to the optimal solution.

The low number of simulations reached during the computation time is explained by the representation of the transition function. In this problem, transitions are not represented by a matrix of probabilities because of the high number of state components. The transitions are based on a set of rules, leading to a longer computation time.

Randomly Generated Environments. These environments allow us to study in a controlled setting the scalability of our algorithms. To create an instance, a number of states n_s, actions n_a and modes n_m have to be defined. Random MDPs are then automatically generated such that, in each state, each action can lead to $\lfloor \frac{|S|}{10} \rfloor$ states and $\lfloor \frac{|S|}{5} \rfloor$ states can yield a positive reward. To enlarge those

Table 5. Results for random environments with $n_s = 50$, $n_a = 5$ and $n_m = 5$

Table 6. Results for random environments with $n_s = 50$, $n_a = 5$ and $n_m = 10$

Table 7. Results for random environments with $n_s = 50$, $n_a = 5$ and $n_m = 20$

Sim.	Orig.	SA	SAER
1	0.41	0.0%	5.6%
2	0.41	4.9%	51.4%
4	0.42	11.5%	140.9%
8	0.44	30.9%	209.6%
16	0.48	34.6%	234.7%
32	0.58	46.0%	223.0%
64	0.77	53.1%	187.2%
128	1.08	45.7%	123.4%
256	1.52	33.5%	70.0%
512	1.98	19.6%	34.5%
1024	2.30	12.5%	17.3%

Sim.	Orig.	SA	SAER
1	0.39	0.1%	8.9%
2	0.39	21.0%	57.5%
4	0.40	9.9%	149.0%
8	0.41	24.0%	224.6%
16	0.43	33.0%	261.3%
32	0.48	58.2%	275.8%
64	0.60	76.2%	248.7%
128	0.83	75.4%	184.5%
256	1.16	64.1%	115.9%
512	1.61	41.5%	61.5%
1024	2.05	2.2%	28.8%

Sim.	Orig.	SA	SAER
1	0.39	0.8%	11.9%
2	0.40	2.6%	51.1%
4	0.40	2.7%	138.9%
8	0.41	11.8%	225.2%
16	0.41	22.3%	270.8%
32	0.45	42.9%	290.3%
64	0.51	77.5%	305.5%
128	0.63	102.2%	261.1%
256	0.85	102.7%	186.8%
512	1.23	73.3%	107.7%
1024	1.66	43.6%	55.3%

environments, we varied the size of the sets of states, actions and modes. We averaged results from 10 different instances with different state/mode transition and reward functions for each parameter set.

Tables 5, 6 and 7 describe results for randomly generated environments with respectively 5, 10 and 20 modes. We were not able to compute the optimal policy for these instances because of their high dimensionality. We can see that our methods significantly outperform the original POMCP method. In fact, the exact representation of belief states always outperforms POMCP versions based on particles filter on sufficiently large environments. Indeed, these methods quickly lack particles to accurately represent the belief state.

Moreover, the computation time of our adaptations are promising for application to large-sized real life problems. For instance, in the random environment with 20 modes (Table 7), one run of 1024 simulations took 1.15 seconds for solving the HS3MDP with structured adapted POMCP and 1.48 seconds for solving the HS3MDP with POMCP and exact representation of the belief state.

6   Conclusion and Discussions

In this paper, we introduced Hidden-Semi-Markov-Mode Markov Decision Processes (HS3MDPs), a new generalization of Hidden-Mode Markov Decision Processes (HM-MDPs) to handle in a more natural and efficient way non-stationary environments. We proposed to use the Partially Observable Monte-Carlo Planning algorithm as a solving method for HS3MDPs. As a subclass of our model, HM-MDPs can be solved efficiently using the same methods. However, this algorithm does not solve large-sized problems modeled with HS3MDPs in the most efficient way. We developed two adaptations of POMCP to improve its performances. The first adaptation exploits the structure of HS3MDPs to alleviate particle deprivation. The second adaptation uses an exact representation of the belief state to reach better results with less simulations than the other two methods. Experimental results on various domains of the literature show that those adaptation significantly improve the performance.

As future work, different research directions could be explored. In HM-MDPs and HS3MDPs, transition functions over modes do not depend on the performed action. This assumption does not hold in environments like stock markets where buying big volumes may influence the market. An extension of HS3MDPs to handle such situations would be interesting. Another research direction is to relax the assumption that the transition function between modes is known and to learn it in a multi-armed bandit or reinforcement learning setting.

Acknowledgments. Funded by the French National Research Agency under grant ANR-10-BLAN-0215.

References

1. Araya-López, M., Thomas, V., Buffet, O., Charpillet, F.: A closer look at MOMDPs. In: ICTAI (2010)
2. Aström, K.: Optimal control of markov decision processes with incomplete state estimation. J. of Math. Analysis and Applications 10, 174–205 (1965)
3. Cassandra, A., Littman, M., Zhang, N.: Incremental Pruning: A simple, fast, exact method for Partially Observable Markov Decision Processes. In: UAI, pp. 54–61 (1997)
4. Cassandra, T.: Pomdp-solve (2003-2013),
   http://www.pomdp.org/code/index.shtml/
5. Chadès, I., Carwardine, J., Martin, T.G., Nicol, S., Sabbadin, R., Buffet, O.: MOMDPs: A solution for modelling adaptive management problems. In: AAAI (2012)
6. Choi, S.: Reinforcement learning in nonstationary environments. Ph.D. thesis, Hong Kong Univ. of Science and Tech. (2000)
7. Choi, S., Yeung, D., Zhang, N.: An environment model for nonstationary reinforcement learning. In: NIPS, pp. 981–993 (2000)
8. Choi, S., Zhang, N., Yeung, D.: Solving Hidden-Mode Markov Decision Problems. In: AISTATS, pp. 19–26 (2001)
9. Doya, K., Samejima, K., Katagiri, K., Kawato, M.: Multiple model-based reinforcement learning. Neural Computing 14(6), 1347–1369 (2002)
10. Kaelbling, L., Littman, M., Cassandra, A.: Planning and acting in partially observable stochastic domains. Artificial Intelligence 101(1-2), 99–134 (1998)
11. Kocsis, L., Szepesvári, C.: Bandit based Monte-Carlo planning. In: Fürnkranz, J., Scheffer, T., Spiliopoulou, M. (eds.) ECML 2006. LNCS (LNAI), vol. 4212, pp. 282–293. Springer, Heidelberg (2006)
12. Ong, S., Png, S., Hsu, D., Lee, W.: POMDPs for robotic tasks with mixed observability. In: Robotics: Science & Syst. (2009)
13. Papadimitriou, C., Tsitsiklis, J.: The complexity of Markov Decision Processes. Math. of OR 12(3), 441–450 (1987)
14. Puterman, M.: Markov Decision Processes: Discrete dynamic stochastic programming. John Wiley Chichester (1994)
15. da Silva, B., Basso, E., Bazzan, A., Engel, P.: Dealing with non-stationary environments using context detection. In: ICML (2006)
16. Silver, D., Veness, J.: Monte-Carlo planning in large POMDPs. In: NIPS, pp. 2164–2172 (2010)
17. Yu, S.: Hidden Semi-Markov Models. Artificial Intelligence 174(2), 215–243 (2010)

Probabilistic Strategies
in Dialogical Argumentation

Anthony Hunter

Department of Computer Science, University College London,
Gower Street, London WC1E 6BT, UK

Abstract. In dialogical argumentation, a participant is often unsure
what moves the other participant(s) might make. If the dialogue is pro-
ceeding according to some accepted protocol, then a participant might
be able to determine what are the possible moves that the other might
make, but the participant might be unsure as to which move will be chosen by the other agent. In this paper, propositional executable logic is
augmented with probabilities that reflect the probability that any given
move will be chosen by the agent. This provides a simple and lucid language that can be executed to generate a dialogue. Furthermore, a set
of such rules for each agent can be represented by a probabilistic finite
state machine (PFSM). For modelling dialogical argumentation, a PFSM
can be used by one agent to model how the other agent may react to
any dialogical move. An agent can then analyze the PFSM to determine
the most likely outcomes of a dialogue given any choices it makes. This
can be used by the agent to determine its choice of moves in order to
optimize its outcomes from the dialogue.

1 Introduction

Dialogical argumentation involves agents exchanging arguments in activities such
as discussion, debate, persuasion, and negotiation. Dialogue games are now a
common approach to characterizing argumentation-based agent dialogues (e.g.
[1, 11]). In order to compare and evaluate dialogical argumentation systems, we
proposed in a previous paper that first-order executable logic could be used as
common theoretical framework to specify and analyze dialogical argumentation
systems [12]. Then in [13], propositional executable logic was presented as a
special case, and for which a finite state machine (FSM) can be generated. An
FSM is a useful structure for investigating various properties of the dialogue,
including conformance to protocols, and application of the minimax strategy.

We can improve on analyzing argumentation strategies by harnessing probabilistic information about an opponent to offer better decision making. In this
paper, we address this need by introducing a probability assignment to the argumentation moves. For this, we introduce probabilistic executable logic and
show how a specification for the protocols for a pair of agents in probabilistic
executable logic can be represented and analyzed in the form of a probabilistic
finite state machine (PFSM). By using Markov chains, we can determine the

U. Straccia and A. Calì (Eds.): SUM 2014, LNAI 8720, pp. 190–202, 2014.
© Springer International Publishing Switzerland 2014

expected utility of such a specification, and we can optimize the expected utility for one of the agents by changing its specification to be deterministic.

2 Probabilistic Executable Logic

We assume a set of atoms $\mathcal{A}$ which we use to form propositional formulae in the usual way using disjunction, conjunction, and negation connectives. We construct modal formulae using the $\boxplus$, $\boxminus$, $\oplus$, and $\ominus$ modal operators. We only allow literals to be in the scope of a modal operator. If α is a literal, then each of $\oplus \alpha$, $\ominus \alpha$, $\boxplus \alpha$, and $\boxminus \alpha$ is an **action unit**.

Informally, we describe the meaning of action units as follows: $\oplus \alpha$ means that the action by an agent is to add the literal α to its next private state; $\ominus \alpha$ means that the action by an agent is to delete the literal α from its next private state; $\boxplus \alpha$ means that the action by an agent is to add the literal α to the next public state; and $\boxminus \alpha$ means that the action by an agent is to delete the literal α from the next public state.

We use the action units to form **action formulae** as follows using the disjunction and conjunction connectives. If $\alpha_1, \ldots, \alpha_n$ are action units, and $v \in [0, 1]$, then $(v : \alpha_1 \wedge \ldots \wedge \alpha_n)$ is an **action option**. As we will see later, v denotes the probability that an agent will undertake the actions $\alpha_1, \ldots, \alpha_n$. We compose the action options into action rules as follows. If β is a classical formula, and $\gamma_1 = (v^1 : \alpha_1^1 \wedge \ldots \wedge \alpha_n^1), \ldots, \gamma_m = (v^m : \alpha_1^m \wedge \ldots \wedge \alpha_n^m)$ are action options, where $v^1 + \ldots + v^m = 1$, then $\beta \Rightarrow \gamma_1 \vee \ldots \vee \gamma_m$ is an **action rule**. So an action rule has a consequent in "disjunctive normal form". Each disjunct is action option, and the sum of the probabilities is 1.

Example 1. Let $\mathcal{A} = \{b(a_1), b(a_2), c(a_1), c(a_2)\}$. So the following is an action rule (which we might use in an example where b denotes belief, and c denotes claim, and a_1 and a_2 are some items of information).

$$b(a_1) \wedge b(a_2) \Rightarrow (0.2 : \boxplus c(a_1) \wedge \boxminus c(a_2)) \vee (0.5 : \boxplus c(a_1)) \vee (0.3 : \boxplus c(a_2))$$

Implicit in the definitions for the language is the fact that we can use it as a meta-language [14]. For this, the object-language will be represented by terms in this meta-language. For instance, the object-level formula $p(a, b) \rightarrow q(a, b)$ can be represented by a term where the object-level literals $p(a, b)$ and $q(a, b)$ are represented by constant symbols, and $\rightarrow$ is represented by a function symbol. Then we can form the atom `belief`$(p(a, b) \rightarrow q(a, b))$ where `belief` is a predicate symbol. Note, in general, no special meaning is ascribed to the predicate symbols or terms. They are used as in classical logic. Also, the terms and predicates are all ground, and so it is essentially a propositional language.

We use a state-based model of dialogical argumentation with the following definition of an execution state. To simplify the presentation, we restrict consideration in this paper to two agents. An execution represents a finite or infinite sequence of execution states. If the sequence is finite, then t denotes the terminal state, otherwise $t = \infty$.

Definition 1. *An* **execution** *e is a tuple $e = (s_1, a_1, p, a_2, s_2, t)$, where for each $n \in \mathbb{N}$ where $0 \leq n \leq t$, $s_1(n)$ is a set of ground literals, $a_1(n)$ is a set of ground action units, $p(n)$ is a set of ground literals, $a_2(n)$ is a set of ground action units, $s_2(n)$ is a set of ground literals, and $t \in \mathbb{N} \cup \{\infty\}$. For each $n \in \mathbb{N}$, if $0 \leq n \leq t$, then an* **execution state** *is $e(n) = (s_1(n), a_1(n), p(n), a_2(n), s_2(n))$ where $e(0)$ is the* **initial state***. We assume $a_1(0) = a_2(0) = \emptyset$. We call $s_1(n)$ the private state of agent 1 at time n, $a_1(n)$ the action state of agent 1 at time n, $p(n)$ the public state at time n, $a_2(n)$ the action state of agent 2 at time n, $s_2(n)$ the private state of agent 2 at time n.*

In general, there is no restriction on the literals that can appear in the private and public state. The choice depends on the specific dialogical argumentation we want to specify. This flexibility means we can capture diverse kinds of information in the private state about agents by assuming predicate symbols for their own beliefs, objectives, preferences, arguments, etc, and for what they know about other agents. The flexibility also means we can capture diverse information in the public state about moves made, commitments made, etc.

Example 2. The first 5 steps of an infinite execution where each row in the table is an execution state where b denotes belief, and c denotes claim.

n	$s_1(n)$	$a_1(n)$	$p(n)$	$a_2(n)$	$s_2(n)$
0	b(a)				b(¬a)
1	b(a)	⊞c(a),⊟c(¬a)			b(¬a)
2	b(a)		c(a)	⊞c(¬a),⊟c(a)	b(¬a)
3	b(a)	⊞c(a),⊟c(¬a)	c(¬a)		b(¬a)
4	b(a)		c(a)	⊞c(¬a),⊟c(a)	b(¬a)
5	. . .	. . .	. . .	. . .	. . .

We define a system in terms of the action rules for each agent, which specify what moves the agent can potentially make based on the current state of the dialogue. In this paper, we assume agents take turns, and at each time point the actions are from the head of just one rule (as defined in the rest of this section). We also assume in this paper that at most one rule can have an antecedent satisfiable at any time.

Definition 2. *A* **system** *is a tuple $(Rules_x, Initials)$ where $Rules_x$ is the set of action rules for agent $x \in \{1, 2\}$, and $Initials$ is the set of initial states.*

For an agent x, the information it has available at any point n in the dialogue is determined by the private state $s_x(n)$ and public state $p(n)$. We augment this set of atoms by the closed world assumption as follows.

Definition 3. *Let $s_x(n)$ be the private state of agent x at time n, and let $p(n)$ be the public state of agent x at time n. The* **knowledge of agent** *x* **at time** *n, denoted $k_x(n)$, is defined as follows.*

$$k_x(n) = s_x(n) \cup p(n) \cup \{\neg \alpha \mid \alpha \in \mathcal{A} \text{ and } \alpha \notin s_x(n) \cup p(n)\}$$

Example 3. Let $\mathcal{A} = \{b(a), b(\neg a), b(b), b(\neg b), c(a), c(\neg a), c(b), c(\neg b)\}$. Also let $s_1(2) = \{b(a)\}$ and $p(2) = \{c(a)\}$. Therefore, $k_1(2) = \{b(a), \neg b(\neg a), \neg b(b), \neg b(\neg b), c(a), \neg c(\neg a), \neg c(b), \neg c(\neg b)\}$.

We give two constraints on an execution to ensure that they are well-behaved. The first (propagated) ensures that each subsequent private state (respectively each subsequent public state) is the current private state (respectively current public state) for the agent updated by the actions given in the action state. The second (engaged) ensures that an execution does not have one state with no actions followed immediately by another state with no actions (otherwise the dialogue can lapse) except at the end of the dialogue where neither agent has further actions.

Definition 4. *An execution* $(s_1, a_1, p, a_2, s_2, t)$ *is* **propagated** *iff for all* $x \in \{1, 2\}$, *for all* $n \in \{0, \dots, t-1\}$, *where* $a(n) = a_1(n) \cup a_2(n)$

1. $s_x(n+1) = (s_x(n) \setminus \{\phi \mid \ominus\phi \in a_x(n)\}) \cup \{\phi \mid \oplus\phi \in a_x(n)\}$
2. $p(n+1) = (p(n) \setminus \{\phi \mid \boxminus\phi \in a(n)\}) \cup \{\phi \mid \boxplus\phi \in a(n)\}$

Definition 5. *Let* $e = (s_1, a_1, p, a_2, s_2, t)$ *be an execution and* $a(n) = a_1(n) \cup a_2(n)$. e *is* **finitely engaged** *iff (1)* $t \neq \infty$; *(2) for all* $n \in \{1, \dots, t-2\}$, *if* $a(n) = \emptyset$, *then* $a(n+1) \neq \emptyset$; *(3)* $a(t-1) = \emptyset$; *and (4)* $a(t) = \emptyset$. e *is* **infinitely engaged** *iff (1)* $t = \infty$; *and (2) for all* $n \in \mathbb{N}$, *if* $a(n) = \emptyset$, *then* $a(n+1) \neq \emptyset$.

The next definition shows how a system provides the initial state of an execution and the actions that can appear in an execution. It also ensures turn taking by the two agents. Given the current state of an execution n, the following definition captures which rules are fired. For agent x, these are the rules that have the condition literals satisfied by the current knowledge of the agent (i.e. $k_x(n)$). We use classical entailment, denoted $\models$, for satisfaction, but other relations could be used (e.g. Belnap's four valued logic). Also recall that in this paper we assume that the antecedents of the rules for an agent are such that at most one rule can fire for any point of the execution. In general, this is not essential, but it makes the definitions simpler.

Definition 6. *Let* $S = (Rules_x, Initials)$ *be a system and* $e = (s_1, a_1, p, a_2, s_2, t)$ *be an execution.* S **generates** e *iff (1)* e *is propagated; (2)* e *is finitely engaged or infinitely engaged; (3)* $e(0) \in Initials$; *and (4) for all* $m \in \{1, \dots, t-1\}$

1. *If* m *is odd, then* $a_2(m) = \emptyset$ *and either* $a_1(m) = \emptyset$ *or there is an* $\phi \Rightarrow (v^1 : \psi^1) \vee \dots \vee (v^j : \psi^j) \in Rules_1$ *where* $k_1(m) \models \phi$ *and there is an* $i \in \{1, \dots, j\}$ *where* $\psi^i = \alpha_1 \wedge \dots \wedge \alpha_p$ *and* $a_1(m) = \{\alpha_1, \dots, \alpha_p\}$
2. *If* m *is even, then* $a_1(m) = \emptyset$ *and either* $a_2(m) = \emptyset$ *or there is an* $\phi \Rightarrow (v^1 : \psi^1) \vee \dots \vee (v^j : \psi^j) \in Rules_2$ *where* $k_2(m) \models \phi$ *and there is an* $i \in \{1, \dots, j\}$ *where* $\psi^i = \alpha_1 \wedge \dots \wedge \alpha_p$ *and* $a_2(m) = \{\alpha_1, \dots, \alpha_p\}$

The rules for each agent constitute the protocol for each agent. So the set of executions generated by a system is the set of executions allowed by the protocols of the two agents.

Example 4. We can obtain the execution in Example 2 with the following rules where the first is for agent 1, and the second is for agent 2. The first action option of the first rule is used in steps 1 and 3, and the first action option of the second rule is used in steps 2 and 4.

- $b(a) \Rightarrow (0.6 : \boxplus c(a) \wedge \boxminus c(\neg a)) \vee (0.4 : \boxplus \top)$
- $b(\neg a) \Rightarrow (0.8 : \boxplus c(\neg a) \wedge \boxminus c(a)) \vee (0.2 : \boxplus \top)$

Note, for the second action option, we have $\boxplus \top$ as the only action. This denotes an "empty action", or a "skip action", since there is no material effect on the public or private states by this choice of action.

So far we have not considered the probabilities. Given a system and an execution state e, for each execution state $e(n)$, we can obtain the probability that this state will result in execution state $e(n+1)$. We obtain this from the action rule, by taking the probability of the action option, as follows.

Definition 7. *Let $S = (Rules_x, Initials)$ be a system and $e = (s_1, a_1, p, a_2, s_2, t)$ be an execution. A **probability function for execution** e, denoted pr, is defined as follows. For all $m \in \{1, \ldots, t-1\}$*

1. *If m is odd, and $a_1(m) = \emptyset$, then $pr(e(m), e(m+1)) = 1$.*
2. *If m is odd, and there is an $\phi \Rightarrow (v^1 : \psi^1) \vee \ldots \vee (v^j : \psi^j) \in Rules_1$ s.t. $k_1(m) \models \phi^i$, then $pr(e(m), e(m+1)) = v^i$, where $i \in \{1, \ldots, j\}$.*
3. *If m is even, and $a_2(m) = \emptyset$, then $pr(e(m), e(m+1)) = 1$.*
4. *If m is even, and there is an $\phi \Rightarrow (v^1 : \psi^1) \vee \ldots \vee (v^j : \psi^j) \in Rules_2$ s.t. $k_2(m) \models \phi^i$, then $pr(e(m), e(m+1)) = v^i$, where $i \in \{1, \ldots, j\}$.*

Example 5. Using the action rules in Example 4 to generate the execution in Example 2, we obtain the following for the steps 1 to 5: $pr(e(1), e(2)) = 0.6$, $pr(e(2), e(3)) = 0.8$, $pr(e(3), e(4)) = 0.6$, and $pr(e(4), e(5)) = 0.8$.

Executable logic can be used to formalize a diverse range of dialogical argumentation where agents exchange arguments and attacks [12, 13]. It is straightforward to formalize the exchange of a wide range of moves and content for both abstract and logical argumentation, and richer notions, such as value-based argumentation [15]), can be captured.

3   Probabilistic Finite State Machines

Probabilistic finite state machines (PFSMs) are an important approach in computer science for modelling behaviours of interacting modules when there is uncertainty in the choices made by those modules. The formalization augments finite state machines with a probability distribution over the transitions coming out of each state [16]. In this section, we show how a executable logic system, together with a choice of initial state, can be used to generate a PFSM.

Definition 8. *A tuple $M = (States, Arcs, Start, Prob)$ is a* **probabilistic finite state machine** *(PFSM) where States is a set of states, $Start \in States$, $Arc \subseteq States \times States$ is a set of arcs, $Prob : Arcs \to [0,1]$ is a probability function such that $\sum_{s' \in States \text{ s.t. } (s,s') \in Arcs} Prob(s,s') = 1$ for each $s \in States$.*

The *Prob* assignment is the probability that that arc is chosen. The transitions out of a state sum to 1.

Definition 9. *Let M be a PFSM, and let $\rho = \sigma_1, \ldots, \sigma_k \subseteq States$ be a sequence of states. ρ is a* **walk** *in M iff (1) if $k = 1$, then σ_1 is the node reached immediately from the Start node and (2) if $k > 1$, then $\sigma_1, \ldots, \sigma_{k-1}$ is a walk in M and $(\sigma_{k-1}, \sigma_k) \in Arcs$.*

Example 6. The following is an PFSM where the probability of each transition labels the arc. For this, $\sigma_1, \sigma_2, \sigma_3, \sigma_2, \sigma_3, \sigma_4$ is a walk.

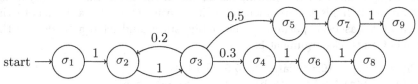

We can construct an FSM that represents the set of executions for an initial state for a system. For this, each state is a tuple $(y, s_1(n), p(n), s_2(n))$, where n is an execution step and y is the agent holding the turn when $n < t$ and r is 0 when $n = t$. The probability that any given actions are executed is given by the probability assignment to the rule. So each execution is a Markov chain.

Definition 10. *A PFSM $M = (States, Arcs, Start, Prob)$* **represents** *a system $S = (Rules_x, Initials)$ for an initial state $I \in Initials$ iff*

(1) $States = \{(y, s_1(n), p(n), s_2(n)) \mid S$ generates $e = (s_1, a_1, p, a_2, s_2, t)$
$$and \ I = (s_1(0), a_1(0), p(0), a_2(0), s_2(0))$$
$$and \ y = 0 \ when \ n = t$$
$$and \ y = 1 \ when \ n < t \ and \ n \ is \ odd$$
$$and \ y = 2 \ when \ n < t \ and \ n \ is \ even \ \}$$

(2) $Start = (1, s_1(0), p(0), s_2(0))$ where $I = (s_1(0), a_1(0), p(0), a_2(0), s_2(0))$

(3) $Arcs$ is the smallest subset of $States \times States$ s.t. for all executions e and for all $n < t$ there is a transition $\tau \in Arcs$ such that

$$\tau = ((x, s_1(n), p(n), s_2(n)), (y, s_1(n+1), p(n+1), s_2(n+1)))$$

where x is 1 when n is odd, x is 2 when n is even, y is 1 when $n+1 < t$ and n is odd, y is 2 when $n+1 < t$ and n is even, and y is 0 when $n+1 = t$.

(4) $Prob$ is the smallest subset of $Arcs \times [0,1]$ s.t. for all $\tau \in Trans$,

$$if \ \tau = ((x, s_1(n), p(n), s_2(n)), (y, s_1(n+1), p(n+1), s_2(n+1)))$$
$$then \ Prob(\tau) = pr(e(n), e(n+1))$$

where $e(n + 1) = (s_1(n + 1), a_1(n + 1), p(n + 1), a_2(n + 1), s_2(n + 1), q(n + 1))$, $e(n) = (s_1(n), a_1(n), p(n), a_2(n), s_2(n), q(n))$, *and pr is the execution probability function.*

Example 7. Consider the PFSM in Example 6 where the states are defined as follows.

$$\sigma_1 = (1, \emptyset, \{a\}, \emptyset) \quad \sigma_4 = (2, \emptyset, \{d\}, \emptyset) \quad \sigma_7 = (1, \emptyset, \{e\}, \emptyset)$$
$$\sigma_2 = (2, \emptyset, \{b\}, \emptyset) \quad \sigma_5 = (2, \emptyset, \{e\}, \emptyset) \quad \sigma_8 = (0, \emptyset, \{d\}, \emptyset)$$
$$\sigma_3 = (1, \emptyset, \{c\}, \emptyset) \quad \sigma_6 = (1, \emptyset, \{d\}, \emptyset) \quad \sigma_9 = (0, \emptyset, \{e\}, \emptyset)$$

This can be obtained from the following set of rules for each agent so that the PFSM represents the system: (1) $\mathsf{a} \Rightarrow (1 : \boxplus\mathsf{b} \wedge \boxminus\mathsf{a})$; (2) $\mathsf{b} \Rightarrow (1 : \boxplus\mathsf{c} \wedge \boxminus\mathsf{b})$; and (3) $\mathsf{c} \Rightarrow (0.2 : \boxplus\mathsf{b} \wedge \boxminus\mathsf{c}) \vee (0.3 : \boxplus\mathsf{d} \wedge \boxminus\mathsf{c}) \vee (0.5 : \boxplus\mathsf{e} \wedge \boxminus\mathsf{c})$.

Example 8. Consider the following set of action rules. For this, let $\mathsf{h}(\mathsf{a})$ denote that an agent holds an argument a in its private state, let $\mathsf{a}(\mathsf{a})$ denote that argument a has been posited in the public state, and let $\mathsf{e}(\mathsf{a}, \mathsf{b})$ denote that argument a attacks argument b.

- $\mathsf{h}(\mathsf{a}) \wedge \mathsf{h}(\mathsf{b}) \Rightarrow (0.5 : \boxplus\mathsf{a}(\mathsf{a})) \vee (0.5 : \boxplus\mathsf{a}(\mathsf{b}))$
- $\mathsf{h}(\mathsf{a}) \wedge \neg\mathsf{a}(\mathsf{a}) \wedge \mathsf{a}(\mathsf{b}) \wedge \mathsf{e}(\mathsf{a}, \mathsf{b}) \Rightarrow (1 : \boxplus\mathsf{a}(\mathsf{a}) \wedge \boxplus\mathsf{e}(\mathsf{a}, \mathsf{b}))$
- $\mathsf{h}(\mathsf{b}) \wedge \neg\mathsf{a}(\mathsf{b}) \wedge \mathsf{a}(\mathsf{a}) \wedge \mathsf{e}(\mathsf{a}, \mathsf{b}) \Rightarrow (1 : \boxplus\mathsf{a}(\mathsf{b}) \wedge \boxplus\mathsf{e}(\mathsf{a}, \mathsf{b}))$

The first action rule adds the argument a to the public state or adds the argument b to the public state. The second action rule adds a to the public state if it has not already been added, and it adds the attack by a on b to the public state. The third action rule adds b to the public state if it has not already been added, and it adds the attack by a on b to the public state.

With this set of action rules, and the initial state $(\{\mathsf{h}(\mathsf{a}), \mathsf{h}(\mathsf{b})\}, \{\}, \{\}, \{\}, \{\})$, we obtain the following PFSM, with the states defined below.

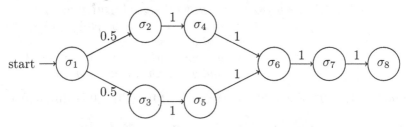

$\sigma_1 = (1, \{\mathsf{h}(\mathsf{a}), \mathsf{h}(\mathsf{b})\}, \{\}, \{\})$ $\sigma_5 = (1, \{\mathsf{h}(\mathsf{a}), \mathsf{h}(\mathsf{b})\}, \{\mathsf{a}(\mathsf{b})\}, \{\})$
$\sigma_2 = (2, \{\mathsf{h}(\mathsf{a}), \mathsf{h}(\mathsf{b})\}, \{\mathsf{a}(\mathsf{a})\}, \{\})$ $\sigma_6 = (2, \{\mathsf{h}(\mathsf{a}), \mathsf{h}(\mathsf{b})\}, \{\mathsf{a}(\mathsf{a}), \mathsf{a}(\mathsf{b}), \mathsf{e}(\mathsf{a}, \mathsf{b})\}, \{\})$
$\sigma_3 = (2, \{\mathsf{h}(\mathsf{a}), \mathsf{h}(\mathsf{b})\}, \{\mathsf{a}(\mathsf{b})\}, \{\})$ $\sigma_7 = (1, \{\mathsf{h}(\mathsf{a}), \mathsf{h}(\mathsf{b})\}, \{\mathsf{a}(\mathsf{a}), \mathsf{a}(\mathsf{b}), \mathsf{e}(\mathsf{a}, \mathsf{b})\}, \{\})$
$\sigma_4 = (1, \{\mathsf{h}(\mathsf{a}), \mathsf{h}(\mathsf{b})\}, \{\mathsf{a}(\mathsf{a})\}, \{\})$ $\sigma_8 = (0, \{\mathsf{h}(\mathsf{a}), \mathsf{h}(\mathsf{b})\}, \{\mathsf{a}(\mathsf{a}), \mathsf{a}(\mathsf{b}), \mathsf{e}(\mathsf{a}, \mathsf{b})\}, \{\})$

So this set of action rules implicitly constructs an abstract argument graph in the public state. The end state is σ_8 and it contains the specification of the abstract argument graph $\mathsf{a} \rightarrow \mathsf{b}$.

Proposition 1. *For each* $S = (Rules_x, Initials)$, *there is a PFSM M such that M represents S for an initial state* $I \in Initials$.

In the next definition, we provide the conditions under which a walk over k states is equivalent to the k steps of an execution after the initial state.

Definition 11. *A sequence of states* $\rho = \sigma_1, \ldots, \sigma_k$ **reflects** *an execution* $e = (s_1, a_1, p, a_2, s_2, t)$ *iff for each* $i \in \{1, \ldots, k-1\}$,

1. $\sigma_i = (s_1(i), p(i), s_2(i))$
2. $Prob((\sigma_i, \sigma_{i+1})) = pr(e(i), e(i+1))$

So for each initial state for a system, we can obtain a PFSM that is a concise representation of the executions of the system following that initial state.

Proposition 2. *Let* $S = (Rules_x, Initials)$ *be a system, and let* M *be a PFSM that represents* S *for* $I \in Initials$:

1. *For all* ρ *s.t.* ρ *is a walk in* M, *there is an execution* e *s.t.* S *generates* e *and* $e(0) = I$ *and* ρ *reflects* e.
2. *For all finite executions* e *s.t.* S *generates* e *and* $e(0) = I$, *then there is a* ρ *such that* ρ *is a walk in* M *and* ρ *reflects* e.

A PFSM provides a more efficient representation of all the possible executions than the set of executions for an initial state. For instance, if there is a set of states that appear in some permutation of each of the executions then this can be more compactly represented by an PFSM.

Furthermore, we can ask simple questions such as is termination possible, is termination guaranteed, and is one system subsumed by another? Then by analyzing the Markov chains, we can answer questions such as what is the probability that we leave a state and never return, or the probability that we visit a state infinitely often? So by translating a system into a PFSM, we can harness substantial theory and tools for analyzing PFSMs.

There are various options we have for dealing with cycles in PFSMs. Here, we consider dropping arcs so as to remove cycles. For this, when there is a walk that is looping back to a state that has already been visited on the walk, that arc is dropped. Then the probability of the arcs that are dropped are redistributed to any remaining arcs.

Definition 12. *Let* M *be a PFSM.* $M' = (States, Arcs', Start, Prob)$ *is **derived** from* $M = (Starts, Arcs, Start, Prob)$ *iff*

1. *Arcs' is a maximal subset of Arcs s.t. for every walk* $\sigma_1, \ldots, \sigma_n$ *in* M, *if* $\sigma_1 = \sigma_n$, *then* $(\sigma_{n-1}, \sigma_n) \notin Arcs'$.
2. *For each* $\sigma_i \in States$, *if* $(\sigma_i, \sigma_j) \in Arcs'$, *let* $Prob'((\sigma_i, \sigma_j)) =$

$$\frac{Prob((\sigma_i, \sigma_j))}{\sum_{(\sigma_i, \sigma_k) \in Arcs'} Prob((\sigma_i, \sigma_k))}$$

Example 9. Consider the PFSM in Example 6. The following is derived from it.

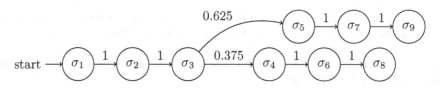

Proposition 3. *If M is a PFSM, and M' is derived from M, then M' is an acyclic PFSM.*

We can regard each walk in M as a Markov chain. Hence, if the graph is acyclic, then we can calculate the probability of walk $\sigma_1, \ldots, \sigma_k$ as follows.

$$\prod_{i \in \{1,\ldots,k-1\}} Prob((\sigma_i, \sigma_{i+1}))$$

For acyclic graphs, the probability that a dialogue is in any particular end state (i.e. state with no arcs out of the state) is the sum of the probabilities of the walks that terminate in that end state. Let $\mathsf{Walks}(M, \sigma) = \{\rho \mid \rho = \sigma_1, \ldots, \sigma_k$ is a walk in M and $\sigma_k = \sigma\}$ be the set of walks with σ as end state.

Proposition 4. *Let M be an PFSM, and let $\mathsf{End}(M)$ be the set of end states in M. If M is acyclic, then $\sum_{\sigma \in \mathsf{End}(M)} \sum_{\rho \in \mathsf{Walks}(M,\sigma)} Prob(\rho) = 1$*

Example 10. For the PFSM in Example 9, $Prob(\sigma_1, \sigma_2, \sigma_3, \sigma_5, \sigma_7, \sigma_9) = 0.625$, and $Prob(\sigma_1, \sigma_2, \sigma_3, \sigma_4, \sigma_6, \sigma_8) = 0.375$.

In the rest of the paper, we assume that we will have an acyclic PFSM. So if a system gives a cycle, we use the above method to get an acyclic PFSM.

4   Analyzing Outcomes through Search

A PFSM generated by a system and an initial state represents the uncertainty that an agent will make any move allowed by the protocol at each point in the dialogue. By exhaustively constructing a search tree, we can analyze an acyclic PFSM. A **search tree** T for an acyclic PFSM M is the smallest tree where every walk ρ in M that terminates in an end state in M (i.e. a state with no transitions out of the state) corresponds to a branch in T. So for each walk $\rho = \tau_1 \ldots \tau_{t-1}$, the source node in τ_1 is the root of the tree, and the destination node in τ_{t-1} is a leaf node. The probability of a leaf is the product of the probability values on the arcs from root to leaf.

We use a **utility function** to measure the value of each leaf in a search tree. For argumentation, there is a wide range of utility functions. In the example below, we have used grounded semantics to determine whether a specified argument (i.e. a goal argument) is in the grounded extension of the argument graph in the public state of the leaf node. A refinement is the **weighted utility function** which weights the utility by $1/d$ where d is the depth of the leaf. This favours shorter dialogues. Further definitions arise from using other semantics and richer formalisms such as valued-based argumentation [15].

Definition 13. *For an PFSM M, where Prob is the probability function, let the search tree be T, let U be a utility function, and let E be a function defined as follows.*

- *If σ is a leaf node in T, then $E(\sigma)$ is $U(\sigma)$.*
- *If σ is a non-leaf node in T, σ has children $\sigma_1, \ldots, \sigma_k$, where for each $i \in \{1, \ldots, k\}$, the transition from σ to σ_i, is τ_i, and the probability of the transition is $Prob(\tau_i)$, then $E(\sigma) = \sum_{i \in \{1, \ldots, k\}} Prob(\tau_i) \times E(\sigma_i)$.*

*If σ is the root of the search tree T, then the **expected tree utility** of T, denoted $E(T, Prob, U)$, is $E(\sigma)$.*

Example 11. In this example, we assume that agent 1 has a goal of making an argument c hold in the grounded extension of the abstract argument graph that exists in the end state of the execution. This goal is represented by the predicate g(c) in the private state of agent 1. In its private state, each agent has zero or more arguments that it holds represented by the predicate h(c), where c is an argument, and zero or more attacks e(d, c) from argument d to argument c. In the public state, each argument c is represented by the predicate a(c). Each agent can add attacks e(d, c) to the public state, if the attacked argument is already in the public state (i.e. a(c) is in the public state), and the agent also has the attacker in its private state (i.e. h(d) is in the private state). Note, r(c) is used to denote that the agent has used its right to present argument c.

$\neg a(a) \wedge h(a) \Rightarrow (1 : \boxplus a(a))$
$\neg r(b) \wedge \neg a(b) \wedge h(b) \wedge a(a) \wedge e(b, a) \Rightarrow (0.5 : \boxplus a(b) \wedge \boxplus e(b, a)) \vee (0.5 : \oplus r(b))$

Let M be the following PFSM representing the system defined by the above action rules.

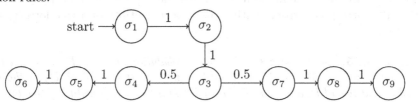

Below we give states for the PFSM. Note, in the end state σ_6 the public state contains the abstract argument graph containing a(a), a(b), and e(b, a) (i.e. the graph is a ← b). and in the end state σ_9 the public state contains the abstract argument graph a(a) (i.e. the graph is a).

$$\sigma_1 = (1, \{h(a), h(b), a(b, a)\}, \{\}, \{\})$$
$$\sigma_2 = (2, \{h(a), h(b), a(b, a)\}, \{a(a)\}, \{\})$$
$$\sigma_3 = (1, \{h(a), h(b), a(b, a)\}, \{a(a)\}, \{\})$$
$$\sigma_4 = (2, \{h(a), h(b), a(b, a)\}, \{a(a), a(b), e(b, a)\}, \{\})$$
$$\sigma_5 = (1, \{h(a), h(b), a(b, a)\}, \{a(a), a(b), e(b, a)\}, \{\})$$
$$\sigma_6 = (0, \{h(a), h(b), a(b, a)\}, \{a(a), a(b), e(b, a)\}, \{\})$$
$$\sigma_7 = (2, \{h(a), h(b), a(b, a), r(b)\}, \{a(a)\}, \{\})$$
$$\sigma_8 = (1, \{h(a), h(b), a(b, a), r(b)\}, \{a(a)\}, \{\})$$
$$\sigma_9 = (0, \{h(a), h(b), a(b, a), r(b)\}, \{a(a)\}, \{\})$$

Suppose the utility of a being in the grounded extension of the graph in the end state is 10, and the utility of a not being in the grounded extension of the graph in the end state is -5. There are two leaves of the search tree to consider: $E(\sigma_6) = -5$ and $E(\sigma_9) = 10$. Hence, the expected tree utility is 2.5.

We can characterize expected tree utility in terms of expected utility of a lottery. We start by briefly reviewing the notion of a lottery. A **lottery** is a probability distribution over a set of possible outcomes. A lottery with possible outcomes $\pi_1,..,\pi_n$, that occur with probabilities $p_1,..,p_n$ respectively, is written as $[p_1, \pi_1; ....; p_n, \pi_n]$. Note, each outcome can be a lottery. For a utility function U, the **expected utility of a lottery** L, denoted $E(L, U)$, is $E(L, U) - \sum_{i=1}^{n} p_i \times U(\pi_i)$.

Definition 14. *Let M be a PFSM, with probability function Prob, and let T be a search tree for M. We define the coding function L as follows:*

- *If σ is a leaf node, then $L(\sigma) = \sigma$;*
- *If σ is a non-leaf node with children $\sigma_1, \ldots, \sigma_k$, and transitions $\tau_1, \ldots, \tau_k$ respectively, then $L(\sigma) = [Prob(\tau_1) : L(\sigma_1), \ldots, Prob(\tau_k) : L(\sigma_k)]$.*

*If σ is the root of the search tree T, then the **tree lottery** for T, denoted $L(T, Prob)$, is $L(\sigma)$.*

Example 12. Let T be a search tree for the PFSM in Example 11. So $L(T, Prob)$ $= [1, [1, [0.5, [1, [1, \sigma_6]; 0.5, [1, [1, \sigma_9]]]]]]$.

The following result shows that evaluating the expected tree utility (i.e. Definition 13) corresponds to evaluating the expected utility of a tree lottery (i.e. Definition 14).

Proposition 5. *Let M be a PFSM, with probability function Prob, and let T be a search tree for M, Let U be a utility function. If $E(T, Prob, U)$ is the expected tree utility of T, and $E(L(T, Prob), U)$ is the expected utility of tree lottery $L(T, Prob)$, then $E(T, Prob, U) = E(L(T, Prob), U)$.*

An agent can then analyze the PFSM to determine the most likely outcomes of a dialogue given any choices it makes. This can be used by the agent to determine its choice of moves in order to optimize its outcomes from the dialogue. It does this by changing its probabilities on its action rules. So that for each action rule, only one action option has probability 1, and the other action options have probability 0 (i.e. we make it deterministic).

Example 13. Consider Example 11. The following set of deterministic rules is the optimal set of rules for agent 1.

$\neg a(a) \wedge h(a) \Rightarrow (1 : \boxplus a(a))$
$\neg r(b) \wedge \neg a(b) \wedge h(b) \wedge a(a) \wedge e(b, a) \Rightarrow (0 : \boxplus a(b) \wedge \boxplus e(b, a)) \vee (1 : \oplus r(b))$

In general for a set of action rules $Rules_1$, let $Rules_1^i$ contain a deterministic version of each rule in $Rules_1$. For each set $Rules_1^i$, together with the set $Rules_2$, and starting state, we obtain the PFSM M^i, and calculate expected tree utility of the search tree T^i. A set $Rules_1^i$ that gives the maximum expected tree utility is an optimal set of deterministic rules for agent 1, and thereby specifies the actions that agent 1 should do to maximize its utility from the dialogue with agent 2 (assuming that $Prob$ is a good estimate of the moves that agent 2 would make).

5 Discussion

In this paper, we have provided a simple, yet expressive language, for specifying the argumentation protocols of agents, and for representing the likelihood that an agent will make any specific move. We have shown how a specification in executable logic can be represented by a PFSM. This can be analyzed to determine expected utility for an agent, and it can be adjusted to optimize the performance of one agent over the other with respect to expected utility.

Some general frameworks for dialogue games have been proposed [17, 7], but they lack sufficient detail to formally analyze or implement specific systems. A more detailed framework, that is based on situation calculus, has been proposed by Brewka [18], though the emphasis is on protocols, and not on the likelihood of moves, or of strategies. Probability theory (for example [19–21]) and utility theory (for example [22–25, 20]) have been considered in other frameworks for multi-agent argumentation though none of these offers a general logical language for specifying diverse protocols for dialogical argumentation with abstract and logic-based arguments, and for representing the uncertainty of moves made by each agent in argumentation. We believe this is the first paper that provides a general framework for specifying probabilistic protocols for dialogical argumentation, and for using them for optimizing the probabilistic strategies in terms of the expected utility of the Markov chains that can be obtained by the protocols. Potentially, this is a powerful tool for assessing agents in dialogical argumentation, and optimizing their outcomes from the argumentation.

References

1. Amgoud, L., Maudet, N., Parsons, S.: Arguments, dialogue and negotiation. In: European Conf. on Artificial Intelligence (ECAI 2000), pp. 338–342. IOS Press (2000)
2. Black, E., Hunter, A.: An inquiry dialogue system. Autonomous Agents and Multi-Agent Systems 19(2), 173–209 (2009)
3. Dignum, F., Dunin-Keplicz, B., Verbrugge, R.: Dialogue in team formation. In: Dignum, F.P.M., Greaves, M. (eds.) Issues in Agent Communication. LNCS, vol. 1916, pp. 264–280. Springer, Heidelberg (2000)
4. Fan, X., Toni, F.: Assumption-based argumentation dialogues. In: Proceedings of International Joint Conference on Artificial Intelligence (IJCAI 2011), pp. 198–203 (2011)

5. Hamblin, C.: Mathematical models of dialogue. Theoria 37, 567–583 (1971)
6. Mackenzie, J.: Question begging in non-cumulative systems. Journal of Philosophical Logic 8, 117–133 (1979)
7. McBurney, P., Parsons, S.: Games that agents play: A formal framework for dialogues between autonomous agents. Journal of Logic, Language and Information 11, 315–334 (2002)
8. McBurney, P., van Eijk, R., Parsons, S., Amgoud, L.: A dialogue-game protocol for agent purchase negotiations. Journal of Autonomous Agents and Multi-Agent Systems 7, 235–273 (2003)
9. Parsons, S., Wooldridge, M., Amgoud, L.: Properties and complexity of some formal inter-agent dialogues. J. of Logic and Comp. 13(3), 347–376 (2003)
10. Prakken, H.: Coherence and flexibility in dialogue games for argumentation. J. of Logic and Comp. 15(6), 1009–1040 (2005)
11. Walton, D., Krabbe, E.: Commitment in Dialogue: Basic Concepts of Interpersonal Reasoning. SUNY Press (1995)
12. Black, E., Hunter, A.: Executable logic for dialogical argumentation. In: European Conf. on Artificial Intelligence (ECAI 2012), pp. 15–20. IOS Press (2012)
13. Hunter, A.: Analysis of dialogical argumentation via finite state machines. In: Liu, W., Subrahmanian, V.S., Wijsen, J. (eds.) SUM 2013. LNCS, vol. 8078, pp. 1–14. Springer, Heidelberg (2013)
14. Wooldridge, M., McBurney, P., Parsons, S.: On the meta-logic of arguments. In: Parsons, S., Maudet, N., Moraitis, P., Rahwan, I. (eds.) ArgMAS 2005. LNCS (LNAI), vol. 4049, pp. 42–56. Springer, Heidelberg (2006)
15. Bench-Capon, T.: Persuasion in practical argument using value based argumentation frameworks. Journal of Logic and Computation 13(3), 429–448 (2003)
16. Rabin, M.: Probabilistic automata. Information and Control 6, 230–245 (1963)
17. Maudet, N., Evrard, F.: A generic framework for dialogue game implementation. In: Proc. 2nd Workshop on Formal Semantics & Pragmatics of Dialogue, University of Twente, pp. 185–198 (1998)
18. Brewka, G.: Dynamic argument systems: A formal model of argumentation processes based on situation calculus. J. Logic & Comp. 11(2), 257–282 (2001)
19. Hunter, A.: Modelling uncertainty in persuasion. In: Liu, W., Subrahmanian, V.S., Wijsen, J. (eds.) SUM 2013. LNCS, vol. 8078, pp. 57–70. Springer, Heidelberg (2013)
20. Rienstra, T., Thimm, M., Oren, N.: Opponent models with uncertainty for strategic argumentation. In: Proceedings of IJCAI 2013. IJCAI/AAAI (2013)
21. Hadjinikolis, C., Siantos, Y., Modgil, S., Black, E., McBurney, P.: Opponent modelling in persuasion dialogues. In: Proceedings of IJCAI (2013)
22. Rahwan, I., Larson, K.: Pareto optimality in abstract argumentation. In: Proceedings of the Twenty-Third AAAI Conference on Artificial Intelligence (AAAI 2008). AAAI Press (2008)
23. Riveret, R., Prakken, H., Rotolo, A., Sartor, G.: Heuristics in argumentation: A game theory investigation. In: Computational Models of Argument (COMMA 2008). Frontiers in Artificial Intelligence and Applications, vol. 172, pp. 324–335. IOS Press (2008)
24. Matt, P., Toni, F.: A game-theoretic measure of argument strength for abstract argumentation. In: Hölldobler, S., Lutz, C., Wansing, H. (eds.) JELIA 2008. LNCS (LNAI), vol. 5293, pp. 285–297. Springer, Heidelberg (2008)
25. Oren, N., Norman, T.: Arguing using opponent models. In: McBurney, P., Rahwan, I., Parsons, S., Maudet, N. (eds.) ArgMAS 2009. LNCS, vol. 6057, pp. 160–174. Springer, Heidelberg (2010)

Analytics over Probabilistic Unmerged Duplicates

Ekaterini Ioannou and Minos Garofalakis

Technical University of Crete, Chania, Greece
{ioannou,minos}@softnet.tuc.gr

Abstract. This paper introduces probabilistic databases with unmerged duplicates (DB^{ud}), i.e., databases containing probabilistic information about instances
found to describe the same real-world objects. We discuss the need for efficiently
querying such databases and for supporting practical query scenarios that require
analytical or summarized information. We also sketch possible methodologies
and techniques that would allow performing efficient processing of queries over
such probabilistic databases, and especially without the need to materialize the
(potentially, huge) collection of all possible deduplication worlds.

1 Introduction

Entity Deduplication is the task of processing a data set in order to create **entities** by
merging the data set **instances** that describe the same real-world **objects**. Traditional
deduplication techniques [4] are based on an a-priori merging of instances: they first detect the possible matches between instances, and then, given a threshold, decide which
instances to merge into entities. The entities resulting from the merges are then used
for replacing the coreference instances in the original data set. Query processing is performed over the updated data set.

To handle the new resolution challenges, the recently introduced approaches (e.g.,
[1], [6], and [9]) moved towards databases that maintain and incorporate *unmerged duplicates*. These approaches perform only the first part of the resolution process, which is
the identification of the possible matches between the instances. This is the deduplication information, and it corresponds to a set of possible linkages between instances. In
some approaches each linkage is accompanied with a probability that reflects the belief
of the deduplication technique that the specific two instances describe the same real-
world object. The resulting information is not used for performing entity merges (using
a given threshold), but is stored alongside the original data. The complete deduplication
is performed during query processing, and thus answers reflect the different real-world
situations that are encoded in the deduplication information. In case the deduplication
information is probabilistic, as for instance in [1] and [6], then the probabilities are used
for computing the overall probability of each query answer.

Although answering queries over unmerged duplicates is important, it is still just
a first step towards a complete solution to the problem. The typical situation is that
the unmerged duplicates are part of a large database that of course contains other tables. Consequently, users would require retrieving information related to all data in the
database, duplicated or not. However, this would require generating and considering all
the possible worlds, which is typically huge [2] and will overwhelm the user instead

U. Straccia and A. Calì (Eds.): SUM 2014, LNAI 8720, pp. 203–208, 2014.

Buyer						Deduplication				Order			
id	name	surname	loc.	gender	year	id	inst_1	inst_2	pr	id	buyer	items	amount
r_1	Marion	Smith	GR	female	2009	l_{r_1,r_2}	r_1	r_2	0.95	t_1	r_1	1	100
r_2	Marion	Smith	DE	female	2010	l_{r_1,r_3}	r_1	r_3	0.55	t_2	r_2	2	300
r_3	Mary	Smith	DE	female	2011					t_3	r_2	4	250
										t_4	r_3	2	250

Fig. 1. A fragment of a probabilistic database with unmerged duplicates

of providing useful information. In addition, users might not care about the exact entities but rather on obtaining insights through analytical and summarizing queries, as for example performed in the online analytical processing.

In this paper, we introduce DB^{ud}: a database containing probabilistic information about instances found to describe the same real-world objects. DB^{ud} adopts the most expressive form of deduplication information (i.e., probabilistic linkages between instances – also accounting for transitivity), and significantly extends its scope by considering the deduplication information as part of a database with other tables providing entity-related data. In the following sections, we first introduce analytical queries for retrieving information of the entities in DB^{ud} (Section 2), and then sketch possible methodologies and techniques for efficiently processing queries over such a probabilistic database with unmerged duplicates (Section 3).

2 Modeling Data and Queries

A probabilistic database with unmerged duplicates DB^{ud} contains deterministic relational tables $T_1, ..., T_n$ as well as tables with duplicates $R_1, ..., R_k$, i.e., some instances of R_i describe the same real-world objects. The deduplication information for table R_i is given in table L_i. More specifically, L_i contains probabilistic linkages over the instances in R_i: $l_{r_\alpha,r_\beta} \in L_i$ means that instances r_α and r_β from R_i describe the same real-world object with probability p^l.

To process queries over DB^{ud} we must be able to support joins between the tables with unmerged duplicates and the deterministic tables. For example, answering queries over the DB^{ud} fragment shown in Figure 1 requires considering the join between table Buyer with Order. Since table Buyer contains duplicates, we must first derive the possible entities using the deduplication information provided in table Deduplication. Each linkage from the Deduplication table can be either accepted or rejected, e.g., we can accept l_{r_1,r_3} with probability 0.55 or reject it with probability (1-0.55). Rejecting the linkage means that the database has two entities, one for each of the instances. Accepting the linkage implies a new entity, with identifier $e_{1,3}$, that replaces both r_1 and r_3. Creating a single entity given these two instances maybe performed using different semantics. For example, if we assume that we keep the instance with the highest value on the year attribute, the tuple for the merge between instances r_1 and r_3 is ⟨$e_{1,3}$, "Mary", "Smith", "DE", "female", "2011"⟩.

For creating the possible entities of a table with unmerged duplicates R_i we need to consider the acceptance and rejection of each linkage of L_i. Deciding which linkages from L_i (e.g., table Deduplication from Figure 1) to accept or reject leads to a huge

Table 1. The possible deduplication worlds with the entities created when requesting the join between Order with Buyer and summation over the Order's "amount" for each entity

	Linkages	Prob.	Entities (with summation over Order's amount)
I_1	l_{r_1,r_2} & l_{r_1,r_3}	0.5225	$\langle e_{1,2,3}, ..., 2009, DE, 900\rangle$
I_2	l_{r_1,r_2} & $\neg l_{r_1,r_3}$	0.4275	$\langle e_{1,2}, ..., 2010, DE, 650\rangle$, $\langle e_3, ..., 2011, DE, 250\rangle$
I_3	$\neg l_{r_1,r_2}$ & l_{r_1,r_3}	0.0275	$\langle e_{1,3}, ..., 2011, DE, 800\rangle$, $\langle e_2, ..., 2010, DE, 550\rangle$
I_4	$\neg l_{r_1,r_2}$ & $\neg l_{r_1,r_3}$	0.0225	$\langle e_1, ..., 2009, GR, 100\rangle$, $\langle e_2, ..., 2010, DE, 550\rangle$, $\langle e_3, ..., 2011, DE, 250\rangle$

(exponentially-large) number of situations, termed *possible deduplication worlds*. Generating all these situations is infeasible. In addition, the huge volume of results that would arise when processing queries over all possible worlds would make it impossible for users to derive any meaningful information.

We suggest to address these issues by applying analytical operators and qualifiers over the possible deduplication worlds. In particularly, we introduce the following two levels of aggregation:

- **First Aggregation Level:** performs aggregation within each possible deduplication world and uses conventional SQL aggregate semantics over the merged entities. For example, consider again the data from Figure 1. Accepting both linkages of table Buyer leads to entity $e_{1,2,3}$, which would join with tuples t_1, t_2, t_3 and t_4 from table Order. The summation over the Order's "amount" is thus 900. Table 1 shows the four deduplication worlds created when requesting the join between Order and Buyer with summation over the Order's "amount" for each entity. Note that we also need to identify and ignore the deduplication worlds in which the entities created by the accepted linkages are not satisfied by the rejected linkages. For example, the deduplication world with entity $e_{\alpha,\beta,\gamma}$ is invalid if it was created by accepting the linkages $l_{\alpha,\beta}$ and $l_{\alpha,\gamma}$ and rejecting linkage $l_{\beta,\gamma}$.
- **Second Aggregation Level:** performs aggregation across all possible deduplication worlds and over all the records created by the first level and based on one (or more) query attributes of interest. The goal is to further reduce the number of information that is created by the first aggregation level, which would help users to reach vital business decisions easier and faster.

As an example, consider again the data from Figure 1 and that a manager wants to retrieve the range of possible total Order amounts per location. The manager poses the following query:

> **SELECT** Buyer.location, range(entity_amount), prob
> **FROM** Order **entity-join** Buyer **based on** Deduplication
> **using** sum(Order.amount) as entity_amount
> **WHERE GROUP BY** Buyer.location

Although not directly expressed in the query, the ENTITY-JOIN implies aggregation of the records corresponding to each entity in the possible worlds by assuming an implicit group-by operator over the entities (i.e., first aggregation level). Evaluating the (explicit) GROUP BY clause over the resulting records gives two locations: "GR" and "DE" (i.e., second aggregation level). Consider now all entities in the possible worlds, i.e., I_{1-4} of Table 1. The amount summation for location "GR" is 100, and for location "DE" it is between 250 and 900, and thus the range is [250-900]. The probability for each location is the summation of the possible worlds in which they participate. The location-range pairs along with their probabilities that compose the answer set are {⟨"GR", [100-100], 0.0225⟩, ⟨"DE", [250-900], 1⟩}.

The manager also wants to retrieve the two most likely aggregate amounts spent by buyers in 2010, along with their respective probabilities. This is basically an iceberg query as it allows users to find the *high-probability deduplication scenarios* satisfying specific selection predicates. The query posed by the manager is now the following:

> **SELECT** top-2 entity_amount, prob
> **FROM** Order **entity-join** Buyer **based on** Deduplication
> **using** sum(Order.amount) as entity_amount
> **WHERE** Buyer.year=2010

The entities satisfying the WHERE conditions are e_2 from possible worlds I_3 and I_4, $e_{1,2}$ from I_2. The probability of each entity is the summation of the probabilities of the worlds in which it participates, i.e., 0.05 for e_2 and 0.4275 for $e_{1,2}$. By default, the entities are ordered by probability, thus, the answer for this query is {⟨650, 0.4275⟩, ⟨550, 0.05⟩}.

Our vision is to provide complex aggregation and iceberg queries that will allow users to efficiently retrieve statistical information about the possible deduplicated entities. As shown in the above examples, a vital operator is a novel ENTITY-JOIN, which will allow expressing joins between a table with unmerged duplicates R_i and deterministic database table T_j. Entities are created using summation, count, minimum, or maximum aggregation over the T_j tuples. The ENTITY-JOIN can be used for query analytics using either aggregation operators (e.g., range, mean and variance[1]) or iceberg operators (e.g., top-k). Instead of top-k, we could also consider simply specifying a lower bound on the probability of the returned aggregate values.

Users might also be interested in retrieving results with more details, probably after executing aggregation queries, which basically implies reversing parts of the performed summarization. This can be performed with a *"drill down"* qualifier, similar to the corresponding qualifier of online analytical processing.

Providing efficient operators for constructing entities given a set of instances is also useful for query processing over DB^{ud}. The majority of the existing deduplication approaches either do not deal with this issue or simply return the most recent instance or the union of all instances. To provide such operators, we could for example consider the [11] approach from information extraction, which constructs entities by detecting a canonical value for each attribute given the corresponding values from all the instances.

3   Possible Mechanisms for Efficient Query Processing

For providing analytics over DB^{ud}, we need to introduce new mechanisms and techniques that exploit processing of aggregation and iceberg queries without the need to

---

[1] Mean can be used for retrieving the average value over the ranges of all possible merges and variance for indicating the typical discrepancy of the expected value.

materialize the possible worlds. Other important aspects that we must consider, include the efficient computation of probabilities over the resulting answers, and the linkage transitivity requirement that, among other things, implies the need for reasoning at query time.

Aggregation Queries. This type of queries has been so far studied only by very few approaches. For example, processing aggregation queries is the main goal of [5]. It is achieved by the structural decompositions of expressions into sub-expressions that are independent and mutually exclusive. DB^{ud} needs to support a more expressive form of aggregation, which captures two aggregation levels.

Another existing approach that targets aggregate operators is [9]. However, there exist crucial differences with the aggregate operators required for DB^{ud}. One difference is that the model followed in [9] assumes that the algorithm is provided with fixed clusters of instances, which allows focusing on basic query-time aggregation. In sharp contrast to [9], DB^{ud} follows a more generic deduplication model that requires dealing also with linkages between instances as well as linkage transitivity. In addition, DB^{ud} also considers probabilistic linkages, in order to capture the relevant entity-linkage uncertainty. Another difference is that DB^{ud} supports a more expressive query syntax in comparison to [9], which includes two aggregation levels and additional aggregation functions.

Processing aggregation queries over DB^{ud} could be efficiently achieved by limiting the number of possible worlds to be materialized or by partially materializing possible worlds. For instance, for minimum and maximum aggregates we do not need to use all the records but rather only one record from T_i for each instance from R_i. As an example, consider again the data of Order from Figure 1. When processing a query with a maximum aggregate, we can safely ignore all tuples related to a specific r_i except the one with the highest amount, i.e., for r_2 we keep only tuple t_2 since this provides the highest amount among all tuples related to r_2.

Iceberg Queries. In contrast to deterministic data, iceberg queries (i.e., top-k) for uncertain data have different interpretations [10]: the top-k tuples from the possible world with the highest probability, the set of k tuples that have the highest aggregated probability to appear together across all possible worlds [8, 10] (called "U-Top$k$"), and the k tuples from any possible world as long as they have the highest probabilities [10] (called "U-kRanks"). For DB^{ud}, this query type corresponds to retrieving the k single-item answers with the highest probabilities (i.e., Top$k$ from [8], k U-Top1 from [10]). Ré et al. [8] process U-Top$k$ through Monte-Carlo simulation. They maintain probability intervals that are then tightened by generating random possible worlds. Soliman et al. [10] introduced a framework that navigates the space of possible worlds in order to generate the top-k tuples. More recent top-k related approaches are [7] and [3]. The approach in [7] shares the probability computation of detected subqueries with several query answer, and further extends for the computation of bounds. The goal of [3] is similar, but here the authors achieve the computation of bounds without materialization.

One option for processing iceberg queries over DB^{ud}, is to create an indexing structure that detects and maintains the entities with the highest probabilities. Ideally, the indexing structure would provide efficient access to the information encoded through the linkages (i.e., potential merges) and allow easy construction of possible worlds (or partial possible worlds), as well as the fast retrieval of their probabilities. Thus, DB^{ud}

would not need to perform a full on-the-fly materialization, but rather directly retrieve query answers, or part of them, from the indexing structure.

4 Summary

In this paper we have presented probabilistic databases with unmerged duplicates, i.e., databases with duplicated instances and probabilistic linkages between duplicated instances. We discussed the need for efficiently supporting practical query scenarios that do not require retrieving the huge collection of all possible deduplication worlds, but rather analytical or summarized information. This primarily involves query analytic, including aggregation and iceberg queries. We have also sketched possible methodologies and techniques that would allow the efficient processing of queries over such probabilistic databases, and especially without the need to materialize the collection of all possible deduplication worlds.

Acknowledgment. This work has been partially funded from the European Unions Seventh Framework Programme under Grant Agreement 619525 (QualiMaster) and Reference number 249217 (HeisenData).

References

[1] Andritsos, P., Fuxman, A., Miller, R.: Clean answers over dirty databases: A probabilistic approach. In: ICDE (2006)

[2] Dalvi, N., Suciu, D.: Efficient query evaluation on probabilistic databases. VLDB 16(4) (2007)

[3] Dylla, M., Miliaraki, I., Theobald, M.: Top-k query processing in probabilistic databases with non-materialized views. In: ICDE (2013)

[4] Elmagarmid, A., Ipeirotis, P., Verykios, V.: Duplicate record detection: A survey. TKDE 19(1) (2007)

[5] Fink, R., Han, L., Olteanu, D.: Aggregation in probabilistic databases via knowledge compilation. PVLDB 5(5) (2012)

[6] Ioannou, E., Nejdl, W., Niederée, C., Velegrakis, Y.: On-the-fly entity-aware query processing in the presence of linkage. PVLDB 3(1) (2010)

[7] Olteanu, D., Wen, H.: Ranking query answers in probabilistic databases: Complexity and efficient algorithms. In: ICDE (2012)

[8] Ré, C., Dalvi, N., Suciu, D.: Efficient top-k query evaluation on probabilistic data. In: ICDE (2007)

[9] Sismanis, Y., Wang, L., Fuxman, A., Haas, P., Reinwald, B.: Resolution-aware query answering for business intelligence. In: ICDE (2009)

[10] Soliman, M., Ilyas, I., Chang, K.: Top-k query processing in uncertain databases. In: ICDE (2007)

[11] Wick, M., Rohanimanesh, K., Schultz, K., McCallum, A.: A unified approach for schema matching, coreference and canonicalization. In: KDD (2008)

A Psychological Analysis of Preference Semantics in Conditional Logics for Preference Representation

Souhila Kaci[1] and Eric Raufaste[2]

[1] LIRMM - UMR 5506, Montpellier, France
kaci@lirmm.fr
[2] CLLE-LTC, CNRS UMR 5263, Toulouse, France
raufaste@univ-tlse2.fr

Abstract. Qualitative and comparative preference statements of the form "prefer α to β" are useful components of many applications. This statement leads to the comparison of two sets of alternatives: the set of alternatives in which α is true and the set of alternatives in which β is true. Different ways are possible to compare two sets of objects leading to what is commonly known as preference semantics. The choice of the semantics to employ is important as they differently rank-order alternatives. Existing semantics are based on philosophical and non-monotonic reasoning grounds. In the meanwhile, they have been widely and mainly investigated by AI researchers from algorithmic point of view. In this paper, we come to this problem from a new angle and complete existing theoretical investigations of the semantics. In particular, we provide a comparison of the semantics on the basis of their psychological plausibility by evaluating their closeness to human behavior.

1 Introduction

Preferences are fundamental in many scientific researches as well as applications. One of the main problems an individual faces when expressing her/his preferences may be due to the number of variables (or attributes) that she/he must take into account to evaluate the different alternatives. This is because the number of alternatives increases exponentially with the number of variables, making the direct assessment of individual preferences over the whole set of alternatives simply infeasible. Fortunately it is commonly acknowledged that individuals generally express preferences over partial descriptions of alternatives. They often take the form of qualitative comparative preference statements, e.g., "I like London more than Paris" and "I prefer tea to coffee". Individuals may express their comparative statements w.r.t. some context, e.g., "If fish is served, then I prefer white wine to red wine". Therefore comparative preference statements allow to express general preferences (e.g., "I prefer fish to meat") and specific preferences in particular contexts (e.g., "If red wine is served, I prefer meat to fish").

AI researchers have developed compact preference representation languages to cope with comparative preference statements. They use more or less strong ways to compare two sets of alternatives. These ways are referred to as preference semantics. So far these semantics have been studied in AI on entirely technical grounds, mainly by characterizing a unique rank-ordering of alternatives. It goes without saying that the choice of

U. Straccia and A. Calì (Eds.): SUM 2014, LNAI 8720, pp. 209–222, 2014.

a particular semantics to employ is important as crucial decisions may be made on the basis of the rank-ordering of the alternatives induced by the semantics. While we know a lot about the technical machinery of existing semantics, we know relatively much less how the latter would be close to individuals preferences if the individuals were able to rank-order the whole set of alternatives. Our aim in this paper is to address the issue of psychological plausibility of the main semantics studied in the literature. Cognitive psychology is a research discipline which aims to understand human cognitive functions such as reasoning, judgment and decision-making. It has not a normative but a descriptive stance. Based on experiments, cognitive psychology provides hints on the validity of hypotheses one may have on the sources of human behavior. It has already been used to evaluate non-monotonic reasoning approaches [1,5,9] and decision theories [10].

2   Background

Let $V = \{X_1, \ldots, X_h\}$ be a set of h variables, each takes its values in a finite domain $Dom(X_i)$. A possible alternative, denoted by ω, is the result of assigning a value in $Dom(X_i)$ to each variable X_i in V. Ω denotes the set of all possible alternatives. We suppose that this set is fixed and finite. Let $\mathcal{L}$ be a language based on V and the symbols $\wedge, \vee$ and $\neg$ which respectively correspond to conjunction, disjunction and negation. Formulas are built on $\mathcal{L}$ using atomic formulas of the form $X_i = A_i$, with $A_i \in Dom(X_i)$. When there is no ambiguity, $X_i = A_i$ is replaced with A_i. $Mod(\alpha)$ denotes the set of alternatives that make the formula α true. It is also called α-*alternatives*.

A preference relation $\succeq$ on Ω is a binary relation. For $\omega, \omega' \in \Omega$, the statement $\omega \succeq \omega'$ stands for "ω is at least as preferred as ω'". The notation $\omega \succ \omega'$ means that ω is strictly preferred to ω'. We have $\omega \succ \omega'$ if $\omega \succeq \omega'$ holds but $\omega' \succeq \omega$ does not. $\succeq$ is total if and only if $\forall \omega, \omega' \in \Omega$, either $\omega \succeq \omega'$ or $\omega' \succeq \omega$ holds. We suppose that a preference relation $\succeq$ is a preorder (reflexive and transitive). $\succeq$ is cyclic if and only if $\exists \omega, \omega', \cdots, \omega'' \in \Omega$ such that $\omega \succ \ldots \succ \omega' \succ \omega'' \succ \omega$ holds. Otherwise it is acyclic.

For convenience, an acyclic total preorder $\succeq$ can also be represented by a well ordered partition of Ω. A set of sets of alternatives of the form $(E_1, \ldots, E_n)$ is a partition of Ω if and only if (i) $\forall i, E_i \neq \emptyset$, (ii) $E_1 \cup \ldots \cup E_n = \Omega$, and (iii) $\forall i, j, E_i \cap E_j = \emptyset$ for $i \neq j$. A sequence $(E_1, \ldots, E_n)$ is an ordered partition of $(\Omega, \succeq)$ if $(E_1, \ldots, E_n)$ is a partition of Ω and ($\forall \omega, \omega' \in \Omega$ with $\omega \in E_i, \omega' \in E_j$ we have $i \leq j$ iff $\omega \succeq \omega'$).

Given an acyclic total preorder $\succeq$ on Ω and its associated ordered partition $(E_1, \cdots, E_n)$, a ranking of outcomes can be defined. Preferred outcomes w.r.t. $\succeq$ receive the rank 1. They are alternatives in E_1. Next preferred ones receive the rank 2 (they are alternatives in E_2), and so on.

3   Comparative Preference Statements

Individuals express their preferences in different forms. Often, these preferences implicitly or explicitly refer to qualitative comparative preference statements of the form "prefer α to β", where α and β are formulas. This statement serves to rank-order the set of alternatives. Let $\succeq$ be the underlying preference relation. Given the statement "prefer α to β", α-alternatives are expected to be preferred to β-alternatives w.r.t. $\succeq$. Dealing with the statement "prefer α to β" is easy when both α and β refer to an alternative. However this task becomes more complex when α and β refer to sets of alternatives and share some alternatives. In order to prevent this situation von Wright [11] interprets the statement "prefer α to β" as a choice problem between $\alpha \wedge \neg\beta$-alternatives and $\beta \wedge \neg\alpha$-alternatives. Particular situations are those when $\alpha \wedge \neg\beta$ (resp. $\beta \wedge \neg\alpha$) is a contradiction or not feasible, i.e. there is no alternative in Ω which satisfies $\alpha \wedge \neg\beta$ (resp. $\beta \wedge \neg\alpha$). In both cases, $\alpha \wedge \neg\beta$ (resp. $\beta \wedge \neg\alpha$) is replaced with α (resp. β) following [11,6]. For simplicity we suppose that both $\alpha \wedge \neg\beta$ and $\beta \wedge \neg\alpha$ are consistent and feasible.

Comparative preference statements may be expressed w.r.t. some context. They are of the form "if γ, prefer α to β". This means that we prefer $\gamma \wedge \alpha \wedge \neg\beta$-alternatives to $\gamma \wedge \beta \wedge \neg\alpha$-alternatives which corresponds to "prefer $\gamma \wedge \alpha$ to $\gamma \wedge \beta$". Indeed for simplicity and without loss of generality, we focus on statements of the form "prefer α to β", denoted $\alpha \rhd \beta$. There are several ways to compare $\alpha \wedge \neg\beta$-alternatives and $\beta \wedge \neg\alpha$-alternatives. They are called preference semantics.

Definition 1. *Let $\succeq$ be a preference relation on Ω, and two formulas α and β. Consider $\alpha \rhd \beta$.*

– **Strong semantics** *[4,12]*
  $\succeq$ *satisfies $\alpha \rhd \beta$ following* strong semantics, *denoted by* $\succeq \models \alpha \rhd_{st} \beta$, *iff* $\forall\omega \in Mod(\alpha \wedge \neg\beta), \forall\omega' \in Mod(\beta \wedge \neg\alpha), \omega \succ \omega'$.

– **Ceteris paribus semantics** *[6]*
  $\succeq$ *satisfies $\alpha \rhd \beta$ following* ceteris paribus semantics, *denoted by* $\succeq \models \alpha \rhd_{cp} \beta$, *iff* $\forall\omega \in Mod(\alpha \wedge \neg\beta), \forall\omega' \in Mod(\beta \wedge \neg\alpha), \omega \succ \omega'$ *if ω and ω' have the same valuation of variables which do not appear in $\alpha \wedge \neg\beta$ and $\beta \wedge \neg\alpha$[1].*

– **Optimistic semantics** *[4]*
  $\succeq$ *satisfies $\alpha \rhd \beta$ following* optimistic semantics, *denoted by* $\succeq \models \alpha \rhd_{opt} \beta$, *iff* $\exists\omega \in Mod(\alpha \wedge \neg\beta), \forall\omega' \in Mod(\beta \wedge \neg\alpha), \omega \succ \omega'$.

– **Pessimistic semantics** *[2]*
  $\succeq$ *satisfies $\alpha \rhd \beta$ following* pessimistic semantics, *denoted by* $\succeq \models \alpha \rhd_{pes} \beta$, *iff* $\exists\omega' \in Mod(\beta \wedge \neg\alpha), \forall\omega \in Mod(\alpha \wedge \neg\beta), \omega \succ \omega'$.

A preference set of type $\rhd_*$, denoted by $\mathcal{P}_{\rhd_*}$, is a set of preferences of the form $\{p_i \rhd_* q_i | i = 1, \ldots, n\}$, with $* \in \{st, cp, opt, pes\}$. An acyclic preorder $\succeq$ is a model of $\mathcal{P}_{\rhd_*}$ if and only if $\succeq$ satisfies each preference $p_i \rhd_* q_i$ in $\mathcal{P}_{\rhd_*}$. A preference set $\mathcal{P}_{\rhd_*}$ is consistent if it has a model.

Generally we have to deal with several comparative preference statements expressed by an individual. There are mainly two kinds of queries in preference representation: either one looks for the maximally preferred alternatives or compares two alternatives. In many applications individuals are more concerned with the preferred alternatives. In the case where these alternatives are not satisfactory (e.g. preferred menus are too expensive), then we need to compute the preferred alternatives among remaining ones, and so on. In order to accommodate these considerations, we associate a total preorder to

---

[1] This is the definition of ceteris paribus the most used in the literature. Other definitions have also been proposed. See [6].

a preference set. Different total preorders may satisfy (i.e., are models of) a preference set given a semantics. However it is widely acknowledged that, for decision purposes, it is more convenient to characterize a *unique* total preorder [4]. Specificity principle has been commonly used to characterize a unique model per semantics.

Definition 2. *[13] Let $\succeq$ and $\succeq'$ be two total preorders on Ω represented by ordered partitions $(E_1, \ldots, E_n)$ and $(E'_1, \ldots, E'_{n'})$ respectively. We say that $\succeq$ is less specific than $\succeq'$, written as $\succeq \sqsubseteq \succeq'$, iff $\forall \omega \in \Omega$, if $\omega \in E_i$ and $\omega \in E'_j$ then $i \leq j$. $\succeq$ belongs to the set of minimally (resp. maximally) specific preorders, among a set of total preorders, if and only if there is no preorder in the set that is strictly less (resp. more) specific than $\succeq$. If $\succeq$ is the unique minimally (resp. maximally) specific total preorder then it is called the least (resp. most) specific preorder.*

It is worth noticing that strict statements "prefer α to β" have been extended with equal preference statements of the form "α and β are equally preferred" [3,8], denoted $\alpha =_* \beta$ with $* \in \{st, cp, opt, pes\}$. Therefore the computation of unique models applies on the set $\mathcal{P}_{\triangleright_*} \cup \mathcal{P}_{=_*}$. Proposition 1 summarizes existing results about the uniqueness of models (which are total preorders) for each semantics.

Proposition 1. *[3,8] Let $\mathcal{P}_{\triangleright_*} \cup \mathcal{P}_{=_*}$ be a consistent preference set.*

- *The least specific model of $\mathcal{P}_{\triangleright_{opt}} \cup \mathcal{P}_{=_{opt}}$ (resp. $\mathcal{P}_{\triangleright_{st}} \cup \mathcal{P}_{=_{st}}$, $\mathcal{P}_{\triangleright_{cp}} \cup \mathcal{P}_{=_{cp}}$) exists.*
- *The most specific model of $\mathcal{P}_{\triangleright_{pes}} \cup \mathcal{P}_{=_{pes}}$ (resp. $\mathcal{P}_{\triangleright_{st}} \cup \mathcal{P}_{=_{st}}$, $\mathcal{P}_{\triangleright_{cp}} \cup \mathcal{P}_{=_{cp}}$) exists.*
- *The most (resp. least) specific model of $\mathcal{P}_{\triangleright_{opt}} \cup \mathcal{P}_{=_{opt}}$ (resp. $\mathcal{P}_{\triangleright_{pes}} \cup \mathcal{P}_{=_{pes}}$) does not always exist.*

Due to space limitation we do not present the algorithms to compute unique models. We refer the reader to [8].

Note that when the set of preferences at hand is inconsistent, the algorithms return a preorder in which not all alternatives are rank-ordered. More specifically, alternatives which are involved in cycles are not rank-ordered (i.e., they are excluded). In this case we speak about a preference relation (and not a model) of the preference set following the principle used. For example the set $\mathcal{P}_{\triangleright_{st}} = \{t_1 : W \triangleright_{st} S, t_2 : B \wedge S \triangleright_{st} B \wedge W, t_3 : W \wedge A \triangleright_{st} W \wedge H\}$ is inconsistent w.r.t. both specificity principles. The preference relation associated with $\mathcal{P}_{\triangleright_{st}}$ following the minimal specificity principle is $\succeq = (\{AMW\})$. We say that the alternatives in $\Omega \backslash \{AMW\}$ are not ranked w.r.t. strong semantics and also w.r.t. the corresponding algorithm. It is worth noticing that alternatives which are not excluded in case of inconsistent preferences are rank-ordered following the underlying principle (minimal or maximal). In the next part of the paper, we will need to deal with both consistent and inconsistent preference sets. In order to encompass both cases, we will speak about the preference relation associated with the set of preferences given a semantics. The preference relation is the model of the preference set if the latter is consistent given the semantics at hand. It is a preorder in which some alternatives are excluded if the preference set is inconsistent given the semantics at hand. In both cases, the preference relation is computed using the corresponding algorithm.

One may observe that, besides the technical machinery of the semantics, we have no indication how they are close to human behavior. In the next part of this paper we aim at providing experimental data to compare the various semantics on the basis of their psychological plausibility.

4 Experimental Study

4.1 Rationale

Recall that strong and ceteris paribus semantics obey both minimal and maximal specificity principles. Therefore we empirically evaluate six semantics:

- optimistic (O),
- pessimistic (P),
- strong-optimistic (SO) when the minimal specificity principle is applied,
- ceteris paribus-optimistic (CPO) when the minimal specificity principle is applied,
- strong-pessimistic (SP) when the maximal specificity principle is applied,
- ceteris paribus-pessimistic (CPP) when the maximal specificity principle is applied.

The main idea is to assess the fit between each semantics and human behavior so as to compare the semantics based on their psychological plausibility. Let us consider three variables X_1, X_2 and X_3 with $Dom(X_1) = \{A_1, A_2\}$, $Dom(X_2) = \{B_1, B_2\}$ and $Dom(X_3) = \{C_1, C_2\}$. Therefore we have eight possible alternatives, called "3-variable alternatives". We construct a set of comparative preference statements involving 1- or 2-variable alternatives, e.g. $\{A_1 \rhd A_2, A_1 \wedge C_1 \rhd A_1 \wedge C_2, C_2 \rhd C_1\}$. Given a semantics this set induces a preference relation (hence a ranking) over the eight alternatives. The experimental study comprised four steps.

- The first three steps are separately considered for each participant:
    1. The participant produces a ranking over the eight 3-variable alternatives;
    2. The participant produces a subset of pairwise comparisons between 1- or 2-variable alternatives. This set is called "a set of comparative preference statements".
    3. For each semantics, the preference relation associated with the set of the participant's comparative statements obtained in Step (2) is computed.
- The fourth step aggregates data obtained from all participants:
    4. The various semantics are compared based on the fit between their predictions computed in Step (3) and actual participants' rankings collected in Step (1).

4.2 Methods

Materials and Design. It has long been known in psychology that humans usually do not process positive valence (agreeable things) and negative valence (things that tend to be avoided) in the same way. Moreover preference reversals sometimes occur between high vs. low intensity of stakes. Intensity refers to how much the problem upon which preferences are expressed is important. In order to control such material effects, participants had to express their preferences about 4 types of scenarios covering a 2×2 between-design: Valence ($\{+, -\}$) × Intensity ($\{$High,Low$\}$) (see Table 1). Each scenario was based on three binary variables X_1, X_2 and X_3

Table 1. The four experimental conditions

Scenario	Studies	Restaurant	Work costs	Diet
Valence	Positive	Positive	Negative	Negative
Intensity	High	Low	High	Low
A_1	long	fish	Moving	press ups
A_2	short	meat	commuting time	jogging
B_1	Bordeaux	white wine	Low salary	sugar-free
B_2	Marseilles	red wine	no bonus	no fat
C_1	short controls	pastry	heavy schedule	intense
C_2	single control	ice cream	Sunday work	long

with $Dom(X_1) = \{A_1, A_2\}$, $Dom(X_2) = \{B_1, B_2\}$ and $Dom(X_3) = \{C_1, C_2\}$. For example the restaurant scenario corresponds to a positive valence and a low intensity. The three variables X_1, X_2 and X_3 respectively correspond to "main dish", "wine" and "dessert" with $Dom(X_1) = \{fish, meat\}$, $Dom(X_2) = \{white, red\}$ and $Dom(X_3) = \{pastry, ice\_cream\}$. Each participant has been asked to express her/his preferences over only one scenario among the four.

Experimental Tasks. The experimental program was run on personal computers.

- In the first part of the experiment, each participant has been asked to rank-order a set of eight alternatives by graphically placing the alternatives in a decreasing order of preference. The preferred alternative had to be placed on the top of the screen, then the next preferred just below, and so on. The least preferred being placed in the lowest part of the list. Between each two subsequent alternatives, a ranking operator, initially set to "?" had to be changed into "=" or ">". Participants could exclude some alternatives of the ranking[2]–but not all. This part of the experiment could only be terminated when all non-excluded alternatives were linked by "=" or ">" operators. Answers from this first step were encoded as follows. With i encoding the downward vertical position of an alternative, the preferred alternative was associated with the rank $rank_1 = 1$. Then $rank_{i+1} = rank_i$ if the ranking operator was "=", and $rank_{i+1} = rank_i + 1$ when the ranking operator was ">". Once this ranking fixed, every alternative received a value representing its rank. The preferred alternatives received the rank 1, next preferred ones received the rank 2, and so on. All excluded alternatives received the value 9.

- In the second part of the experiment, a set of comparative preference statements was constructed. It was composed of pairwise comparisons over 1- or 2- variable alternatives. Let $\mathbb{L}$ (for left) and $\mathbb{R}$ (for right) be two 1- or 2- variable alternatives to be compared. Participants were asked to choose one of the following four possibilities: (1) I would prefer $\mathbb{L}$ over $\mathbb{R}$, (2) I am indifferent to both $\mathbb{L}$ and $\mathbb{R}$, (3) I would prefer $\mathbb{R}$ over $\mathbb{L}$, and (4) neither of the previous choices is acceptable to me. Answers (1) to (3) added the corresponding preferences, namely $\mathbb{L} \triangleright \mathbb{R}$, $\mathbb{L} = \mathbb{R}$,

---

[2] e.g., if they judge the alternatives to be totally unacceptable.

and $\mathbb{R} \triangleright \mathbb{L}$ respectively, to the set of preferences under construction. The fourth choice made the pairwise comparison simply ignored.

We distinguish between 81 possible pairwise comparisons. Fortunately not all of them are relevant. We constructed the set of comparisons in the following way. First the participant was presented with the three 1-variable homogeneous comparisons (i.e., $A_1 \Box A_2$, $B_1 \Box B_2$, $C_1 \Box C_2$, in random order). These comparisons are called homogeneous since they compare the values of a same variable. Depending on the outcome of these comparisons, 12 to 21 additional comparisons were proposed.

A first set of additional comparisons concerned 2-variable alternatives. More precisely, we added context to 1-variable homogeneous comparisons. For example regarding $A_1 \Box A_2$, we added the following comparisons: $(A_1 \wedge B_1) \Box (A_2 \wedge B_1)$, $(A_1 \wedge B_2) \Box (A_2 \wedge B_2)$, $(A_1 \wedge C_1) \Box (A_2 \wedge C_1)$ and $(A_1 \wedge C_2) \Box (A_2 \wedge C_2)$. These are conditional (called also contextual) preferences. For example $(A_1 \wedge B_1) \Box (A_2 \wedge B_1)$ means that the participant is asked to compare A_1 and A_2 with the hypothesis that B takes the value B_1. In a similar way, we constructed 2-variable comparisons from $B_1 \Box B_2$ and $C_1 \Box C_2$. This resulted in 12 additional pairwise comparisons.

Next, depending on the outcome of the first three comparisons (i.e., $A_1 \Box A_2$, $B_1 \Box B_2$, $C_1 \Box C_2$) we added importance statements. These statements occur when, for two different variables, a strict preference is expressed over the values of each variable. For example if a participant returned $A_1 \triangleright A_2$ and $B_1 \triangleright B_2$ then we added the following comparison $A_1 \Box B_1$ which expresses the relative importance of A and B. Such information is important in case we have to make an exclusive choice between A_1 and B_1. Therefore the participant was asked to express whether (i) she/he preferred A getting its preferred value, i.e. A is more important than B (in which case the participant choses $A_1 \triangleright B_1$ interpreted as $A_1 \wedge B_2 \triangleright A_2 \wedge B_1$) or (ii) she/he preferred B getting its preferred value, i.e. B is more important than A (in which case the participant returns $B_1 \triangleright A_1$ interpreted as $A_2 \wedge B_1 \triangleright A_1 \wedge B_2$). The participant may express indifference between the two alternatives, i.e. $A_1 = B_1$ interpreted as $A_1 \wedge B_2 = A_2 \wedge B_1$ which means that A and B have equal importance. Lastly, the participant may also refuse to make a choice in which case the comparison between the two alternatives is no longer considered. It is worth noticing that importance statements are meaningful only when a strict comparison is expressed over the values of two variables at least. Indeed we added at most 3 importance statements. Lastly we added contexts to importance statements. This set is composed of 6 comparisons at most.

To summarize, for each participant we have a fixed set composed of 15 comparisons and a variable set composed of at most 9 2-variable comparisons. In the remainder of this paper, the constructed set of pairwise comparisons is referred to as the set of comparative preference statements.

Participants. $N = 60$ persons (45 women, 15 men) participated in the experiment, all with college education, 15 in each of the four conditions.

Analyses. To compare the semantics and human data, we used the following procedures.

1. *Number of dissonant alternatives: Clustering and comparison.* Each human participant provides a ranking of a given number of alternatives. Similarly, and for a given set of preference statements provided by a participant, each algorithm can rank a number of alternatives. For each semantics and participant, the total number of alternatives ranked by the participant or by the algorithm –but not by both– was used for the first analysis of human-semantics compatibility.
2. Comparisons based on a *distance measure* computed between the ranking over the eight alternatives provided by a participant (produced in Step (1), subsection 4.1) and the ranking (or preference relation) associated with a set of comparative preference statements (produced in Step (2), Subsection 4.1) given a semantics. A semantics is considered plausible insofar as the distance is low.
3. Finally, we compared *Covariations* between semantics and data: for a good semantics as humans' measures become high (resp. low), predicted measures must also become high (resp. low), that is, there should be a positive covariation between human and semantics. A semantics is poorly plausible when predicted measures decrease as humans' measures become higher (null or negative covariation).

Let us now give the technical details of the analyses conducted at each of the three levels above-cited.

Number of Dissonant Alternatives. For a given participant and a given semantics applied to her/his set of comparative preference statements, an alternative could either be ranked or not by the participant or the semantics. We computed the number of dissonant alternatives, that is, the total number of alternatives that were ranked by the participant but not by the semantics, or ranked by the semantics but not by the participant. As 8 alternatives were proposed, this number could range from 0 (i.e., no dissonance or perfect consistency) to 8 (i.e, maximal dissonance).

We sorted participants into two clusters based on their minimum number of dissonant alternatives computed over the six semantics : Participants of Cluster #1 ($N_1 = 29$) had their minimum number of dissonant alternatives > 0 (over the 6 semantics) whereas participants of Cluster #2 ($N_2 = 31$) had their minimum number of dissonant alternatives $= 0$. The potential bias of gender on the clustering process was checked using a χ^2 test on the contingency table formed by crossing clusters and gender. After clusters were constructed, mean absolute deviations were statistically compared using a repeated-measures ANOVA, with the cluster as fixed factor. Post-hoc comparisons were done using Student-Newman-Keuls (S-N-K) procedure.

Computing and Comparing a Distance Measure. For each participant we computed the preference relation returned by each semantics given the set of compact preference statements she/he provided. Recall that a ranking can be associated with a preference relation. Therefore, for each participant we got six rankings (O,P,SO,SP,CPO,CPP). Let us denote by $j = 0$ to 7 the eight alternatives, from $\omega_0 = A_1 B_1 C_1$ to $\omega_7 = A_2 B_2 C_2$. For each alternative, we denote by $r_p(j)$ its rank according to participant p, and $r_s(j)$ its rank according to the semantics s under consideration. Thus, for each alternative six distances could be computed between a participant's ranking and the semantics's ranking. In our analysis, we used Spearman's footrule distance, which we denote d_{fr}. This distance is the sum of absolute differences between the two rankings, $d_{fr} = \sum_j |r_p(j) -$

$r_s(j)|$. For example, the footrule distance between "32114599" and "32145999" is $|3 - 3| + |2 - 2| + |1 - 1| + |1 - 4| + |4 - 5| + |5 - 9| + |9 - 9| + |9 - 9| = 8$.

Given the results of the previous step, indicating that Cluster #1 participants could not be accounted for by any of the semantics at hand, the distance were statistically analyzed only for Cluster #2 participants.

Covariations. Several indexes can be used to estimate covariation among nonparametric measures. We used the most classical one, Spearman's Rho. Given two vectors of ranks, Spearman's Rho first consists in recoding each vector so that each component of the vector is replaced with the average of the ranks having the value of this component. For example $(1; 1; 2; 5; 5; 3; 2; 3)$ is replaced with $(1.5; 1.5; 5; 4.5; 4.5; 7; 5; 7)$. Once the data is recoded, ρ is simply the Pearson correlation over the recoded data. Here the two vectors were $(r_p(0), \cdots, r_p(7))$ for ranks derived from the pth participant's alternative ranking, and $(r_{ps}(0), \cdots, r_{ps}(7))$ for ranks that were derived from the pth participant's compact preferences using the algorithm of the semantics s. In our case, ρ must be positive and statistically significant. Negative or non significant values reflect a bad fit.

4.3   Results

Number of Dissonant Alternatives. The two clusters were clearly differentiated with all the semantics having over 7.33 dissonant alternatives in Cluster #1 vs. 0.92 to 6.75 dissonant alternatives in Cluster #2 (see descriptive statistics in Table 2). Given the disparity in the sample–15 men for 45 women–gender might have affected the clustering. In fact, it was not the case and the male to female ratio was similar in both clusters $(7/22$ in Cluster #1 vs. $8/23$ in Cluster #2, $\chi^2(1, N = 60) = 0.022, p > .88)$.

Table 2. Descriptive statistics (numerical variables)

	SO	SP	O	P	CPO	CPP
Cluster #1 ($N = 29$)						
Number of dissonant alternatives	7.71 (0.16)	7.63 (0.13)	7.75 (0.55)	7.33 (0.49)	7.54 (0.48)	7.71 (0.29)
Cluster #2 ($N = 31$)						
Number of dissonant alternatives	6.67 (0.13)	6.75 (0.10)	3.19 (0.45)	1.61 (0.40)	1.47 (0.39)	0.92 (0.24)
Footrule distance	38.67 (1.23)	45.89 (0.68)	21.79 (3.23)	15.66 (2.05)	15.61 (2.81)	14.01 (2.18)

Analysis of Cluster #1 Results. Not surprisingly, Cluster #1 participants did not behave as predicted by any of the semantics (upper curve in Figure 1). Furthermore, with these participants, the six semantics could not be statistically distinguished, $F(5, 140) = 1.605, p > .16$.

Analysis of Cluster #2 Results. The results suggest that some semantics capture (at least partially) Cluster #2 participants' behaviors (lower curve in Figure 1). With these participants, the various semantics were significantly different, $F(5, 135) = 67.5, p < .001$. More precisely, post-hoc tests distinguished between three groups of homogeneous semantics. As one can easily figure out from Figure 1, the most plausible group included CPO, CPP and P. Within this group, the three semantics were not significantly different from each other. The second group contained O only, which generated significantly more dissonant alternatives than CPO, CPP, and P (All $p_s < .001$), and significantly

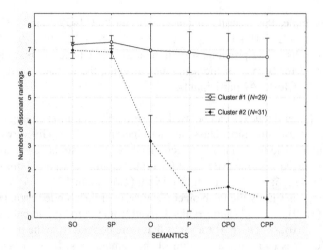

Fig. 1. Number of dissonant alternatives as a function of semantics and Cluster. Vertical bars reflect 95% confidence intervals.

less dissonant alternatives than SO and SP (All $p_s < .001$). The latter two could not be significantly distinguished ($p > .89$).

Interestingly, in Cluster #2, even though there was no primary effect of the experimental conditions ($p > .15$ for valence, $p > .72$ for intensity) there was a significant three-way interaction between the type of semantics and the four experimental conditions ($F(5, 135) = 4.32$, $p = .001$). Although we had no a priori hypothesis about such effect, it corroborates our expectation that the plausibility of the various semantics might vary as a function of the scenario framing. This fact should deserve attention in subsequent investigations of the semantics psychological plausibility.

Let us now conclude about the number of ranked alternatives:

- The first clear result is the fact that 29 participants out of 60 (i.e., 48%) ranked the alternatives in a way that was completely inconsistent with every single semantics. Of course, further investigations will be needed to understand specifically why those subjects behave in such incompatible way with rational standards (see Conclusion).
- The second clear result is the fact that the other half of participants (31 out of 60, i.e., 52%) behaved in a completely different way, much more in accordance with standards promoted in Artificial Intelligence.
- With regard to the purpose of this article, which is to evaluate the psychological plausibility of the various semantics, there is no hope to learn more from participants in Cluster #1 at this point. Consequently, in the next section we will focus on participants from Cluster #2 only.
- From the number of dissonant alternatives observed in by Cluster #2 participants, the six semantics can be ranked according to their psychological plausibility in the following way: 1. CPO $\simeq$ CPP $\simeq$ P, 2. O, 3. SO $\simeq$ SP, where $\simeq$ stands for current impossibility to statistically distinguish between the semantics given our experimental data.

Distance Measure: Spearman Footrule Distance. See the main descriptive statistics in Table 2. The various semantics were clearly different in terms of footrule distance, $F(5, 135) = 52.6$, $p < .001$. Post-hoc tests distinguished between four homogeneous groups. The most plausible semantics were CPP, P, CPO, with mean distances in the range $[14.01; 15.66]$ but no significant differences between them. Then, significantly more distant, came O ($m = 21.79$; $p_s < .05$ for all comparisons with CPP, P and CPO), SO ($m = 38.7$; $p < .001$ for the comparison O - SO), and finally SP ($m = 45.9$; $p < .01$ for the comparison SP - SP).

In summary, from footrule distances in Cluster #2 participants, the six semantics can be ranked according to their psychological plausibility in the following way: 1. CPO $\simeq$ CPP $\simeq$ P, 2. O, 3. SO, 4. SP.

One may also be doubtful as whether standard repeated measures ANOVA can be applied to footrule distances, because such distances come from rankings not real numbers. Thus, assuming that some readers would prefer using more conservative nonparametric tests, we computed Friedman's ANOVA and appropriate post-hoc tests –at the cost of statistical power and the possibility to test interactions. Overall, the difference between semantics was still significant ($p < .001$). However, the loss of statistical power gave a less sharp picture. The distinctions between SP and SO on one hand, versus all other semantics on the other hand were all significant at $p < .001$, except for the comparison between O and SO which was significant at $p = .017$). Within each of these two groups, post-hoc tests for Friedman's ANOVA were not significant. Thus, if we adopt a conservative standpoint, psychological plausibility of the six semantics can be ranked as follows: 1. CPO $\simeq$ CPP $\simeq$ P $\simeq$ O, 2. SO $\simeq$ SP.

Table 3. Descriptive statistics for covariation between participants rankings and semantics' rankings. Means and standard errors were computed from Fisher transformed correlations, then transformed back into the $]-1; 1[$ interval.

	N	Mean ρ	Standard error
SO	28	.51	.05
SP	30	-.49	.05
O	25	.77	.13
P	30	.69	.17
CPO	25	.86	.08
CPP	29	.78	.14

Covariations. Covariations display a complementary picture (descriptive statistics in Table 3). As one can see in Figure 2, SO outperformed SP, O outperformed P, and CPO outperformed CPP. The top graph indicates that 28 out of 30 SP rankings were *negatively* correlated with participants rankings! The middle and bottom graphs indicate that O and CPO produced fewer negative correlations than P and CPP respectively. Finally, ceteris paribus semantics appeared promising. Particularly CPO with all correlations being above .40 and even 17 being above .80.

One sample t-tests showed that correlations with participants' rankings were significantly above 0 (all $t_s > 5.04$, all $p_s < .001$) for all semantics except for SP where ρ

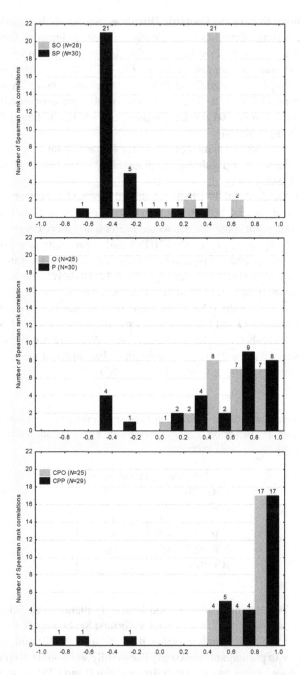

Fig. 2. Distributions of Spearman's rho. Clear bars exhibit optimistic variants of the semantics (SO, O, CPO), dark bars exhibit pessimistic variants of semantics (SP, P, CPP). Values are compatible with human data insofar as they tend to be close to +1. Negative values, on the left parts of the graphs represent particularly incompatible data.

values were significantly *below* 0 ($t(30) = -10.32$; $p_s < .001$), which clearly discards SP as a candidate for claiming psychological plausibility!

Then we performed a repeated-measure ANOVA on Fisher transformed values of the six ρ_s. As it is usual in ANOVA analysis, missing values were replaced with the grand mean of the variable. Intensity and Gender appeared to have no effect of any sort, so we kept Valence as the only between-factor.

As one would expect from the distributions of Spearman's Rho, there was a significant effect of the semantics ($F(5, 145) = 49.58$, $p < .001$). S-N-K post-hoc tests showed that CPO, CPP, P, and O could hardly be distinguished. CPO could outperform P ($p = .006$). SO and SP were outperformed by all other semantics (all $p_s < .001$) but SO outperformed SP ($p < .001$). Thus, from Spearman's correlations in Cluster #2 participants, the six semantics can be ranked according to their psychological plausibility in the following way: 1. CPO $\simeq$ CPP $\simeq$ O $\simeq$ P, 2. SO, 3. SP.

Aggregating the Different Plausibility Orders of the Semantics. We have used several statistical tests to compare the semantics. Despite some minor differences in the ordering of psychological plausibility of the semantics, our various criteria let emerge a global view that is relatively clear in the sense that CPO, CPP and P hold the best. The remaining semantics are rank-ordered differently by the semantics. We distinguish between two extreme strategies to conclude from these results:

- *We avoid differentiating wrongly between two semantics:* this means that if at least one of our measures cannot distinguish between two semantics then they are considered as not distinguishable. Accordingly the ordering of the semantics would be as follows: 1. CPO $\simeq$ CPP $\simeq$ P $\simeq$ O, 2. SO $\simeq$ SP.
- *We avoid failing to differentiate between two semantics:* this means that if two semantics are distinguishable by at least one of our measures then they are considered as distinguishable. Accordingly the ordering of the semantics would be as follows: 1. CPO $\simeq$ CPP $\simeq$ P, 2. O, 3. SO, 4. SP.

Interestingly both interpretations corroborate AI researchers claims. In fact the first interpretation tells us that ceteris paribus, pessimistic and optimistic semantics are more plausible than strong semantics. The latter has been widely criticized by AI researchers as it is too "strong" [4]. On the other hand the former have been promoted as they weaken strong semantics. This interpretation also tells us that humans are likely not to distinguish between minimal and maximal specificity principles, respectively corresponding to negative and positive reading of preferences when handling comparative statements [2], a claim recently given in [7]. The second interpretation tells us that humans do not distinguish between the two readings with ceteris paribus semantics. However it seems that humans are likely to have a positive reading of preferences (P) when the latter are interpreted as defeasible preferences. In fact P and O have been originally proposed in order to cope with defeasible preferences. They respectively correspond to a positive and negative reading of preferences. Regarding strong semantics the second interpretation completely corroborates AI view of this semantics. In fact strong semantics represents constraints whose interpretation is "what doesn't violate constraints is accepted". This is minimal specificity principle which corresponds to SO.

5 Conclusion

We reported first promising results which allow comparing preference semantics on the basis of their closeness to human behavior. We analyzed six semantics given 60 participants (among which 31 participants have been considered more in details) and four scenarios based on three variables. Our results suggest that ceteris paribus (optimistic and pessimistic) and pessimistic semantics hold the best. They however provide various plausibility ordering of the remaining semantics. An experimental analysis, not presented in this paper due to space limitation, showed that the remaining participants (those of Cluster #1) didn't comply with any semantics because they referred to different semantics simultaneously. This result corroborates AI claims that individuals may refer to different semantics at the same time [12,8].

Readers may wonder whether the use of three variables is sufficient to validate our results. In fact the existence of a phenomenon can empirically be asserted as soon as it is observed, which was clearly the case with three variables in our analyses.

Finally, it is worth noticing that cognitive psychology provides descriptive but not normative hints. Our results should not be understood as suggesting that the ordering we observed over the semantics is always followed by humans. They rather tell us which semantics are likely to be used more than others.

References

1. Benferhat, S., Bonnefon, J.F., Da Silva Neves, R.: An overview of possibilistic handling of default reasoning: an experimental study. Synthese 146(1), 53–70 (2005)
2. Benferhat, S., Dubois, D., Kaci, S., Prade, H.: Bipolar representation and fusion of preferences in the possibilistic logic framework. In: KR 2002, pp. 421–432 (2002)
3. Benferhat, S., Dubois, D., Prade, H.: Towards a possibilistic logic handling of preferences. Applied Intelligence 14(3), 303–317 (2001)
4. Boutilier, C.: Toward a logic for qualitative decision theory. In: KR 1994, pp. 75–86 (1994)
5. Dubois, D., Fargier, H., Bonnefon, J.F.: On the qualitative comparison of decisions having positive and negative features. JAIR 32, 385–417 (2008)
6. Hansson, S.O.: The structure of values and norms. Cambridge University Press (2001)
7. Kaci, S.: Characterization of positive and negative information in comparative preference representation. In: ECAI 2012, pp. 450–455 (2012)
8. Kaci, S., van der Torre, L.: Reasoning with various kinds of preferences: Logic, non-monotonicity and algorithms. Annals of Operations Research 163(1), 89–114 (2008)
9. Da Silva Neves, R., Bonnefon, J.F., Raufaste, E.: An empirical test of patterns for nonmonotonic inference. AMAI 34(1-3), 107–130 (2002)
10. Raufaste, E., Da Silva Neves, R., Mariné, C.: Testing the descriptive validity of possibility theory in human judgments of uncertainty. Artificial Intelligence 148(1-2), 197–218 (2003)
11. von Wright, G.H.: The Logic of Preference. University of Edinburgh Press (1963)
12. Wilson, N.: Extending CP-nets with stronger conditional preference statements. In: AAAI 2004, pp. 735–741 (2004)
13. Yager, R.R.: Entropy and specificity in a mathematical theory of evidence. International Journal of General Systems 9, 249–260 (1983)

Answering Ontological Ranking Queries
Based on Subjective Reports

Thomas Lukasiewicz[1], Maria Vanina Martínez[1], Cristian Molinaro[2],
Livia Predoiu[1], and Gerardo I. Simari[1]

[1] Department of Computer Science, University of Oxford, UK
{thomas.lukasiewicz,vanina.martinez,
livia.predoiu,gerardo.simari}@cs.ox.ac.uk
[2] DIMES, Università della Calabria, Italy
cmolinaro@dimes.unical.it

Abstract. The use of preferences in query answering, both in traditional databases and in ontology-based data access, has recently received much attention,
due to its many real-world applications. In this paper, we tackle the problem of
query answering in Datalog+/– ontologies subject to the querying user's preferences and a collection of subjective reports (i.e., scores for a list of features) of
other users, who have their own preferences as well. All these pieces of information are combined to rank the query results. We first focus on the problem of
ranking atoms in a database by leveraging reports and customizing their content
according to the user's preferences. Then, we extend this approach to deal with
ontological query answering using provenance information. Though the general
problem is shown to have an exponential-time data complexity upper bound, we
propose a special case that has polynomial time data complexity.

1 Introduction

The use of preferences in query answering, both in traditional databases and in ontology-
based data access, has recently received much attention due to its many real-world applications. In particular, in recent times, there has been a huge change in the way data
is created and consumed, and users have largely moved to the Social Web, a system of
platforms used to socially interact by sharing data and collaborating on tasks.

In this paper, we tackle the problem of preference-based query answering in Data-
log+/– ontologies assuming that the user must rely on subjective reports to get a complete picture and make a decision. This kind of situation arises all the time on the Web;
for instance, when searching for a hotel, users provide some basic information and receive a list of answers to choose from, each associated with a set of subjective reports
(often called reviews) written by other users to tell everyone about their experience. The
main problem with this setup, however, is that users are often overwhelmed and frus-
trated, because they cannot decide which reviews to focus on and which ones to ignore,
since it is likely that, for instance, a very negative (or positive) review may have been
produced on the basis of a feature that is completely irrelevant to the querying user.

We study a formalization of this process and its incorporation into preference-based
query answering in Datalog+/– ontologies, proposing a framework for user-tailored
query answers on the basis of the ontology reports and users' preferences.

U. Straccia and A. Calì (Eds.): SUM 2014, LNAI 8720, pp. 223–236, 2014.

The following are the main contributions of this paper:

- We present an approach to preference-based query answering in Datalog+/− ontologies, given a collection of subjective reports. Here, each report contains scores for a list of features.
- We first propose a basic approach to rank atoms in the ontology's database, which combines reports, their authors' preferences among the features, and the querying user's preferences among the features.
- We then extend our framework to ontologies with dependencies and propose a method for dealing with query answering by leveraging provenance information for propagating reports from the database atoms to newly inferred ones.
- As we are using a kind of "how"-provenance modeled using semirings, we can map the general semiring to more specific ones that capture different ways in which the information contained in reports can be leveraged. Though the general case is exponential, we explore one such mapping for which ranking query answers can be done in polynomial time data complexity.

The rest of this paper is organized as follows. In Section 2, we provide some preliminaries on Datalog+/− and the used preference models. Section 3 then defines subjective reports and proposes a basic framework to deal with simple (no dependencies) Datalog+/− ontologies. Section 4 considers ontologies with dependencies and addresses query answering. Section 5 discusses related work, and Section 6 concludes.

2 Preliminaries

We first briefly recall some basics on Datalog+/− [8], namely, on relational databases and (Boolean) conjunctive queries, along with tuple- and equality-generating dependencies and negative constraints, the chase, and ontologies in Datalog+/−. We also define the used preference models.

Databases and Queries. We assume (i) an infinite universe of *(data) constants* Δ (which constitute the "normal" domain of a database), (ii) an infinite set of *(labeled) nulls* Δ_N (used as "fresh" Skolem terms, which are placeholders for unknown values, and can thus be seen as variables), and (iii) an infinite set of variables $\mathcal{V}$ (used in queries, dependencies, and constraints). Different constants represent different values (*unique name assumption*), while different nulls may represent the same value. We denote by $\mathbf{X}$ sequences of variables $X_1, \ldots, X_k$ with $k \geqslant 0$. We assume a *relational schema* $\mathcal{R}$, which is a finite set of *predicate symbols* (or simply *predicates*). A *term t* is a constant, null, or variable. An *atomic formula* (or *atom*) A has the form $p(t_1, ..., t_n)$, where p is an n-ary predicate, and $t_1, ..., t_n$ are terms. It is *ground* (resp., *existentially closed*) iff every t_i belongs to Δ (resp., $\Delta \cup \Delta_N$). Every ground atom A is uniquely identified with an id, denoted $id(A)$; when it is clear from the context, an atom and its id are used interchangeably. We use $\mathcal{H}$ to denote the set of all possible ground and existentially closed atoms.

A *database (instance)* D for a relational schema $\mathcal{R}$ is a set of ground atoms with predicates from $\mathcal{R}$. A *conjunctive query (CQ)* Q over $\mathcal{R}$ has the form $q(\mathbf{X}) = \exists \mathbf{Y}\, \Phi$

$(\mathbf{X}, \mathbf{Y})$, where q is a predicate, and $\Phi(\mathbf{X}, \mathbf{Y})$ is a nonempty conjunction of atoms (possibly equalities, but not inequalities) with the variables $\mathbf{X}$ and $\mathbf{Y}$, and possibly constants, but no nulls. With a slight abuse of notation, we will sometimes treat $\Phi(\mathbf{X}, \mathbf{Y})$ as a *set* of atoms. A *Boolean CQ* (BCQ) over $\mathcal{R}$ is a CQ where all variables are existentially quantified.

Answers to CQs and BCQs are defined via *homomorphisms*, which are mappings $\mu: \Delta \cup \Delta_N \cup \mathcal{V} \to \Delta \cup \Delta_N \cup \mathcal{V}$ such that (i) $c \in \Delta$ implies $\mu(c) = c$, and (ii) $c \in \Delta_N$ implies $\mu(c) \in \Delta \cup \Delta_N$. Moreover, μ is naturally extended to atoms, sets of atoms, and conjunctions of atoms. The set of all *answers* to a CQ Q of the form $q(\mathbf{X}) = \exists \mathbf{Y} \, \Phi(\mathbf{X}, \mathbf{Y})$ over a database D, denoted $Q(D)$, is the set of all ground atoms $q(t)$ with tuples t over Δ for which a homomorphism $\mu: \mathbf{X} \cup \mathbf{Y} \to \Delta \cup \Delta_N$ exists such that $\mu(\Phi(\mathbf{X}, \mathbf{Y})) \subseteq D$ and $\mu(\mathbf{X}) = t$. The *answer* to a BCQ Q over a database D is *Yes*, denoted $D \models Q$, iff $Q(D) \neq \emptyset$.

Given a relational schema $\mathcal{R}$, a *tuple-generating dependency* (TGD) σ is a first-order formula $\forall \mathbf{X} \forall \mathbf{Y} \, \Phi(\mathbf{X}, \mathbf{Y}) \to \exists \mathbf{Z} \, \Psi(\mathbf{X}, \mathbf{Z})$, where $\Phi(\mathbf{X}, \mathbf{Y})$ and $\Psi(\mathbf{X}, \mathbf{Z})$ are conjunctions of atoms over $\mathcal{R}$ (without nulls), called the *body* and the *head* of σ, denoted $body(\sigma)$ and $head(\sigma)$, respectively. Such σ is satisfied in a database D for $\mathcal{R}$ iff, whenever there exists a homomorphism h that maps the atoms of $\Phi(\mathbf{X}, \mathbf{Y})$ to atoms of D, there exists an extension h' of h that maps the atoms of $\Psi(\mathbf{X}, \mathbf{Z})$ to atoms of D. All sets of TGDs are finite here. Since TGDs can be reduced to TGDs with only single atoms in their heads, in the sequel, every TGD has w.l.o.g. a single atom in its head. A TGD σ is *guarded* iff it contains an atom in its body that contains all universally quantified variables of σ. The leftmost such atom is the *guard atom* (or *guard*) of σ. A set of TGDs is guarded iff all its TGDs are guarded.

Query answering under TGDs, i.e., the evaluation of CQs and BCQs on databases under a set of TGDs is defined as follows. For a database D for $\mathcal{R}$, and a set of TGDs Σ on $\mathcal{R}$, the set of *models* of D and Σ, denoted $mods(D, \Sigma)$, is the set of all (possibly infinite) databases B such that (i) $D \subseteq B$ and (ii) every $\sigma \in \Sigma$ is satisfied in B. The set of *answers* for a CQ Q of the form $q(\mathbf{X}) = \exists \mathbf{Y} \, \Phi(\mathbf{X}, \mathbf{Y})$ to D and Σ, denoted $ans(Q, D, \Sigma)$ (or, for $KB = (D, \Sigma)$, $ans(Q, KB)$), is the set of all ground atoms $q(t)$ such that $q(t) \in Q(B)$ for all $B \in mods(D, \Sigma)$. The *answer* for a BCQ Q to D and Σ is *Yes*, denoted $D \cup \Sigma \models Q$, iff $ans(Q, D, \Sigma) \neq \emptyset$. Query answering under general TGDs is undecidable [2], even when $\mathcal{R}$ and Σ are fixed [7]. Decidability and tractability in the data complexity of query answering for the guarded case follows from a bounded tree-width property.

A *negative constraint* (or simply *constraint*) γ is a first-order formula of the form $\forall \mathbf{X} \Phi(\mathbf{X}) \to \bot$, where $\Phi(\mathbf{X})$ (called the *body* of γ) is a conjunction of atoms over $\mathcal{R}$ (without nulls). Under the standard semantics of query answering of BCQs in Datalog+/– with TGDs, adding negative constraints is computationally easy, as for each constraint $\forall \mathbf{X} \Phi(\mathbf{X}) \to \bot$, we only have to check that the BCQ $\exists \mathbf{X} \, \Phi(\mathbf{X})$ evaluates to false in D under Σ; if one of these checks fails, then the answer to the original BCQ Q is true, otherwise the constraints can simply be ignored when answering the BCQ Q.

As another component, the Datalog+/– language allows for special types of *equality-generating dependencies* (EGDs). Since they can also be modeled via negative constraints, we refer to [8] for their details. We usually omit the universal quantifiers in

TGDs, negative constraints, and EGDs, and we implicitly assume that all sets of dependencies and/or constraints are finite.

The Chase. The *chase* was first introduced to enable checking implication of dependencies, and later also for checking query containment. By "chase", we refer both to the chase procedure and to its output. The TGD chase works on a database via so-called TGD *chase rules* (see [8] for further details and for an extended chase with also EGD chase rules). The (possibly infinite) chase of a database D relative to a set of TGDs Σ, denoted $chase(D, \Sigma)$, is a *universal model*, i.e., there is a homomorphism from $chase(D, \Sigma)$ onto every $B \in mods(D, \Sigma)$ [8]. Thus, BCQs Q over D and Σ can be evaluated on the chase for D and Σ, i.e., $D \cup \Sigma \models Q$ is equivalent to $chase(D, \Sigma) \models Q$. For guarded TGDs Σ, such BCQs Q can be evaluated on an initial fragment of $chase(D, \Sigma)$ of constant depth $k \cdot |Q|$, which is possible in polynomial time in the data complexity.

Datalog+/– Ontologies. A *Datalog+/– ontology* $KB = (D, \Sigma)$, where $\Sigma = \Sigma_T \cup \Sigma_E \cup \Sigma_{NC}$, consists of a database D, a set of TGDs Σ_T, a set of EGDs Σ_E, and a set of negative constraints Σ_{NC}. In order to ensure decidability and tractability of query answering, we assume that EGDs are *separable*, which means that the interaction between TGDs and EGDs is controlled—this condition can be ensured by the syntactic criterion of *non-conflicting keys*; for details on these conditions, we refer the reader to [8]. Finally, we say that KB is *guarded* iff Σ_T is guarded. The following example illustrates a simple Datalog+/– ontology, used in the sequel as a running example.

Example 1. Consider the following ontology $KB = (D, \Sigma)$:

$$D = \{id_1 : apthotel(h_1), \quad id_2 : hotel(h_2),$$
$$id_3 : bb(bb_1), \quad id_4 : hostel(hs_1)\},$$

$$\Sigma = \{\sigma_1 : hotel(H) \rightarrow accom(H),$$
$$\sigma_2 : apartment(A) \rightarrow accom(A),$$
$$\sigma_3 : bb(B) \rightarrow accom(B),$$
$$\sigma_4 : apthotel(A) \rightarrow hotel(A),$$
$$\sigma_5 : apthotel(A) \rightarrow apartment(A),$$
$$\sigma_6 : hostel(H) \rightarrow accom(H)\}.$$

This ontology models a very simple accommodation booking domain, which can be used as the underlying model in an online system. Accommodations can be either hotels, apartments, bed and breakfasts, hostels, or aparthotel. Moreover, an aparthotel is both a hotel and an apartment. The database D provides some instances for some kinds of accommodation. ∎

Preference Models. A *preference relation* $\succ$ over a set S is a strict partial order (SPO) over S, i.e., an irreflexive and transitive binary relation over S—we consider these to be the minimal requirements for a useful preference relation. If $a \succ b$, we say that a is *preferred to* b. The *indifference relation* $\sim$ induced by $\succ$ is defined as follows: for any $a, b \in S, a \sim b$ iff $a \nsucc b$ and $b \nsucc a$.

A *stratification* of S relative to $\succ$ is an ordered sequence $\langle S_1, \ldots, S_k \rangle$, where each S_i is a maximal subset of S such that for every $a \in S_i$ there is no $b \in \bigcup_{j=i}^{k} S_j$ with

$b \succ a$. Intuitively, S_1 contains the most preferred elements in S relative to $\succ$, i.e., those elements $a \in S$ for which there is no $b \in S$ such that $b \succ a$. Then, S_2 contains the most preferred elements of $S - S_1$, and so on so forth. Notice that a stratification always exists and is unique—moreover, it is a partition of S. Each S_i is called a *stratum*.

3   A Framework for Ranking Atoms Based on Subjective Reports

In this section, we propose a framework that allows us to produce a ranking of atoms in a knowledge base by leveraging subjective reports. For now, we restrict our attention to the case where the set of dependencies Σ is empty—we generalize our framework to deal with dependencies and ontological query answering in the next section.

More specifically, we consider the following setting. We are given a Datalog+/− ontology $KB = (D, \Sigma)$, with $\Sigma = \emptyset$, where ground atoms in D are associated with *reports* provided by *observers*. A report is an evaluation for an entity of interest that specifies a score for different *features*, which are the dimensions along which entities can be evaluated. Also, each observer expresses a preference relation over the features. Finally, a user looking at the ontology also has her own preference relation over the features and wishes to rank the atoms in the ontology on the basis of the reports, her preferences, as well as the observers' preferences.

Example 2. Consider $KB = (D, \emptyset)$ where D is the database from Example 1—recall that predicate symbols refer to different types of accommodations. Accommodations might be rated relative to location, cleanliness, price, breakfast, and internet; in this case, these would be the features of interest. Observers can leave reports for the accommodations in the knowledge base, with each report specifying one score per feature. Each observer has also a preference relation over the features specifying their importance from the observer's point of view. A user, who also has her own preference relation over the features, wishes to rank the atoms in the knowledge base (that is, all the atoms in D), taking into account all the elements discussed above.   ∎

Features and Reports. We assume the existence of an ordered sequence of *features* $\langle f_1, \ldots, f_m \rangle$, each of which has a domain $dom(f_i) = [0, 1] \cup \{-\}$. We use $\mathcal{F}$ to denote the set of features $\{f_1, \ldots, f_m\}$.[1] The values in a feature domain are possible scores that can be given for the feature, with "$-$" meaning that no score is given. The binary relation $>$ over $dom(f_i)$ is defined in the standard way with the additional requirement that $v > -$ for every $v \in [0, 1]$.

A *report* is an element of $dom(f_1) \times \cdots \times dom(f_m)$. We use $Reports$ to denote the set of all possible reports, i.e., $Reports = dom(f_1) \times \cdots \times dom(f_m)$. A report given by a certain observer for a ground atom A specifies m "scores" representing an evaluation of A for each feature.

Multiple reports, coming from different observers, can be associated with the same ground atom—each report expresses the rating given to the ground atom by a specific observer. Thus, we assume we have N observers and N partial functions $\Gamma_i : D \rightarrow$

---

[1] Throughout this section, we consider a single set of features shared by all predicate symbols; this limitation is removed in the following section.

Reports, called *report functions*, for $1 \leqslant i \leqslant N$. Given an observer $1 \leqslant i \leqslant N$ and a ground atom $A \in D$, if $\Gamma_i(A)$ is defined, it denotes the report given by observer i to A; if undefined, this means that observer i has no report for A (note that we assume that an observer can have at most one report for a given ground atom).

Example 3. Consider the accommodation booking domain of Example 2. Suppose the features with respect to which accommodations are rated are $\langle location, cleanliness,$ $price, breakfast, internet \rangle$; in the following, we abbreviate these features as *loc*, *cl*, *pri*, *br*, *net*, respectively. Suppose we have 3 observers and that the Γ_i's are as follows:

	loc	cl	pri	br	net
$\Gamma_1(id_1)$	0.8	0	0.4	–	0.6
$\Gamma_1(id_2)$	0.6	0.7	0.4	–	0.2
$\Gamma_1(id_3)$	0.9	1	0.6	0.5	0.1
$\Gamma_1(id_4)$	1	0.1	0.2	0.3	0.4

	loc	cl	pri	br	net
$\Gamma_2(id_1)$	0.7	0.5	0.3	0.1	0.5
$\Gamma_2(id_2)$	0.5	0.2	0.2	0.3	0.3
$\Gamma_2(id_3)$	0.4	0.1	0.9	0.5	0.7
$\Gamma_2(id_4)$	0.3	0.7	–	0.4	0.6

	loc	cl	pri	br	net
$\Gamma_3(id_1)$	0.5	0	0.4	–	0.1
$\Gamma_3(id_2)$	0.7	–	0.6	0.9	0.4
$\Gamma_3(id_3)$	0.9	0.3	0.3	0.1	0
$\Gamma_3(id_4)$	0.8	0.5	0.6	–	0.2

For instance, Γ_1 says that the first observer assigned the report $\langle 0.8, 0, 0.4, -, 0.6 \rangle$ to atom id_1, which means that the score given to id_1 relative to *location* (resp., *cleanliness*, *price*, *internet*) is 0.8 (resp., 0, 0.4, 0.6), while no score is given for *breakfast*.

Comparing Reports. Given a report $r = \langle s_1, \ldots, s_m \rangle$, we use $r[i]$ to refer to s_i. Given two reports r_1, r_2 and a set of features $\mathcal{F}' \subseteq \mathcal{F}$, we write

1. $r_1[\mathcal{F}'] = r_2[\mathcal{F}']$ iff $r_1[i] = r_2[i]$ for every $f_i \in \mathcal{F}'$;
2. $r_1[\mathcal{F}'] \geqslant r_2[\mathcal{F}']$ iff $r_1[i] \geqslant r_2[i]$ for every $f_i \in \mathcal{F}'$;
3. $r_1[\mathcal{F}'] > r_2[\mathcal{F}']$ iff $r_1[\mathcal{F}'] \geqslant r_2[\mathcal{F}']$ and $r_1[j] > r_2[j]$ for some $f_j \in \mathcal{F}'$.

Now, a user inspecting the ontology might have preferences among features—for instance, location and price might be the most preferred. We thus allow a user to specify a preference relation over the features in order to influence the comparison of reports and eventual ranking of atoms. The following definition formalizes the comparison between two reports r_1 and r_2 relative to a preference relation $\succ$ over $\mathcal{F}$. Intuitively, if $\mathcal{F}_1$ is the first stratum of the stratification of $\mathcal{F}$ with respect to $\succ$, and $r_1[\mathcal{F}_1] > r_2[\mathcal{F}_1]$, then $r_1 > r_2$. If $r_1[\mathcal{F}_1] = r_2[\mathcal{F}_1]$, we compare r_1 and r_2 relative to the second stratum $\mathcal{F}_2$ and if $r_1[\mathcal{F}_2] > r_2[\mathcal{F}_2]$, then $r_1 > r_2$. If $r_1[\mathcal{F}_2] = r_2[\mathcal{F}_2]$, then we compare r_1 and r_2 over the third stratum, and so on.

Definition 1. Consider two reports r_1, r_2 and a preference relation $\succ$ over $\mathcal{F}$. Let $\mathcal{F}_1, \ldots, \mathcal{F}_k$ be the stratification of $\mathcal{F}$. We say $r_1 > r_2$ *relative to* $\succ$ iff there exists $1 \leqslant i \leqslant k$ such that: (i) $r_1[\mathcal{F}_i] > r_2[\mathcal{F}_i]$, and (ii) $r_1[\mathcal{F}_j] = r_2[\mathcal{F}_j]$ for $1 \leqslant j < i$.

Example 4. Consider the reports of Example 3 and let $\mathcal{F}$ be the set of features therein. Suppose a user has the preference relation $\succ_u$ over $\mathcal{F}$ of Figure 1, where an edge from a to b means that $a \succ_u b$. The stratification of $\mathcal{F}$ relative to $\succ_u$ consists of three strata: $\mathcal{F}_1 = \{loc, net\}$, $\mathcal{F}_2 = \{cl\}$, and $\mathcal{F}_3 = \{pri, br\}$. Consider now two reports: $r_1 = \Gamma_1(id_1)$ and $r_2 = \Gamma_1(id_2)$. Then, we have $r_1 > r_2$, since $r_1[\mathcal{F}_1] > r_2[\mathcal{F}_1]$. ∎

Each observer can have a preference relation over the set of features in much the same way as the user looking at the ontology. Thus, we assume the existence of N such preference relations, denoted with $\succ_1, \ldots, \succ_N$ (one for each observer).

Ranking Atoms. Our goal is to obtain a ranking of the ground atoms in the knowledge base on the basis of: *(i)* N report functions $\Gamma_1, \ldots, \Gamma_N$ (one for each observer), *(ii)* N preference relations $\succ_1, \ldots, \succ_N$ over $\mathcal{F}$ (one for each observer), and *(iii)* a preference relation $\succ_u$ over $\mathcal{F}$ (the user's preferences). For this, we adopt a two-phase approach:

1. Generate a preference relation over D for each observer i on the basis of Γ_i and $\succ_u$, thereby obtaining a preference relation over ground atoms that takes into account the i-th observer's scores and is "customized" according to the user's preferences.
2. Then, the N preference relations obtained in the previous step are combined into one by taking into account how "relevant" the observers' feature preferences are relative to the user's preferences.

Definition 2. Let Γ_i be a report function and $\succ_u$ the user's preference relation over $\mathcal{F}$. For all $A_1, A_2 \in D$, $A_1 >_i A_2$ iff

- $\Gamma_i(A_1)$ is defined and $\Gamma_i(A_2)$ is undefined; or
- both $r_1 = \Gamma_i(A_1)$ and $r_2 = \Gamma_i(A_2)$ are defined and $r_1 > r_2$ relative to $\succ_u$.

The following example shows how atoms in the database can be ranked in our running scenario by each observer when the user's preferences are also incorporated.

Example 5. Consider the reports of Example 3 and the user's preference relation $\succ_u$ over the features of Figure 1. Then, for each observer, we can determine an order among ground atoms based on the observer's reports, customizing it relative to the user's feature preferences. Specifically, we get

$$
\begin{array}{llll}
\textit{Observer 1:} & id_1 >_1 id_2; & id_4 >_1 id_2; & id_4 >_1 id_3; \\
\textit{Observer 2:} & id_1 >_2 id_2; & id_3 >_2 id_4; & \\
\textit{Observer 3:} & id_2 >_3 id_1; & id_4 >_3 id_1. &
\end{array}
$$
∎

After obtaining a preference relation $>_i$ over D for each observer i, we need to combine them into a single one. In this second step, we want to take into account how "relevant" the observers' feature preferences are given the user's feature preferences. Thus, we assume the existence of a *relevance function* ρ which takes as input two preference relations over $\mathcal{F}$ (the user preference relation and the preference relation of an observer) and gives as output a value in $[0, 1]$, measuring how similar the two preference relations are (e.g., various distance measures over fully and partially specified preference structures have been proposed in [13]).

Below, we provide a general definition of an operator that combines a set of preference relations where each is associated with a relevance value.

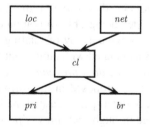

Fig. 1. A user preference relation over features

Definition 3. A *preference aggregation operator* is a function that takes as input a set $\mathcal{P}$ of pairs $\langle >_i, \rho_i \rangle$ where $>_i$ is a preference relation and $\rho_i \in [0,1]$, and returns a preference relation $>$ such that if $a >_i b$ for every $>_i$ appearing in $\mathcal{P}$ then $a > b$.

For instance, when combining the $>_i$'s we may want to give more importance to the more relevant ones, i.e., those with higher ρ_i value, and give the same importance to $>_i$'s with the same ρ_i value.

Example 6. Consider again the setting from Example 5, and suppose the observers' preference relations over the features are such that the relevance function yields 0.8 for observers 1 and 2, and 0.3 for observer 3. An example of preference aggregation operator can be the one that first combines $>_1$ and $>_2$ using, e.g., pareto composition, thereby obtaining $id_1 >_{1 \otimes 2} id_2$ and $id_4 >_{1 \otimes 2} id_2$. Then, it combines $>_{1 \otimes 2}$ with $>_3$ with prioritized composition giving more importance to $>_{1 \otimes 2}$ (because of the higher relevance of the first two observers) thereby obtaining a final order among atoms $>$ as follows: $id_1 > id_2$, $id_4 > id_2$, and $id_4 > id_1$.   ∎

Definition 4. Given a Datalog+/− ontology $KB = (D, \Sigma)$, N report functions $\Gamma_1, \ldots,$ Γ_N, N preference relations $\succ_1, \ldots, \succ_N$ over $\mathcal{F}$, a preference relation $\succ_u$ over $\mathcal{F}$, a relevance function ρ, and a preference aggregation operator agg, we define a ranking for KB as $agg(\{\langle >_1, \rho(\succ_1, \succ_u)\rangle, \ldots, \langle >_N, \rho(\succ_N, \succ_u)\rangle\})$.

Notice that in the previous definition each $>_i$ is obtained as per Definition 2. Once a partial order over the ground atoms is obtained, a total order over this set can easily be derived by computing one of its topological sortings.

Proposition 1. *The worst-case time complexity of computing a ranking for a Datalog+/− ontology (D, Σ) is $km^2 + N(mn^2 + f_\rho + logN) + f_{agg}^N$, where m is the number of features, $n = |D|$, N is the number of observers, $k = |\succ_u|$, f_ρ is the worst-case time complexity of the relevance function, and f_{agg}^N is the worst-case time complexity of the adopted preference aggregation operator over N preference relations.*

4   Ontological Query Answering Based on Subjective Reports

Up to now, we have considered simple knowledge bases $(D, \emptyset)$, without directly addressing ontological query answering. We now generalize our framework to deal with conjunctive query answering over ontologies containing tuple-generating dependencies.

   The application of such dependencies in Σ can generate new atoms, to which we need to "propagate" reports associated with the ground atoms in D that participated in the creation. Therefore, besides being able to compute answers to queries in the classical manner, we must also provide a way of relating query answers to the information contained in the reports associated with atoms that "contributed" to them—this suggests the need to use an adequate *data provenance representation*. There is a variety of approaches that have been proposed in the formalization of provenance and lineage [12,6]. In this work, we adopt a special case of the framework from [12], where "how"-provenance is formalized through *semirings of polynomials*; this is a very general and flexible formalism that we adapt to our needs, as discussed next.

A Data Provenance Model. Consider an ontology $KB = (D, \Sigma)$. Recall that every ground atom $A \in D$ has an id, denoted $id(A)$; let $X = \{id(A) \mid A \in D\}$ and $S = \{0, 1\}$, we define a semiring of polynomials $(S[X], +, \times, 0, 1)$, with variables from X and coefficients from S, with operations $+$ and $\times$ being idempotent, associative, and commutative, and $\times$ distributing over $+$. Note that 1 (resp. 0) is the identity element of $\times$ (resp. $+$). The provenance information is modeled as annotations to atoms in $\mathcal{H}$, defined by a function $Ann : \mathcal{H} \mapsto S[X]$, i.e., Ann maps ground atoms to polynomials in the semiring. This approach allows us to record the provenance information by means of symbolic expressions over the ids of atoms using the semiring operations, which in turn allows us to model not only information about *which* atoms contributed to the answer's computation but also *how* they contributed.

The Guarded Chase Forest. Note that, as shown in [12], commutative semirings are not enough to correctly model the propagation of provenance annotations that arise from derivations in Datalog; this is the case because derivations in Datalog (and therefore in Datalog+/− ontologies) can be infinite, which requires semirings with infinite sums. For this reason, we focus on guarded Datalog+/− ontologies, and instead of computing the provenance annotations over the chase itself, we compute them over the necessary finite part of the *guarded chase forest* (introduced in [8]). It is sufficient to consider only the finite part of the chase, as for every derivation outside, there is an isomorphic one inside the finite part (which follows from the results in [8]). In the following, we show how the guarded chase forest can be extended to consider provenance annotations.

The guarded chase forest for D and Σ contains *(i)* a node n_A for each atom $A \in D$, having $label(n_A) = A$, and *(ii)* for any two atoms $A, B \in chase(D, \Sigma)$ there are two nodes n_A and n_B with $label(n_A) = A$ and $label(n_B) = B$ along with an arrow from n_A to n_B iff B is obtained from A and possibly other atoms by a one-step application of a TGD $\sigma \in \Sigma$ with A as guard. In [8], it has been shown that whenever homomorphic images of a CQ $Q(\mathbf{X})$ are contained in $chase(D, \Sigma)$, then they are also contained in a finite, initial portion of the guarded chase forest, whose size is determined only by the query and the schema. Furthermore, for guarded Datalog+/− ontologies, the whole derivation of the query atoms is also contained in such a portion of the forest. This means that we can construct the provenance annotations based on this finite portion of the chase forest. For this, we associate with each node n a provenance annotation $ann\_ch(n)$; the forest is computed in the following way: first, for all nodes n labeled with atoms $A \in D$, we have $ann\_ch(n) = id(A)$. Then, each time the chase rule is applied via a TGD $\sigma : \Upsilon(\mathbf{X}, \mathbf{Y}) \rightarrow \exists \mathbf{Z} \Psi(\mathbf{X}, \mathbf{Z})$ to atoms labeling the nodes of the guarded chase forest constructed thus far (let these nodes be $n_1, \ldots, n_k$), the annotation of the new node n' labeled with the atom to which $\Psi(\mathbf{X}, \mathbf{Z})$ is mapped is defined as $ann\_ch(n') = \prod_{1 \leqslant i \leqslant k} ann\_ch(n_i)$.

The guarded depth of an atom A in the guarded chase forest for D and Σ, denoted $depth(A)$, is the minimum length of a path from D to a node labeled with A in the forest. Then, the guarded chase of level up to $k > 0$ for D and Σ, denoted $g\_chase^k(KB)$, is the set of all atoms in the guarded chase forest of depth at most k. Given a CQ Q, we denote with $g\_chase^\gamma(KB, Q)$ the part of the chase forest needed to answer Q; as pointed out above, it has been shown that for guarded Datalog+/− ontologies $g\_chase^\gamma(KB, Q)$ is finite and its size (i.e., γ) depends only on Q and KB.

Definition 5. Let $KB = (D, \Sigma)$ be a Datalog+/– ontology and $(S[X], +, \times, 0, 1)$ a commutative doubly idempotent semiring of polynomials. We define annotation function *Ann*: $\mathcal{H} \rightarrow S[X]$ as follows: for each atom $A \in \mathcal{H}$, if $KB \models A$ then $Ann(A) = \sum_{n \in g\_chase^\gamma(KB,A) s.t. \ label(n)=A} ann\_ch(n)$, otherwise $Ann(A) = 0$.

Note that $\mathcal{H}$ can be infinite as can the guarded chase forest; however, the support for function *Ann* is defined as $supp(Ann) = \{A \mid Ann(A) \neq 0\}$, which is clearly finite in $g\_chase^\gamma(KB, Q)$. Note that the annotations depend on the query and the schema.

A Note about Idempotent Operators. Though in general the assumption of idempotent $+$ and $\times$ operators is not made in the context of lineage expressions, for our purposes it makes sense to adopt it since: (i) The $+$ operator is used to express the fact that an answer can be derived in more than one way—if two terms in the polynomial are identical, we only need to keep track of one of them since they both represent the same information regarding which reports should be considered (cf. the derivation of atom $accom(h_1)$ in Example 7). (ii) The $\times$ operator is used to keep track of which atoms participated in a single derivation of an answer—if identical atoms appear in an expression, we only need to keep track of the atom's presence since, as before, we only need to know which reports are associated with the atom and not how many times the atom participated in the derivation; thus, we have that $a * a = a$. Therefore, all exponents and coefficients in our lineage expressions will be equal to 1.

Example 7. Consider the ontology $KB = (D, \Sigma)$ obtained from the one of Example 1 by adding the following atoms to the database:

$id_5 : pub(p_1), \quad id_8 : locIn(h_1, oxford, st\_giles), \quad id_{11} : locIn(p_2, oxford, st\_giles),$
$id_6 : pub(p_2), \quad id_9 : locIn(p_1, oxford, st\_giles), \quad id_{12} : locIn(p_3, oxford, queen\_st).$
$id_7 : pub(p_3), \quad id_{10} : locIn(bb_1, oxford, queen\_st),$

Then, each ground atom in D is annotated with its id, i.e. $Ann(apthotel(h_1)) = id_1$, $Ann(hotel(h_2)) = id_2$, etc. The following atoms, along with their annotations, are derived from D and Σ:

$$Ann(accom(h_1)) = id_1 + id_1 = id_1 \quad Ann(accom(h_2)) = id_2$$
$$Ann(accom(bb_1)) = id_3 \quad Ann(accom(hs_1)) = id_4$$
$$Ann(hotel(h_1)) = id_1 \quad Ann(apartment(h_1)) = id_1 \quad \blacksquare$$

Query Answers with Lineage. Given the formalism described above, we can now associate useful information about how query answers are computed.

Definition 6. Let $KB = (D, \Sigma)$ be a Datalog+/– ontology and Q be a CQ of the form $q(\mathbf{X}) = \exists \mathbf{Y} \bigwedge_{i=1}^{k} p_i(\mathbf{X_i}, \mathbf{Y_i})$. The *annotation* of a query answer $A \in ans(Q, KB)$ is defined as follows:

$$Ann(A) = \sum_{\theta \in subs(Q, KB) \wedge A = \theta q(\mathbf{X})} \prod_{i=1}^{k} Ann(\theta p_i(\mathbf{X_i}, \mathbf{Y_i})),$$

where $subs(Q, KB)$ denotes the set of all substitutions θ such that $KB \models \theta p_i(\mathbf{X_i}, \mathbf{Y_i})$ for all $i \in \{1, \ldots, k\}$.

Example 8. Consider the ontology from Example 7 and the query $myaccom(X) = \exists S, Y \, accom(X) \wedge locIn(X, oxford, S) \wedge pub(Y) \wedge locIn(Y, oxford, S)$, asking for accommodations in Oxford such that there is a pub on the same street. There are two query answers, namely $myaccom(h_1)$ with annotation $(id_1 \times id_8 \times id_5 \times id_9) + (id_1 \times id_8 \times id_6 \times id_{11})$ and $myaccom(bb_1)$ with annotation $(id_3 \times id_{10} \times id_7 \times id_{12})$. ∎

Thus, each query answer A is associated with an annotation computed as follows: $Ann = \sum_{i=1}^{n} \prod_{j=1}^{k} Ann_{ij}$. Each expression $\prod_{j=1}^{k} Ann_{ij}$ $(1 \leqslant i \leqslant n)$ keeps track of one way of deriving A; then, all possible ways are combined with a summation. We define $S\text{-}Ann(Ann) = \{ \prod_{j=1}^{k} Ann_{ij} \mid 1 \leqslant i \leqslant n \}$. We also define the set of *annotated answers* as $\overline{ans}(Q, KB) = \{ \langle A_1, Ann(A_1) \rangle, \ldots, \langle A_n, Ann(A_n) \rangle \}$, where $ans(Q, KB) = \{ A_1, \ldots, A_n \}$. The set of *possible set of answers* is defined as:

$$\{ \{ \langle A_1, Ann'_1 \rangle, \ldots, \langle A_n, Ann'_n \rangle \} \mid Ann'_i \in S\text{-}Ann(Ann(A_i)) \text{ for } 1 \leqslant i \leqslant n \}$$

Example 9. Returning to Example 8, There are two possible sets of answers, namely:

$$\{ \langle myaccom(h_1), (id_1 \times id_8 \times id_5 \times id_9) \rangle, \langle myaccom(bb_1), (id_3 \times id_{10} \times id_7 \times id_{12}) \rangle \}$$
$$\{ \langle myaccom(h_1), (id_1 \times id_8 \times id_6 \times id_{11}) \rangle, \langle myaccom(bb_1), (id_3 \times id_{10} \times id_7 \times id_{12}) \rangle \}. \quad ∎$$

We can now apply the framework proposed in the previous section to each possible set of answers, and then combine the results obtained for all of them. But before that, we need a way of *determining the features* of the query answers and the reports associated with them. Of course, these should be obtained on the basis of the query structure and the available provenance information.

Each predicate symbol can be associated with a sequence of features. In the following, given a conjunctive query Q, we assume there is an arbitrary but fixed criterion to determine a sequence of features $\mathcal{F}_Q$ for the query answers and assume a way of generating a report for each query answer A on the basis of $Ann(A)$ (in Propositions 2 and 3, we assume that both can be accomplished in polynomial time). A concrete approach is illustrated in the following example.

Example 10. Consider the set of possible answers of Example 9:

$$\{ \langle myaccom(h_1), (id_1 \times id_8 \times id_5 \times id_9) \rangle, \quad \langle myaccom(bb_1), (id_3 \times id_{10} \times id_7 \times id_{12}) \rangle \}$$

Suppose the features of *accom* are *loc, cl, pri, br, net*, while those of *pub* are *food, drink*. The features of the query answers might be defined as the union of the aforementioned ones, i.e., *loc, cl, pri, br, net, food, drink*—in general, there can be different reasonable ways of combining the features of the predicate symbols in the query (in our case, we considered only predicates *accom* and *pub* and took the union of their features). The reports for the query answers may be derived by merging the reports of the *accom*- and *pub*-atoms that contributed to each query answer. For instance, suppose we have two observers and their reports are as follows:

$$\Gamma_1(id_1) = \langle 1, 0.7, 0.5, 1, - \rangle \qquad \Gamma_2(id_1) = \langle 0.3, 0.4, 1, -, - \rangle$$
$$\Gamma_1(id_5) = \langle 0.5, 1 \rangle \qquad \Gamma_2(id_5) = \langle 0.7, 0.2 \rangle$$
$$\Gamma_1(id_3) = \langle 0.8, 0.3, 0.2, 0.5, 0.7 \rangle \quad \Gamma_2(id_3) = \langle 0.4, 0.5, 0.5, 0.6, 0.1 \rangle$$
$$\Gamma_1(id_7) \text{ not defined} \qquad \Gamma_2(id_7) = \langle 0.1, 0.8 \rangle$$
$$\cdots \qquad \qquad \cdots$$

Thus, the reports of the first and second observers for the query answers are:

$$\Gamma_1(myaccom(h_1)) = \langle 1, 0.7, 0.5, 1, -, 0.5, 1 \rangle$$
$$\Gamma_1(myaccom(bb_1)) = \langle 0.8, 0.3, 0.2, 0.5, 0.7, -, - \rangle$$
$$\Gamma_2(myaccom(h_1)) = \langle 0.3, 0.4, 1, -, -, 0.7, 0.2 \rangle$$
$$\Gamma_2(myaccom(bb_1)) = \langle 0.4, 0.5, 0.5, 0.6, 0.1, 0.1, 0.8 \rangle$$

At this point, we can apply the framework of the previous section.   ∎

Once we get a preference relation for each possible set of answers, they can be combined using a preference aggregation operator, and a total order over this set can easily be derived (e.g., as before, via topological sorting). Thus, eventually, we obtain a ranking over the query answers.

The following proposition provides an upper bound on the (data) complexity of computing query answer rankings. The exponential time upper bound is due to the fact that the number of possible sets of answers can be exponential in the worst case.

Proposition 2. *Computing a query answer ranking can be done in exponential time in the data complexity.*

Alternatively, we can use the annotations in a different way. For instance, we can use a function that maps the id of an atom to a value in the $[0, 1]$ interval such that it compiles all reports (from different observers) associated with the atom into a single score (in Proposition 3 below, we assume that such a function can be computed in polynomial time); this score may represent, for instance, the overall relevance of the atom for the user based on the reports and the relative importance of each dimension. The scores can be combined through the min (resp., max) operator when $\times$ (resp., $+$) is encountered in the annotation. This yields the application of a commutative semiring $(\mathbb{R}^+_{\leq 1}, min, max, 1, 0)$. This evaluation of reports is called *extensional*.

The following proposition provides an upper bound on the (data) complexity of computing query answer rankings under the extensional evaluation of reports.

Proposition 3. *Computing a query answer ranking under the extensional evaluation of reports can be done in polynomial time in the data complexity.*

The polynomial time complexity of the extensional approach follows from the fact that each of the following tasks can be accomplished in polynomial time: computing the provenance information, mapping atoms' ids to scores, and combining the scores.

5   Related Work

The study of preferences has been carried out in many disciplines; in computer science, the developments that are most relevant to our work is in the incorporation of preferences into query answering mechanisms. To date (and to our knowledge), the state of the art in this respect is centered around relational databases and, recently, in ontological languages for the Semantic Web [15]. The seminal work in preference-based query answering was that of [14], in which SQL is extended to incorporate user preferences. The preference formula formalism was introduced in [9] as a way to embed a user's preferences into SQL. An important development in this line of research is the well-known *skyline* operator, which was introduced in [4]. A recent survey of preference-based query answering formalisms is provided in [17]. The problem of evaluating ranked top-k queries in the context of ontology-based access over relational databases was considered in [18]. Studies of preferences related to our approach have also been done in classical logic programming [11] as well as answer set programming frameworks [5].

The present work can be considered as a further development of the PrefDatalog+/– framework presented in [15], where we develop algorithms to answer skyline queries, and their generalization to k-rank queries, over classical Datalog+/– ontologies. The main difference between PrefDatalog+/– and the work presented here is that PrefDatalog+/– assumes that a model of the user's preferences is given at the time the query is issued. On the other hand, we make no such assumption here; instead, we assume that the user only provides some very basic information regarding her preferences over certain features, and has access to a set of reports provided by other users in the past. In a sense, this approach is akin to building an ad hoc model *on the fly* at query time and using it to provide a ranked list of results.

Related to our approach is the work in [10], where preferences are created from provenance annotations of RDF data. There, different provenance annotation dimensions are considered, each yielding a total preference order represented by a semiring. These preferences are then aggregated according to common social choice theory preference aggregation methods. Another line of related research are constraint-based formalisms for modeling preferences based on semirings, like the one presented in [3].

This work is also related to the study and use of provenance in information systems and, in particular, the Semantic Web and social media [16,10,1]. Here, we use a kind of "how"-provenance [12] to study how to propagate report annotations of atoms in an ontology to create preference-based ranked results of ontological queries. Representing the "how"-provenance in this manner allows us to map the semiring into others depending on the manipulations that we wish to perform during the propagation of reports. To our knowledge, this is the first study of a direct application of provenance of reports of this kind found in online reviews to query answering.

6   Summary and Outlook

In this paper, we have studied the problem of preference-based query answering in Datalog+/– ontologies under the assumption that the user's preferences are informed by a set of subjective reports representing opinions of others—such reports model the kind of information found, for instance, in online reviews of products, places, and services.

We first introduced a basic approach to rank atoms in an ontology by combining reports, report authors' preferences, and the user's preferences. Then, we extended our framework to deal with dependencies and query answering using provenance information to keep track of which reports should be considered to evaluate query answers, as well as new information derived from dependencies. Representing the provenance by means of a semiring enabled us to adapt our framework depending on the kind of report propagation that we wish to carry out. One direction for future work involves investigating further semirings for manipulating reports during their propagation.

Much work remains to be done in this line of research, such as expressing general reports that apply to sets of tuples, and exploring the application of existing techniques to gather reports from actual information available in Web reviews. We also plan to experimentally evaluate our framework over synthetic and real-world data.

Acknowledgments. This work was supported by the UK EPSRC grant EP/J008346/1 ("PrOQAW"), an EU (FP7/2007-2013) Marie-Curie Intra-European Fellowship, the ERC grant 246858 ("DIADEM"), and a Yahoo! Research Fellowship.

References

1. Barbier, G., Feng, Z., Gundecha, P., Liu, H.: Provenance Data in Social Media. Morgan and Claypool (2013)
2. Beeri, C., Vardi, M.Y.: The implication problem for data dependencies. In: Even, S., Kariv, O. (eds.) ICALP 1981. LNCS, vol. 115, pp. 73–85. Springer, Heidelberg (1981)
3. Bistarelli, S., Pini, M.S., Rossi, F., Venable, K.B.: Bipolar preference problems: Framework, properties and solving techniques. In: Azevedo, F., Barahona, P., Fages, F., Rossi, F. (eds.) CSCLP. LNCS (LNAI), vol. 4651, pp. 78–92. Springer, Heidelberg (2007)
4. Börzsönyi, S., Kossmann, D., Stocker, K.: The skyline operator. In: Proc. ICDE, pp. 421–430 (2001)
5. Brewka, G.: Preferences, contexts and answer sets. In: Dahl, V., Niemelä, I. (eds.) ICLP 2007. LNCS, vol. 4670, p. 22. Springer, Heidelberg (2007)
6. Buneman, P.: The providence of provenance. In: Gottlob, G., Grasso, G., Olteanu, D., Schallhart, C. (eds.) BNCOD 2013. LNCS, vol. 7968, pp. 7–12. Springer, Heidelberg (2013)
7. Calì, A., Gottlob, G., Kifer, M.: Taming the infinite chase: Query answering under expressive relational constraints. In: Proc. KR, pp. 70–80 (2008)
8. Calì, A., Gottlob, G., Lukasiewicz, T.: A general Datalog-based framework for tractable query answering over ontologies. J. Web Sem. 14, 57–83 (2012)
9. Chomicki, J.: Preference formulas in relational queries. TODS 28(4), 427–466 (2003)
10. Dividino, R., Gröner, G., Scheglmann, S., Thimm, M.: Ranking RDF with provenance via preference aggregation. In: ten Teije, A., Völker, J., Handschuh, S., Stuckenschmidt, H., d'Acquin, M., Nikolov, A., Aussenac-Gilles, N., Hernandez, N. (eds.) EKAW 2012. LNCS, vol. 7603, pp. 154–163. Springer, Heidelberg (2012)
11. Govindarajan, K., Jayaraman, B., Mantha, S.: Preference queries in deductive databases. New Generat. Comput. 19(1), 57–86 (2001)
12. Green, T.J., Karvounarakis, G., Tannen, V.: Provenance semirings. In: Proc. PODS, pp. 31–40 (2007)
13. Ha, V., Haddawy, P.: Toward case-based preference elicitation: Similarity measures on preference structures. In: Proc. UAI, pp. 193–201 (1998)
14. Lacroix, M., Lavency, P.: Preferences: Putting more knowledge into queries. In: Proc. VLDB, vol. 87, pp. 1–4 (1987)
15. Lukasiewicz, T., Martinez, M.V., Simari, G.I.: Preference-based query answering in Datalog+/– ontologies. In: Proc. IJCAI, pp. 1017–1023 (2013)
16. Moreau, L.: The foundations for provenance on the Web. Found. Trends Web Sci. 2(2/3), 99–241 (2010)
17. Stefanidis, K., Koutrika, G., Pitoura, E.: A survey on representation, composition and application of preferences in database systems. TODS 36(3), 19:1–19:45 (2011)
18. Straccia, U.: On the top-k retrieval problem for ontology-based access to databases. In: Pivert, O., Zadrożny, S. (eds.) Flexible Approaches in Data, Information and Knowledge Management. SCI, vol. 497, pp. 95–114. Springer, Heidelberg (2013)

A Petri Net Model of Argumentation Dynamics

Diego C. Martinez[1,2], Maria Laura Cobo[2], and Guillermo Ricardo Simari[2]

[1] Consejo Nacional de Investigaciones Cientficas y Tcnicas - CONICET
Artificial Intelligence Research and Development Laboratory (LIDIA)
[2] Department of Computer Science and Engineering, Universidad Nacional del Sur
Av. Alem 1253 - (8000) Bahía Blanca - Bs. As. - Argentina
{mlc,dcm,grs}@cs.uns.edu.ar
http://www.cs.uns.edu.ar/lidia

Abstract. Petri nets are a mathematical modelling tool suitable for describing dynamic computational systems. In this work we present a formalization of abstract argumentation frameworks using Petri nets, where arguments and attacks are represented as places and transitions. This provides a formalism to study the semantic consequences of a procedural evaluation of argument attacks. The relation between markings of the net and argument extensions is analysed.

1 Introduction

Roughly speaking, *argumentation* is the study of arguments and their relationships. It is a form of reasoning suitable to deal with incomplete and contradictory information in dynamic domains. Although several proposals of argumentation systems are available, it is possible to study pure semantic notions in a general framework with a high level of abstraction. In abstract argumentation formalisms some components remain unspecified, being the structure of an argument the main abstraction. In this kind of systems, the emphasis is put on the semantic notion of finding the set of accepted arguments. Most of these *abstract argumentation frameworks* are based on the single concept of information conflict called *attack*, represented as an abstract relation, and extensions are defined as sets of possibly accepted arguments. The study of *dynamics* of argumentation has been an important topic in this area. An initial proposal of dynamic argumentation is presented in [3] using situation calculus. In recent formalisms [5,4,2,11] the semantic study of arguments and attacks is addressed under a temporal perspective, where arguments and attacks are progressively considered as the framework evolves through time. This is important since argumentation is intrinsically tied to dialectic activities, like dialogues, debates and even introspection.

Then it is possible to study how the process of argumentation advances while arguments and attacks are selected or provided in a sequential manner, which is relevant in systems with a large amounts of arguments. For instance, a DeLP program [7] of few defasible rules may produce hundreds of defeasible arguments. Even more, in some contexts it is not necessary to consider all of these arguments as a whole while reasoning. Gradual consideration of arguments and attacks is interesting. However, the fact that some controversies between arguments may be addressed in a sequential, distributed and concurrent operation with different semantic consequences was not previously studied under a suitable mathematical model in the argumentation community.

U. Straccia and A. Calì (Eds.): SUM 2014, LNAI 8720, pp. 237–250, 2014.

Argumentation processes can be naturally complex. For instance, when a set of topics is discussed by several participants, some of the debates may occur at the same time. This is how many legislative corps works, such as in the United Nations or several state Congresses around the world. A formal model for the study of a procedural, timed interpretation of the act of attacking an argument is novel and interesting in the context of timed argumentation, where it is uncertain what arguments are going to be addressed next. In this work we are not interested in the logical aspects of argumentation as in [3], but in the abstract characterization of potentially distributed, asynchronous argumentation.

Petri nets are a mathematical modelling tool suitable for describing concurrent, asynchronous, non-deterministic computational systems [1,10,8]. The basic formalism is simple and it provides a sound framework for the study of properties of discrete events systems, yet in the last years several extensions were proposed in order to provide models for different characterizations of dynamic systems. Petri nets provide a description of independence and causal relations between system actions, which allows to reason about partially ordered sets of actions without having to consider their interleavings [8]. Properties of Petri nets were studied for half a century around the globe and it is widely considered a mature discipline.

In this work we propose a Petri net representation of an abstract argumentation framework. This provides a formalism to study the semantic consequences of a procedural evaluation of individual argument attacks. Being a classic engineering-oriented formalism, Petri nets are appropriate for *argumentation process analysis* which is an important direction of this line of research. In this paper we introduce the formalism and we show there is a correspondence between the evolution of the net and the underlying argumentation semantics.

This paper is organized as follows. In Section 2 we recall the basic notions of classical abstract argumentation frameworks. In Section 3 a brief, general description of Petri nets is included. In Section 4 we present a Petri net model for abstract argumentation and related semantic notions are introduced in Section 5. In Section 6 the dialectical interpretation of the argumentation net is discussed. Finally, conclusions and future work is discussed.

2   Classic Abstract Argumentation

Dung defines several argument extensions that are used as a reference for many authors. The formal definition of the classic argumentation framework follows.

Definition 1 [6] *An argumentation framework is a pair $AF = \langle AR, attacks \rangle$ where AR is a set of arguments, and attacks is a binary relation on AR, i.e. attacks $\subseteq$ AR $\times$ AR.*

Arguments are denoted by labels starting with an upper-case letter, leaving the underlying logic unspecified. A set of accepted arguments is characterized in [6] using the concept of *acceptability*, which is a central notion in argumentation, formalized by Dung in the following definition.

Definition 2 [6] *An argument $A \in AR$ is acceptable with respect to a set of arguments S if and only if every argument B attacking A is attacked by an argument in S.*

If an argument A is acceptable with respect to a set of arguments S then it is also said that S *defends* A. Also, the attackers of the attackers of A are called *defenders* of A. We will use these terms throughout this paper.

Acceptability is the main property of Dung's semantic notions, some of them summarized in the following definition.

Definition 3 *A set of arguments S is said to be*
– conflict-free if there are no arguments A, B in S such that A attacks B.
– admissible if it is conflict-free and defends all its elements.
– a preferred extension if S is a maximal (for set inclusion) admissible set.

In [6], theorems stating conditions of existence and equivalence between these and other extensions are also introduced.

Example 1 *Consider the argumentation framework $AF_1 = \langle AR, attacks \rangle$, where $AR = \{A, B, C, D, E, F, G, H\}$ and attacks $= \{(B, A), (C, B), (D, A), (E, D), (G, H), (H, G)\}$. Then*
– $\{A, C, E\}$ is an admissible set of arguments.
– $\{A, C, E, F, G\}$ is a preferred extension.

In the following section we recall the basic definitions of Petri nets, as needed later. For a more detailed introduction to Petri nets, the reader may refer to [10].

3 Petri Nets

A Petri net is a directed, weighted, bipartite graph consisting of two kind of nodes called *places* and *transitions*. Usually places are represented as circles and transitions are represented as boxes or bars. Arcs connect transitions and places and have a weight (positive integers). A *marking M* of the net assigns a nonnegative integer to each place in the net. If a marking M assigns to place p an integer k, it is said that p has (or is marked with) k *tokens*. This tokens are graphically represented as dots inside a place, or just simply a number.

Definition 4 *A Petri net is a 5-tuple $PN = (P, T, F, W, M_0)$ where*

- $P = \{p_1, p_2, \ldots, p_n\}$ *is a finite set of places and $T = \{t_1, t_2, \ldots, t_n\}$ is a finite set of transitions, with $P \cap T = \emptyset$ and $P \cup T = \emptyset$,*
- $F \subseteq (P \times T) \cup (T \times P)$ *is a set of arcs,*
- $W : F \to \{1, 2, 3, \ldots\}$ *is a weight function,*
- $M_0 : P \to \{1, 2, 3, \ldots\}$ *is the initial marking.*

A transition t is said to be *enabled* at marking M if each input place p of t is such that $M(p) \geq W(p, t)$. Enabled transitions may be *fired*. A firing of a transition t removes $W(p, T)$ tokens from each input place p and adds $W(t, p)$ tokens to each output place. A sequence of firings leads to a sequence of markings. A marking M_i is said to be *reachable* from marking M_j if there exists a sequence of firings that transforms M_j to M_i. The set of all reachable markings of a net P is called the *reachability space* of P. A *vanishing state* of the net is a marking that can be changed since it enable some transitions. A *tangible state* is a marking in which no transition is enabled. The number of tokens in a place p at marking M_k is denoted as $M_k(p)$. A Petri net is said to be *k-bounded* if the number of tokens in each place does not exceed a finite number k for any marking reachable from M_0. A Petri net is said to be *safe* if it is 1-bounded. A *self loop* is a transition with an output and an input from the same place. Two transitions that output to the same place are said to be in *backward conflict*. Two transitions are in *forward conflict* if, being both of them enabled by a common place, only one can be fired.

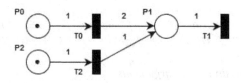

Fig. 1. A simple Petri net

In Figure 1 a simple Petri net is depicted, where places are circles and transitions are black bars. The marking is represented by black dots inside places, showing $M(P0) = 1$, $M(P1) = 0$ and $M(P2) = 1$. Transitions $T0$ and $T2$ are enabled and can be fired. If transition $T0$ is fired, it consumes the only token of place $P0$ and produces two tokens in place $P1$. After that, transition $T1$ is enabled and can be fired twice, each one consuming one token of place $P1$. Some firing sequences on this net are $\{T0, T1, T1, T2, T1\}$ and $\{T2, T1, T0, T1, T1\}$. Both of them lead to the only *tangible state* of the net, with all the places free of tokens. Between the initial state and this tangible state, it is possible to generate eight vanishing states. This is because the firing order of some transitions is interchangeable, as $T0$ and $T2$. Transition $T1$ consumes all the tokens produced by these two transitions and can only be fired after them.

An extension of Petri nets distinguishes a special kind of arcs called *inhibitor arcs*. This set of arcs appears as a new component in the formal definition of the Petri net, being then a 6-tuple. An inhibitor arc connects a place to a transition and it is graphically represented as a line with a white circle in its end. This kind of arc disables the transition when the input place has a token, and enables the transition when the input place has no token and any other normal input has the required tokens. This extension allows the test for absence of tokens and this simple fact makes Petri nets as expressive as Turing Machines.

In Petri nets there are several definitions of fairness. Two transitions t_1 and t_2 are said to be in a bounded-fair relation if the maximum number of times that either one can fire while the other is not firing is bounded. Hence, one transition cannot block the other by firing infinitely. A Petri net is said to be a bounded-fair net if every pair of transitions in the net are in a bounded-fair relation. A firing sequence T is said to be unconditionally fair if it is finite or every transition in the net appears infinitely often in T. A Petri net is said to be an unconditionally fair net if every firing sequence from the initial state is unconditionally fair.

In the following section we present a formalization of argumentation frameworks using Petri nets with inhibitor arcs. Semantic notions are discussed in subsequent sections.

4   Argumentation Nets

An abstract argumentation framework as in Definition 1 induces a Petri net where places are arguments and transitions represent the conflict between arguments. This is formalized in the following definition.

Definition 5 *Let $\Phi = \langle AR, attacks \rangle$ be an argumentation framework. The argumentation net of Φ, or simply argnet, is a Petri net $V_\Phi = (AR, T, F, H, W, M)$ where*

- *AR is the set of places*
- *$T = \{t_1, t_2, \ldots, t_n\}$ is a finite set of transitions.*
- *$F \subseteq (AR \times T) \cup (T \times AR)$ is a set of arcs.*
- *$H \subseteq (AR \times T) \cup (T \times AR)$ is a set of inhibitors arcs.*
- *$W : F \to \{1, 2, 3 \ldots\}$ is the weight function.*
- *$M : AR \to \{0, 1, 2, 3 \ldots\}$ is the initial marking.*

such that

- *for any attack $(\mathcal{A}, \mathcal{B}) \in attacks$, there exist a transition t_{AB} and the arcs $(\mathcal{A}, t_{AB}), (t_{AB}, \mathcal{A}), (\mathcal{B}, t_{AB}) \subseteq F$. This transition is called an attack transition.*
- *for any argument $\mathcal{A}$ with attackers $\mathcal{X}_1, \mathcal{X}_2, \ldots, \mathcal{X}_n$ there exists the transition t_A with the inhibitor arcs $(\mathcal{A}, t_A)$ and $(\mathcal{X}_i, t_A)$ for $1 \leq i \leq n$, and the arc $(t_A, \mathcal{A})$. This transition is called a restoring transition.*
- *$M(\mathcal{X}) = 1$ for any argument $\mathcal{X} \in AR$.*

Tokens represent potential strength of arguments for attacking each other. There is an *attack* transition whenever an argument $\mathcal{A}$ attacks another argument $\mathcal{B}$. Sometimes in this text arguments and places are treated as equivalents. In this paper the weight of every transition is 1, and then transitions remove or add only one token at a time. When referring to attack transitions, the corresponding place for $\mathcal{A}$ will be called the *attacking* or input place and the corresponding place for $\mathcal{B}$ will be called the *attacked* or output place. Such a transition can be fired when a token is available in both the *attacking* and *attacked* place. The attack transition consumes the tokens of both places, and restores the consumed token in the attacking place. Restoring transitions links attacked arguments with all of its attackers. Such a transition places a token in an empty attacked place $\mathcal{X}$ whenever all the attacker places are empty. This models the fact that, since every attacker of $\mathcal{X}$ has no strength, then the strength of $\mathcal{X}$ can be reinstated.

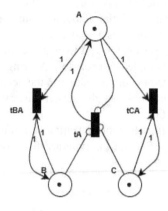

Fig. 2. Arguments $\mathcal{B}$ and $\mathcal{C}$ attack argument $\mathcal{A}$

Example 2 *Let* $\Phi = (\{\mathcal{A}, \mathcal{B}, \mathcal{C}\}, \{(\mathcal{B}, \mathcal{A}), (\mathcal{C}, \mathcal{A})\})$ *be an argumentation framework. The corresponding argnet* $\mathcal{V}_\Phi$ *is depicted in Figure 2. Arguments are represented as places and the attacks are represented by transitions* t_{BA} *and* t_{CA}. *The restoring transition* t_A *adds a token to the place* $\mathcal{A}$ *only when* $\mathcal{A}$ *and its attackers have no tokens.*

Petri nets are mainly a model of computation. In this argumentation net our *units of computation* are the transitions, either attacking or restoring ones. Therefore, we are interested in the evolution of the strength of arguments as transitions are fired, *i.e.* when tokens are consumed and restored in argument places. In the initial marking M_0 every place has a token, since no attack is considered yet and then every argument has the potential strength of affecting other arguments. Note that transitions can put tokens in a place whenever (a) a token is removed from the same place as in attacking transitions or (b) when there is no token in the place as in restoring transitions. Hence no place can hold more than one token, as stated in the following Proposition.

Proposition 1 *For any argumentation framework* Φ, *the argnet* $\mathcal{V}_\Phi$ *is safe (1-bounded).*

If the Petri net is k-bounded, then the reachability space is finite. In this work we are interested in the connection between the firing of transitions, the evolution of markings and the underlying argumentation semantics. Consider the net of Figure 2. The enabled transitions are tBA and tCA. Restoring transition tA is not enabled since places A, B and C are not free of tokens. There are two possible firing sequences in this net: $\{tBA\}$ and $\{tCA\}$. Since the firing of an attacking transition removes the token in the attacker argument, then the firing of tCA inhibits the firing of tBA and viceversa. This happens with attacking transitions for the same arguments, as stated in the following Proposition.

Proposition 2 *Let* $\mathcal{A}$ *be an argument attacked by arguments* $\mathcal{D}_1, \mathcal{D}_2 \ldots \mathcal{D}_{n},$. *The transitions* $t\mathcal{D}_1\mathcal{A}, t\mathcal{D}_2\mathcal{A} \ldots t\mathcal{D}_n\mathcal{A}$ *are all in forward conflict.*

However, whether tBA or tCA are fired in the net of Figure 2 the final marking M of $\mathcal{V}_\Phi$ is the same: $M(A) = 0$, $M(B) = 1$ and $M(C) = 1$. In other words, any of these attacks can be applied and the final outcome is the same, but only one of them is firable since these transitions are not independent of each other. Note that the maximal admissible set of Φ is $\{\mathcal{B}, \mathcal{C}\}$, the only places with a token in the tangible state.

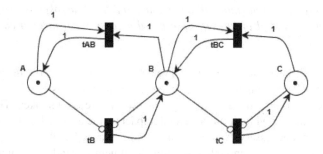

Fig. 3. Argument $\mathcal{A}$ attacks $\mathcal{B}$. Argument $\mathcal{B}$ attacks $\mathcal{C}$.

Consider the net of Figure 3 where a situation of argument defense is presented. The initial enabled transitions are tAB and tBC. This means that both attacks are fireable at the beginning. If transition tAB is fired first, then no transition is later enabled, leading to a tangible state. If transition tBC is fired first, transition tAB is still enabled. After firing transition tAB, the restoring transition tC becomes enabled, since neither B nor C have tokens. After firing transition tC, place (argument) C gains a token and then no transition is enabled after that. Thus, a tangible state can be reached by firing tAB or by firing the sequence $\{tBC, tAB, tC\}$. Moreover, there is only one tangible state with marking $M(A) = 1$, $M(B) = 0$ and $M(C) = 1$. Note that the maximal admissible set of the corresponding abstract framework is $\{\mathcal{A}, \mathcal{C}\}$.

In the following section we present semantic notions for argumentation Petri nets, based on markings and sequence of transitions.

5   Argumentation Semantics

As stated before, Petri nets are a model of computation for concurrent and distributed systems, where the emphasis is put in the firing of transitions and how the marking of the net evolves as a consequence. Argnets provide an interesting model for procedural argumentation semantics. In this section we consider nets without isolated parts. As shown in the example of Figure 3 a sequence of firing of transitions leads to a sequence of markings, which can be interpreted as an evolution of the strength of arguments as attacks take place. As long as there are enabled transitions, an attack or a restoration can occur and then there are still arguments able to loose or gain strength. There is, however, a set of arguments that never loose its tokens. A *trap* of a Petri net is a set of places S such that any transition with an input in S has also an output in S and if S is marked under some marking M, it is still marked under any succesor marking of M.

Proposition 3 *Let* $\Phi = \langle AR, attacks \rangle$ *be an argumentation framework with corresponding argnet* $\mathcal{V}_\Phi$ *and let* $D_f \subseteq AR$ *be the set of all defeater-free arguments in AF. Then* D_f *is a trap of* $\mathcal{V}_\Phi$.

Proof: Defeater-free arguments can only attack other arguments and then every outgoing transition of the corresponding place is an attack transition, conforming a self-loop. These are the only transitions that will be enabled and then the token is never lost. In some formalisms, this kind of loop if represented as a single transition called read-transition. □

Hence, defeater-free arguments always have a token in the corresponding place. Because of this, any attack transition from a defeater-free argument is enabled at M_0 and it will be enabled as long as the attacked argument still possesses its token. That means that the attack is enabled while it has an actual impact on the attacked argument (otherwise the attack is not necessary). In the net of Figure 2, places $\mathcal{B}$ and $\mathcal{C}$ never loose their tokens, yet only one attack is enough to suppress argument $\mathcal{A}$. Under the interpretation of attacks as actions, in this example only one attack is sufficient to reach a tangible state.

An admissible extension is basically a set of arguments defending each other. In the Petri net this is interpreted as a distribution of tokens over the net, with a particular condition. This marking can be reached by firing transitions until no transition is enabled, as stated in the following proposition.

Proposition 4 *Let* $\Phi = \langle AR, attacks \rangle$ *be an argumentation framework with argnet* $\mathcal{V}_\Phi = (P, T, F, H, W, M)$. *Let* T *be a sequence of firing transitions* $\{t_1, t_2, \ldots, t_n\}$ *that transforms* M *to* M_n *such that every transition in* $\mathcal{V}_\Phi$ *after* M_n *is not enabled. Then the set of arguments* $S = \{\mathcal{A} \in AR | M_n(\mathcal{A}) = 1\}$ *is an admissible set of* Φ.

Proof: Suppose no transitions are enabled and S is not admissible. Then either (a) it is not free of conflict or (b) at least one argument $\mathcal{A} \in S$ is not defended by S. If (a) is the case, then at least two arguments $\mathcal{X}$ and $\mathcal{Y}$ are such that $M(\mathcal{X}) = M(\mathcal{Y}) = 1$ are in conflict. But then the attack transition between them is enabled, which is a contradiction. If (b) is the case, then at least an argument $\mathcal{X} \in S$ is attacked by an argument $\mathcal{Y}$, but not defended by an argument in S. But $\mathcal{Y}$ must be free of tokens, otherwise (a) is the case. Since $M(\mathcal{Y}) = 0$, then $\mathcal{Y}$ is not free of attackers and then there is at least one argument attacking $\mathcal{Y}$. But, since $\mathcal{Y}$ has no token and no transitions are enabled, then it is not possible for all the attackers of $\mathcal{Y}$ to have no tokens. Then at least one argument $\mathcal{Z}$ attacks $\mathcal{Y}$ such that $M(\mathcal{Z}) = 1$. But then $\mathcal{X}$ is actually defended by S, which is a contradiction. Since (a) and (b) cannot be the case, then S is admissible whenever the transitions are not enabled. □

Hence, a tangible state in the net corresponds to a distribution of tokens signalling arguments in an admissible extension. If no tangible state can be reached through a sequence of firings of transitions, then there is always an enabled attack or restoring transition, *i.e.* a token can be placed or removed somewhere in the net.

Proposition 5 *If at least one transition in $\mathcal{V}_\Phi$ is potentially fireable in any marking M_k then AF is not well-formed.*

Proof: If at least one transition is potentially fireable, then every state of the net is a vanishing state. Since there is a finite set of possible states and the net is safe, then at least one argument is gaining and loosing a token repeatedly (although not necessarily in consecutive markings).   □

Consider the net of Figure 4 where an odd cycle of attacks is present. Starting from the initial marking, no tangible state can be produced. In fact, there is an infinite sequence of attacking and restoration transitions. Whenever a token is restored in a place $\mathcal{X}$, it enables an attack transition tXY from that place. There are six vanishing states in this net and the only admissible set in the corresponding argumentation framework is the empty set.

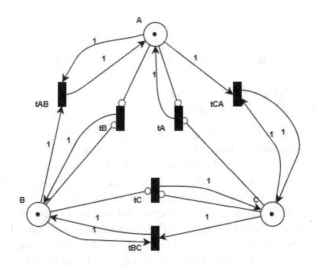

Fig. 4. Argument cycle between $\mathcal{A}$, $\mathcal{B}$ and $\mathcal{C}$

The converse of Proposition 5 is not true, as shown in the net of Figure 5, corresponding to the argumentation framework $\Phi = \langle \{\mathcal{A}, \mathcal{B}, \mathcal{C}, \mathcal{D}\}, \{(\mathcal{A}, \mathcal{B}), (\mathcal{B}, \mathcal{C}), (\mathcal{C}, \mathcal{D}), (\mathcal{D}, \mathcal{A})\} \rangle$ with a cycle of attacks between the four arguments. Starting from the initial marking, it is possible to reach thirteen different states, with only two of them being tangible. As stated in Proposition 4 these tangible states correspond to the admissible sets $S_1 = \{\mathcal{A}, \mathcal{C}\}$ and $S_2 = \{\mathcal{B}, \mathcal{D}\}$. The set S_1 can be reached, for instance, by the sequence of firings $T_1 = \{tAB, tCD\}$ but also by $T_1' = \{tCD, tBC, tAB, tC\}$. The set S_2 can be reached, for instance, by $T_2 = \{tBC, tDA\}$ or $T_2' = \{tBC, tAB, tDA, tB\}$. Note that in T_1' and T_2' a restoration transition is needed.

Consider the following sequences of transitions in the net of Figure 5:

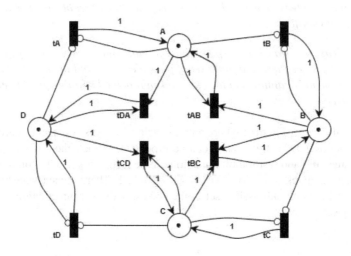

Fig. 5. Four arguments $\mathcal{A}$, $\mathcal{B}$, $\mathcal{C}$ and $\mathcal{D}$ in a cycle

- $T = \{tDA, tCD, tA, tBC, tD, tDA\}$
- $T' = \{tDA, tCD, tBC, tA, tAB, tC\}$

Although both sequences start firing transitions $\{tDA, tCD\}$, they lead to different tangible states. Sequence T leaves tokens in $\mathcal{D}$ and $\mathcal{B}$ while the sequence T' leaves tokens in $\mathcal{A}$ and $\mathcal{C}$. The former decides to reinstate $\mathcal{A}$ (by firing tA) before triggering an attack to $\mathcal{C}$ from $\mathcal{B}$ (by firing tBC). The sequence T', however, decides to attack $\mathcal{C}$ before reinstate $\mathcal{A}$. This shows, of course, that the choice of transitions to fire may completely change the outcome of the process. But the most interesting aspect is that the reinstatement of tA and the attack tBC are interchangeable, since both T and T' lead to a state of the net in which only tAB and tD are the enabled transitions after the fourth transition. In other words, $\{tDA, tCD, tA, tBC\}$ and $\{tDA, tCD, tBC, tA\}$ lead to the same vanishing state. After that, both sequences take different paths: T reinstantes $\mathcal{D}$ to attack $\mathcal{A}$ later, while T' attacks $\mathcal{B}$ to later reinstate $\mathcal{C}$. As expected in argumentation, a restoring transition can be fired only after other transitions are fired. For a given sequence of transitions, we will denote $t_1 \ll t_2$ if transition t_1 occurs before transition t_2. In order to reinstate an argument $\mathcal{X}$, all of its attackers must loose their tokens. This is formalized in the following proposition.

Proposition 6 *Let Φ be an argumentation framework with argnet $\mathcal{V}_\Phi$. For any sequence of firings of transitions T, then*

1. *$tXY \ll tY$ in T and*
2. *for every argument W_i attacking Y, $tZ_iW_i \ll tY$ in T.*

Proof: Trivial. Place Y must loose its token in order to fire its reinstate transition. This can only be achieved by an attack transition with input Y. The same is true for all of its attackers. □

Moreover, if $tZ_iW_i \ll tY$ in a sequence T then there is a sub-sequence $T' = [tZ_iW_i \ldots tY]$ such that there are no extra occurrences of tZ_iW_i in T' and $tW_i \nll tY$. In other words, although W_i loose its token after the attacking transition, it is not reinstated before Y is. Note that in Proposition 6 there is no specific restriction about the order of tXY and every tZ_iW_i. This is consistent with the notion of attack and defense in argumentation frameworks.

Another important aspect of Petri nets is the *reachability problem*. This is a *decision problem* about deciding, for a given marking M, whether it is reachable in a particular net. In our formalism, the reachability graph is finite since the net is safe. What is interesting is to prove that certain relevant markings are in the reachability graph induced by a net.

Proposition 7 *Let* $\Phi = \langle AR, attacks \rangle$ *be an argumentation framework with argnet* $\mathcal{V}_\Phi = (P, T, F, H, W, M)$. *If* $S \subseteq AR$ *is a preferred extension of* Φ, $S \neq \emptyset$, *then there exists a sequence of firings* $\{t_1, t_2, \ldots, t_n\}$ *that transforms* M *to* M_n *such that* $M_n(\mathcal{A}) = 1$ *if* $\mathcal{A} \in S$ *and* $M_n(\mathcal{A}) = 0$ *if* $\mathcal{A} \notin S$.

Proof: Let M_S *be the marking such that only arguments in the preferred extension* S *have tokens. Suppose* M_S *is not in the reachability space of the net* $\mathcal{V}_\Phi$. *This means that there exists at least one argument* $\mathcal{A}$, *such that* $\mathcal{A}$ *cannot (a) acquire a token if* $M_S(\mathcal{A}) = 1$ *or (b) loose a token if* $M_S(\mathcal{A}) = 0$ *in any sequence of firing transitions. Suppose (a) is the case. Since the initial marking assigns tokens for every place, then* $\mathcal{A}$ *looses its token and it is not able to recover it in any sequence of transitions. However, since* $\mathcal{A}$ *is in the preferred extension, then it is defended by arguments in* S. *But if every defender* $\mathcal{D}_i$ *is such that* $M(\mathcal{D}_i) = 1$, *then after firing the outgoing attack transitions, every attacker of* $\mathcal{A}$ *looses its token, which after the restoring transition makes* $M(\mathcal{A}) = 1$. *Then clearly at least one defender* $\mathcal{D}_k$ *of* $\mathcal{A}$ *has no token, otherwise* $\mathcal{A}$ *could recover its token. Now* $\mathcal{D}_k$ *and* $\mathcal{A}$ *are two arguments that* M_S *assigns tokens to, but they cannot acquire them. The same analysis can be made for* $\mathcal{D}_k$. $\qquad\square$

It could be the case that some controversies are present in the framework, as shown in Example 4, where an infinite sequence of transitions can be fired. In this particular case of a three-argument cycle, every transition leads to a vanishing state. The same is true for longer odd cycles, when an argument attacks its own (indirect) defender. This is sometimes called a *contradictory argumentation line*, since every argument indirectly attacks itself. An argument $\mathcal{A}$ is said to be *controversial* with respect to another argument $\mathcal{B}$ if $\mathcal{B}$ indirectly attacks and indirectly defendes $\mathcal{A}$ [6].

Proposition 8 *Let* Φ *be an argumentation framework with argnet* $\mathcal{V}_\Phi$. *If there exists an infinite sequence of firings of transitions in* $\mathcal{V}_\Phi$, *then* Φ *is controversial.*

Proof: If there exists an infinite sequence of transitions, then some arguments are repeatedly loosing and gaining their tokens. Thus, there is a cycle of attacking and restoring transitions. It means that at least two transitions tXY *and* tY *are involved a bounded-fair relation. Hence the attack of* X *on* Y *indirectly causes the restoration of* Y. *It means that* X *attacks and indirectly defends* Y, *and then* Φ *is controversial.* $\qquad\square$

A Petri net may have an isolated sub-graph with a cycle causing an infinite sequences of firings of transitions. Since there is always a transition that is potentially fireable, then the net cannot reach a tangible state. There may be, however, a subnet such that no transition is enabled and there are no other, external transitions that can change that fact. A subnet generated by a set of transitions T is another net formed by T and all of its input and output places with its corresponding arcs. In argumentation nets, every transition of a net V_Φ that is not potentially fireable at a given marking M, forms a subnet V'_Φ such that the restriction of M to V'_Φ is a potential admissible subset of arguments. Although there are no tangible states in the whole net, some transitions will be never fired and thus some places are not receiving or loosing tokens any more.

In the following section we discuss a dialectical interpretation of sequences of firings of transitions in an argumentation net.

6 Transitions as Dialogue Acts

An argumentation system may produce thousands of arguments from a knowledge base. In Defeasible Logic Programming [7], the addition of a simple defeasible rule may cause new derivation trees, thus incrementing the set of arguments. The size and complexity of argumentation makes procedural evaluation of arguments and its relationship an important topic. It is interesting to evaluate the role of transitions under procedural models of argumentation. A sequence of firings can be considered as a sequence of moves in an argumentation game, where two participants (*agents*) decide what attack is considered next. This is basically a dialogue that last until some particular condition is reached. Several forms of dialogue games may be defined, for instance:

– *Single-topic*: an agent P proposes an argument $\mathcal{A}$ to agent O. The goal of P is to defend $\mathcal{A}$, *i.e.* to keep the corresponding place tokenized. The goal of O is to de-tokenize $\mathcal{A}$.
– *Set-of-beliefs*: both agents propose a set of arguments S as a set of beliefs in contention. The goal of the dialogue is to collaboratively analyse the acceptance of arguments in S by highlighting attacks and restorations until some condition is reached.

The first dialogue is competitive, while the other is collaborative. Both dialogues may run until no transitions are enabled, or for a finite period of time, or until a maximal number of transitions were fired. In any case, a restriction of valid sequences of firings can be considered. For instance, an agent that proposes tXY cannot propose tY later. This means that an agent that causes the disqualification of an argument in the dialogue (by deleting its token) cannot provoke the restoration of the same argument. This restriction is probably more reasonable in single-topic dialogues. Also some transitions may be completely forbidden. This may be the case when some arguments and some relations are previously agreed to be off-topic.

It is also possible to consider restoration transitions as an automatic consequence of the last transition that enables such a restoration, an then restoring transitions are not a move by itself in the dialogue. Thus, a restoration transition is not a legal single move. It must be always preceded by an attack transition. According to this form of dialogue, there are only two possible moves for an agent: (tXY) and $(tXY; tZ)$ for any arguments $\mathcal{X}$, $\mathcal{Y}$ and $\mathcal{Z}$ such that $\mathcal{Y}$ is an attacker of $\mathcal{Z}$

Another interpretation may be to provide restoring transitions with special, additional conditions for firing. Here restoration is not automatic, but reserved for particular moments in the dialogue. Then there is a priority between transitions, being attack ones preferred over restoring ones. When no attack is possible, a restoration of a place may be considered by any agent, even when that agent previously removed the token of the same place. This is a sort of belief revision made by the agent, by consenting the validity of a previously challenged argument.

In the Petri nets model of argumentation this restrictions to the dialogue can be achieved by the notion of *supervisors* [9], as shown in the following Definition.

Definition 6 [9] *Let $\mathcal{V}_\Phi = (AR, T, F, W)$ be a Petri net, $\mathcal{M}$ the set of all markings of $\mathcal{V}_\Phi$ and $U \subseteq \mathcal{M}$. A **supervisor** Ξ is a function $\Xi : U \rightarrow 2^T$ that maps to every marking a set of transitions that the Petri net is allowed to fire.*

The notion of supervisor is usually associated with the task of preventing deadlock in Petri nets. However, the same formalism may be used to direct the dialogue to specific purposes.

Consider the net of Figure 4, where there is no tangible state. A dialogue of transitions engaged in this net requires additional controls to avoid circular argumentation. A possible supervisor for this situation may be $\Xi(M) = \emptyset$ for any marking $M \subseteq \{(1, 0, 0), (0, 1, 0), (0, 0, 1)\}$ *i.e.* markings with only one token. This means that no transition is enabled after removing two tokens. In other words, the last argument to survive leads to a special kind of tangible state since no transition is legally fireable. The notion of tangible state is now contextual to the overall state of enabled transitions and the supervisor restrictions. Even more, it is possible to use more than one supervisor, with priorities. One supervisor define legal attack transitions and the other define legal restorations. The change of supervisor takes place when the net reaches a tangible state under supervisors restriction.

7   Conclusions and Future Work

Petri nets are a model of computation for concurrent and distributed systems, where the emphasis is put in the firing of transitions and how the marking of the net evolves as a consequence. In this work we proposed a Petri net representation of abstract argumentation frameworks as an approach to the study of procedural interpretation of attacks, *i.e.* the consideration of argument attacks as actions in an argumentation system. Given this new Petri net model, we have proved that there is a relation between tangible states of the net and admissible sets of the corresponding framework. We also discussed that the procedural profile of Petri nets makes this formalism suitable to provide a framework for the study of argumentation dialogues, under the formal regulation of a supervisor.

Future work has several directions, as intended in this seminal proposal. A Petri net model of argumentation frameworks that allow places to have more than one token is being studied. Tokens here are intended to represent the strength of an argument, and a single attack weakens such an argument by removing one token. An argument is considered rejected if it has no tokens. In other direction, it is important to study the relation between partial repetitive nets and the existence of admissible extensions in the corresponding argumentation framework. Dialectical strategies to avoid cyclic transitions in an argumentation dialogue are interesting. Finally, the addition of *timed* transitions is important to model the dynamics of timed argumentation formalisms [5].

References

1. van der Aalst, W.M.P., Stahl, C., Westergaard, M.: Strategies for modeling complex processes using colored petri nets. In: Jensen, K., van der Aalst, W.M.P., Balbo, G., Koutny, M., Wolf, K. (eds.) ToPNoc VII. LNCS, vol. 7480, pp. 6 55. Springer, Heidelberg (2013)
2. Barringer, H., Gabbay, D.M., Woods, J.: Temporal, numerical and meta-level dynamics in argumentation networks. Argument & Computation 3(2-3), 143–202 (2012)
3. Brewka, G.: Dynamic argument systems: A formal model of argumentation processes based on situation calculus. J. Log. Comput. 11(2), 257–282 (2001)
4. Cobo, M.L., Martinez, D.C., Simari, G.R.: Stable extensions in timed argumentation frameworks. In: Modgil, S., Oren, N., Toni, F. (eds.) TAFA 2011. LNCS, vol. 7132, pp. 181–196. Springer, Heidelberg (2012)
5. Cobo, M.L., Martinez, D.C., Simari, G.R.: An approach to timed abstract argumentation. In: Proc. of Int. Workshop of Non-Monotonic Reasoning 2010 (2010)
6. Dung, P.M.: On the acceptability of arguments and its fundamental role in nonmonotonic reasoning, logic programming and n-person games. Artificial Intelligence 77(2), 321–358
7. García, A.J., Simari, G.R.: Defeasible logic programming: An argumentative approach. Theory and Practice of Logic Programming 4(1-2), 95–138 (2004)
8. Rintanen, J., Thiébaux, S., White, L., Hickmott, S.: Planning via petri net unfolding. In: Proc. of 20th Int. Joint Conference on Artificial Intelligence, pp. 1904–1911. AAAI Press, Menlo Park (2007)
9. Iordache, M.V., Antsaklis, P.J.: Generalized conditions for liveness enforcement and deadlock prevention in petri nets. In: Colom, J.-M., Koutny, M. (eds.) ICATPN 2001. LNCS, vol. 2075, pp. 184–203. Springer, Heidelberg (2001)
10. Murata, T.: Petri nets: properties, analysis and applications. Proc. of IEEE 77(4) (1989)
11. Pardo, P., Godo, L.: t-delp: an argumentation-based temporal defeasible logic programming framework. Annals of Mathematics and Artificial Intelligence, 1–33 (2013)

Integrity Constraints
for Probabilistic Spatio-Temporal Knowledgebases

Francesco Parisi[1] and John Grant[2]

[1] Department of Informatics, Modeling, Electronics and System Engineering,
University of Calabria, Italy
fparisi@dimes.unical.it
[2] Department of Computer Science and UMIACS,
University of Maryland, College Park, USA
grant@cs.umd.edu

Abstract. We formally introduce integrity constraints for probabilistic spatio-temporal knowledgebases. We start by defining the syntax and semantics of PST knowledgebases. This definition generalizes the SPOT framework which is a declarative framework for the representation and processing of probabilistic spatio-temporal data where probability is represented as an interval because of uncertainty. We augment the previous definition by adding a type of non-atomic formula that expresses integrity constraints. The result is a highly expressive formalism for knowledge representation dealing with probabilistic spatio-temporal data. Our main results concern the complexity of checking the consistency of PST knowledgebases.

1 Introduction

Recent years have seen a great deal of interest in tracking moving objects. For this reason, researchers have investigated in detail the representation and processing of spatio-temporal data, both in AI [3,8,32,33,13] and databases [1,27]. However, in many cases the location of objects is uncertain: such cases can be handled by using probabilities [26,29]. Sometimes the probabilities themselves are uncertain. The SPOT (Spatial PrObabilistic Temporal) framework was introduced in [23] to provide a declarative framework for the representation and processing of probabilistic spatio-temporal data with uncertain probabilities.

The SPOT framework is able to represent atomic statements of the form "object id is/was/will be inside region r at time t with probability in the interval $[\ell, u]$". This allows the representation of information concerning moving objects in several application domains. For instance, a military agency is interested in modelling enemy vehicles that may be in a region at a given time point and with a given probability (in order to adequately arrange its defense line) [16,12]. A cell phone provider is interested in knowing which cell phones will be in the range of some towers at a given time and with what probability. A transportation company is interested in predicting the vehicles that will be on a given road at a given time (and with what probability) in order avoid congestion.

The framework introduced in [23] was then extended in [25,10] to include the specific integrity constraint that, for a given moving object, only some points are reachable from a given starting point in one time unit. This captures the scenario where objects

U. Straccia and A. Calì (Eds.): SUM 2014, LNAI 8720, pp. 251–264, 2014.

have speed limits and only some points are reachable by objects depending on the distance between the points. However, even such an extended SPOT framework is not yet general enough to represent additional knowledge concerning the movements of objects. Examples of facts we may be aware of but cannot represent in the SPOT framework are, for instance, the fact that (*i*) there cannot be two distinct objects in a given region in a given time interval; (*ii*) some object cannot reach a given region starting from a given location in less than a given time (not necessarily one time unit); (*iii*) an object can go away from a given region only if it stayed there for at least a given time. To overcome such limitation and allow this kind of knowledge to be represented, we define probabilistic spatio-temporal (PST) knowledgebases (KBs) consisting of atomic statements, such as those representable in the SPOT framework and *spatio-temporal denial formulas*, a general class of formulas that account for all the three cases above, and many more (including the reachability constraint of [25,10]).

The focus of this paper is the systematic study of knowledge representation in probabilistic spatio-temporal data. We start by defining the concept of a PST KB and provide its formal semantics, which is given in terms of worlds and interpretations. We define the concept of a consistent PST KB, and characterize the complexity of checking consistency, showing that it is NP-complete in general. Then we present a mixed-binary linear programming algorithm for dealing with the consistency checking problem, and identify tractable cases. Finally, we discuss how checking consistency can be exploited to answer queries in PST KBs.

2   The PST Framework

This section introduces the syntax and semantics of PST KBs generalizing the SPOT framework introduced in [23] and further extended in [25,10]. Basically, we define a PST KB by augmenting the previous framework with non-atomic formulas (i.e., spatio-temporal denial formulas) that represent integrity constraints. This way we can make statements whose meaning is that certain object trajectories cannot occur.

2.1   Syntax

We assume the existence of a finite set ID of object ids, a finite set $T = [0, 1, \ldots, tmax]$ of time points (where $tmax$ in an integer), and a finite set $Space$ of points.

A *spatio-temporal atom* (*st-atom*) is an expression of the form $loc(X, Y, Z)$, where:

(i) X is a variable ranging over ID, or a constant $id \in ID$;
(ii) Y is a variable ranging over the power set $\mathcal{P}(Space)$, or a constant $r \subseteq Space$
(iii) Z is a variable ranging over T, or a constant $t \in T$.

We say that st-atom $loc(X, Y, Z)$ is *ground* if all of its arguments X, Y, Z are constants. For instance, $loc(id, r, t)$ is a ground st-atom. The intuitive meaning of $loc(id, r, t)$ is that object id is/was/will be inside region r at time t.

Definition 1 (PST atom). *A PST atom is a ground st-atom* $loc(id, r, t)$ *annotated with a probability interval* $[\ell, u] \subseteq [0, 1]$ *(with both ℓ and u rational numbers), and denoted as* $loc(id, r, t)[\ell, u]$.

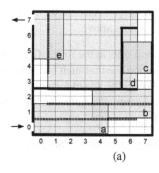

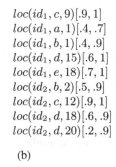

$loc(id_1, c, 9)[.9, 1]$
$loc(id_1, a, 1)[.4, .7]$
$loc(id_1, b, 1)[.4, .9]$
$loc(id_1, d, 15)[.6, 1]$
$loc(id_1, e, 18)[.7, 1]$
$loc(id_2, b, 2)[.5, .9]$
$loc(id_2, c, 12)[.9, 1]$
$loc(id_2, d, 18)[.6, .9]$
$loc(id_2, d, 20)[.2, .9]$

(a)   (b)

Fig. 1. (a) A map of an airport area (names of regions are on their bottom-right corner) (b) PST atoms

Intuitively, the PST atom $loc(id, r, t)[\ell, u]$ says that object id is/was/will be inside region r at time t with probability in the interval $[\ell, u]$. Hence, PST atoms can represent information about the past and the present (such as from techniques for interpreting RFID readings [5,6]), but also information about the future, such as from methods for predicting the destination of moving objects [16,12,28], or from querying predictive databases [2,21,22].

In the original SPOT definition, for ease of implementation, *Space* was a grid within which only rectangular regions were considered; however, in our general framework, *Space* is arbitrary and a region is any nonempty subset of *Space*. Still, for convenience we use such rectangular regions in our running example.

Example 1. Consider an airport security system which collects data from biometric sensors as well as from Bluetooth or WiFi enabled devices. Biometric data such as faces recognized by sensors [14] are matched against given profiles (such as those of checked-in passports, or of wanted criminals). Similarly, device identifiers (e.g., MAC addresses) recognized in the areas covered by network antennas are matched against profiles collected by the airport hotspots (such as logins, possibly associated with passport numbers). A simplified plan of an airport area is reported in Fig. 1(a), where regions a, b, c, d, e covered by sensors and/or antennas are highlighted. Once entered in this area, passengers typically move through the path delimited by queue dividers (represented by dotted lines in the figure, and overlapping with regions a and b), and reach the room on the upper-half right side where security checks are performed (region c is included in this room). Next, passengers can spend some time in the hall room (overlapping with region d), and finally go towards the exit (near region e).

Suppose that the security system uses a PST KB to represent the information where every PST atom consists of the profile id resulting from the matching phase, the region where the sensor/antenna recognizing the profile is operating, the time point at which the profile is recognized, and the lower and upper probability bounds of the recognizing process. For instance, PST atom $loc(id_1, c, 9)[.9, 1]$ says that a profile having id id_1 was in region c at time 9 with probability in the interval $[.9, 1]$ (the high-accuracy sensors used at security check points located in region c entail a narrow probability interval with upper bound equal to 1). Atom $loc(id_1, a, 1)[.4, .7]$ says that id_1 was recognized in region a at the earlier time 1 with probability in $[.4, .7]$. Assume the PST KB consists of the atoms in Fig. 1(b), which includes the two atoms above.   □

In order to form **PST** KBs we add integrity constraints in the form of *spatio-temporal denial* formulas (*std* formulas for short). We will soon see that such formulas are expressive enough to capture a large set of conditions. Basically, an std formula is the negation of the conjunctions of st-atoms and built-in predicates. We note that *std* formulas are closely related to the first-order formulas of [4].

Definition 2 (Std- formula). *An std-formula is an expression of the form*

$$\vee \, X, Y, Z \, \neg \left[\left(\bigwedge_{i=1}^{k} loc(X_i, Y_i, Z_i) \right) \wedge \alpha(X) \wedge \beta(Y) \wedge \gamma(Z) \right]$$

where:

- X, Y, *and* Z *are sets whose variables range over* ID, $\mathcal{P}(Space)$, *and* T, *respectively;*
- $loc(X_i, Y_i, Z_i)$, *with* $i \in [1..k]$, *are st-atoms, where the* X_i, Y_i, Z_i *may be variables or constants of the appropriate type, such that, if* X_i *(resp.,* Y_i, Z_i*) is a variable, then it occurs in* X *(resp,* Y, Z*). Moreover, each variable in* X, Y, *and* Z *occurs in at least one st-atom* $loc(X_i, Y_i, Z_i)$, *with* $i \in [1..k]$;
- $\alpha(X)$ *is a conjunction of built-in predicates of the form* $X_i \diamond X_j$, *where* X_i *and* X_j *are either variables occurring in* X *or ids in* ID, *and* $\diamond$ *is an operator in* $\{=, \neq\}$;
- $\beta(Y)$ *is a conjunction of built-in predicates of the form* $Y_i \diamond Y_j$, *where* Y_i *and* Y_j *are either variables occurring in* Y *or regions (i.e., non-empty subsets of* $Space$*), and* $\diamond$ *is a comparison operator in* $\{=, \neq, \subseteq, \supset, ov, nov\}$ *(where ov stands for "overlaps" and nov stands for "does not overlap");*
- $\gamma(Z)$ *is a conjunction of built-in predicates of the form* $Z_i \diamond Z_j$, *where each* Z_i *and* Z_j *is either a time point in* T *or a variable in* Z *that may be followed by* $+n$ *where* n *is a positive integer, and* $\diamond$ *is an operator in* $\{=, \neq, <, \geq\}$.

Example 2. In our running example, in region c security checks on one individual at a time are performed. The constraint *"there cannot be two distinct objects in region c at any time point between 1 and 20"* can be expressed by the following std-formula:
$f_1 = \forall X_1, X_2, Z_1 \, \neg[loc(X_1, c, Z_1) \wedge loc(X_2, c, Z_1) \wedge X_1 \neq X_2 \wedge Z_1 \geq 1 \wedge 20 \geq Z_1]$.

Due to the distance and the several obstacles between the entrance and the exit, we also have the constraint *"no object can reach region e starting from region a in less than 10 time points"*, that can be expressed as:
$f_2 = \forall X_1, Z_1, Z_2 \, \neg[loc(X_1, a, Z_1) \wedge loc(X_1, e, Z_2) \wedge Z_1 < Z_2 \wedge Z_2 < Z_1 + 10]$.

Moreover, as the security check on each individual takes at least 2 time units, we know that *"object id can go away from region c only if it stayed there for at least 2 time points"*, that can be expressed as:
$f_3 = \forall Y_1, Y_2, Z_1, Z_2, Z_3 \, \neg[loc(id, Y_1, Z_1) \wedge loc(id, c, Z_2) \wedge loc(id, Y_2, Z_3) \wedge Y_1 nov \, c \wedge Y_2 nov \, c \wedge Z_2 = Z_1 + 1 \wedge Z_2 < Z_3 \wedge Z_2 + 2 \geq Z_3]$.   □

In the initial **SPOT** framework [23] only PST atoms were considered. Moreover, it was assumed that all points in $Space$ are reachable from all other points by all objects.

To overcome this limitation, in [10] the SPOT framework was extended by introducing *reachability definitions*. A *reachability atom* is written as $reachable_{id}(p, q)$ where $id \in ID$ is an object id, and $p, q \in Space$. Intuitively, the reachability atom says that it is possible for the object id to reach location q from location p in one unit of time. Hence, what is reachable in one time point depends not only on the locations p and q, but also the object id. As we now show, reachability can be expressed in our formalism as an integrity constraint. However, in order to formulate reachability in our framework of denial formulas, we need to deal with what is not reachable, rather than what is reachable.

Example 3. Let r be the region consisting of all points q that are not reachable from p in one time unit. The corresponding std-formula is:
$$\forall X_1, Z_1, Z_2 \neg [loc(X_1, p, Z_1) \wedge loc(X_1, r, Z_2) \wedge Z_2 = Z_1 + 1].$$ □

Note how using std-formulas we can also express which points can not be reached from p in any number of time units, not just 1.

Example 4. In our running example, the following std-formula states that the points in region $r = \{(x, y) | 0 \leq x \leq 5 \wedge y = 3\}$ (i.e., those close to the upper-side of the wall dividing the hall room and the one where there are queue dividers) are not reachable in less than 3 time units from any point in $r' = \{(x, y) | 0 \leq x \leq 5 \wedge y = 2\}$ (i.e., the points close to the other side of that wall):
$$f_4 = \forall X_1, Z_1, Z_2 \neg [loc(X_1, r', Z_1) \wedge loc(X_1, r, Z_2) \wedge Z_1 < Z_2 \wedge Z_2 < Z_1 + 3].$$ □

Definition 3 (PST KB). *A PST KB $\mathcal{K}$ is pair $\langle \mathcal{A}, \mathcal{F} \rangle$, where $\mathcal{A}$ is a finite set of PST atoms and $\mathcal{F}$ is finite set of std-formulas.*

Example 5. In our running example, PST KB $\mathcal{K}_{ex}$ is the pair $\langle \mathcal{A}_{ex}, \mathcal{F}_{ex} \rangle$, where $\mathcal{A}_{ex}$ is the set consisting of the PST atoms in Fig. 1(b), and $\mathcal{F}_{ex}$ is the set $\{f_1, f_2, f_3, f_4\}$ of std-formulas defined in Examples 2 and 4. □

2.2 Semantics

The semantics of a PST KB is defined through the concept of *worlds*. Before introducing this concept, we define *ground* std-formulas.

Given an std-formula f having the form in Definition 2, we denote by Θ_f the set of all substitutions of variables in $\mathbf{X}$, $\mathbf{Y}$, and $\mathbf{Z}$ with constants in ID, $\mathcal{S}$, and T, respectively, where $\mathcal{S}$ is the set of all sets of *Space* that contain a single point[1]. Moreover, given substitution $\theta \in \Theta_f$, we denote as $\theta(f)$ the *ground* std-formula resulting from applying θ to f: $\theta(f) = \neg [(\bigwedge_{i=1}^{k} loc(\theta(X_i), \theta(Y_i), \theta(Z_i))) \wedge \alpha(\theta(\mathbf{X})) \wedge \beta(\theta(\mathbf{Y})) \wedge \gamma(\theta(\mathbf{Z}))]$. As the ground conjunction of built-in predicates $\alpha(\theta(\mathbf{X})) \wedge \beta(\theta(\mathbf{Y})) \wedge \gamma(\theta(\mathbf{Z}))$ evaluates to either *true* or *false*, $\theta(f)$ is either the negation of a conjunction of ground st-atoms or the truth value *true* (when the conjunction of built-in predicates evaluates to *false*).

---

[1] We use only such singleton subsets of *Space* in order to reduce the number of possible instantiations of variables Y from exponential to linear in the size of *Space*, without serious effect on the meanings of the std-formulas.

Example 6. Consider the formula $f_1 = \forall X_1, X_2, Z_1 \neg[loc(X_1, c, Z_1) \wedge loc(X_2, c, Z_1) \wedge X_1 \neq X_2 \wedge Z_1 \geq 1 \wedge 20 \geq Z_1]$ introduced in Example 2, and the substitution $\theta = \{X_1/id_1, X_2/id_2, Z_1/6\}$, where $id_1, id_2 \in ID$ and time point 6 is in T. Thus, $\theta(f_1) = \neg[loc(id_1, c, 6) \wedge loc(id_2, c, 6)]$, where the conjunction of ground built-in predicates $id_1 \neq id_2 \wedge 6 \geq 1 \wedge 6 \leq 20$, evaluating to *true*, is not reported in $\theta(f_1)$. □

Definition 4 (World). *A world w is a function, $w : ID \times T \to Space$.*

Basically, a world w specifies a trajectory for each $id \in ID$. That is, for each $id \in ID$, w says where in *Space* object id was/is/will be at each time $t \in T$. In particular, this means that an object can be in only one location at a time. However, a location may contain multiple objects. It is easy to see that world w can be represented by the set $\{loc(id, \{p\}, t) \mid w(id, t) = p\}$ of ground st-atoms.

Example 7. World w_1 describing the trajectories of id_1 and id_2 for time points in $[0, 20]$ is $w_1(id_1, t) = (4, 1)$ for $t \in [0, 5]$, $w_1(id_1, t) = (7, 2)$ for $t \in [6, 7]$, $w_1(id_1, t) = (7, 4)$ for $t \in [8, 10]$, $w_1(id_1, t) = (4, 4)$ for $t \in [11, 16]$, $w_1(id_1, t) = (1, 6)$ for $t \in [17, 20]$, $w_1(id_2, t) = (4, 1)$ for $t \in [0, 11]$, $w_1(id_2, t) = (7, 5)$ for $t \in [12, 15]$, $w_1(id_2, t) = (7, 7)$ for $t \in [16, 16]$, $w_1(id_2, t) = (4, 5)$ for $t \in [17, 20]$.

Given a world w and a ground st-atom $a = loc(id, r, t)$, we say that w *satisfies* a (denoted as $w \models a$) iff $w(id, t) \in r$. Moreover, we say that w satisfies a conjunction of ground st-atoms $\bigwedge_{i=1}^k a_i$ (denoted as $w \models \bigwedge_{i=1}^k a_i$) iff $w \models a_i \; \forall i \in [1..k]$. Finally, world w *satisfies std-formula f* (denoted as $w \models f$) iff for each substitution $\theta \in \Theta_f$, $w \models \theta(f)$. Note that, as there is a negation in front of f, $w \models \theta(f)$ iff w does not satisfy a ground st-atom in $\theta(f)$ or the conjunction of ground built-in predicates in $\theta(f)$ evaluates to false.

Example 8. World w_1 of Example 7 satisfies the st-atom $loc(id_1, b, 0)$, as $w_1(id_1, 0) = (4, 1)$ belongs to region b (see Fig. 1(a)). Moreover, $w_1 \models \neg[loc(id_1, b, 0) \wedge loc(id_1, e, 15)]$ as $w_1 \not\models loc(id_1, e, 15)$, since $w_1(id_1, 15) = (4, 4) \notin e$. □

In the following, we will denote as $\mathcal{W}(\mathcal{K})$ the set of all worlds of the PST KB $\mathcal{K}$. Moreover, in order to simplify formulas, we will assume that w ranges over $\mathcal{W}(\mathcal{K})$.

An interpretation I for a PST KB $\mathcal{K}$ is a probability distribution function (PDF) over $\mathcal{W}(\mathcal{K})$, that is, a function assigning a probability value to each world in $\mathcal{W}(\mathcal{K})$. $I(w)$ is the probability that w describes the actual trajectories of all the objects. Some interpretations are models of $\mathcal{K}$ in which case we write M instead of I.

Definition 5 (Model). *A model M for a PST KB $\mathcal{K} = \langle \mathcal{A}, \mathcal{F} \rangle$ is an interpretation for $\mathcal{K}$ such that:*

- $\forall \, loc(id, r, t)[\ell, u] \in \mathcal{A}, \quad \sum_{w \mid w \models loc(id, r, t)} M(w) \in [\ell, u];$
- $\forall f \in \mathcal{F}, \quad \sum_{w \mid w \not\models f} M(w) = 0.$

The first condition in the definition above means that, for each $a = loc(id, r, t)[\ell, u] \in \mathcal{A}$, the sum of the probabilities assigned by M to the worlds satisfying the st-atom

$loc(id, r, t)$ have to belong to the probability interval $[\ell, u]$ specified by a. The second condition means that every world not satisfying a formula $f \in \mathcal{F}$ must be assigned by M probability equal to 0.

Example 9. Let w_1 be the world introduced in Example 7. Let w_2 be as w_1 except that $w_2(id_1, 1) = (3, 2)$, and let w_3 be as w_2 except that $w_3(id_2, 2) = (2, 2)$, $w_3(id_2, t) = (0, 3)$ for $t \in [18..20]$. Let M be such that $M(w_1) = .7$ $M(w_2) = .2$ $M(w_3) = .1$, and $M(w) = 0$ for all the other worlds in $\mathcal{W}(\mathcal{K}_{ex})$. It can be checked that M satisfies both the conditions of Definition 5 for the PST KB $\mathcal{K}_{ex}$ of our running example. For instance, for atom $loc(id_1, a, 1)[.4, .7] \in A_{ex}$, $\sum_{w|w \models loc(id_1, a, 1)} M(w) = M(w_1) = .7 \in [.4, .7]$ (note that, at time 1, $w_2(id_1, 1) = w_3(id_1, 1) = (3, 2)$ that is not in region a). Moreover, it is easy to check that w_1, w_2, w_3 satisfy every sdt-formula in $\mathcal{F}_{ex}$. Thus, M is a model for $\mathcal{K}_{ex}$. □

We say that PST KB $\mathcal{K}$ is consistent iff if there is a model for it. The set of models for $\mathcal{K}$ will be denoted as $\mathbf{M}(\mathcal{K})$.

Definition 6 (Consistency). PST *KB* $\mathcal{K}$ *is consistent iff* $\mathbf{M}(\mathcal{K}) \neq \emptyset$.

Example 10. PST KB $\mathcal{K}_{ex}$ of our running example is consistent, as there exists the model M of Example 9 for it. □

3 Checking the Consistency of PST KBs

We now address the fundamental problem of checking the consistency of PST KBs. Given a PST KB $\mathcal{K} = \langle \mathcal{A}, \mathcal{F} \rangle$, the consistency checking problem is deciding whether $\mathbf{M}(\mathcal{K}) \neq \emptyset$, that is, whether there is a model for $\mathcal{K}$.

Theorem 1. *Given a* PST *KB* $\mathcal{K} = \langle \mathcal{A}, \mathcal{F} \rangle$, *deciding whether* $\mathbf{M}(\mathcal{K}) \neq \emptyset$ *is NP-complete.*

Proof. (Membership). For any world $w_i \in \mathcal{W}(\mathcal{K})$, let v_i be a variable ranging over the domain of rational numbers. The variable v_i will be used to represent the probability $M(w_i)$ assigned to w_i by a model $M \in \mathbf{M}(\mathcal{K})$. Deciding whether $\mathcal{K}$ is consistent, that is $\mathbf{M}(\mathcal{K}) \neq \emptyset$, can be done by checking the feasibility of the following system $LP(\mathcal{K})$ of linear (in)equalities:

(1) $\forall\, loc(id, r, t)[\ell, u] \in \mathcal{A}$,   (2) $\forall f \in \mathcal{F}$, $\displaystyle\sum_{w_i \,|\, w_i \not\models f} v_i = 0$;

    (a) $\ell \le \displaystyle\sum_{w_i | w_i \models loc(id, r, t)} v_i$;   (3) $\displaystyle\sum_{w_i \,|\, w_i \in \mathcal{W}(\mathcal{K})} v_i = 1$;

    (b) $\displaystyle\sum_{w_i | w_i \models loc(id, r, t)} v_i \le u$;   (4) $\forall w_i \in \mathcal{W}(\mathcal{K})$, $v_i \ge 0$.

It is easy to see that every solution s of $LP(\mathcal{K})$ corresponds one-to-one to a model $M \in \mathbf{M}(\mathcal{K})$ such that $M(w_i)$ is equal to the value of v_i in s. Therefore, deciding whether $\mathcal{K}$ is consistent is equivalent to deciding the feasibility of $LP(\mathcal{K})$. However, it turns out that representing $LP(\mathcal{K})$ whose number of variables is $|\mathcal{W}(\mathcal{K})| = |Space|^{|ID| \cdot |T|}$ is not necessary, since we can exploit a guess-and-check strategy based on the following result: "*if a system of m linear (in)equalities*[2] *is feasible, then it admits at least one solution with at most m non-zero variables*" [18]. In our setting, this means that $LP(\mathcal{K})$ is feasible iff there is a solution for $LP(\mathcal{K})$ consisting of at most $2 \cdot |\mathcal{A}| + |\mathcal{F}| + 1$ non-zero variables. Hence, a guess-and-check strategy for deciding the feasibility of $LP(\mathcal{K})$ is the following. First, guess an assignment s' consisting of $2 \cdot |\mathcal{A}| + |\mathcal{F}| + 1$ pairs $\langle v_i, \overline{v}_i \rangle$, where v_i is a variable in $LP(\mathcal{K})$, and $\overline{v}_i$ a value to be assigned to v_i. Then, check whether s' is a solution of the system $LP^*(\mathcal{K})$, which is the system of linear inequalities obtained from $LP(\mathcal{K})$ by keeping in it only the occurrences of the variables in the guessed assignment s'. Thus, the size of $LP^*(\mathcal{K})$ is polynomial w.r.t. the size of $\mathcal{K}$, and checking whether s' is a solution of $LP^*(\mathcal{K})$ can be accomplished in polynomial time. This guess-and-check strategy is correct since if s' turns out to be a solution of $LP^*(\mathcal{K})$, then $LP(\mathcal{K})$ is feasible. Its completeness derives from a result in [18].

(*Hardness*). We show a reduction to our problem from the NP-hard Hamiltonian path problem, that is, the problem of checking whether there is a path π in a directed graph G such that π visits each vertex of G exactly once.

Given a directed graph $G = \langle V, E \rangle$, where $V = \{v_0, \ldots, v_k\}$ is the set of its vertexes, and E is a set of pairs (v_i, v_j) with $v_i, v_j \in V$, we construct an instance of our problem as follows. Let $ID = \{id\}$, $Space = V$, and $T = [0, \ldots, k]$. $\mathcal{K}$ is the pair $\langle \mathcal{A}, \mathcal{F} \rangle$ such that $\mathcal{A}$ consists of the **PST** atom $loc(id, v_0, 0)[1, 1]$, and $\mathcal{F}$ consists of std-formulas f_1^i (with $i \in [0..k]$) and f_2 such that:

- $f_1^i = \forall Z_1, Z_2 \, \neg[loc(id, \{v_i\}, Z_1) \wedge loc(id, Space \backslash V', Z_2) \wedge Z_2 = Z_1 + 1]$ where V' is the set of vertexes v_j s.t. $(v_i, v_j) \in E$. This formula says that the only points id can reach starting from v_i in one time step are those in V'.
- $f_2 = \forall Y_1, Z_1, Z_2 \, \neg[loc(id, Y_1, Z_1) \wedge loc(id, Y_1, Z_2) \wedge Z_1 \neq Z_2]$, saying that id can not be on the same location at distinct time points.

We show that $\mathbf{M}(\mathcal{K}) \neq \emptyset$ iff there is a Hamiltonian path in G.

($\Rightarrow$) As there is only one id in $\mathcal{A}$, every world $w \in \mathcal{W}(\mathcal{K})$ is such that w places id on a vertex in V at each time point $t \in T$. As $\mathbf{M}(\mathcal{K}) \neq \emptyset$, there is a model $M \in \mathbf{M}(\mathcal{K})$ such that M assigns probability greater than zero only to worlds w such that $\forall f \in \mathcal{F}, w \models f$. In particular, let w be one such world. The fact that $w \models f_1^i$ entails that $\forall t \in [0, k-1], w(id, t) = v_i$ and $w(id, t+1) = v_j$ iff $(v_i, v_j) \in E$. Moreover, the fact that $w \models f_2$ entails that $\forall t, t' \in [0, k], t \neq t', w(id, t) \neq w(id, t')$, meaning that id is never placed by w on the same vertex at different time points. Since $loc(id, v_0, 0)[1, 1] \in \mathcal{A}$, every world which is assigned probability greater than zero by M is such that $w(id, 0) = v_0$. It follows that every world $w \in \mathcal{W}(\mathcal{K})$ which is assigned by $M \in \mathbf{M}(\mathcal{K})$ a probability greater than zero encodes a Hamiltonian path of G whose first vertex is v_0. In fact, $\forall w \in \mathcal{W}(\mathcal{K})$ such that $M(w) > 0$ the following properties hold: (i) $w(id, 0) = v_0$, (ii) $\forall t \in [0, k-1], w(id, t) = v_i, w(id, t+1) = v_j$ iff $(v_i, v_j) \in E$. (iii) $\forall t, t' \in [0, k], t \neq t', w(id, t) \neq w(id, t')$. Conditions (i) and (ii) entail that $\pi = w(id, 0), w(id, 1), \ldots, w(id, k)$ is a path on G starting from vertex v_0, while condition (iii) entails that each vertex $v \in V$ occurs exactly once in π.

($\Leftarrow$) Let π be a Hamiltonian path of G. We denote by $\pi[i]$ (with $i \in [0..k]$) the i-th

---

[2] Inequalities imposing that variables are non-negative must not be considered.

vertex of π. W.l.o.g. we assume that the first vertex of π is v_0, that is, $\pi[0] = v_0$. We now show that $\mathbf{M}(\mathcal{K})$ is not empty. Let M be a function over $\mathcal{W}$ such that for all worlds $w \in \mathcal{W}$, $M(w) = 0$, except for the world w^* which is such that: $w^*(id, 0) = \pi[0] = v_0$, $\forall t \in [1, k]$, $w^*(id, t) = \pi[t]$. It is easy to see that $w^* \models \mathcal{F}$. In fact, for each $i \in [0..k]$, f_1^i is satisfied by w^*, since the fact that π a path on G entails that $\forall t \in [0, k-1]$, $w^*(id, t) = v_i$ and $w^*(id, t+1) = v_j$ only if edge (v_i, v_j) is an edge of G. Moreover, f_2 is satisfied by w^*, since the fact that π is a Hamiltonian path entails that w^* places id on different locations (i.e., vertexes of G) at different time points. Since $w^* \models \mathcal{F}$, it can be assigned by M a probability different from 0. Let $M(w^*) = 1$. Therefore, as $\sum_{w|w \models loc(id, v_0, 0)} M(w) = M(w^*) + \sum_{w|w \neq w^* \wedge w \models loc(id, v_0, 0)} M(w) = 1$, the condition required by atom $loc(id, v_0, 0)[1, 1] \in \mathcal{A}$ holds too. Thus, M is a model for $\mathcal{K}$. □

3.1 Sufficient Condition for Checking Consistency

We present a mixed-binary linear programming problem whose feasibility entails the consistency of PST KB $\mathcal{K} = \langle \mathcal{A}, \mathcal{F} \rangle$. As shown in [23], the consistency of a PST KB $\mathcal{K} = \langle \mathcal{A}, \emptyset \rangle$ can be checked in polynomial time w.r.t. the size of $\mathcal{K}$ by solving a linear programming problem whose variables $v_{id,t,p}$ represent the probability that object id is at point p at time t. Here, we start from this linear programming problem and augment the set of its (in)equalities with some inequalities ensuring that if the so-obtained set of linear inequalities is feasible then every ground std-formula derived from $\mathcal{F}$ is satisfied. To achieve this, we need to introduce the binary variables δ, thus obtaining a mixed-binary linear programming problem.

Definition 7 (MBLP($\mathcal{K}$)). *Let $\mathcal{K} = \langle \mathcal{A}, \mathcal{F} \rangle$. The linear program MBLP($\mathcal{K}$) consists of the following (in)equalities:*

(1) $\forall loc(id, r, t)[\ell, u] \in \mathcal{A}$: $\ell \leq \sum_{p \in r} v_{id,t,p} \leq u$;
(2) $\forall id \in ID, t \in T$: $\sum_{p \in Space} v_{id,t,p} = 1$;
(3) $\forall p \in Space, id \in ID, t \in T$: $v_{id,t,p} \geq 0$;
(4) for each $f \in \mathcal{F}$ and $\theta \in \Theta_f$ such that $\theta(f)$ is logically equivalent to the negation of the conjunction of st-atoms $\bigwedge_{i=1}^{k} loc(\theta(X_i), \theta(Y_i), \theta(Z_i))$, the inequalities:
    (a) $\forall i \in [1..k]$: $\sum_{p \in \theta(Y_i)} v_{\theta(X_i), \theta(Z_i), p} \leq \delta_i$;
    (b) $\sum_{i=1}^{k} \delta_i = k - 1$;
    (c) $\forall i \in [1..k]$: $\delta_i \in \{0, 1\}$.

Basically, inequalities (1) ensure that a solution of *MBLP($\mathcal{K}$)* places the object in r with a probability between ℓ and u, as required by the atom $(id, r, t, [\ell, u])$. Inequalities (2) and (3) ensure that for each id and t, the $v_{id,t,p}$ variables jointly represent a probability distribution. Moreover, for each ground st-atom $loc(\theta(X_i), \theta(Y_i), \theta(Z_i))$ of the ground std-formula $\theta(f)$, inequalities (4)(a) and (4)(c) entail that the probability $v_{\theta(X_i), \theta(Z_i), p}$ that object $\theta(X_i)$ is in any point p in region $\theta(Y_i)$ at time $\theta(Z_i)$ is either constrained to be 0 or free to take any value not greater than 1. Intuitively enough, if $v_{\theta(X_i), \theta(Z_i), p}$ is enforced to be zero (i.e., $\delta_i = 0$), then object $\theta(X_i)$ can not be in region $\theta(Y_i)$ at time $\theta(Z_i)$. On the other hand, if $v_{\theta(X_i), \theta(Z_i), p}$ is left free to take any value less than or equal to one (i.e., $\delta_i = 1$), then $\theta(X_i)$ may or may not be in region $\theta(Y_i)$ at time

$\theta(Z_i)$. Finally, equality $(4)(b)$ entails that there is at least one of the k ground st-atoms $loc(\theta(X_i), \theta(Y_i), \theta(Z_i))$ of $\theta(f)$ such that $\theta(X_i)$ is not placed in a point in $\theta(Y_i)$ at time $\theta(Z_i)$.

Example 11. Consider the ground std-formula $\theta(f_1) = \neg[loc(id_1, c, 6) \wedge loc(id_2, c, 6)]$ of Example 6. Then, the inequalities in $MBLP(\mathcal{K})$ corresponding to $\theta(f_1)$ are:

(4a) $\sum_{p \in c} v_{id_1, 6, p} \leq \delta_1$; $\sum_{p \in c} v_{id_2, 6, p} \leq \delta_2$; (4b) $\delta_1 + \delta_2 = 1$; (4c) $\delta_1, \delta_2 \in \{0, 1\}$. $\square$

The following theorem states that $MBLP(\mathcal{K})$ can be used to check if $\mathcal{K}$ is consistent.

Theorem 2. *If $MBLP(\mathcal{K})$ is feasible then $\mathbf{M}(\mathcal{K}) \neq \emptyset$.*

Proof. Let σ be a solution of $MBLP(\mathcal{K})$, and $\sigma(v_{id,t,p})$ be the value assigned to variable $v_{id,t,p}$ by σ. We define the function M over $\mathcal{W}(\mathcal{K})$ such that, for each world $w \in \mathcal{W}(\mathcal{K})$, $M(w) = \prod_{id \in ID, t \in T, w(id,t)=p} \sigma(v_{id,t,p})$, that is $M(w)$ is the product of the values assigned by solution σ to variables $v_{id,t,p}$ such that $w(id, t) = p$. It can be shown that, (in)equalities (2) and (3) of the definition of $MBLP(\mathcal{K})$ entail that M is a PDF over $\mathcal{W}(\mathcal{K})$. Moreover, since $\sigma(v_{id,t,p})$ is equal to $\sum_{w | w \models loc(id,t,p)} M(w)$, for each atom $loc(id, r, t)[\ell, u] \in \mathcal{A}$, $\sum_{w|w \models loc(id,r,t)} M(w) = \sum_{p \in r} \sum_{w|w \models loc(id,t,p)} M(w) = \sum_{p \in r} \sigma(v_{id,t,p}) \in [\ell, u]$. Given $f \in \mathcal{F}$ and $\theta \in \Theta_f$ such that $\theta(f)$ is logically equivalent to the negation of the conjunction of the st-atoms $\bigwedge_{i=1}^k loc(\theta(X_i), \theta(Y_i), \theta(Z_i))$, the inequalities (4)(a-c) entail that there is $i \in [1..k]$ s.t. $\sum_{p \in \theta(Y_i)} \sigma(v_{\theta(X_i), \theta(Z_i), p}) = 0$. Thus $\forall p \in \theta(Y_i), \sigma(v_{\theta(X_i), \theta(Z_i), p}) = 0$. Hence, for each world $w \in \mathcal{W}(\mathcal{K})$ such that $w(\theta(X_i), \theta(Z_i)) = p$, $M(w) = 0$ due to the presence of the factor $\sigma(v_{\theta(X_i), \theta(Z_i), p}) = 0$ in the product defining $M(w)$. Therefore, for each std-formula $f \in \mathcal{F}$, it holds that $\sum_{w | w \not\models f} M(w) = 0$; hence M is a model for $\mathcal{K}$. $\square$

A consequence of Theorem 2 is that well-known techniques for solving linear optimization problems can be adopted to address the consistency checking problem.

The following example shows that the converse of Theorem 2 does not hold ($\mathcal{K}$ may be consistent even if $MBLP(\mathcal{K})$ is not feasible).

Example 12. Let $ID = \{id\}$, $T = [0, 1]$, $Space = \{p_0, p_1\}$, $\mathcal{K} = \langle \mathcal{A}, \mathcal{F} \rangle$ where $\mathcal{A} = \{loc(id, p_0, 0)[0.5, 0.5], loc(id, p_1, 1)[0.5, 0.5], \}$ and $\mathcal{F} = \{\neg[loc(id, \{p_0\}, 0) \wedge loc(id, \{p_1\}, 1)]\}$. Thus, $\mathcal{W} = \{w_1, w_2, w_3, w_4\}$ where: $w_1(id, 0) = p_0$, $w_1(id, 1) = p_0$, $w_2(id, 0) = p_0$, $w_2(id, 1) = p_1$, $w_3(id, 0) = p_1$, $w_3(id, 1) = p_0$, $w_4(id, 0) = p_1$, $w_4(id, 1) = p_1$. It is easy to check that M is such that $M(w_1) = 0.5$, $M(w_2) = 0$, $M(w_3) = 0$, $M(w_4) = 0.5$ is a model for $\mathcal{K}$. However, $MBLP(\mathcal{K})$ is not feasible as it includes the following inequalities: $0.5 \leq v_{id,0,p_0} \leq 0.5$; $0.5 \leq v_{id,1,p_1} \leq 0.5$; $v_{id,0,p_0} + v_{id,0,p_1} = 1$; $v_{id,1,p_0} + v_{id,1,p_1} = 1$; $v_{id,0,p_0} \leq \delta_1$; $v_{id,1,p_1} \leq \delta_2$; $\delta_1 + \delta_2 = 1$; $\delta_1, \delta_2 \in \{0, 1\}$; $v_{id,0,p_0} \geq 0, v_{id,0,p_1} \geq 0, v_{id,1,p_0} \geq 0, v_{id,1,p_1} \geq 0$. $\square$

3.2   A Tractable Case

We now identify a tractable case of the consistency checking problem, one that holds when all std-formulas are *unary*, that is, each formula in $\mathcal{F}$ consists of only one st-atom and possibly a conjunction of built-in predicates (i.e., in Definition 2, $k = 1$).

Example 13. The constraint "*there is no object in region r at any time between* 5 *and* 10" can be expressed by the following unary std-formula:

$\forall X_1, Z_1 \neg [loc(X_1, r, Z_1) \wedge Z_1 \geq 5 \wedge 10 \geq Z_1]$.

The constraint "*object id is always in region r*" can be expressed as:

$\forall Y_1, Z_1 \neg [loc(id, Y_1, Z_1) \wedge Y_1 nov\ r]$. □

Checking consistency is tractable if only *unary* std-formulas are considered.

Theorem 3. *Let* $\mathcal{K} = \langle \mathcal{A}, \mathcal{F} \rangle$ *be a* PST *KB such that* $\mathcal{F}$ *consists of unary std-formulas only. Then, deciding whether* $\mathcal{K}$ *is consistent is in* $PTIME$.

Proof. We reduce this case to a result from [23]. The statement follows from the fact that if $\mathcal{F}$ consists of *unary* std-formulas only, $\mathcal{K} = \langle \mathcal{A}, \mathcal{F} \rangle$ is equivalent to (i.e., it has exactly the same set of models as) $\mathcal{K}' = \langle \mathcal{A}', \emptyset \rangle$, where $\mathcal{A}'$ consists of the atoms in $\mathcal{A}$ plus atom $loc(\theta(X_i), \theta(Y_i), \theta(Z_i))[0, 0]$ for each ground std-formula $\theta(f) = \neg [loc(\theta(X_i), \theta(Y_i), \theta(Z_i))]$, where $f \in \mathcal{F}$ and $\theta \in \Theta_f$. Since, $\bigcup_{f \in \mathcal{F}} \Theta_f$ is polynomial w.r.t. the size of $\mathcal{K}$, the size of $\mathcal{A}'$ (and thus of $\mathcal{K}'$) increases by a polynomial number of atoms. Hence, we can apply the result of [23], entailing that the consistency of PST KBs with $\mathcal{F} = \emptyset$ can be decided in $PTIME$. □

Additional tractable cases involving reachability definitions were identified in [10].

4 Using Consistency Checking to Answer Queries

In this section, we consider the problem of answering *selection queries* in PST KBs, and show that consistency checking can be used to address this problem.

A selection query is an expression of the form $(?id, q, ?t, [\ell, u])$, where q is a region and $[\ell, u]$ is a probability interval. Intuitively, a selection query says: "Given a region q and a probability interval $[\ell, u]$, find all objects id and times t such that id is inside q at time t with a probability in the interval $[\ell, u]$." There are two semantics for interpreting this statement, leading to two types of answers to selection queries. *Optimistic answers* are objects and time points that *may* be in the query region with probability in the specified interval, whereas *cautious answers* consist only of those objects and time points that are *guaranteed* to be in that region with probability in the given interval. Thus, the cautious answers are a subset of the optimistic ones.

Definition 8 (Optimistic/Cautious Query Answers). *Let* $\mathcal{K}$ *be a consistent* PST *KB, and* $Q = (?id, q, ?t, [\ell, u])$ *a selection query. Then,* $\langle id, t \rangle$ *is*

- *an optimistic answer to Q w.r.t.* $\mathcal{K}$ *iff* $\exists M \in \mathbf{M}(\mathcal{K})$ *s.t.* $\sum_{w | w \models loc(id,q,t)} M(w) \in [\ell, u]$.

- *a cautious answer to Q w.r.t.* $\mathcal{K}$ *iff* $\forall M \in \mathbf{M}(\mathcal{K})$, $\sum_{w | w \models loc(id,q,t)} M(w) \in [\ell, u]$.

Example 14. Let $q_1 = \{(7, 3), (7, 4)\}$ (q_1 overlaps with region c, see Fig. 1(a)). Model M of Example 9 entails that $\langle id_1, 9 \rangle$ is an optimistic answer to $Q = (?id, q_1, ?t, [.7, 1])$, as $w_1(id_1, 9) = w_2(id_1, 9) = w_3(id_1, 9) = (7, 4) \in q_1$ and $M(w_1) + M(w_2)$

$+ M(w_3) = 1 \in [.7, 1]$. Now, let q_2 be any region including region c. $\langle id_1, 9 \rangle$ is a cautious answer to $Q' = (?id, q_2, ?t, [.7, 1])$, as according to any model for $\mathcal{K}_{ex}$, id_1 must be in region c (and thus in q_2) at time 9 with probability in $[.9, 1]$ (due to $loc(id_1, c, 9)[.9, 1] \in \mathcal{A}_{ex}$). Clearly, $\langle id_1, 9 \rangle$ is also an optimistic answer to Q'.   □

The following proposition says how consistency checking can be used to answer selection queries under both optimistic and cautious semantics.

Proposition 1. *Let $\mathcal{K}$ be a consistent PST KB, and $Q = (?id, q, ?t, [\ell, u])$. Then,*

- $\langle id, t \rangle$ *is an optimistic answer to Q w.r.t. $\mathcal{K}$ iff $\langle \mathcal{A} \cup \{loc(id, q, t)[\ell, u]\}, \mathcal{F} \rangle$ is consistent.*
- $\langle id, t \rangle$ *is a cautious answer to Q w.r.t. $\mathcal{K}$ iff $\langle \mathcal{A} \cup \{loc(id, q, t)[0, \ell - \epsilon]\}, \mathcal{F} \rangle$ and $\langle \mathcal{A} \cup \{loc(id, q, t)[u + \epsilon, 1]\}, \mathcal{F} \rangle$ are not consistent, where $\epsilon - 1/(ma)^m$ where $m = 2 \cdot |\mathcal{A}| + |F| + 1$ and a is the maximum among the numerators and denominators of the probabilities in $\mathcal{K}^3$.*

5   Related Work

A comprehensive survey on the SPOT framework can be found in [11], where related research is also reviewed. Here we mention just a few papers.

[15] proposes an important probabilistic logic programming approach where conditional rules that can express denial formulas are studied. The problem of checking the consistency of (relational) probabilistic databases in the presence of denial constraints is addressed in [7]. However, these frameworks do not explicitly deal with space and time. Substantial work has been done on spatio-temporal logics [8,13] which combine spatial and temporal formalisms. This includes important contributions on qualitative spatio-temporal representation and reasoning [17,30,3], which focus on describing entities and qualitative relationships between them while dealing with discrete time. However, these works are not intended for reasoning about moving objects whose location at a given time is uncertain (they do not put probabilities into the mix).

[32,31,33] focus on spatio-temporal logical theories that describe known plans of moving objects by sets of *go* atoms, each of them stating that an object go from location L_1 to L_2, leaving L_1 and reaching L_2 at some time points in some intervals, and travelling with a speed in a given interval. Later, [26] extends this logic to include some probabilistic information about such plans. The SPOT framework in [23] further extends this work to uncertainty about where objects might be at a given time. As SPOT data provide information on moving objects, one issue addressed in [25] and then further investigated in [10] is that of revising SPOT data so that information on these objects may be changed as objects move. Other efforts focused on the processing of selection [24,20] and aggregate queries [9].

While there is much work on spatio-temporal databases [1,27] and probabilistic spatio-temporal databases [29,34,35], these works mainly focus on devising indexing mechanisms and scaling query computation, instead of representing knowledge in a declarative fashion. None of these works systematically addresses the issue of considering integrity constraints over probabilistic spatio-temporal data.

---

[3] The size of ϵ is polynomial w.r.t. the size of $\mathcal{K}$. Its value can be determined by applying a well-known result [18] on boundedness of solutions of linear programming problems.

6 Conclusion and Future Work

We believe that this is the first work that focuses systematically on knowledge representation in the form of integrity constraints for probabilistic spatio-temporal data. The knowledge is represented both in the form of spatio-temporal atoms describing the location of objects in time with a probability interval as well as spatio-temporal denial formulas describing the integrity constraints the system must satisfy. Within this framework we showed that consistency checking is NP-complete. However, we also identified a class of formulas for which consistency checking is feasible.

There are further issues that we plan to investigate. We will show how to use the results of this paper to study the complexity of query answering in probabilistic spatio-temporal knowledgebases. Using [19], we will further consider the problem of repairing an inconsistent probabilistic spatio-temporal knowledgebase. We will look into the possibility of semantic query optimization for probabilistic spatio-temporal knowledgebases. We will investigate probabilistic std-formulas for expressing constraints that hold with a probability in a given interval. Finally, we also plan to study the use of previous knowledge to efficiently check for consistency and process queries after updates.

References

1. Agarwal, P.K., Arge, L., Erickson, J.: Indexing moving points. J. Comput. Syst. Sci. 66(1), 207–243 (2003)
2. Akdere, M., Çetintemel, U., Riondato, M., Upfal, E., Zdonik, S.B.: The case for predictive database systems: Opportunities and challenges. In: CIDR, pp. 167–174 (2011)
3. Cohn, A.G., Hazarika, S.M.: Qualitative spatial representation and reasoning: An overview. Fundam. Inform. 46(1-2), 1–29 (2001)
4. Doder, D., Grant, J., Ognjanović, Z.: Probabilistic logics for objects located in space and time. J. of Logic and Computation 23(3), 487–515 (2013)
5. Fazzinga, B., Flesca, S., Furfaro, F., Parisi, F.: Cleaning trajectory data of RFID-monitored objects through conditioning under integrity constraints. In: Int. Conf. on Extending Database Technology (EDBT), pp. 379–390 (2014)
6. Fazzinga, B., Flesca, S., Furfaro, F., Parisi, F.: Offline cleaning of RFID trajectory data. In: Int. Conf. on Scientific and Statistical Database Management (SSDBM), p. 5 (2014)
7. Flesca, S., Furfaro, F., Parisi, F.: Consistency checking and querying in probabilistic databases under integrity constraints. J. Comput. Syst. Sci. 80(7), 1448–1489 (2014)
8. Gabelaia, D., Kontchakov, R., Kurucz, Á., Wolter, F., Zakharyaschev, M.: Combining spatial and temporal logics: Expressiveness vs. complexity. J. Artif. Intell. Res. 23, 167–243 (2005)
9. Grant, J., Molinaro, C., Parisi, F.: Aggregate count queries in probabilistic spatio-temporal databases. In: Liu, W., Subrahmanian, V.S., Wijsen, J. (eds.) SUM 2013. LNCS, vol. 8078, pp. 255–268. Springer, Heidelberg (2013)
10. Grant, J., Parisi, F., Parker, A., Subrahmanian, V.S.: An agm-style belief revision mechanism for probabilistic spatio-temporal logics. Artif. Intell. 174(1), 72–104 (2010)
11. Grant, J., Parisi, F., Subrahmanian, V.S.: Research in probabilistic spatiotemporal databases: The spot framework. In: Ma, Z., Yan, L. (eds.) Advances in Probabilistic Databases. STUDFUZZ, vol. 304, pp. 1–22. Springer, Heidelberg (2013)
12. Hammel, T., Rogers, T.J., Yetso, B.: Fusing live sensor data into situational multimedia views. In: Multimedia Information Systems, pp. 145–156 (2003)

13. Knapp, A., Merz, S., Wirsing, M., Zappe, J.: Specification and refinement of mobile systems in mtla and mobile uml. Theor. Comput. Sci. 351(2), 184–202 (2006)
14. Li, S.Z., Jain, A.K. (eds.): Handbook of Face Recognition, 2nd edn. Springer (2011)
15. Lukasiewicz, T.: Probabilistic logic programming with conditional constraints. ACM Trans. on Computational Logic 2(3), 289–339 (2001)
16. Mittu, R., Ross, R.: Building upon the coalitions agent experiment (CoAx) - integration of multimedia information in gccs-m using impact. In: Multimedia Information Systems, pp. 35–44 (2003)
17. Muller, P.: A qualitative theory of motion based on spatio-temporal primitives. In: Int. Conf. on Principles of Knowledge Representation and Reasoning (KR), pp. 131–143 (1998)
18. Papadimitriou, C.H., Steiglitz, K.: Combinatorial optimization: algorithms and complexity. Prentice-Hall, Inc., Upper Saddle River (1982)
19. Parisi, F., Grant, J.: Repairs and consistent answers for inconsistent probabilistic spatio-temporal databases. In: Straccia, U., Cali, A. (eds.) SUM 2014. LNCS, vol. 8720, pp. 265–279. Springer, Heidelberg (2014)
20. Parisi, F., Parker, A., Grant, J., Subrahmanian, V.S.: Scaling cautious selection in spatial probabilistic temporal databases. In: Jeansoulin, R., Papini, O., Prade, H., Schockaert, S. (eds.) Methods for Handling Imperfect Spatial Information. STUDFUZZ, vol. 256, pp. 307–340. Springer, Heidelberg (2010)
21. Parisi, F., Sliva, A., Subrahmanian, V.S.: Embedding forecast operators in databases. In: Benferhat, S., Grant, J. (eds.) SUM 2011. LNCS, vol. 6929, pp. 373–386. Springer, Heidelberg (2011)
22. Parisi, F., Sliva, A., Subrahmanian, V.S.: A temporal database forecasting algebra. Int. J. of Approximate Reasoning 54(7), 827–860 (2013)
23. Parker, A., Subrahmanian, V.S., Grant, J.: A logical formulation of probabilistic spatial databases. IEEE TKDE, 1541–1556 (2007)
24. Parker, A., Infantes, G., Grant, J., Subrahmanian, V.S.: Spot databases: Efficient consistency checking and optimistic selection in probabilistic spatial databases. IEEE TKDE 21(1), 92–107 (2009)
25. Parker, A., Infantes, G., Subrahmanian, V.S., Grant, J.: An agm-based belief revision mechanism for probabilistic spatio-temporal logics. In: AAAI, pp. 511–516 (2008)
26. Parker, A., Yaman, F., Nau, D.S., Subrahmanian, V.S.: Probabilistic go theories. In: Int. Joint Conf. on Artificial Intelligence (IJCAI), pp. 501–506 (2007)
27. Pelanis, M., Saltenis, S., Jensen, C.S.: Indexing the past, present, and anticipated future positions of moving objects. ACM Trans. Database Syst. 31(1), 255–298 (2006)
28. Southey, F., Loh, W., Wilkinson, D.F.: Inferring complex agent motions from partial trajectory observations. In: IJCAI, pp. 2631–2637 (2007)
29. Tao, Y., Cheng, R., Xiao, X., Ngai, W.K., Kao, B., Prabhakar, S.: Indexing multi-dimensional uncertain data with arbitrary probability density functions. In: VLDB, pp. 922–933 (2005)
30. Wolter, F., Zakharyaschev, M.: Spatio-temporal representation and reasoning based on RCC-8. In: KR, pp. 3–14 (2000)
31. Yaman, F., Nau, D., Subrahmanian, V.: A motion closed world assumption. In: Int. Joint Conf. on Artificial Intelligence (IJCAI), pp. 621–626 (2005)
32. Yaman, F., Nau, D.S., Subrahmanian, V.S.: A logic of motion. In: KR, pp. 85–94 (2004)
33. Yaman, F., Nau, D.S., Subrahmanian, V.S.: Going far, logically. In: Int. Joint Conf. on Artificial Intelligence (IJCAI), pp. 615–620 (2005)
34. Zhang, M., Chen, S., Jensen, C.S., Ooi, B.C., Zhang, Z.: Effectively indexing uncertain moving objects for predictive queries. PVLDB 2(1), 1198–1209 (2009)
35. Zheng, K., Trajcevski, G., Zhou, X., Scheuermann, P.: Probabilistic range queries for uncertain trajectories on road networks. In: EDBT, pp. 283–294 (2011)

Repairs and Consistent Answers for Inconsistent Probabilistic Spatio-Temporal Databases

Francesco Parisi[1] and John Grant[2]

[1] Department of Informatics, Modeling, Electronics and System Engineering,
University of Calabria, Italy
fparisi@dimes.unical.it
[2] Department of Computer Science and UMIACS,
University of Maryland, College Park, USA
grant@cs.umd.edu

Abstract. We formally introduce the concept of repair and consistent answer for inconsistent probabilistic spatio-temporal databases. We start by defining the syntax and semantics of SPOT databases, a declarative framework that has been explored in recent years for the representation of spatio-temporal data with uncertainty expressed as probability intervals. In this framework we study two types of repairs, that is, minimal modifications that lead to consistent databases: maximal consistent subsets and probability interval expansion. We also extend the concept of consistent answer to this framework and find that this can be done in several different ways. In emphasizing tractable cases we propose polynomial-time algorithms for computing consistent answers and repairs based on probability interval expansion, and experimentally validate our approach.

1 Introduction

Recent years have seen a great deal of interest in tracking moving objects. For this reason, researchers have investigated in detail the representation and processing of spatio-temporal databases [32,36,22,2,33,20]. However, in many cases the location of objects is uncertain: such cases can be handled by using probabilities [35,10,8,5]. Sometimes the probabilities themselves are uncertain. The SPOT (Spatial PrObabilistic Temporal) database concept was introduced in [30] to provide a declarative framework for the representation and processing of probabilistic spatio-temporal databases with uncertain probabilities.

A SPOT database represents atomic statements of the form "object id is/was/will be inside region r at time t with probability in the interval $[\ell, u]$". This allows the representation of information concerning moving objects in several application domains. For instance, a military agency is interested in modelling enemy vehicles that may be in a region at a given time point and with a given probability (in order to adequately arrange its defense line) [25,21]. A cell phone provider is interested in knowing which cell phones will be in the range of some towers at a given time and with what probability. A transportation company is interested in predicting the vehicles that will be on a given road at a given time (and with what probability) in order avoid congestion.

U. Straccia and A. Calì (Eds.): SUM 2014, LNAI 8720, pp. 265–279, 2014.

Previous work on SPOT databases included a formal syntax and semantics as well as checking for consistency: an object can not be in two places at the same time. Additional research focused on the efficient processing of selection queries [31,27], aggregate queries [17], and database updates [18]. However, all of these works assume that a consistent version of the database is somehow available before processing queries or performing updates. This is often not the case.

To overcome such limitation and allow the user to profitably use inconsistent probabilistic spatio-temporal data, in this paper, we investigate a principled method for querying inconsistent SPOT database. Our approach relies on the well-known consistent query answering (CQA) approach [4], which has been deeply investigated in the context of relational databases in the last fifteen years.

Starting from the seminal paper [4], substantial work has been developed on repairing and consistently answering queries in inconsistent relational databases (see [6] for a comprehensive survey). While repairing aims at restoring consistency by "minimally" changing the inconsistent database instance, CQA characterizes meaningful answers from possibly inconsistent databases: a consistent answer to a query is an answer that is invariant regarding the way the database is repaired.

In this paper, we present a consistent query answering approach to manage inconsistency in probabilistic spatio-temporal databases. In particular, we first introduce two natural strategies for restoring consistency in SPOT databases. The first strategy is based on "minimally" removing a subset of atomic statements from the original SPOT database, leading to repairs consisting of maximal consistent subsets that we call S-repairs. Basically, this repairing strategy corresponds to the idea of repairing a relational database by minimally removing tuples, such as done in [9]. The other strategy is based on "minimally" updating the probability bounds of the atomic statements of the SPOT database that we call PU-repairs. Analogue repairing strategies in the context of relational databases are those based on numerical values updates, such as those proposed in [7,15,13,16,14]. Both value-updates and tuple-eliminations are suggested as basic primitives for managing inconsistent relational databases in [23,24]. However, here we focus on probability interval updates and on spatio-temporal atom eliminations for dealing with inconsistent SPOT databases.

After presenting S- and PU-repairs, we define the corresponding semantics for answering selection queries in possibly inconsistent SPOT databases, that is, S-consistent and PU-consistent answers. We show that computing repairs is tractable for both kinds of repairing strategies we propose. In contrast, checking PU-consistent answers can be done in PTIME, but checking S-consistent answers is coNP-complete. In emphasizing tractable cases, we implemented a prototype for computing PU-repairs and PU-consistent answers, and experimentally validated our technique.

2   SPOT Databases

This section reviews the syntax and semantics of SPOT databases given in [30].

2.1   Syntax

We assume the existence of a set *ID* of objects ids, a set T of time points ranging over the integers, and a finite set *Space* of points. We assume that *Space* is a grid of size $N \times N$ where we only consider integer coordinates (the framework is easily extensible to higher dimensions). We assume that an object can be in only one location at a time, but that a single location may contain more than one object.

Definition 1 (SPOT **atom/database**). *A* SPOT *atom is a tuple* $(id, r, t, [\ell, u])$, *where* $id \in ID$ *is an object id*, $r \subseteq Space$ *is a region in the space*, $t \in T$ *is a time point, and* $[\ell, u] \subseteq [0, 1]$ *is a probability interval (both ℓ and u are rational numbers).*
A SPOT *database is a finite set of* SPOT *atoms.*

Intuitively, the SPOT atom $(id, r, t, [\ell, u])$ says that object *id* is/was/will be inside region r at time t with probability in the interval $[\ell, u]$. Hence, SPOT atoms can represent information about the past and the present (such as from techniques for interpreting RFID readings [11,12]), but also information about the future, such as that deriving from methods for predicting the destination of moving objects [25,21,34], or from querying predictive databases [3,1,28,29].

The initial SPOT definition used only rectangular regions; however, we allow a region to be any non-empty set of points. Still, for convenience we use such rectangular regions in our running example and in the implementation of our prototype.

Example 1. Consider a lab where data coming from biometric sensors are collected and analyzed. Biometric data such as faces, voices, and fingerprints recognized by sensors are matched against given profiles (such as those of people having access to the lab) and tuples like those in Fig. 1(a) are obtained. Every tuple consists of the profile id resulting from the matching phase, the area of the lab where the sensor recognizing the profile is operating, the time point at which the profile has been recognized, and the lower and upper probability bounds of the recognizing process getting the tuple. For instance, the tuple in the first row of the table in Fig. 1(a), representing the SPOT atom $(id_1, d, 1, [0.9, 1])$, says that the profile having id id_1 was in region d at time 1 with probability in the interval $[0.9, 1]$. In Fig. 1(b), the plan of the lab and the areas covered by sensors are shown. In area d a fingerprint sensor is located, whose high accuracy

Id	Area	Time	Lower Probability	Upper Probability
id_1	d	1	0.9	1
id_1	b	3	0.6	1
id_1	c	3	0.7	0.8
id_2	b	1	0.5	0.9
id_2	e	2	0.2	0.5
id_3	e	1	0.6	0.9

(a)

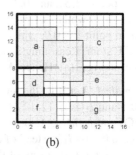

(b)

Fig. 1. (a) SPOT database $\mathscr{S}_{lab}$ (b) Areas of the lab

entails a narrow probability interval with upper bound equal to 1. After fingerprint authentication, id_1 was recognized at time 3 in areas b and c with probability in $[0.6, 1]$ and $[0.7, 0.8]$, respectively.   □

Given a SPOT database $\mathscr{S}$, an object id, and a time t, we use the notation $\mathscr{S}_{id,t}$ to refer to the set $\mathscr{S}_{id,t} = \{(id', r', t', [\ell', u']) \in \mathscr{S} \mid id' = id \wedge t' = t\}$.

2.2 Semantics

The meaning of a SPOT database is given by the set of interpretations that satisfy it.

Definition 2 (SPOT interpretation). *A* SPOT *interpretation is a function* $I : ID \times Space \times T \rightarrow [0, 1]$ *such that for each* $id \in ID$ *and* $t \in T$, $\sum_{p \in Space} I(id, p, t) = 1$.

Observe that $I^{id,t}(p) = I(id, p, t)$ is a probability distribution function (PDF). The set of all interpretations for database $\mathscr{S}$ will be denoted as $\mathbf{I}(\mathscr{S})$.

Example 2. Interpretation M for the SPOT database $\mathscr{S}_{lab}$ of Example 1 is as follows.

$M(id_1, (3,6), 1) = 0.4$
$M(id_1, (3,5), 1) = 0.3$
$M(id_1, (2,5), 1) = 0.2$
$M(id_1, (7,7), 1) = 0.1$
$M(id_2, (5,7), 1) = 0.7$
$M(id_2, (12,12), 1) = 0.3$

$M(id_3, (10,5), 1) = 0.8$
$M(id_3, (5,6), 1) = 0.2$
$M(id_1, (7,5), 2) = 0.5$
$M(id_1, (4,2), 2) = 0.5$
$M(id_2, (9,7), 2) = 0.3$
$M(id_2, (12,13), 2) = 0.7$
$M(id_3, (5,5), 2) = 0.5$

$M(id_3, (6,5), 2) = 0.5$
$M(id_1, (10,10), 3) = 0.7$
$M(id_1, (7,5), 3) = 0.3$
$M(id_2, (8,7), 3) = 0.9$
$M(id_2, (11,15), 3) = 0.1$
$M(id_3, (5,3), 3) = 0.6$
$M(id_3, (5,6), 3) = 0.4$

Moreover, $M(id, p, t) = 0$ for all triplets (id, p, t) not mentioned above.   □

Given an interpretation I and region r, the probability that object id is in r at time t *according to* I is $\sum_{p \in r} I(id, p, t)$. We now define satisfaction and SPOT models.

Definition 3 (Satisfaction and SPOT model). *Let* $A = (id, r, t, [\ell, u])$ *be a* SPOT *atom and let* I *be a* SPOT *interpretation. We say that* I *satisfies* A *(denoted* $I \models A$*) iff* $\sum_{p \in r} I(id, p, t) \in [\ell, u]$. *I satisfies a* SPOT *database* $\mathscr{S}$ *(denoted* $I \models \mathscr{S}$*) iff* $\forall A \in \mathscr{S}$, $I \models A$. *If* I *satisfies a* SPOT *atom* A *(resp.* SPOT *database* $\mathscr{S}$*), we say that* I *is a* model *for* A *(resp.* $\mathscr{S}$*).*

Example 3. In our running example, interpretation M is a model for the SPOT atom $(id_1, d, 1, [0.9, 1])$ as, for id id_1 and time point 1, M assigns probability 0.4 to point $(3,6)$, 0.2 to point $(2,5)$, and 0.3 to point $(3,5)$ (which are points in area d), and probability 0.1 to $(7,7)$ which is a point outside area d. Hence, the probability that id_1 is in area d at time point 1 is 0.9, which belongs to the interval $[0.9, 1]$ specified by the considered SPOT atom. Reasoning analogously, it is easy to see that M is a model for all of the atoms in Fig. 1(a). Hence, M is a model for $\mathscr{S}_{lab}$.   □

Example 4. Let I_1 be the interpretation which is equal to M except that $I_1(id_2, (5,7), 1) = 0.1$ and $I_1(id_2, (12, 12), 1) = 0.9$. It is easy to check that I_1 is not a model for the SPOT atom $(id_2, b, 1, [0.5, 0.9])$, as the probability to be in area b at time 1 for id id_2 is set to 0.1 by I_1, instead of a value in $[0.5, 0.9]$. Hence, since $(id_2, b, 1, [0.5, 0.9])$ is in $\mathscr{S}_{lab}$ (see the fourth row of the table in Fig. 1(a)), it follows that I_1 is a not model for $\mathscr{S}_{lab}$.   □

We use $\mathbf{M}(\mathcal{S})$ to denote the set of models for a SPOT database $\mathcal{S}$, that is, $\mathbf{M}(\mathcal{S}) = \{I \mid I \in \mathbf{I}(\mathcal{S}) \wedge I \models \mathcal{S}\}$. In the following we will use the symbol M to refer to interpretations that are models, that is, elements in $\mathbf{M}(\mathcal{S})$.

Definition 4 (Consistency). *A* SPOT *database $\mathcal{S}$ is consistent iff* $\mathbf{M}(\mathcal{S}) \neq \emptyset$.

Example 5. Model M of Example 3 proves that the database $\mathcal{S}_{lab}$ is consistent.   □

As shown in [30], the consistency of a SPOT database can be checked by means of a linear programming algorithm whose complexity is $O(|ID(\mathcal{S})| \cdot |T| \cdot (|Space| \cdot |\mathcal{S}|)^3)$.

Next we give an example of an inconsistent SPOT database that we will also use later to illustrate database repair and consistent answers.

Example 6. Consider the SPOT database shown below, which is just a slight modification of database $\mathcal{S}_{lab}$ of our original example (we number the atoms so we can refer back to them later as needed). It's easy to see that there is an inconsistency for object id_3 at time 1, object id_2 at time 2, and object id_1 at time 3.

Atom	Id	Area	Time	Lower Pr.	Upper Pr.
at_1	id_1	d	1	0.9	1
at_2	id_1	a	3	0.5	0.9
at_3	id_1	b	3	0.6	1
at_4	id_1	c	3	0.7	0.8
...	...	...	...	...	...

Atom	Id	Area	Time	Lower Pr.	Upper Pr.
...	...	...	...	...	...
at_5	id_2	b	1	0.5	0.9
at_6	id_2	e	2	0.3	0.5
at_7	id_2	f	2	0.5	0.7
at_8	id_2	g	2	0.9	1.0
at_9	id_3	c	1	0.5	0.8
at_{10}	id_3	e	1	0.6	0.9

3   Repairs

We now introduce two strategies for repairing SPOT databases, each of them aiming at minimally modifying the original database in order to restore consistency. Although the repair concept is really useful only for inconsistent databases, we define repairs for all databases. Hence a consistent database is its own repair. The criterion for the first strategy is to mimimally modify the original database by finding maximal consistent subsets.

Definition 5 (S-repairs). *Given a* SPOT *database $\mathcal{S}$, an S-repair for $\mathcal{S}$ is a maximal (under $\subseteq$) consistent subset of $\mathcal{S}$.*

Thus, according to this repairing strategy, an atom belonging to the inconsistent SPOT database either is deleted from the database or kept in the repair.

Example 7. Continuing with Example 6 we note that there are three problems, one per time period. At time 1 the inconsistency is due to atoms at_9 and at_{10}. At time 2 the inconsistency is due to three atoms at_6, at_7, and at_8. Finally, at time 3 the problem is due to atoms at_2, at_3, and at_4.

The possible S-repairs are as follows. For time 1 we must choose either at_9 or at_{10}. For time 2 we must pick either at_6 and at_7 or just at_8. Then, for time 3 the two choices are at_2 and at_3 or at_3 and at_4. As there are 2 possible choices for each time period, the total number of maximal consistent subsets is 8. Hence there are 8 S-repairs.   □

The second repairing strategy we propose is instead based on minimally updating the extreme values of the probability interval of each of the SPOT atom in the database in order to achieve consistency.

Given a SPOT atom $a = (id, r, t, [\ell, u])$, a *probability-interval updated atom* for a is a SPOT atom a' obtained from a by enlarging its probability interval, that is, $a' = (id, r, t, [\ell', u'])$ where $[\ell', u'] \supseteq [\ell, u]$. Given $a = (id, r, t, [\ell, u])$ and a probability-interval updated atom $a' = (id, r, t, [\ell', u'])$ for a, we denote as $low(a, a')$ and $up(a, a')$ the absolute values of the differences between the lower and upper probability bounds of a and a', respectively, that is, $low(a, a') = \ell - \ell'$ and $up(a, a') = u' - u$. Basically, $low(a, a')$ and $up(a, a')$ are a measure of the distance between the original atom a and its probability update a'. A *PU*-repair aims at minimizing this distance for each atom in the SPOT database.

Definition 6 (PU-repairs). *A PU-repair for a SPOT database $\mathscr{S}$ is a consistent SPOT database $\mathscr{S}'$ consisting of a probability-interval update atom a' for each $a \in \mathscr{S}$ and such that $\sum_{a \in \mathscr{S}} low(a, a') + up(a, a')$ is minimum.*

Example 8. Continuing with Example 6, *PU*-repairs are as follows. For time 1 the problem is that the lower bounds 0.6 and 0.5 add to 1.1. A *PU*-repair must lower that number to 1. This can be accomplished in infinitely many ways; for example by lowering the lower bound of at_9 to 0.4, or lowering the lower bounds to 0.45 and 0.55 respectively. In any case, the minimal probability change must be 0.1. Next, for time 2 the lower bounds add to 1.7, so similarly as for time 1 the sum of the lower bounds must be lowered to 1 by a minimal probability change of 0.7 that can be distributed among at_6, at_7, and at_8 in any manner. Finally, for time 3 the sum of the lower bounds of at_2 and at_4 must be lowered to 1 by a minimal probability change of 0.2. □

Note that in this example for all *PU*-repairs we needed to work only with the lower probability bounds. This will be the case in general when the regions do not cover all of *Space*. But consider a simple case where 2 regions cover all of *Space*. If the sum of the upper bounds is less than 1, for a *PU*-repair we must work with the upper bounds.

The following proposition states an important property that holds for both kinds of repairs we have introduced above: a repair for a SPOT database $\mathscr{S}$ can be obtained by combining the repairs for the single-id, single-time-point SPOT databases $\mathscr{S}_{id,t}$ with $id \in ID$ and $t \in T$. We already exploited this result in our example.

Proposition 1. *Given a SPOT database $\mathscr{S}$, a SPOT database $\mathscr{S}'$ is an S-repair (resp. PU-repair) for $\mathscr{S}$, iff $\mathscr{S}' = \bigcup_{id \in ID, t \in T} \mathscr{S}'_{id,t}$, where $\mathscr{S}'_{id,t}$ is an S-repair (resp. PU-repair) for $\mathscr{S}_{id,t}$.*

It is easy to see that, for a SPOT database $\mathscr{S}$, both *S*-repairs and *PU*-repairs always exist. In fact, in the worst case, an *S*-repair for $\mathscr{S}$ can be obtained by keeping in it only one atom for each id, t pair. As regards *PU*-repairs, it is easy to see that relaxing the probability intervals of all the SPOT atoms to the interval $[0, 1]$ results in a consistent database, which makes the problem of deciding whether a *PU*-repair exist trivial.

In the following, we will denote as $Rep_S(\mathscr{S})$ (resp. $Rep_{PU}(\mathscr{S})$) the set of *S*-repairs (resp. Rep_{PU}-repairs) for $\mathscr{S}$.

3.1 Checking Repairs

We characterize the complexity of the problem of checking whether a given SPOT database $\mathscr{S}'$ is a repair for a SPOT database $\mathscr{S}$. We consider the two kinds of repairs introduced above, and show that this problem is tractable for both S- and PU-repairs. For both repairing strategy, the problem of computing a repair turns out to be in *PTIME*.

Theorem 1. *Let $\mathscr{S}, \mathscr{S}'$ be two* SPOT *databases. Deciding whether $\mathscr{S}'$ is an S-repair for $\mathscr{S}$ is in PTIME.*

Proof. If $\mathscr{S}$ is consistent, then $\mathscr{S}'$ is an S-repair for $\mathscr{S}$ iff $\mathscr{S}'$ coincides with $\mathscr{S}$. Recall that checking whether $\mathscr{S}$ is not consistent can be accomplished in polynomial time [30]. Now assume that $\mathscr{S}$ is not consistent. If $\mathscr{S}' \not\subseteq \mathscr{S}$ or $\mathscr{S}'$ is not consistent, then $\mathscr{S}'$ is not an S-repair for $\mathscr{S}$. Otherwise, for each $a \in \mathscr{S} \setminus \mathscr{S}'$ check whether $\mathscr{S}' \cup \{a\}$ is not consistent. If so, then $\mathscr{S}'$ is an S-repair for $\mathscr{S}$, as it is a maximal consistent subset of $\mathscr{S}$ — in fact, no strict superset of $\mathscr{S}'$ is consistent if $\forall a \in \mathscr{S} \setminus \mathscr{S}'$, $\mathscr{S}' \cup \{a\}$ is not consistent. If there is $a \in \mathscr{S} \setminus \mathscr{S}'$ such that $\mathscr{S}' \cup \{a\}$ is consistent, it follows that $\mathscr{S}'$ is not an S-repair for $\mathscr{S}$, as it is not a maximal consistent subset of $\mathscr{S}$. $\qquad\square$

The proof of Theorem 1, along with the result of Proposition 1, suggests that an S-repair for SPOT database $\mathscr{S}$ can be computed as the union of S-repairs for $\mathscr{S}_{id,t}$ (with $id \in ID$ and $t \in T$), each of them incrementally computed by scanning the atoms in $\mathscr{S}_{id,t}$ according to any total ordering, and then adding atom $a \in S_{id,t}$ to $\mathscr{S}'_{id,t}$ iff $\mathscr{S}'_{id,t} \cup \{a\}$ is consistent. It is easy to see that the so obtained database $\mathscr{S}'_{id,t}$ is an S-repair for $\mathscr{S}_{id,t}$.

Corollary 1. *An S-repair for* SPOT *database $\mathscr{S}$ can be computed in PTIME.*

We now consider PU-repairs. Given a SPOT database $\mathscr{S}$, an object $id \in ID$, and a time point $t \in T$, we define a linear programming problem, called $PULP(\mathscr{S}, id, t)$, such that each of its optimal solutions one-to-one corresponds to a PU-repair for $\mathscr{S}_{id,t}$. $PULP(\mathscr{S}, id, t)$ uses variables v_p to denote the probability that object id will be at point $p \in Space$ at time t. Moreover, $PULP(\mathscr{S}, id, t)$ uses variables low_i and up_i, for each atom $a_i \in \mathscr{S}_{id,t}$, to denote the absolute values of the differences between the lower and upper probability bounds of atom a_i and its probability-interval updated atom a'_i.

Definition 7 $(PULP(\cdot))$. *For* SPOT *database $\mathscr{S}$, $id \in ID$, and $t \in T$, $PULP(\mathscr{S}, id, t)$ is defined as the following linear programming problem:*
minimize $\sum_{a_i \in \mathscr{S}_{id,t}} low_i + up_i$ **subject to:**

$$
\begin{cases}
1) \ \forall a_i = (id, r_i, t, [\ell_i, u_i]) \in \mathscr{S}_{id,t} \\
\quad \ell_i - low_i \leq \sum_{p \in r_i} v_p \leq u_i + up_i \\
\quad 0 \leq low_i \leq \ell_i \\
\quad 0 \leq up_i \leq 1 - u_i \\
2) \ \sum_{p \in Space} v_p = 1 \\
3) \ \forall p \in Space \quad v_p \geq 0
\end{cases}
$$

Given a solution σ of $PULP(\mathscr{S}, id, t)$, we define $\mathscr{S}_{id,t}(\sigma)$ as the SPOT database obtained from $\mathscr{S}_{id,t}$ by replacing each atom $a_i = (id, r_i, t, [\ell_i, u_i]) \in \mathscr{S}_{id,t}$ with the probability-interval updated atom $a'_i = (id, r_i, t, [\ell_i - \sigma[low_i], u_i + \sigma[up_i]])$, where $\sigma[low_i]$ and $\sigma[up_i]$ denote the values assigned to variables low_i and up_i by solution σ.

Theorem 2. *For each optimal solution σ of $PULP(\mathscr{S}, id, t)$, $\mathscr{S}_{id,t}(\sigma)$ is a PU-repair for $\mathscr{S}_{id,t}$. Moreover, every optimal solution σ for $PULP(\mathscr{S}, id, t)$ one-to-one corresponds to a model for PU-repair $\mathscr{S}_{id,t}(\sigma)$ for $\mathscr{S}_{id,t}$, and vice versa.*

A consequence of the above theorem is that deciding whether $\mathscr{S}'_{id,t}$ is a *PU*-repair for $\mathscr{S}_{id,t}$ can be accomplished in polynomial time. In fact, $\mathscr{S}'_{id,t}$ is a *PU*-repair for $\mathscr{S}_{id,t}$ iff by replacing each occurrence of variables low_i and up_i in $PULP(\mathscr{S}, id, t)$ with the values of the corresponding probability updates described by $\mathscr{S}'_{id,t}$ results in a solution of $PULP(\mathscr{S}, id, t)$ whose objective value is the optimal one. This result along with Proposition 1 entail that checking whether a SPOT database $\mathscr{S}'$ is a *PU*-repair for SPOT database $\mathscr{S}$ can be accomplished in polynomial time.

Corollary 2. *Deciding whether $\mathscr{S}'$ is a PU-repair for SPOT database $\mathscr{S}$ is in PTIME.*

Theorem 2 and Proposition 1 also entail that PU-repairs can be computed in *PTIME*, by finding optimal solutions of $PULP(\mathscr{S}, id, t)$, for each $\langle id, t \rangle$ pair occurring in $\mathscr{S}$.

Corollary 3. *A PU-repair for SPOT database $\mathscr{S}$ can be computed in PTIME.*

4 Consistent Query Answers

Selection queries are the most investigated kind of query in the SPOT framework [30,31,27]. A selection query is an expression of the form $(?id, r, ?t, [\ell, u])$, where r is a region and $[\ell, u]$ is a probability interval. Intuitively, a selection query says: "Given a region r and a probability interval $[\ell, u]$, find all objects id and times t such that id is inside r at time t with a probability in the interval $[\ell, u]$."

Definition 8 (Selection Query Answers). *Given a selection query $Q = (?id, r, ?t, [\ell, u])$ and a consistent SPOT database $\mathscr{S}$, $\langle id, t \rangle$ is an answer to Q w.r.t. $\mathscr{S}$, iff for every model $M \in \mathbf{M}(\mathscr{S})$, $M \models (id, r, t, [\ell, u])$.*

Example 9. Consider the consistent SPOT database of Fig. 1. One may be interested in knowing the ids and time points of profiles that were in the room where the fingerprint sensor is located, with probability greater than .75. This can be expressed by the selection query $(?id, r, ?t, [.75, 1])$, where r is the rectangle defined by constraints $0 \le x \le 6$ and $4 \le y \le 8$ (this query region includes the whole area d, a portion of area b, and some other points). The answer of this query is the set consisting of the pair $\langle id_1, 1 \rangle$.

Selection query answers are defined for consistent SPOT databases only. So, in order to query inconsistent SPOT databases, now we adapt the well-known consistent query answering (CQA) approach originally proposed for relational databases [4]. To make the following simpler we use the terminology *X-CQA*, where X is either S or PU.

Definition 9 (X-Consistent Selection Query Answers). *Given a SPOT database $\mathscr{S}$ and a selection query $Q = (?id, r, ?t, [\ell, u])$, $\langle id, t \rangle$ is an X-consistent answer to Q w.r.t. $\mathscr{S}$ iff for each $\mathscr{S}' \in Rep_X(\mathscr{S})$, $\langle id, t \rangle$ is an answer to Q w.r.t. $\mathscr{S}'$.*

Example 10. Continuing with our example of inconsistent SPOT database, previously in examples 7 and 8 we described all the *S*-repairs and *PU*-repairs, respectively. Consider now the large region R that contains the right-half of *Space* so that it extends horizontally from 8 to 16 and vertically from 0 to 16. In particular, R contains the regions c, e, g, and a portion of b. Our sample query asks for all objects and times such that the object at that time was in the region R with probability at least 0.6: $Q = (?id, R, ?t, [0.6, 1])$. Then, the *S*-consistent answer is $\emptyset$, while the *PU*-consistent answer is $\{(id_3, 1)\}$.

4.1 Complexity

We characterize the complexity of the problem of deciding whether a given $\langle id, t \rangle$ pair is a consistent answer. It turns out that checking if a pair $\langle id, t \rangle$ is an *S*-consistent answer is coNP-complete. However, checking if a pair is a *PU*-consistent answer is in PTIME.

Theorem 3. *Given a* SPOT *database* $\mathscr{S}$, *a query* $Q = (?id, r, ?t, [\ell, u])$, *and a pair* $\langle id, t \rangle$. *Deciding whether* $\langle id, t \rangle$ *is an S-consistent answer to Q w.r.t.* $\mathscr{S}$ *is coNP-complete.*

Proof. (Membership). A guess-and-check strategy for deciding the complement of our problem is as follows. First, guess a subset $\mathscr{S}'$ of $\mathscr{S}$. Then, check whether (i) $\mathscr{S}'$ is an *S*-repair for $\mathscr{S}$ and (ii) $\langle id, t \rangle$ is not an answer to Q. The fact that both (i) and (ii) can be checked in polynomial time follows from Theorem 1 and a result of [30], where it is shown that deciding whether $\langle id, t \rangle$ is an answer to Q is polynomial time.
(Hardness). [*Sketch*]. We can show a LOGSPACE reduction to the complement of our problem from the NP-hard SUBSET SUM problem, which is defined as follows: given a set of positive integers $S = \{s_1, \ldots, s_n\}$ and a positive integer constant C, decide whether there is a subset $S' \subseteq S$ such that $\sum_{s_i \in S'} s_i = C$. W.l.o.g. we assume that $\forall i \in [1..n], s_i \leq C$.

Given S and C, we construct an instance of our problem as follows. Let *Space* = $\{p_1, \cdots, p_n, p_{n+1}\}$, where n is the cardinality of S. Let T consist of the single time point 0. We define the SPOT database $\mathscr{S}$ as follows. $\mathscr{S}$ consists of two SPOT atoms $a_i^0 = (id, \{p_i\}, 0, [0, 0])$ and $a_i^1 = (id, \{p_i\}, 0, [s_i/C, s_i/C])$, for each $s_i \in S$. Finally, let $Q = (?id, \{p_{n+1}\}, ?t, [1/C, 1])$, and $\langle id, 0 \rangle$ be the pair to be checked being an *S*-consistent answer to Q w.r.t. $\mathscr{S}$.

Observe that every repair $\mathscr{S}' \in Reps(\mathscr{S})$ is such that, for each $i \in [1..n]$, either a_i^0 or a_i^1 is in $\mathscr{S}'$. This entails that $\mathscr{S}'$ has exactly one model. Then, it can be shown that $\langle id, 0 \rangle$ is not an *S*-consistent answer to Q w.r.t. $\mathscr{S}$ iff there is $S' \subseteq S$ s.t. $\sum_{s_i \in S'} s_i = C$ $\square$

We now address the problem of consistently answering selection queries in inconsistent SPOT databases under the *PU*-repair semantics. We introduce a polynomial time method based on first solving an instance of $PULP(\mathscr{S}, id, t)$ to obtain the minimum cost o^* of *PU*-repairs for $\mathscr{S}_{id,t}$, and then using this value to construct and solving the two additional instances of the linear programming problems defined below (basically, these problems are obtained by adding equality (4) and objective function to $PULP$).

Definition 10 (PU-CQA$^\ell(\cdot)$ and PU-CQA$^u(\cdot)$). *Let* o^* *be the optimal value returned by* $PULP(\mathscr{S}, id, t)$. *For* SPOT *database* $\mathscr{S}$, *query* $Q = (?id, r, ?t, [\ell, u])$, $id \in ID$, *and* $t \in T$, *we define* $PU\text{-}CQA^\ell(\mathscr{S}, Q, id, t)$ *(resp.* $PU\text{-}CQA^u(\mathscr{S}, Q, id, t)$) *as the following linear programming problem:*

*minimize (resp., **maximize**)* $\sum_{p \in r} v_p$ *subject to:*

$$\begin{cases} 1) \; 2) \; and \; 3) \; of \; Definition \; 7, \; and \\ 4) \; \sum_{a_i \in \mathscr{S}_{id,t}} low_i + up_i = o^* \end{cases}$$

The relationship between solutions of *PU-CQA$^\ell$* (resp., *PU-CQA^u*)) and the *PU*-repairs for $\mathscr{S}$ is as follows.

Lemma 1. *Let $\mathscr{S}_{id,t}$ be a* SPOT *database, and $Q = (?id, r, ?t, [\ell, u])$. Every optimal solution σ for PU-CQA$^\ell(\mathscr{S}, Q, id, t)$ (resp., PU-CQA$^u(\mathscr{S}, Q, id, t)$) one-to-one corresponds to a model M for PU-repair $\mathscr{S}_{id,t}(\sigma)$ such that $\sum_{p \in r} M(id, p, t)$ is minimum (resp., maximum), and vice versa.*

The following theorem provides a method for checking whether an $\langle id, t \rangle$ pair is a *PU*-consistent answer to query Q w.r.t $\mathscr{S}_{id,t}$ by comparing the optimal values returned by *PU-CQA$^\ell(\mathscr{S}, Q, id, t)$* and *PU-CQA$^u(\mathscr{S}, Q, id, t)$* with the probability bounds of Q.

Theorem 4. *Let $\mathscr{S}_{id,t}$ be a* SPOT *database, and $Q = (?id, r, ?t, [\ell, u])$. Let ℓ^* and u^* be the optimal values returned by PU-CQA$^\ell(\mathscr{S}, Q, id, t)$ and PU-CQA$^u(\mathscr{S}, Q, id, t)$, respectively. A pair $\langle id, t \rangle$ is a PU-consistent answer to Q w.r.t. $\mathscr{S}$ iff $\ell \leq \ell^*$ and $u^* \leq u$.*

Hence, computing the optimal values of *PU-CQA$^\ell(\mathscr{S}, Q, id, t)$* and *PU-CQA$^u(\mathscr{S}, Q, id, t)$* suffices to decide if $\langle id, t \rangle$ is a *PU*-consistent answer to Q w.r.t. $\mathscr{S}_{id,t}$. As this can be done in polynomial time, and since $\langle id, t \rangle$ is a *PU*-consistent answer to Q w.r.t. $\mathscr{S}$ iff $\langle id, t \rangle$ is a *PU*-consistent answer to Q w.r.t. $\mathscr{S}_{id,t}$, we obtain the following result.

Theorem 5. *Given a* SPOT *database $\mathscr{S}$, a query $Q = (?id, r, ?t, [\ell, u])$, and a pair $\langle id, t \rangle$. Deciding whether $\langle id, t \rangle$ is a PU-consistent answer to Q w.r.t. $\mathscr{S}$ is in PTIME.*

4.2 Other Semantics

We note that in Definition 8 selection queries are interpreted under a cautious semantics. However, selection queries can be also interpreted under optimistic semantics: a pair $\langle id, t \rangle$ is an optimistic answer to query $Q = (?id, r, ?t, [\ell, u])$ w.r.t. $\mathscr{S}$ iff *there exists* a model $M \in \mathbf{M}(\mathscr{S})$ s.t. $M \models (id, r, t, [\ell, u])$. Thus, cautious answers (Definition 8) are a subset of the optimistic ones. For instance, the cautious answer to the query of Example 9 is contained in the optimistic one, that is $\{\langle id_1, 1 \rangle, \langle id_1, 2 \rangle, \langle id_2, 1 \rangle, \langle id_2, 2 \rangle\}$.

Basically, under the optimistic semantics the query applies to at least one model, while under the cautious one the query applies to all models. Now, as there are usually multiple repairs for a SPOT database, we can choose to consider whether the query applies to all repairs or at least one repair. Based on this, it is natural to consider also other semantics for X-consistent query answers. Let us denote the type of consistent answers introduced in Section 4 as *X-consistent universal cautious answers*. Given a SPOT database $\mathscr{S}$ and a selection query Q, we say that $\langle id, t \rangle$ is an *X-consistent*

- *existential cautious answer* to Q w.r.t. $\mathscr{S}$ iff there exists $\mathscr{S}' \in Rep_X(\mathscr{S})$, such that $\langle id, t \rangle$ is a cautious answer to Q w.r.t. $\mathscr{S}'$.
- *universal optimistic answer* to Q w.r.t. $\mathscr{S}$ iff for each $\mathscr{S}' \in Rep_X(\mathscr{S})$, $\langle id, t \rangle$ is an optimistic answer to Q w.r.t. $\mathscr{S}'$.

- *existential optimistic answer* to Q w.r.t. $\mathscr{S}$ iff there exists $\mathscr{S}' \in Rep_X(\mathscr{S})$, such that $\langle id, t \rangle$ is an optimistic answer to Q w.r.t. $\mathscr{S}'$.

For a consistent database there is no difference between the universal and existential versions. Moreover, every X-consistent universal cautious (resp. optimistic) answer is an X-consistent existential cautious (resp. optimistic) answer. Also, every X-consistent universal (resp. existential) cautious answer is an X-consistent universal (resp. existential) optimistic answer. Next we give an example to illustrate these concepts.

Example 11. Continuing Example 10, the answers for the above-defined types of consistent query answers are:

S-consistent existential cautious: $\{(id_3, 1), (id_2, 2), (id_1, 3)\}$
S-consistent universal optimistic: $\{(id_3, 1), (id_2, 1)\}$
S-consistent existential optimistic: $\{(id_3, 1), (id_2, 1), (id_2, 2), (id_1, 3)\}$
PU-consistent existential cautious: $\{(id_3, 1)\}$
PU-consistent universal optimistic: $\{(id_3, 1), (id_2, 2), (id_1, 3), (id_2, 1)\}$
PU-consistent existential optimistic: $\{(id_3, 1), (id_2, 1), (id_2, 2), (id_1, 3)\}$ □

5 Experiments

We experimentally validated our approach for computing PU-repairs and PU-consistent (universal cautious) answers on randomly generated SPOT databases, each of them characterized by the following parameters: the sizes $|ID|$, $|T|$, and $|Space|$ of ID, T, and $Space$, respectively; the average size ω of one side of the atom's rectangles; and the average cardinality of $\mathscr{S}_{id,t}$ for $id \in ID$ and $t \in T$, which we call the *density* of the database. Basically, the density of the database represents the average number of times that an object was detected at a time point, while ω^2 is the average number of points in the detection's regions.

Each SPOT database in the dataset was generated as follows. ID and T was set to $\{0, .., |ID| - 1\}$ and $\{0, .., |T| - 1\}$. Next, for each $\langle id, t \rangle$ pair, with $id \in ID$ and $t \in T$, a random number of atoms were generated in the interval of integers between 1 and $2d$, where d is the density of the database (thus on average d atoms were generated for each $\langle id, t \rangle$ pair). Each of these atoms was generated by randomly choosing the width and the height of the atom's rectangle, as well as the lower and upper bound probabilities. Specifically, the rectangle's width and height of the region of each atom were randomly chosen in the interval of integers between 1 and 2ω. For the first generated atom in $\mathscr{S}_{id,0}$, the rectangle's upper left corner was chosen uniformly in the set of points in $Space$. For the subsequent generated atoms, the x- and y- coordinates of the rectangle's upper left corners were chosen by perturbing that of the previously generated atom by adding/subtracting an integer randomly chosen in the interval $[0, 2\omega]$, according to randomly choosing a direction of travel (that is compatible with the x- and y- bounds of the grid representing $Space$). Moreover, the rectangle's upper left corner of the first generated atom in $\mathscr{S}_{id,t}$, was obtained by perturbing the rectangle's upper left corner of the last atom generated in $\mathscr{S}_{id,t-1}$. This way, adjacent, possibly intersecting, regions were iteratively generated. For each generated atom, the probability interval was set by choosing a random number in $[0, 1]$ for ℓ, and then a random number in $[\ell, 1]$ for u.

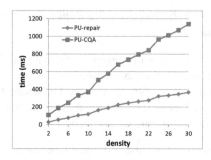
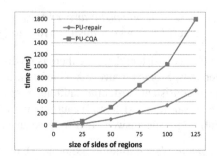

Fig. 2. Repair and CQA time vs. the density $(\omega = 75, |Space| = 1000 \times 1000)$

Fig. 3. Repair and CQA time vs. ω $(d = 16, |Space| = 1000 \times 1000)$

If, for an $\langle id, t \rangle$ pair, the database $\mathscr{S}_{id,t}$ generated using the above-described procedure was consistent, we iteratively generated a new instance until an inconsistent database was generated (this was typically achieved in less than 2 steps).

The procedure used for generating the region as well as the lower and upper bounds of queries was the same as that used to generate the region and the lower and upper bounds of the first atom in the database. In the experiments, when considering a query issued on a database having average size of the width and height of the atom's rectangles equal to ω, the width and height of the query region were randomly chosen in $[0, 2\omega]$.

All experiments have been carried out on an Intel Core Duo CPU 2.10GHz with 4GB RAM running Ubuntu 12.04 64bit. Our prototype calls the linear programming solver CPLEX for finding solutions of $PULP(\cdot)$, PU-CQA$^{\ell}(\cdot)$ and PU-CQA$^{u}(\cdot)$. All data points reported on the figures are averages over 50 trials.

Fig. 2 reports the running time needed to decide if a randomly generated $\langle id, t \rangle$ pair (with $id \in ID$ and $t \in T$) is a PU-consistent (universal cautious) answer to a randomly generated query versus the density of the database, where $\omega = 75$ and $Space$ consists of one million points. Fig. 2 also reports the running time needed to compute a PU-repair for $\mathscr{S}_{id,t}$ (a repair for $\mathscr{S}$ can be obtained by assembling repairs for $\mathscr{S}_{id,t}$, see Proposition 1). For the same size of $Space$, and density equal to 16, Fig. 3 reports the running time needed to decide if a randomly generated $\langle id, t \rangle$ pair is a PU-consistent (universal cautious) answer to a randomly generated query versus the size of the side of atom's regions. The running time needed to compute a PU-repair for $\mathscr{S}_{id,t}$ is reported as well. These experiments show that

(i) the running time for deciding whether an $\langle id, t \rangle$ pair is a PU-consistent answer is about three times that needed for computing a repair for database $\mathscr{S}_{id,t}$;

(ii) the running time for deciding if a given $\langle id, t \rangle$ pair is a PU-consistent answer (as well as that for computing repairs for $\mathscr{S}_{id,t}$) increases linearly with the density;

(iii) the running time for deciding if a given $\langle id, t \rangle$ pair is a PU-consistent answer (as well as that for computing repairs for $\mathscr{S}_{id,t}$) increases quadratically with the size of one side of the atoms' regions. That is, the running time increases linearly with the number of points in the atoms' regions, which quadratically depends on ω.

We expected (i), as to decide if $\langle id, t \rangle$ pair is a PU-consistent answer using Theorem 4, the linear programming problems $PU\text{-}CQA^\ell(\mathscr{S}, Q, id, t)$ and $PU\text{-}CQA$ $^u(\mathscr{S}, Q, id, t)$ are solved after solving $PULP(\mathscr{S}, id, t)$ to obtain o^* used in the definitions of $PU\text{-}CQA$, and these problems are of comparable sizes. Items (ii) and (iii) show that the running times linearly increase with the number of atoms in $\mathscr{S}_{id,t}$, as well as with the number of points in the atoms' regions. However, checking if an $\langle id, t \rangle$ pair is a PU-consistent answer takes less than 1 second for databases whose density is 16 and atom's regions consist of 75×75 points on average.

We also performed the kinds of experiments reported above for $|Space| = 10000 \times 10000$. However, the results we obtained are substantially the same as those shown in figures 2 and 3. Basically, this is due to the fact that in our implementation, variable v_p corresponding to point $p \in Space$ is actually added to the instances of $PULP(\mathscr{S}, id, t)$, $PU\text{-}CQA^\ell(\mathscr{S}, Q, id, t)$ and $PU\text{-}CQA^u(\mathscr{S}, Q, id, t)$ only if there is some atom in $\mathscr{S}_{id,t}$ whose region includes p. Thus, $|Space|$ does not affect the running times by itself.

6 Conclusion and Topics for Further Research

A comprehensive survey of the results on the SPOT framework can be found in [19], where several research problems awaiting further investigation were identified. However, all the previous works on SPOT assume that the database is consistent. In fact, we are not aware of any work involving inconsistent probabilistic spatio-temporal databases in any framework. In this paper we use the SPOT framework to define both the concept of database repair and consistent selection query answers. We show that some cases can be solved in PTIME. We also experimentally show the feasibility of our approach.

Future work will involve a detailed investigation of the complexity of different types of repairs as well as the complexity of checking different types of query answers. We also plan to investigate repairs and consistent answers in the presence of the spatio-temporal integrity constraints proposed in [26].

References

1. Agarwal, D., Chen, D., Lin, L.J., Shanmugasundaram, J., Vee, E.: Forecasting high-dimensional data. In: SIGMOD Conference, pp. 1003–1012 (2010)
2. Agarwal, P.K., Arge, L., Erickson, J.: Indexing moving points. J. Comput. Syst. Sci. 66(1), 207–243 (2003)
3. Akdere, M., Çetintemel, U., Riondato, M., Upfal, E., Zdonik, S.B.: The case for predictive database systems: Opportunities and challenges. In: CIDR, pp. 167–174 (2011)
4. Arenas, M., Bertossi, L.E., Chomicki, J.: Consistent query answers in inconsistent databases. In: Int. Symposium on Principles of Database Systems (PODS), pp. 68–79 (1999)
5. Benjelloun, O., Sarma, A.D., Halevy, A.Y., Widom, J.: Uldbs: Databases with uncertainty and lineage. In: VLDB, pp. 953–964 (2006)
6. Bertossi, L.: Database Repairing and Consistent Query Answering. Morgan & Claypool Publishers (2011)
7. Bertossi, L.E., Bravo, L., Franconi, E., Lopatenko, A.: The complexity and approximation of fixing numerical attributes in databases under integrity constraints. Inf. Syst. 33(4-5), 407–434 (2008)

8. Cao, H., Wolfson, O., Trajcevski, G.: Spatio-temporal data reduction with deterministic error bounds. VLDB J. 15(3), 211–228 (2006)

9. Chomicki, J., Marcinkowski, J.: Minimal-change integrity maintenance using tuple deletions. Inf. Comput. 197(1-2), 90–121 (2005)

10. Dai, X., Yiu, M.L., Mamoulis, N., Tao, Y., Vaitis, M.: Probabilistic spatial queries on existentially uncertain data. In: Medeiros, C.B., Egenhofer, M., Bertino, E. (eds.) SSTD 2005. LNCS, vol. 3633, pp. 400–417. Springer, Heidelberg (2005)

11. Fazzinga, B., Flesca, S., Furfaro, F., Parisi, F.: Cleaning trajectory data of RFID-monitored objects through conditioning under integrity constraints. In: Int. Conf. on Extending Database Technology (EDBT), pp. 379–390 (2014)

12. Fazzinga, B., Flesca, S., Furfaro, F., Parisi, F.: Offline cleaning of RFID trajectory data. In: Int. Conf. on Scientific and Statistical Database Management (SSDBM), p. 5 (2014)

13. Flesca, S., Furfaro, F., Parisi, F.: Preferred database repairs under aggregate constraints. In: Prade, H., Subrahmanian, V.S. (eds.) SUM 2007. LNCS (LNAI), vol. 4772, pp. 215–229. Springer, Heidelberg (2007)

14. Flesca, S., Furfaro, F., Parisi, F.: Consistent answers to boolean aggregate queries under aggregate constraints. In: Bringas, P.G., Hameurlain, A., Quirchmayr, G. (eds.) DEXA 2010, Part II. LNCS, vol. 6262, pp. 285–299. Springer, Heidelberg (2010)

15. Flesca, S., Furfaro, F., Parisi, F.: Querying and repairing inconsistent numerical databases. ACM Trans. Database Syst. 35(2) (2010)

16. Flesca, S., Furfaro, F., Parisi, F.: Range-consistent answers of aggregate queries under aggregate constraints. In: Deshpande, A., Hunter, A. (eds.) SUM 2010. LNCS, vol. 6379, pp. 163–176. Springer, Heidelberg (2010)

17. Grant, J., Molinaro, C., Parisi, F.: Aggregate count queries in probabilistic spatio-temporal databases. In: Liu, W., Subrahmanian, V.S., Wijsen, J. (eds.) SUM 2013. LNCS, vol. 8078, pp. 255–268. Springer, Heidelberg (2013)

18. Grant, J., Parisi, F., Parker, A., Subrahmanian, V.S.: An agm-style belief revision mechanism for probabilistic spatio-temporal logics. Artif. Intell. 174(1), 72–104 (2010)

19. Grant, J., Parisi, F., Subrahmanian, V.S.: Research in probabilistic spatiotemporal databases: The SPOT framework. In: Ma, Z., Yan, L. (eds.) Advances in Probabilistic Databases. STUDFUZZ, vol. 304, pp. 1–22. Springer, Heidelberg (2013)

20. Hadjieleftheriou, M., Kollios, G., Tsotras, V.J., Gunopulos, D.: Efficient indexing of spatiotemporal objects. In: Jensen, C.S., Jeffery, K., Pokorný, J., Šaltenis, S., Bertino, E., Böhm, K., Jarke, M. (eds.) EDBT 2002. LNCS, vol. 2287, pp. 251–268. Springer, Heidelberg (2002)

21. Hammel, T., Rogers, T.J., Yetso, B.: Fusing live sensor data into situational multimedia views. In: Multimedia Information Systems, pp. 145–156 (2003)

22. Kollios, G., Gunopulos, D., Tsotras, V.J.: On indexing mobile objects. In: Int. Symposium on Principles of Database Systems (PODS), pp. 261–272 (1999)

23. Martinez, M.V., Parisi, F., Pugliese, A., Simari, G.I., Subrahmanian, V.S.: Inconsistency management policies. In: KR, pp. 367–377 (2008)

24. Martinez, M.V., Parisi, F., Pugliese, A., Simari, G.I., Subrahmanian, V.S.: Policy-based inconsistency management in relational databases. Int. J. Approx. Reas. 55(2), 501–526 (2014)

25. Mittu, R., Ross, R.: Building upon the coalitions agent experiment (CoAx) - integration of multimedia information in gccs-m using impact. In: Multimedia Inf. Syst., pp. 35–44 (2003)

26. Parisi, F., Grant, J.: Integrity constraints for probabilistic spatio-temporal knowledgebases. In: Straccia, U., Cali, A. (eds.) SUM 2014. LNCS, vol. 8720, pp. 251–264. Springer, Heidelberg (2014)

27. Parisi, F., Parker, A., Grant, J., Subrahmanian, V.S.: Scaling cautious selection in spatial probabilistic temporal databases. In: Jeansoulin, R., Papini, O., Prade, H., Schockaert, S. (eds.) Methods for Handling Imperfect Spatial Information. STUDFUZZ, vol. 256, pp. 307–340. Springer, Heidelberg (2010)

28. Parisi, F., Sliva, A., Subrahmanian, V.S.: Embedding forecast operators in databases. In: Benferhat, S., Grant, J. (eds.) SUM 2011. LNCS, vol. 6929, pp. 373–386. Springer, Heidelberg (2011)
29. Parisi, F., Sliva, A., Subrahmanian, V.S.: A temporal database forecasting algebra. Int. J. of Approximate Reasoning 54(7), 827–860 (2013)
30. Parker, A., Subrahmanian, V.S., Grant, J.: A logical formulation of probabilistic spatial databases. IEEE TKDE, 1541–1556 (2007)
31. Parker, A., Infantes, G., Grant, J., Subrahmanian, V.S.: Spot databases: Efficient consistency checking and optimistic selection in probabilistic spatial databases. IEEE TKDE 21(1), 92–107 (2009)
32. Pelanis, M., Saltenis, S., Jensen, C.S.: Indexing the past, present, and anticipated future positions of moving objects. ACM Trans. Database Syst. 31(1), 255–298 (2006)
33. Pfoser, D., Jensen, C.S., Theodoridis, Y.: Novel approaches in query processing for moving object trajectories. In: VLDB, pp. 395–406 (2000)
34. Southey, F., Loh, W., Wilkinson, D.F.: Inferring complex agent motions from partial trajectory observations. In: IJCAI, pp. 2631–2637 (2007)
35. Tao, Y., Cheng, R., Xiao, X., Ngai, W.K., Kao, B., Prabhakar, S.: Indexing multi-dimensional uncertain data with arbitrary probability density functions. In: VLDB, pp. 922–933 (2005)
36. Tao, Y., Papadias, D., Sun, J.: The TPR*-tree: an optimized spatio-temporal access method for predictive queries. In: VLDB, pp. 790–801 (2003)

Skyline Queries in an Uncertain Database Model Based on Possibilistic Certainty

Olivier Pivert[1] and Henri Prade[2]

[1] University of Rennes 1 – Irisa, Lannion, France
[2] CNRS/IRIT – University of Toulouse, Toulouse, France
pivert@enssat.fr, prade@irit.fr

Abstract. This paper deals with skyline queries in the context of an uncertain database model where the notion of necessity is used to qualify the certainty that an ill-known piece of data takes a given value or belongs to a given subset. In this framework, skyline queries aim at computing the extent to which any tuple from a given relation is certainly not dominated by any other tuple from that relation.

1 Introduction

The last two decades have witnessed a profusion of research works on uncertain databases. Even though most authors consider probability theory as the underlying uncertainty model, some approaches rather rely on possibility theory [11], see e.g. [3]. In contrast with probability theory, one expects the following advantages when using possibility theory: i) the qualitative nature of the model makes easier the elicitation of the degrees; ii) in probability theory, the fact that the sum of the degrees from a distribution must equal 1 makes it difficult to deal with incompletely known distributions.

Recently, Bosc *et al.* [5] introduced a new model based on *possibilistic certainty*. The idea is to use the notion of necessity to qualify the certainty that an ill-known piece of data takes a given value or is in a given subset. In contrast with both probabilistic databases *and* possibilistic ones in the sense of [3], the main advantage of the certainty-based model lies in the fact that operations from relational algebra can be extended in a simple way and with a data complexity that is the same as in a classical database context.

In this paper, we move beyond relational agebra and consider a popular approach to preference queries, namely the skyline model [2] that aims to retrieve the items that are not Pareto-dominated by any other. Skyline queries have already been extended to the frameworks of databases with missing values [7], probabilistic databases [9,10], and possibilistic ones [4]. Here, we consider skyline queries in the context of the certainty-based model introduced in [5] and the objective is to assess the extent to which any tuple from a given relation is certainly not dominated by any other.

The remainder of this paper is structured as follows. Section 2 gives a brief overview of the certainty-based model. A refresher on skyline queries is provided

U. Straccia and A. Calì (Eds.): SUM 2014, LNAI 8720, pp. 280–285, 2014.

in Section 3. Section 4 defines the form that a skyline takes in the certainty-based model and the way it can be computed. Finally, Section 5 concludes the paper and outlines perspectives for future research.

2   A Short Overview of the Certainty-Based Model

As the possibilistic model described in [3], the certainty-based model [5] relies on possibility theory [11]. However, it only keeps pieces of information that are more or less certain and leaves aside what is just possible. This corresponds to the most important part of information (a possibility distribution is "summarized" by keeping its most plausible elements, associated with a certainty level). Certainty is modeled as a lower bound of a necessity measure. For instance, $\langle 037, John, (40, \alpha)\rangle$ denotes the existence of a person named John, whose age is 40 with certainty α. Then the possibility that his age differs from 40 is upper bounded by $1 - \alpha$ without further information.

The underlying possibility distribution associated with an uncertain attribute value (a, α) is $\{1/a, (1 - \alpha)/\omega\}$ where ω denotes $domain(\mathcal{A}) - \{a\}$, $\mathcal{A}$ being the attribute considered (due to the duality necessity (certainty) / possibility: $N(a) \geq \alpha \Leftrightarrow \Pi(a) \leq 1 - \alpha$). For instance, let us assume that the domain of attribute $City$ is $\{Boston, Newton, Quincy\}$. The uncertain attribute value $(Boston, \alpha)$ is assumed to correspond to the possibility distribution $\{1/Boston, (1 - \alpha)/Newton, (1 - \alpha)/Quincy\}$. The model can also deal with disjunctive uncertain values. For instance, $\langle 3, Peter, (Newton \vee Quincy, 0.8)\rangle$ represents the fact that it is 0.8-certain that the person number 3 named Peter lives in Newton or in Quincy. Then, the underlying possibility distributions π are of the form $\pi(u) = \max(A(u), 1 - \alpha)$ where A is an α-certain subset of the attribute domain and $A(u)$ equals 1 if $u \in A$, 0 otherwise.

Moreover, since some operations (e.g., the selection) may create "maybe tuples", each tuple t from an uncertain relation r has to be associated with a degree N expressing the certainty that t exists in r. It will be denoted by N/t.

Example 1. Let us consider a relation r of schema $(\#id, Name, City)$ containing tuple $t_1 = \langle 1, John, (Quincy, 0.8)\rangle$, and the query "find the people who live in Quincy". Let the domain of attribute $City$ be $\{Boston, Newton, Quincy\}$. The answer contains $0.8/t_1$ since it is 0.8 certain that t_1 satisfies the requirement, while the result of the query "find the people who live in Boston, Newton or Quincy" contains $1/t_1$ since it is totally certain that t_1 satisfies the condition. ◇

To sum up, a tuple $\alpha/\langle 037, John, (Quincy, \beta)\rangle$ from relation r means that it is α certain that person 037 exists in the relation, and that it is β certain that 037 lives in Quincy (independently from the fact that it is or not in relation r).

3   Refresher About Skyline Queries

The notion of a skyline in a set of tuples is easy to state (since it amounts to exhibit non dominated points in the sense of Pareto ordering). Assume we have:

- a given set of criteria $C = \{c_1, \ldots, c_n\}(n \geq 2)$ associated respectively with a set of attributes A_i, $i = 1, \ldots, n$;
- a complete ordering $\succeq_i$ given for each criterion i expressing preference between attribute values[1] (the case of non comparable values is left aside).

A tuple $u = (u_1, \cdots, u_n)$ in a relation r *dominates* (in the sense of Pareto) another tuple $u' = (u'_1, \cdots, u'_n)$ in r, denoted by $u \succ_C u'$, iff u is at least as good as u' in all dimensions and strictly better than u' in at least one dimension:

$$u \succ_C u' \Leftrightarrow \forall i \in \{1, \ldots, n\}, u_i \succeq_i u'_i \text{ and } \exists i \in \{1, \ldots, n\} \text{ s.t. } u_i \succ_i u'_i. \quad (1)$$

A tuple $u = (u_1, \cdots, u_n)$ in a relation r belongs to the skyline S, denoted by $u \in S$, if there is no other tuple $u' = (u'_1, \cdots, u'_n)$ in r which dominates it (the skyline query returns the Pareto frontier):

$$u \in S \Leftrightarrow \forall u', \neg(u' \succ_C u). \quad (2)$$

Then any tuple u' is either dominated by u, or is non comparable with u. The following example uses the syntax of the language *Preference SQL* [8], which is a typical representative of a Pareto-order-based approach.

Table 1. An extension of relation *car*

	make	*category*	*price*	*color*	*mileage*
t_1	Opel	roadster	4500	blue	20,000
t_2	Ford	SUV	4000	red	20,000
t_3	VW	roadster	5000	red	10,000
t_4	Opel	roadster	5000	red	8000
t_5	Fiat	roadster	4500	red	16,000
t_6	Kia	coupe	5500	blue	24,000
t_7	Seat	sedan	4000	green	12,000
t_8	VW	sedan	3500	black	7500

Example 2. Let us consider a relation *car* of schema (*make, category, price, color, mileage*) whose extension is given in Table 1, and the query:

select * **from** *car* **where** *color* $\neq$ 'black'
preferring
(*make* = 'VW' **else** *make* = 'Seat' **else** *make* = 'Opel' **else** *make* = 'Ford') **and**
(*category* = 'sedan' **else** *category* = 'roadster' **else** *category* = 'coupe') **and**
(**least** *price*) **and** (**least** *mileage*);

In this query, "$A_i = v_{1,1}$ *else* $A_i = v_{1,2}$" means that value $v_{1,1}$ is strictly preferred to value $v_{1,2}$ for attribute A_i. It is assumed that any domain value which is absent from a preference clause is less preferred than any value explicitly specified in the clause (but it is not absolutely rejected). Here, the tuples that are not dominated in the sense of the *preferring* clause are $\{t_3, t_4, t_7\}$. Indeed, t_7 dominates t_1, t_2, and t_5, whereas every tuple dominates t_6 except t_2. Notice that t_8 is discarded from the start as it does not satisfy the condition from the *where* clause. $\diamond$

---

[1] $u \succ v$ means u is preferred to v. $u \succeq v$ means u is at least as good as v, i.e., $u \succeq v \Leftrightarrow u \succ v \lor u \approx v$, where $\approx$ denotes indifference.

4 Certainty-Based Skyline

4.1 Case of an Initial Relation

Let us first assume that the relation concerned, denoted by r, is an initial one, which implies that $\forall t \in r$, $N(t) = 1$. Let us consider two tuples u and u' of r. We denote: $u.A_i = (u_i, \rho_i)$ and $u'.A_i = (u'_i, \rho'_i)$ where u_i (resp. u'_i) is a candidate value (or a subset of values) and ρ_i (resp. ρ'_i) is the associated certainty degree. For instance, if $u.A_i = (Newton \lor Quincy, 0.8)$, then $u_i = \{Newton, Quincy\}$ and $\rho_i = 0.8$. Let us consider the event $(u.A_i \, \theta_i \, u'.A_i)$ where θ_i is $\succ_i$, $\succeq_i$ or $\approx_i$. We denote by $c(E)$ the certainty degree associated with event E. According to the min-decomposability of necessity in possibility theory, we have:

$$c(u.A_i \, \theta_i \, u'.A_i) = \min(\rho_i, \, \rho'_i, \, \min_{(x,y) \in u_i \times u'_i} \mu_{\theta_i}(x, y)) \tag{3}$$

where $\mu_{\theta_i}(x, y) = 1$ if $x \, \theta_i \, y$, 0 otherwise.

Example 3. With the query of Example 2, we have: $c((VW \lor Opel, 0.7) \succ_{make} (Ford, 0.4)) = \min(0.7, 0.4) = 0.4$ as both VW and Opel are preferred to Ford. On the other hand, $c((VW \lor Opel, 0.7) \succ_{make} (Seat, 0.4)) = 0$ as $(Opel \succ_{make} Seat)$ is false. $\diamond$

The expression of $c(u \succ_C u')$ stems straightforwardly from Formula (1):

$$c(u \succ_C u') = \min(\min_i c(u.A_i \succeq_i u'.A_i), \, \max_i c(u.A_i \succ_i u'.A_i)) \tag{4}$$

Example 4. Let us consider the preferences from Example 2 and the tuples t_3 and t_4 from Table 2. We have: $\min_i c(t_3.A_i \succeq_i t_4.A_i) = c((10{,}000, 0.8) \succ_{mileage} (8000, 1)) = 0$, and $\max_i c(t_3.A_i \succ_i t_4.A_i) = c((VW, 1) \succ_{make} (Opel, 0.7)) = 0.7$. Thus, $c(t_3 \succ_C t_4) = \min(0, 0.7) = 0$. $\diamond$

As for $c(u \approx_C u')$, which corresponds to the certainty that u is either incomparable or equally preferred to u', it stems from:

$$u \approx_C u' \Leftrightarrow ((\exists i \text{ such that } u.A_i \succ_i u'.A_i) \land (\exists i \text{ such that } u'.A_i \succ_i u.A_i)) \\ \lor (\forall i, \, u.A_i \approx_i u'.A_i) \tag{5}$$

and we get:

$$c(u \approx_C u') = \max(\min(\max_i c(u.A_i \succ_i u'.A_i), \, \max_i c(u'.A_i \succ_i u.A_i)), \\ \min_i c(u.A_i \approx_i u'.A_i)). \tag{6}$$

Example 5. Consider again t_3 and t_4 from Table 2. We have: $\max_i c(t_3.A_i \succ_i t_4.A_i) = c((VW, 1) \succ_{make} (Opel, 0.7)) = 0.7$, $\max_i c(t_4.A_i \succ_i t_3.A_i) = c((8000, 1) \succ_{mileage} (10{,}000, 0.8)) = 0.8$, and $\min_i c(t_3.A_i \approx_i t_4.A_i) = c((VW, 1) \approx_{make} (Opel, 0.7)) = 0$. Thus, $c(u \approx_C u') = \max(0.7, 0) = 0.7$. $\diamond$

The skyline obtained is represented as a fuzzy set of points, let us denote it by S'. The degree of certainty of the event "u belongs to S'" is defined as:

$$\mu_{S'(r)}(u) = \min_{u' \in r,\, u' \neq u} \max(c(u \succ_C u'),\ c(u \approx_C u')). \tag{7}$$

Formula (7) implies that data complexity is in $\theta(n^2)$ — the tuples of r have to be compared pairwise — as in a classical database context.

Remark 1. We have: $support(S'(r)) = S(r_{0+})$ (where $S(r_{0+})$ denotes the regular skyline computed on r where all the certainty degrees have been replaced by 1). Indeed, due to (3), $(\exists u' \mid c(u \succ_C u') = 0 \vee (c(u \approx_C u') = 0)$ is equivalent to $(\exists u' \mid u' \succ_C u)$, i.e., $u \notin S$.

Table 2. Relation *car* with uncertain attribute values

	make	category	price	mileage
t_1	(Opel, 1)	(roadster, 0.8)	(4500, 1)	(20,000, 0.8)
t_2	(Ford, 0.9)	(SUV, 1)	(4000, 0.7)	(20,000, 0.6)
t_3	(VW, 1)	(roadster, 0.4)	(5000, 0.9)	(10,000, 0.8)
t_4	(Opel, 0.7)	(roadster, 1)	(5000, 0.3)	(8000, 1)
t_5	(Fiat, 0.6)	(roadster, 1)	(4500, 0.8)	(16,000, 0.9)
t_6	(Kia $\vee$ Ford, 1)	(coupe, 1)	(5500, 1)	(24,000, 1)
t_7	(Seat, 1)	(sedan, 1)	(4000 $\vee$ 3800, 0.8)	(12,000, 0.7)

Example 6. Let us consider the query from Example 2 (minus the *where* clause) along with the data from Table 2 (where the values in brackets correspond to certainty degrees). The skyline obtained is: $\{0.4/t_3, 0.3/t_4, 0.6/t_7\}$. ◇

Remark 2. Let us denote by r_α the α-cut of relation r. For instance, considering relation *car* depicted in Table 2, $car_{0.7}$ contains the tuples ⟨Opel, roadster, 4500, 20,000⟩, ⟨Kia $\vee$ Ford, coupe, 5500, 24,000⟩, and ⟨Seat, sedan, 4000 $\vee$ 3800, 12,000⟩ stemming from t_1, t_6 and t_7 respectively. Notice that a user may be interested in computing $S(r_\alpha)$ if s/he cares only about the nondominated tuples among those whose values are at least α certain. Here, for instance, we have $S(car_{0.7}) = \{t_7\}$. An interesting issue would be to compare the approach presented above (certainty-based skyline) with the computation of the classical skyline on different certainty level cuts of the database.

4.2   Case of a Relation Resulting from a Selection

In *Preference SQL* [8], queries may involve a *where* clause (cf. Example 1), which is used to filter out some tuples before computing the skyline. In the context considered here, this may produce *maybe tuples* (cf. the degree N introduced in Section 2). In this case, the certainty degrees attached to the tuples must be taken into account in the computation of the skyline. We have:

$$c(\alpha/u \succ_C \beta/u') = c(u \text{ exists and } v \text{ exists and } u \text{ dominates } u')$$

which leads to $c(\alpha/u \succ_C \beta/u') = \min(\alpha, \beta, \mu)$ where μ is computed by means of Equation (4). Similarly, we have:

$c\left(\alpha/u \approx_C \beta/u'\right) = c\left(u \text{ exists and } v \text{ exists and}\right.$
$$u \text{ is incomparable or equally preferred to } u')$$

which leads to $c\left(\alpha/u \approx_C \beta/u'\right) = \min(\alpha,\ \beta,\ \mu')$ where μ' is computed using Equation (6). As for the computation of the skyline, it is based on the following expression that extends Formula (7):

$$\mu_{S'(r)}(u) = \min_{u' \in r,\, u' \neq u} \max(c\left(\alpha/u \succ_C \beta/u'\right),\ c\left(\alpha/u \approx_C \beta/u'\right)) \qquad (8)$$

where α (resp. β) denotes the certainty degree attached to tuple u (resp. u') in relation r before computing the skyline.

5 Conclusion

In this paper, we have investigated skyline queries in the framework of an uncertain database model where the notion of necessity is used to qualify the certainty that an ill-known piece of data takes a given value or belongs to a given subset. In this framework, skyline queries aim at computing the extent to which any tuple from a given relation is certainly not dominated by any other tuple.

Among the perspectives opened by this work, let us mention: i) the study of a possible adaptation of classical optimization techniques (see e.g. [1]) to the processing of skyline queries in the certainty-based model; ii) the extension of the approach to a *graded* dominance relation, see e.g. [6].

References

1. Bartolini, I., Ciaccia, P., Patella, M.: Efficient sort-based skyline evaluation. ACM Trans. Database Syst. 33(4), 1–49 (2008)
2. Börzsönyi, S., Kossmann, D., Stocker, K.: The skyline operator. In: Proc. of ICDE 2001, pp. 421–430 (2001)
3. Bosc, P., Pivert, O.: About projection-selection-join queries addressed to possibilistic relational databases. IEEE Trans. on Fuzzy Systems 13(1), 124–139 (2005)
4. Bosc, P., Hadjali, A., Pivert, O.: On possibilistic skyline queries. In: Christiansen, H., De Tré, G., Yazici, A., Zadrozny, S., Andreasen, T., Larsen, H.L. (eds.) FQAS 2011. LNCS, vol. 7022, pp. 412–423. Springer, Heidelberg (2011)
5. Bosc, P., Pivert, O., Prade, H.: A model based on possibilistic certainty levels for incomplete databases. In: Godo, L., Pugliese, A. (eds.) SUM 2009. LNCS, vol. 5785, pp. 80–94. Springer, Heidelberg (2009)
6. Hadjali, A., Pivert, O., Prade, H.: On different types of fuzzy skylines. In: Kryszkiewicz, M., Rybinski, H., Skowron, A., Raś, Z.W. (eds.) ISMIS 2011. LNCS, vol. 6804, pp. 581–591. Springer, Heidelberg (2011)
7. Khalefa, M.E., Mokbel, M.F., Levandoski, J.J.: Skyline query processing for incomplete data. In: Proc. of ICDE 2008, pp. 556–565 (2008)
8. Kießling, W., Köstler, G.: Preference SQL – design, implementation, experiences. In: Proc. of VLDB 2002, pp. 990–1001 (2002)
9. Pei, J., Jiang, B., Lin, X., Yuan, Y.: Probabilistic skylines on uncertain data. In: Proc. of VLDB 2007, pp. 15–26 (2007)
10. Yong, H., Kim, J., Hwang, S.: Skyline ranking for uncertain data with maybe confidence. In: ICDE Workshops, pp. 572–579. IEEE Computer Society (2008)
11. Zadeh, L.: Fuzzy sets as a basis for a theory of possibility. Fuzzy Sets and Systems 1(1), 3–28 (1978)

Querying Uncertain Multiple Sources

Olivier Pivert[1] and Henri Prade[2]

[1] University of Rennes 1, Irisa
Lannion France
[2] IRIT-CNRS, University of Toulouse
Toulouse, France
pivert@enssat.fr, prade@irit.fr

Abstract. Information often comes from multiple sources that may be conflicting. This makes uncertain the answers of a query to a set of sources. Possibility theory-based approaches to the handling of uncertainty in databases have been proposed and developed for a long time, in the case of a unique source of information. A multiple source counterpart of possibility theory has been recently proposed. Possibility and necessity set functions are then valued in terms of (possibly fuzzy) subsets of sources. Uncertainty may be assessed here either in terms of global reliability levels of sources or of tuples inside a source. When each source contains precise attribute values, each tuple is associated with the subset of sources that supports it as being an answer to a considered query, and with the subset of sources according to which the tuple is not an answer. In fact, these subsets of sources are fuzzy as they reflect the reliability levels. The benefit of the approach is to rank-order the answers to a query on a qualitative basis, in terms of subsets of sources and reliability levels.

Keywords: Distributed data sources, uncertainty, database querying.

1 Introduction

Managing inconsistent information has become a very important issue at a time where the amount and the variety of pieces of information to be handled considerably increase the possibilities of inconsistency. For this reason, inconsistency, after uncertainty, has appeared on the agenda of both artificial intelligence and database research for more than a decade now. Still, this issue is addressed in a different perspective by the two research fields, even if in both cases, the proposed approaches can be roughly divided between those which try to restore or maintain consistency and those which accept inconsistent information trying to "draw the best" from it.

In this paper, we take a database point of view and investigate the issue of querying distributed sources that may be mutually inconsistent. Indeed, it is well-known that when a database results from the integration of multiple data sources, or (case considered here) when the data comes from unverified sources or is uncertain, the resulting database generally contains inconsistencies [10].

U. Straccia and A. Calì (Eds.): SUM 2014, LNAI 8720, pp. 286–291, 2014.

Extracting consistent and useful pieces of information from various sources is the major task of information fusion [6]. Moreover, information delivered by sources may be uncertain or imprecise, and then these aspects have to be taken into account in information fusion. Several information fusion operators have been proposed in the literature for combining potentially conflicting pieces of information [12]. However, in [2], the authors use formal concept analysis and pattern structures to associate subsets of sources to combination results obtainable from consistent subsets of pieces of information. This may be better than computing a unique answer as the union of these combination results (which is what fusion operators would do).

In the present work, we keep the idea of associating possibly conflicting answers with subsets of sources. We choose to keep the data as they are (no fusion of any kind is performed) and we aim to inform the user about inconsistencies when answering a query (according to a philosophy somewhat similar to Consistent Query Answering [4] in the sense that one accepts to "live with dirty data" and try to draw the best from them). The presence of inconsistent data may of course lead to "suspect answers" when querying such distributed data sources. Let us assume for instance that we have available two sources S_1 and S_2 that contain a relation Emp of schema (*ss_number, name, age, city, job*) describing employees of given company. If Emp in S_1 contains the tuple ⟨17, John, 35, Boston, clerk⟩ and Emp in S_2 contains ⟨17, John, 37, Quincy, clerk⟩, then 17 is a suspect answer to the query "find the people who live in Boston" since S_2 believes that this employee rather lives in Quincy. Notice that, on the other hand, 17 would not be considered a suspect answer to the query "find the employees whose age is greater than 30", even though the sources do not agree on John's age since in both sources employee 17 is described as being older than 30. In the following, we assume that in each source, each tuple is associated with a reliability level (either directly attached to it or inherited from the source), corresponding to a certainty degree assessed on a qualitative scale.

The remainder of the paper is structured as follows. In Section 2, we present the principle of the approach. Section 3 discusses related work. Section 4 concludes the paper and outlines some perspectives for future work.

2 Principle of the Approach

The principle of the approach is inspired by [15] where suspect answers (i.e. answers that may be involved in the violation of a functional dependency) to a query Q adressed to a centralized database are detected by checking if they are both in the result of Q and in the result of the "opposite query" $\overline{Q}$.

Let Q be a query addressed to a set of sources $S = \{S_1, \ldots, S_n\}$. Data mediation is not the issue here, so we assume that each source includes a relation r with the same attribute names (but there may exist inconsistencies between the contents of r in the different sources). Query Q is assumed to be of the form $\pi_X(\sigma_\varphi(r))$ where π denotes the projection operation, σ the selection operation, φ is a condition on Y, and X and Y denote two subsets of attributes of r.

It is also assumed that the functional dependency $X \to Y$ is valid on r. Query Q aims to retrieve the X-component of the tuples of r that satisfy condition φ. In the following, for the sake of simplicity, we assume that X is a key of r or a superset of such a key. Let $\overline{Q} = \pi_X(\sigma_{\neg\varphi}(r))$ be the opposite query (the selection condition is negated). In each source S_i, every t is assumed to be associated with a certainty degree $c_i(t)$ expressed on a qualitative scale, for instance

$$\mathcal{L} = \; < \tau_1 = \textit{very uncertain} < \tau_2 = \textit{rather uncertain}$$
$$< \tau_3 = \textit{rather certain} < \tau_4 = \textit{very certain} < \tau_5 = \textit{totally certain}\rangle.$$

One gets the sets of answers to Q stemming from each source: $A_1, \ldots, A_n$, and one does the same thing for the answers to $\overline{Q}$: $\overline{A_1}, \ldots, \overline{A_n}$:

$$A_i = \{c_i(t)/x \mid t \in S_i \wedge x = t.X \wedge Q(x)\}$$
$$\overline{A_i} = \{c_i(t)/x \mid t \in S_i \wedge x = t.X \wedge \overline{Q}(x)\}.$$

In other words, an answer set is a fuzzy set of the form $\{c_i(t)/x\}$ where t is the tuple from r whose key value is x ($t.X = x$) and $c_i(t) \in \mathcal{L}$. In the following, we use $c_i(x)$ in place of $c_i(t)$ (let us recall that X is assumed to include a key of r).

For every tuple t of the union of the A_i's, one determines two fuzzy sets:

$$V^+(x) = \{\tau_1/F_1^+(x), \ldots, \tau_m/F_m^+(x)\}$$
$$V^-(x) = \{\tau_1/F_1^-(x), \ldots, \tau_m/F_m^-(x)\}$$

where m is the number of levels in $\mathcal{L}$ (i.e., τ_m corresponds to "totally certain") and

$$F_i^+ = \{S_j \in S \mid x \in A_j \wedge c_j(x) = \tau_i\},$$
$$F_i^- = \{S_j \in S \mid x \in \overline{A_j} \wedge c_j(x) = \tau_i\}.$$

This can be related to the approach proposed in [3] where the authors outline a joint extension of a multiple agent logic and possibilistic logic [11]. In this extended logic, propositions are associated with both sets of agents and certainty levels. A formula $(a, \alpha/A)$ then expresses that "all agents in set A are certain at least at level α that a is true". More generally, the language is then made of formulas of the form (a, F) where F is a fuzzy set of agents. The degree of membership $\mu_F(k)$ is the minimal degree of certainty of a for agent k. The set of all agents is partitioned by F into subgroups of agents F_i having the same certainty level α_i associated with proposition a; then F can be viewed as a weighted union $\bigcup_i \alpha_i/F_i$, where all the α_i's are distinct and strictly positive, and the F_i's are classical, mutually disjoint subsets.

One must then rank-order the answers from the most certain to the least certain. Let us denote:

- $F^+(x) = \langle n_m^+, \ldots, n_1^+ \rangle$, and
- $F^-(x) = \langle n_m^-, \ldots, n_1^- \rangle$,

where $n_i^+ = card(F_i^+(x))$, and $n_i^- = card(F_i^-(x))$.

Example 1. Let us consider ten sources S_1, ..., S_{10}. Let us assume that $m = 5$ and consider a query Q and three answers x_1, x_2 and x_3 such that:

- $V^+(x_1) = \{\tau_5/\{S_3\}, \tau_3/\{S_1, S_4\}, \tau_2/\{S_7\}\}$,
- $V^-(x_1) = \{\tau_5/\{S_2, S_6, S_{10}\}, \tau_4/\{S_8\}, \tau_1/\{S_9\}\}$,
- $V^+(x_2) = \{\tau_5/\{S_1\}, \tau_4/\{S_8\}, \tau_2/\{S_7\}, \tau_1/\{S_3, S_6\}\}$,
- $V^-(x_2) = \{\tau_3/\{S_{10}\}, \tau_2/\{S_2, S_9\}\}$,
- $V^+(x_3) = \{\tau_5/\{S_{10}, S_5\}, \tau_3/\{S_4\}, \tau_2/\{S_1\}\}$,
- $V^-(x_3) = \{\tau_5/\{S_2, S_3\}, \tau_4/\{S_6\}, \tau_1/\{S_7\}\}$.

One has:

- $F^+(x_1) = \langle 1, 0, 2, 1, 0\rangle$,
- $F^-(x_1) = \langle 3, 1, 0, 0, 1\rangle$,
- $F^+(x_2) = \langle 1, 1, 0, 1, 2\rangle$,
- $F^-(x_2) = \langle 0, 0, 1, 2, 0\rangle$,
- $F^+(x_3) = \langle 2, 0, 1, 1, 0\rangle$,
- $F^-(x_3) = \langle 2, 1, 0, 0, 1\rangle$. ◇

The idea for rank-ordering two items x and x' is to compare $F^+(x)$ with $F^+(x')$ on the one hand, and $F^-(x)$ with $F^-(x')$ on the other hand. Formally, the statement "x is a better (more reliable) answer than x'" is defined as follows:

$$x \succ x' \Leftrightarrow ((F^+(x) >^{lex} F^+(x')) \wedge (F^-(x) \leq^{lex} F^-(x'))) \vee$$
$$((F^+(x') \geq^{lex} F^+(x)) \wedge (F^-(x) <^{lex} F^-(x'))). \qquad (1)$$

where $>^{lex}$ is the lexicographic ordering. Obviously, we only get a partial order and some answers may be incomparable.

Example 2. Let us consider again the context and data of Example 1. We get the ordering: $\{x_2, x_3\} \succ x_1$. Tuples x_2 and x_3 are incomparable since one has both $F^+(x_3) >^{lex} F^+(x_2)$ and $F^-(x_3) >^{lex} F^-(x_2)$. ◇

The partial order obtained may always be linearized by, for instance, applying the maximum specificity principle (i.e. by ranking each item as high as permitted by the order constraints). This amounts in the above example to putting x_1 and x_2 at the same level.

3 Related Work

As mentioned in [14] — which presents a system for integrating multiple heterogeneous and autonomous information sources that uses *data fusion* to resolve factual inconsistencies among the individual sources —, several approaches have attempted to resolve extensional conflicts based on the content of the conflicting data and possibly some probabilistic information that is assumed to be available. They either detect the existence of data inconsistencies and provide their users with some additional information on their nature (e.g., [1]), or they try to resolve such conflicts by returning a probabilistic value: a set of alternative values with attached probabilities, see e.g. [16,13,17].

Another family of approaches is based on the concept of consistent query answering (CQA) [4] that aims at retrieving only the *certain* answers to a given query, see e.g. [5] for a recent work on the topic that considers ontology-based data access (OBDA) systems and proposes two families of inconsistency-tolerant semantics which approximate the CQA semantics. In [9], the authors use defeasible logic programming and argumentative reasoning for computing consistent answers to yes/no questions in the presence of conflicting data sources.

To the best of our knowledge, the approach presented here is the first one that makes it possible to return all the candidate answers to a query, rank-ordered according to their level of reliability.

4   Conclusion

In this paper, we have investigated the issue of querying several data sources that may be conflicting. We have considered the case where a level of reliability is associated with each source, or with each tuple inside a source. Instead of performing a fusion of the data sources, the approach described aims to detect the suspect answers to a selection-projection query and to rank-order the answers according to i) the tuples from the base relations that support them and ii) the tuples from the base relations that contradict them. More precisely, each answer is associated with the (fuzzy) subset of sources that supports it as being an answer to the considered query, and with the (fuzzy) subset of sources according to which the tuple is not an answer. A technique based on a double lexicographic ordering is then used to rank-order the answers.

Among the perspectives for future work, let us mention the two following ones. First, it would be interesting to extend the approach to the case where the sources may contain *imprecise* or missing attribute values. These imprecise values may also be uncertain, i.e., such an imprecise value is associated with a subset that is not a singleton, and there is a nonzero possibility that the value is outside the subset. One could then make use of the possibilistic database model presented in [7], or alternatively that introduced in [8] where a certainty level (interpreted as a lower bound of a necessity degree) is associated with each attribute value. Then, one would have to consider another (fuzzy) subset of sources that only support that it is possible that a considered tuple is an answer. A second challenge is to extend the approach in order to deal with a larger range of queries.

References

1. Agarwal, S., Keller, A.M., Wiederhold, G., Saraswat, K.: Flexible relation: An approach for integrating data from multiple, possibly inconsistent databases. In: Yu, P.S., Chen, A.L.P. (eds.) ICDE, pp. 495–504. IEEE Computer Society (1995)

2. Assaghir, Z., Napoli, A., Kaytoue, M., Dubois, D., Prade, H.: Numerical information fusion: Lattice of answers with supporting arguments. In: ICTAI, pp. 621–628. IEEE (2011)

3. Belhadi, A., Dubois, D., Khellaf-Haned, F., Prade, H.: Multiple agent possibilistic logic. Journal of Applied Non-Classical Logics 23(4), 299–320 (2013)

4. Bertossi, L.E.: Consistent query answering in databases. SIGMOD Record 35(2), 68–76 (2006)

5. Bienvenu, M., Rosati, R.: Tractable approximations of consistent query answering for robust ontology-based data access. In: Rossi, F. (ed.) IJCAI. IJCAI/AAAI (2013)

6. Bloch, I., Hunter, A., Appriou, A., Ayoun, A., Benferhat, S., Besnard, P., Cholvy, L., Cooke, R.M., Cuppens, F., Dubois, D., Fargier, H., Grabisch, M., Kruse, R., Lang, J., Moral, S., Prade, H., Saffiotti, A., Smets, P., Sossai, C.: Fusion: General concepts and characteristics. Int. J. Intell. Syst. 16(10), 1107–1134 (2001)

7. Bosc, P., Pivert, O.: About projection-selection-join queries addressed to possibilistic relational databases. IEEE T. Fuzzy Systems 13(1), 124–139 (2005)

8. Bosc, P., Pivert, O., Prade, H.: A model based on possibilistic certainty levels for incomplete databases. In: Godo, L., Pugliese, A. (eds.) SUM 2009. LNCS, vol. 5785, pp. 80–94. Springer, Heidelberg (2009)

9. Deagustini, C.A.D., Fulladoza Dalibón, S.E., Gottifredi, S., Falappa, M.A., Simari, G.R.: Consistent query answering using relational databases through argumentation. In: Liddle, S.W., Schewe, K.-D., Tjoa, A.M., Zhou, X. (eds.) DEXA 2012, Part II. LNCS, vol. 7447, pp. 1–15. Springer, Heidelberg (2012)

10. Decker, H., Martinenghi, D.: Avenues to flexible data integrity checking. In: DEXA Workshops, pp. 425–429. IEEE Computer Society (2006)

11. Dubois, D., Lang, J., Prade, H.: Possibilistic logic. In: Gabbay, D., Hogger, C., Robinson, J. (eds.) Handbook of Logic in Artificial Intelligence and Logic Programming, vol. 3. Oxford University Press, Oxford (1994)

12. Dubois, D., Prade, H.: Possibility theory and data fusion in poorly informed environments. Control Eng. Practice 2, 811–823 (1994)

13. Genesereth, M.R., Keller, A.M., Duschka, O.M.: Infomaster: An information integration system. In: Peckham, J. (ed.) SIGMOD Conference, pp. 539–542. ACM Press (1997)

14. Motro, A., Anokhin, P.: Fusionplex: resolution of data inconsistencies in the integration of heterogeneous information sources. Information Fusion 7(2), 176–196 (2006)

15. Pivert, O., Prade, H.: Detecting suspect answers in the presence of inconsistent information. In: Lukasiewicz, T., Sali, A. (eds.) FoIKS 2012. LNCS, vol. 7153, pp. 278–297. Springer, Heidelberg (2012)

16. Tseng, F.S.C., Chen, A.L.P., Yang, W.P.: A probabilistic approach to query processing in heterogeneous database systems. In: Yu, P.S. (ed.) RIDE-TQP, pp. 176–183. IEEE Computer Society (1992)

17. Yu, J., Wessels, D., Li, Y.Q.: Fuzzy algorithm for selection of reliable information from a collection of different data sources. In: Klein, B.D., Rossin, D.F. (eds.) IQ, pp. 76–82. MIT (2000)

Population Size Extrapolation in Relational Probabilistic Modelling[*]

David Poole[1], David Buchman[1], Seyed Mehran Kazemi[1], Kristian Kersting[2],
and Sriraam Natarajan[3]

[1] University of British Columbia
{poole,davidbuc,smkazemi}@cs.ubc.ca
www.cs.ubc.ca/~{poole,davidbuc,smkazemi}/
[2] Technical University of Dortmund
http://www-ai.cs.uni-dortmund.de/PERSONAL/kersting.html
[3] Indiana University
http://homes.soic.indiana.edu/natarasr/

Abstract. When building probabilistic relational models it is often difficult to determine what formulae or factors to include in a model. Different models make quite different predictions about how probabilities are affected by population size. We show some general patterns that hold in some classes of models for all numerical parametrizations. Given a data set, it is often easy to plot the dependence of probabilities on population size, which, together with prior knowledge, can be used to rule out classes of models, where just assessing or fitting numerical parameters will be misleading. In this paper we analyze the dependence on population for relational undirected models (in particular Markov logic networks) and relational directed models (for relational logistic regression). Finally we show how probabilities for real data sets depend on the population size.

1 Introduction

Relational probabilistic models [4,17] or template-based models [10] represent the probabilistic dependencies between relations of individuals. In these models, individuals about which we have the same information are exchangeable (i.e. the individuals are treated identically when we have no evidence to distinguish them) and the probabilities are about relations among individuals, which can be specified independently of actual individuals.

In a relational probabilistic model, the predictions of the model may depend on the number of individuals (the population size). For instance, whether someone enjoys a party or not may depend on the number of people they know at that party, and each person at a party may know a different number of people.

Even simple models make strong predictions about the effect of population size on probabilities. If we want to extrapolate from data (as opposed to interpolating), it is important to know how the models handle changes in population

---

[*] Parts of this paper appeared in the UAI-2012 StarAI workshop [16].

U. Straccia and A. Calì (Eds.): SUM 2014, LNAI 8720, pp. 292–305, 2014.

size. Extrapolating from small sample sizes to large ones can be very presumptu-ous, e.g., people act very differently in small groups than in mobs. The structure of the model reflects implicit prior knowledge and assumptions, which are impor-tant to understand. We advocate that we should choose from the models where the extrapolation is reasonable given the data and prior knowledge.

We consider two classes of relational models, undirected models exemplified by Markov logic networks (MLNs) [18,2], and directed models with aggregators exemplified by relational logistic regression (RLR) [9], the directed analogue of MLNs.

This work is complementary to the work of Jain et al. [8,7], who allow weights to vary with the population. Varying weights may be necessary for a particular domain, but from a modeling perspective it is first important to understand what happens when weights are not varied. This paper mainly considers what happens as the population varies, rather that just the limiting probabilities [6].

In the rest of the paper, we first introduce some basic definitions and describe MLNs and RLR. Then we consider a simple model and explain how RLR models and MLNs are influenced by population size and how they behave differently even for this simple model. We then expand these results to more complicated cases, and give some general theoretical results, some empirical data and many open problems.

2   Some Basic Definitions

A **population** is a set of **individuals**. A population corresponds to a domain in logic. The **population size** is the cardinality of the population which can be any non-negative integer. For this paper we assume the populations are disjoint; each individual is only in one population. When there is a single population, we use n for the population size, and write the population as $A_1 \ldots A_n$.

Each **logical variable**, written in lower case, is typed with a population. $pop(x)$ is the population associated with the logical variable x, and $|x| = |pop(x)|$. Constants, denoting individuals, start with an upper case letter. We assume there is a constant for each individual, and there is no uncertainty about the identity of the individuals.

A **parametrized random variable (PRV)** is of the form $F(t_1, \ldots, t_k)$ where F is a k-ary predicate symbol and each t_i is a logical variable or a constant. For example, $At(x, y)$, $At(x, Home)$, $At(Sam, Home)$ are PRVs. The range of the random variables is $\{False, True\}$. (It is possible to have PRVs with more general domains, but the points of the paper can already be made in this simpler setting.) A **ground random variable** is a PRV where all t_i are constants.

An **atom** is an assignment of a value to a PRV. For example, $At(x, Home) = True$ is an atom. We will write assignments in lower case; $R(x) = True$ is written as $r(x)$, and $R(x) = False$ is written as $\neg r(x)$. A **formula** is made up of atoms with logical connectives (we ignore quantification in this paper.) An **instance of a formula** is obtained by replacing logical variables with constants.

A **world** is an assignment of a value to each ground random variable. The number of worlds is exponential in the number of ground random variables.

3   Markov Logic Networks and Relational Logistic Regression

Markov logic networks (MLNs) [18,2] and relational logistic regression (RLR) [9] are defined in terms of weighted formulae. In MLNs the formulae are used to define joint probability distributions. In RLR the formulae are used to define conditional probabilities.

A **weighted formula** (WF) is a triple $\langle L, F, w \rangle$ where L is a set of logical variables, F is a formula where all of the free logical variables in F are in L, and w is a real-valued weight.

An MLN is a set of weighted formulae[1], where the probability of any world is proportional to the exponent of the sum of the weights of the instances of the formulae that are true in the world.

RLR is a form of aggregation, defining conditional probabilities in terms of weighted formulae. We assume a directed acyclic graph on PRVs (where the PRVs of different nodes do not unify), which defines a Bayesian network on the corresponding ground random variables. For each PRV, there are weighted formulae involving an instance of that PRV and PRVs involving instances of (a subset of) the parent PRVs. The conditional probability of each ground random variable given an assignment of values to each of its parent ground random variables is proportional[2] to the exponential of the sum of the weights of the instances of the formulae that are true for that assignment.

Example 1. Suppose we have the weighted formulae:

$$\langle \{\}, q, \alpha_0 \rangle$$
$$\langle \{x\}, q \wedge \neg r(x), \alpha_1 \rangle$$
$$\langle \{x\}, q \wedge r(x), \alpha_2 \rangle$$
$$\langle \{x\}, r(x), \alpha_3 \rangle$$

Treating this as an MLN, if the truth value for $r(x)$ for every individual x is observed:

$$P(q \mid obs) = sigmoid(\alpha_0 + n_F \alpha_1 + n_T \alpha_2) \tag{1}$$

where *obs* has $R(x)$ true for n_T individuals, and false for n_F individuals out of a population of $n = n_F + n_T$ individuals. $sigmoid(x)$ is $1/(1 + e^{-x})$.

Note that, in the MLN, α_3 is not required for representing the conditional probability (because it cancels out), but can be used to affect $P(r(A_i))$.

---

[1] MLNs typically do not explicitly include the set of logical variables as part of the weighted formulae, but use the free variables in F. If one wanted to add an extra logical variable, x, one could conjoin $true(x)$ to F where $true$ is a property that is true for all individuals.

[2] In MLNs there is a single normalizing constant, guaranteeing the probabilities of the worlds sum to 1. In RLR, normalization is done separately for each possible assignment to the parents.

In [9], the sigmoid, as in Equation (1), is used as the definition of RLR. ([9] assumed all formulae were conjoined with $q\wedge$, and omitted $q\wedge$ from the formulae.) When not all $R(A_i)$ are observed, RLR uses Equation (1) for the conditional probability of q given each combination of assignments to the $R(x)$, and requires a separate model for the probability of the $R(x)$.

In summary: RLR uses the weighted formulae to define the conditional probabilities, and MLNs use them to define the joint probability distribution.

Example 2. Suppose people want to go to a party, and the party is fun for them if they know at least one social person in the party. In this case, a PRV $funFor(x)$ is a child of PRVs $knows(x,y)$ and $social(y)$. The following weighted formulae can be used to model the dependence of $funFor(x)$ on its parents:

$$\langle\{x\}, funFor(x), -5\rangle$$
$$\langle\{x,y\}, funFor(x) \wedge knows(x,y) \wedge social(y), 10\rangle$$

RLR sums over the above weighted formulae and takes the *sigmoid*, giving:

$$P(funFor(x) \mid \Pi) = sigmoid(sum), \qquad \text{where } sum = -5 + 10n_T$$

where, for each x, Π is an assignment of values to $knows(x,y)$ and $social(y)$, and n_T represents the number of individuals y for which $knows(x,y) \wedge social(y)$ is *True* in Π. When $n_T = 0$, $sum < 0$ and the probability is closer to 0; when $n_T > 0$, $sum > 0$ and the probability is closer to 1.

Example 3. This example is similar to Example 1, but uses only positive conjunctions[3], and also involves multiple logical variables of the same population.

$$\langle\{\}, q, \alpha_0\rangle$$
$$\langle\{x\}, q \wedge true(x), \alpha_1\rangle$$
$$\langle\{x\}, q \wedge r(x), \alpha_2\rangle$$
$$\langle\{x\}, true(x), \alpha_3\rangle$$
$$\langle\{x\}, r(x), \alpha_4\rangle$$
$$\langle\{x,y\}, q \wedge true(x) \wedge true(y), \alpha_5\rangle$$
$$\langle\{x,y\}, q \wedge r(x) \wedge true(y), \alpha_6\rangle$$
$$\langle\{x,y\}, q \wedge r(x) \wedge r(y), \alpha_7\rangle$$

In RLR and in MLN, if all $R(A_i)$ are observed:

$$P(q \mid obs) = sigmoid(\alpha_0 + n\alpha_1 + n_T\alpha_2 + n^2\alpha_5 + n_T n\alpha_6 + n_T^2\alpha_7)$$

where obs has $R(x)$ true for n_T individuals, and false for n_F individuals out of a population of n. The use of two logical variables (x, y) of the same population gives a squared dependency in the population.

---

[3] Here $true(x)$ is true of every x. This notation is redundant. If you want the traditional MLN notation, you can remove the explicit set of logical variables and keep the $true(\cdot)$ relations. If you are happy with the explicit logical variables, you can remove the $true(\cdot)$ predicates. Removing both is incorrect. Keeping both is harmless. Formulae that involve negation are redundant; any set of weighted formulae involving negation can be replaced by weighted formulae that don't involve negation [9].

4   Three Elementary Models

Consider the simplest case of aggregating over populations, with a PRV Q connected to a PRV $R(x)$ containing an extra logical variable, x, as in Figure 1. In the grounding, Q is connected to $n = |pop(x)|$ instances of $R(x)$. We assume the model is defined before n is known; it is applicable for all values of n.

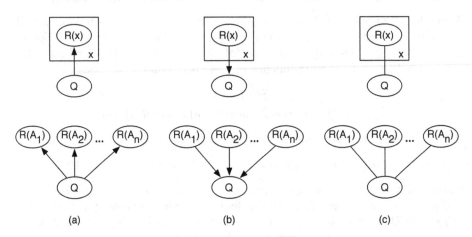

Fig. 1. Running example as (a) naïve Bayes (b) logistic regression with independent priors for each $R(x)$ and (c) Markov network. On the top are the networks using plate notation, where plates [1], drawn as rectangles, correspond to logical variables. On the bottom are the groundings for the population $\{A_1, A_2, \ldots, A_n\}$.

For this situation, Fig. 1(c) shows an undirected model with a factor for Q and a pairwise factor for Q with each individual. Fig. 1(a) shows a directed model where $R(x)$ is a child of Q. In the grounding it produces a naïve Bayes model with a factor for $P(Q)$ and a separate factor for $P(R(A_i) \mid Q)$ for each individual. In both of these models the joint probability is the product of factors. In terms of MLNs and RLR, factors corresponds to weighted formulae.

The naïve Bayes model of Figure 1(a) is an instance of the Markov network of 1(c). Every naïve Bayes model can be represented by a Markov network, but the converse is not true. In some sense the naïve Bayes model is the Markov network with the constraint that the factors represent conditional probabilities (sum to 1, given Q).

For a directed model with $R(x)$ as a parent of Q (Fig. 1(b)), Q has an unbounded number of parents in the grounding, so we need some way to aggregate the parents. Common ways to aggregate in relational domains, e.g. [5,3,12,14,11], include logical operators such as *noisy-or*, *noisy-and*, as well as ways to combine probabilities. This requirement for aggregation occurs in a directed model whenever a parent contains an extra logical variable.

While it may seem that these models are syntactic variants, the models involve very different independence assumptions [13]:

- In the naïve Bayes and the MLN (Figure 1(a) and 1(c)), the variables $R(x)$ and $R(y)$ (for $x \neq y$) are independent given Q, and dependent not given Q.
- In the directed model with aggregation (Figure 1(b)) the variables $R(x)$ and $R(y)$ (for $x \neq y$) are dependent given Q, and independent not given Q.

These dependencies do not depend on what aggregation is used for the directed model. For the rest of this paper we assume that RLR is used as the aggregator. Note that RLR can use the same formulae as the MLN, in which case, when all $R(A_i)$ are observed, the posterior probability of Q would be the same in the MLN and RLR models; however, the posterior probabilities of Q are different when not all of the $R(A_i)$ are observed.

The difference in the dependency structure means that we cannot represent a logistic regression model where the $R(A_i)$ are dependent when Q is observed using an MLN, because in such an MLN the $R(A_i)$ are independent given Q. It is an open problem whether introducing new formulae that involve multiple individuals may allow an MLN to represent the regression model. Similarly, an RLR model cannot represent the MLN where the $R(A_i)$'s are dependent not given Q, without introducing other relations or dependencies among the variables. It is an open question as to whether any finite set of formulae is adequate to make them able to represent the same distributions.

5   Effects of Population Sizes

In this section we investigate the behaviour of MLNs and RLR as the population size n varies.

5.1   A Comparison of MLN, RLR and MF for the Simplest Case

We now compare MLN, RLR, and a simple mean-field (MF) approximation of RLR, for the elementary models in Figure 1. For MLN (Figure 1 (c)), we use the MLN parametrization of Example 1 as the joint distribution. For RLR (Figure 1 (b)), we use p_r as the i.i.d. prior probability of each $r(x)$, and use the RLR parametrization of Example 1 for $P(q \mid R(A_1), \ldots, R(A_n))$. (Note that $P(r(x)) = p_r$ can be represented by RLR model for $R(x)$ using the single formula $\langle \{x\}, r(x), \alpha_3 \rangle$, where $sigmoid(\alpha_3) = p_r$.) We can now sum out the unobserved variables $R(x)$, and get $P(q \mid n)$. The dependency of $P(q)$ on n is an effect of population size.

For the MLN, when Q is conditioned on, the graph is disconnected, with each component $R(x)$ having the same probability. So to compute $P_{MLN}(q \mid n)$, we can compute the probability of one of them and raise it to the power of n [15]:

$$P_{MLN}(q \mid n) = sigmoid(\ \alpha_0 + n \log(e^{\alpha_2} + e^{\alpha_1 - \alpha_3})\ ) \qquad (2)$$

Note this is a logistic function (the sigmoid of a linear function) of n and α_0, but not a logistic function of the other parameters.

For the RLR model, summing out the unobserved variables $R(x)$ gives:

$$P_{RLR}(q \mid n) = \sum_{i=0}^{n} \binom{n}{i} sigmoid(\alpha_0 + i\alpha_1 + (n-i)\alpha_2)(1 - p_r)^i p_r^{n-i}$$

where i is the number of individuals for which $R(x)$ is false. This inference is an instance of first-order variable elimination [19].

Finally, the simple mean-field approximation to the RLR model is:

$$P_{MF}(q \mid n) = sigmoid(\alpha_0 + np_r\alpha_1 + n(1 - p_r)\alpha_2)$$

Note that np_r is the expected number of $R(x)$'s that are true, and $n(1 - p_r)$ is the expected number of $R(x)$'s that are false.

Example 4. Fig. 2 compares $P(q \mid n)$ for RLR, MLN and the mean-field approximation of RLR, using $\alpha_0 = -4.5$, $\alpha_1 = 1$, $\alpha_2 = -1$, and $p_r = 0.7$ (thus $P_{MF}(q \mid n) = sigmoid(-4.5 + 0.4n)$). The MLN uses $\alpha_3 = 2.82$, chosen to give it the same probability as the RLR for $n = 1$.

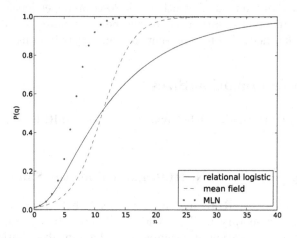

Fig. 2. $P(q \mid n)$ in Example 4

$P_{MLN}(q \mid n)$ is a logistic function (the sigmoid of a linear function) of n, and so is monotonic with n. It might be conjectured that the MLN and RLR models are qualitatively similar. It is therefore intuitive to make the following conjecture:

Conjecture 1. $P_{RLR}(q \mid n)$ (in the RLR model for Fig. 1 (b)) is monotonic in n.

It turns out that this conjecture is false.

Example 5. Fig. 3 demonstrates the setting: $\alpha_0 = -2$, $\alpha_1 = 2$, $\alpha_2 = -1$, $p_r = 0.3$. Whereas the mean-field approximation of RLR, $P_{MF}(q \mid n) = sigmoid(-2 - 0.1n)$, is monotonic, $P_{RLR}(q \mid n)$ is not, having a maximum at $n = 18$. (Example 6 shows $P_{MLN}(q \mid n)$ for this setting.)

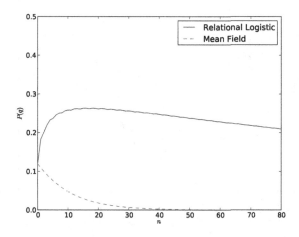

Fig. 3. $P(q \mid n)$ in Example 5

5.2   Phase Transitions in MLNs

A phase transition in physics arises when a value flips from one state to another. In this section we show how a probability can flip from one value to another (e.g, close to 1 or close to 0) as either a parameter varies or a population varies. These interact, as rate of change can depend on the population and on parameter values.

One of the properties of the directed model of Figure 1(b) is that $P_{RLR}(R(A_i) \mid n)$ does not depend on n and can be given as input to the model. In MLNs, however, $P_{MLN}(R(A_i) \mid n)$ depends on n, except for the special case of a naïve Bayes model represented using an MLN. We show that for some MLNs, there is a phase transition where $P_{MLN}(R(A_i) \mid n)$ cannot be arbitrarily set in the limit as the population increases.

Example 6. Consider the same parametrization as Example 5, and the mapping to MLNs given in Example 1. Under this mapping, the MLN and the RLR both represent the same conditional probability $P(q \mid R(A_1), \ldots, R(A_n))$. To fully specify the model, RLR requires p_r, representing $P(r(x))$ for all x. The MLN requires α_3.

Fig. 4 shows $P_{MLN}(q \mid \alpha_3)$ for different population sizes n. All of these slopes are logistic functions. As n increases the slope becomes steeper.

There is a phase transition at approximately $\alpha_3 = 0.7$. For $\alpha_3 < 0.7$, $P_{MLN}(q \mid n)$ decreases with n, and for $\alpha_3 > 0.7$, $P_{MLN}(q \mid n)$ increases with n. At the phase transition point, $P_{MLN}(q \mid n)$ does not depend on n. The phase transition occurs when the coefficient of n in Equation (2) is 0.

Fig. 5 shows $P_{MLN}(r(A_1) \mid \alpha_3)$ for different population sizes n ($P_{MLN}(r(A_i))$ is identical for all individuals A_i). Similarly to Figure 4, the slope becomes steeper with increasing n's.

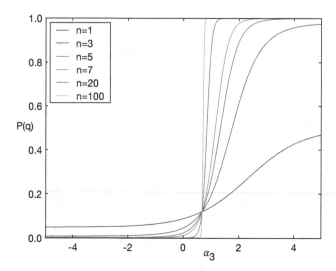

Fig. 4. $P_{MLN}(q \mid \alpha_3)$ in an MLN for various population sizes n, for Example 6

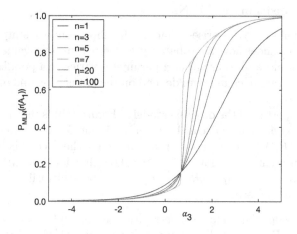

Fig. 5. $P_{MLN}(r(A_1) \mid \alpha_3)$ in an MLN for various population sizes n, for Example 6

Notice the way the parameter α_3 affects $P_{MLN}(q)$ or $P_{MLN}(r(A_i))$ depends on n. We cannot set the parameters so that the MLN represents arbitrary values for $P_{MLN}(r(A_i))$ as the population varies, as we show:

At the phase transition, there is an approximately vertical line segment for large populations. The corresponding probabilities for $r(A_1)$ cannot be represented in the limit $n \to \infty$. In the limit, $P_{MLN}(q \mid n)$ approaches either 0 or 1 (or is not affected by n). Suppose in the limit $P_{MLN}(q \mid n) \to 1$ and we tried to adjust α_3 to fit $P_{MLN}(r(A_1) \mid n) = 0.3$ when $P_{MLN}(q \mid n) = 1$. The new value found for α_3 implies that $P_{MLN}(q \mid n) \to 0$ in the limit. Similarly, suppose

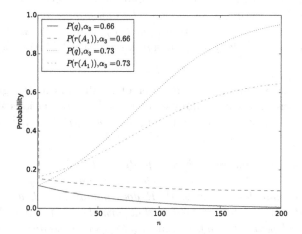

Fig. 6. $P_{MLN}(q \mid n)$ and $P_{MLN}(r(A_1) \mid n)$ for $\alpha_3 = 0.66$ and 0.73, for Example 6

$P_{MLN}(q \mid n) \to 0$ and we tried to adjust α_3 to fit $P_{MLN}(r(A_1) \mid n) = 0.3$ when $P_{MLN}(q \mid n) = 0$, the new value found for α_3 implies that $P_{MLN}(q \mid n) \to 1$. Thus α_3 cannot be set to make $P_{MLN}(r(A_1) \mid n) \to 0.3$ as $n \to \infty$.

Fig. 6 shows how $P(q)$ and $P(r(A_1))$ vary with population size for two different parameterizations, $\alpha_3 = 0.66$ and 0.73. The monotonically increasing lines are for $\alpha_3 = 0.73$ and the decreasing lines are for $\alpha_3 = 0.66$. As α_3 gets closer to the phase transition, the graphs approach the extremes at a slower rate.

5.3   Behavior of MLNs on More General Cases

In general it is a complex inference problem to determine the probability of a random variable as a function of n. However, we can characterize some of the cases where the probability is bounded away from 0 and 1, or approaches 0 or 1 in the limit as a population approaches infinity.

Proposition 1. *Consider an MLN with finite weights. Let n be the size of some population and V be a ground random variable. If the number of formula instantiations that depend on V's value is independent of n, then $P_{MLN}(V \mid n)$ is bounded away from 0 and 1, i.e., exists $c > 0$ such that $0 < c \leq P_{MLN}(V \mid n) \leq 1 - c < 1$ for all n's.*

Proof. The number of such formula instantiations was guaranteed to be fixed (independent of n). The weights are finite, so each such contribution is bounded. Define the neighbours of V to be the grounding of the other PRVs in the weighted formulae that V appears in. Let c be the minimum of the conditional probability of V given its neighbours, and $\neg V$ given its neighbours. This c has the property specified in the proposition, as $P(V \mid n)$ is a linear interpolation of the probabilities of V given its neighbours.   □

Proposition 2. *Consider an MLN with finite weights. Let pop be some population, $n = |pop|$, V be any PRV, and V' be any ground instance of V. If V' does not unify with a PRV that is in a weighted formula with another PRV that has an extra logical variable typed with pop, then $P_{MLN}(V' \mid n)$ is bounded away from 0 and 1.*

Proof. In this case V' has a fixed number of neighbours in the grounding as n varies, and there are a fixed number of formula instantiations that depend on V''s value. Therefore, Proposition 1 guarantees $P_{MLN}(V')$ is bounded away from 0 and 1.   □

Proposition 3. *Consider an MLN with finite weights. If PRV V is in a formula with PRV R that includes a logical variable of a population of size n that does not appear in V, and for any such R, R does not unify with a PRV in other formulae or with an instance of itself in that formula, then either $P_{MLN}(V \mid n)$ is a constant (independent of n), or $\lim_{n \to \infty} P_{MLN}(V)$ is either 1 or 0.*

Proof. Such cases are locally isomorphic to the simple case analyzed earlier.   □

It is an open problem to characterize other cases of what happens in the limit.

5.4 Real Data and Prior Knowledge

Figure 7 show $P(25 < Age(p) < 45 \mid n)$ for a person p, given the number n of movies they rated, for the Movielens 1M dataset (http://grouplens.org/datasets/movielens/), averaged over all people. This is calculated by bucketing over n, with at least 20 people in each bucket.

When trying to fit models to such data, we first need to choose what model class to use. We might want to not only fit to the data, but to fit what we expect

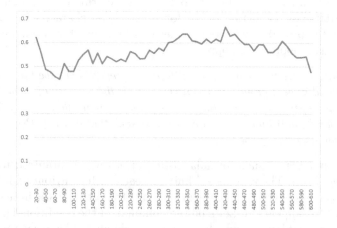

Fig. 7. Observed $P(25 < Age(p) < 45 \mid n)$ from the Movielens dataset

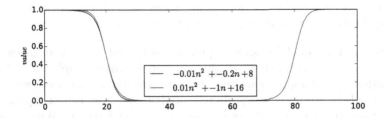

Fig. 8. Sigmoids of polynomials of n. The population size, n, is on the x-axis.

in the limit. We can design the structure of the model to either go to 0 or 1 in the limit or to be bounded away from 0 and 1. In this particular example, we would not expect the probability to go to 0 or 1, and we would also not expect the age to be independent of the number of movies a person has rated (the population size n for each person). So in the model we would not just have weighted formulae that contain $Age(person)$ and $Rated(person, movie)$, for if we did, by Proposition 3, either the age does not depend on the number of movies rated or the Age becomes deterministic (is 1 or 0) in the limit. This does not preclude more complicated formulae, but a preference for simpler models might.

5.5   Fitting Polynomials

In Example 3, $P(q \mid n)$ is a (sigmoid of a) degree-2 polynomial of n. One might innocently write weighted formulae like in Example 3 without realizing the implications of such statements and get very surprising results. In this section we show by example what can happen unexpectedly.

Consider fitting a degree-2 polynomial to data in which the population size n is in the range $0 \leq n \leq 50$. Suppose we find that the closest fit is $0.01n^2 - n + 16$. Suppose in another run, we fit $-0.01n^2 - 0.2n + 8$. Figure 8 plots these, but in the range $0 \leq n \leq 100$. The polynomials are very close in the training range, but the first polynomial goes up soon after, even though we have no evidence of this in the data set.

This is not an isolated occurrence. A degree-k polynomial may have up to $k-1$ points where it changes between increasing and decreasing. If the polynomial we fit has one or more of these points beyond the region of the training set, we are likely to get very unintuitive predictions.

The sign of the coefficient of the leading power in the polynomial determines whether the probability approaches 0 or 1. However, this is often difficult to determine, particularly if we are close to phase transitions.

6   Conclusion

In this paper we investigated the dependence on population size for relational models. Even for simple models that are well understood at the non-relational level, there are complex interactions of the parameters with population size. The results of this paper are important for a number of reasons:

- If we learn a model for some population sizes and apply it to other population sizes, it is important to know what the model implies about such extrapolation of population sizes. Here we have shown some cases where the details of the model makes particular predictions about the extrapolation.
- We want to know the effect of choosing particular formulae. What assumptions are we making? For example, adding an adding an extra variable to a formula adds a dependency on population size.
- If one model fits some data better than another, it is important to understand why. We have investigated the effects of some design decisions for directed and undirected models.
- If we want to extrapolate from data, how can prior information affect the formulae used. The prior information we have considered is what how should the probability change as the population grows.

The other message is that undirected models such as MLNs are different to directed models, such as those that use RLR. It is important to understand these differences if we are to choose an appropriate model for a domain. In particular, when fitting a model to data, we should consider both models, and not assume that one works better than the other independently of the domain.

This paper has exposed more questions than it has answered. Determining dependencies on population sizes for more complicated models is an open question, which may allow a modeler to rule out some models for their specific application. Ideally, we would like ways to generate qualitative descriptions about the model from the model's formulae.

References

1. Buntine, W.L.: Operations for learning with graphical models. arXiv preprint cs/9412102 (1994)
2. Domingos, P., Kok, S., Lowd, D., Poon, H., Richardson, M., Singla, P.: Markov logic. In: De Raedt, L., Frasconi, P., Kersting, K., Muggleton, S.H. (eds.) Probabilistic ILP. LNCS (LNAI), vol. 4911, pp. 92–117. Springer, Heidelberg (2008)
3. Friedman, N., Getoor, L., Koller, D., Pfeffer, A.: Learning probabilistic relational models. In: Proc. IJCAI 1999, pp. 1300–1309 (1999)
4. Getoor, L., Taskar, B.: Introduction to Statistical Relational Learning. MIT Press, Cambridge (2007)
5. Horsch, M., Poole, D.: A dynamic approach to probabilistic inference using Bayesian networks. In: Proc. Sixth Conference on Uncertainty in AI, pp. 155–161 (1990)
6. Jaeger, M.: Convergence results for relational Bayesian networks. In: Proceedings of LICS 1998 (1998)
7. Jain, D., Barthels, A., Beetz, M.: Adaptive Markov logic networks: Learning statistical relational models with dynamic parameters. In: 9th European Conference on Artificial Intelligence (ECAI), pp. 937–942 (2010)

8. Jain, D., Kirchlechner, B., Beetz, M.: Extending markov logic to model probability distributions in relational domains. In: Hertzberg, J., Beetz, M., Englert, R. (eds.) KI 2007. LNCS (LNAI), vol. 4667, pp. 129–143. Springer, Heidelberg (2007)

9. Kazemi, S.M., Buchman, D., Kersting, K., Natarajan, S., Poole, D.: Relational logistic regression. In: Proc. 14th International Conference on Principles of Knowledge Representation and Reasoning (KR 2014) (2014)

10. Koller, D., Friedman, N.: Probabilistic Graphical Models: Principles and Techniques. MIT Press, Cambridge (2009)

11. Natarajan, S., Khot, T., Lowd, D., Tadepalli, P., Kersting, K., Shavlik, J.: Exploiting causal independence in Markov logic networks: Combining undirected and directed models. In: Balcázar, J.L., Bonchi, F., Gionis, A., Sebag, M. (eds.) ECML PKDD 2010, Part II. LNCS, vol. 6322, pp. 434–450. Springer, Heidelberg (2010)

12. Neville, J., Simsek, O., Jensen, D., Komoroske, J., Palmer, K., Goldberg, H.: Using relational knowledge discovery to prevent securities fraud. In: Proceedings of the 11th ACM SIGKDD International Conference on Knowledge Discovery and Data Mining. ACM Press (2005)

13. Pearl, J.: Probabilistic Reasoning in Intelligent Systems: Networks of Plausible inference. Morgan Kaufmann, San Mateo (1988)

14. Perlich, C., Provost, F.: Distribution-based aggregation for relational learning with identifier attributes. Machine Learning 62(1-2), 65–105 (2006)

15. Poole, D.: First-order probabilistic inference. In: Proceedings of the 18th International Joint Conference on Artificial Intelligence (IJCAI 2003), Acapulco, pp. 985–991 (2003)

16. Poole, D., Buchman, D., Natarajan, S., Kersting, K.: Aggregation and population growth: The relational logistic regression and Markov logic cases. In: UAI 2012 Workshop on Statistical Relational AI (2012)

17. De Raedt, L., Kersting, K.: Probabilistic Inductive Logic Programming: Theory and Applications. In: De Raedt, L., Frasconi, P., Kersting, K., Muggleton, S.H. (eds.) Probabilistic Inductive Logic Programming. LNCS (LNAI), vol. 4911, pp. 1–27. Springer, Heidelberg (2008)

18. Richardson, M., Domingos, P.: Markov logic networks. Machine Learning 42, 107–136 (2006)

19. de Salvo Braz, R., Amir, E., Roth, D.: Lifted first-order probabilistic inference. In: Getoor, L., Taskar, B. (eds.) Introduction to Statistical Relational Learning. MIT Press (2007)

Improving Personalized Search on the Social Web Based on Similarities between Users

Zhenghua Xu, Thomas Lukasiewicz, and Oana Tifrea-Marciuska

Department of Computer Science, University of Oxford, UK
{zhenghua.xu,thomas.lukasiewicz,oana.tifrea}@cs.ox.ac.uk

Abstract. To characterize a user's preferences and the social summary of a document, the user profile and the general document profile are widely adopted in existing folksonomy-based personalization solutions. However, in many real-world situations, using only these two profiles cannot personalize well the search results on the Social Web, because (i) different people usually have different perceptions about the same document, and (ii) the information contained in the user profile is usually not comprehensive enough to characterize a user's preference. Therefore, in this work, in order to improve personalized search on the Social Web, we propose a dual personalized ranking (D-PR) function, which adopts two novel profiles: an extended user profile and a personalized document profile. For each document, instead of using a general document profile for all users, our method computes for each individual user a personalized document profile to better summarize his/her perception about this document. A solution is proposed to estimate this profile based on the perception similarities between users. Moreover, we define an extended user profile as the sum of all of the user's personalized document profiles to better characterize a user's preferences. Experimental results show that our D-PR ranking function achieves better personalized ranking on the Social Web than the state-of-the-art baseline method.

1 Introduction

Recently, with the rise of Web 2.0 applications, such as social bookmarking systems, electronic commerce websites, blogs, and social network sites, the Web has evolved towards the so-called *Social Web*, where users can freely provide social annotations to online documents (i.e., Web pages or resources on the Social Web) via bookmarking, tagging, rating, commenting, and so on. Social annotations are valuable resources for personalized search on the Social Web. On the one hand, annotations provided by different Web users from different perspectives are usually good summaries of the corresponding documents. On the other hand, social annotations are also ideal data for privacy-enhanced personalization: first, they are provided by a user directly, so these annotations can be treated as a user's individual opinion about a document; these interests and preferences of the user can be harvested by the aggregation of his/her social annotations; second, these social annotations are usually publicly available and contain little sensitive information about users, so they can be safely utilized without violating user privacy. In this paper, we refer to social annotations as social tags assigned to documents by users in bookmarking systems, but relevant techniques can be easily adapted to other social metadata (e.g., comments, blogs, etc.) as well.

U. Straccia and A. Calì (Eds.): SUM 2014, LNAI 8720, pp. 306–319, 2014.

Consequently, more and more research activities focus on personalizing the search on the Social Web using social tags [3,4,25,26]. Generally, given a query issued by a user, the existing methods rank the online documents by the corresponding *ranking scores*, which are normally comprised of two parts: a query-related part, measuring the textual similarity between the given query and each document, and a personalization part, measuring the similarity between the user's preferences (in the user profile) and the social summary of each document (in the general document profile).

A user profile is a weighted vector, whose dimensions are tags and whose values in each dimensions are the corresponding tag weights. In the user profile, the tag weight is influenced by the number of times that this user uses the tag for bookmarking. Similarly, the general document profile is also a weighted vector and its tag weight is influenced by the number of times that the document is bookmarked with the tag.

However, in many real-world situations, using these two profiles cannot personalize well the search results on the Social Web. On one hand, users usually have different perceptions about the same document, so, for a specific user, not all tags assigned by all the other users are equally helpful to summarize his real perception about the document (some of them are actually harmful). Therefore, the general document profile, which treats tags from all users with equal importance, cannot properly summarize a special user's personal perception about the document. On the other hand, in practice, there are tens of billions of documents on the Web and even a long-time Social Web user can only annotate a very small portion of them. Therefore, the user profile, based on only the tags assigned by the corresponding user, usually does not contain sufficient information to comprehensively characterize the user's preferences.

To solve these problems, we propose a *dual personalized ranking (D-PR)* function which utilizes two novel profiles, called *personalized document profile* and *extended user profile*, to better characterize a user's preferences and better summarize his/her personal perception about a document, respectively. Instead of using the same general document profile for all users, for each of the documents, our method computes for each individual user a personalized document profile to characterize his/her personal perception about this document. Furthermore, the extended user profile is defined as the sum of all of the user's personalized document profiles. As each user has a personalized document profile for each of the documents, the extended user profile contains more information to comprehensively characterize a user's preferences.

However, how to obtain the user's personalized document profiles for all online documents is a challenge. The tags assigned by a user to a document may be a good outline of his personal perception about this document. But, in fact, this is unpractical: on the one hand, a user normally uses only a few (typically 1 or 2) tags to annotate a document, so these tags contain too little information to comprehensively summarize the document; on the other hand, only a small portion of online documents are annotated by a user, but we need a personalized document profile for each document.

Therefore, we propose to estimate the personalized document profile of a user u by using the perception similarity between u and the other users as weights to sum up tags assigned to the relevant document by the users having high perception similarity with u. The underlying intuition is that users having similar perceptions about the existing documents are very likely to also share similar perceptions about future documents; so, for a user u, tags assigned by users having high perception similarity with u are more helpful to characterize u's personal perception about the document than tags assigned by users having low perception similarity with u. Intuitively, the higher perception similarity between two users, the higher their tags are weighted for each other.

In summary, we make the following contributions in this paper.

- We propose a *dual personalized ranking (D-PR)* function to improve personalized search on the Social Web by introducing two novel profiles: the extended user profile and the personalized document profile, to better characterize a user's preferences and better summarize his/her personal perception about a document.
- We formally define the extended user profile as the sum of all the user's personalized document profiles; and we further propose to estimate a user's personalized document profile using the perception similarity between users. Finally, a method used to quantify the perception similarity is also presented.
- We conduct extensive experimental studies based on a public real-world large scale research dataset [15]. The results validate the effectiveness of our *D-PR* function: it outperforms the state-of-the-art *SoPRa* function [3].

The rest of this paper is organized as follows. In Section 2, we present some preliminaries. Section 3 formally defines two state-of-the-art personalized ranking solutions and illustrates their potential problems. In Section 4, we propose a novel D-PR function to solve these problems; while the approaches of estimating the user's personalized document profile and constructing the extended user profile are also presented in this section. Experiments are discussed in Section 5. Section 6 reviews some closely related works. Finally, Section 7 concludes this work and provides some future directions.

2   Preliminaries

Social bookmarking systems are based on the techniques of social tagging. The main idea behind them is to provide the user with a means to freely annotate resources on the Web (e.g., URIs in delicious[1] or images in Flickr[2]) with tags. Since the annotations can be shared with others, this practice of collaboratively creating and translating tags to annotate and categorize online content is usually called *collaborative tagging* or *social tagging*, and the resulting tag-based classification is called a *folksonomy*.

Definition 1. Let U, T, and D be the sets of *users*, *tags*, and *documents*. A *bookmark* is a triple $(u, t, d) \in U \times T \times D$, which represents the fact that the user u has annotated the document d with the tag t. A *folksonomy* $\mathcal{F}(U, T, D)$ is a subset of $U \times T \times D$.

The following example illustrates the above concepts, including folksonomies and bookmarks; it will be used in the sequel as a running example.

Example 1. Consider the set of users $U = \{Alice, Bob, Carl\}$, their set of tags $T = \{English, Chinese, Comedy, Action, Interesting, Boring\}$, and a set of documents $D = \{d_1, d_2, d_3\}$. A folksonomy $\mathcal{F}$ may then express the following knowledge:

---

[1] https://delicious.com/
[2] https://www.flickr.com/

Table 1. Tags used by users to annotate documents

	tags in d_1	tags in d_2	tags in d_3
Alice	English, Comedy, Interesting	Boring	Chinese, Comedy, Interesting
Bob	Boring	Chinese, Action, Interesting	Boring
Carl	English, Comedy, Interesting	Boring	(Null)

(i) *Alice* and *Carl* are interested in all comedies and dislike action movies, while *Bob* has the right opposite preferences; and (ii) d_1 is an introduction page of an English comedy movie, d_2 is an introduction page of a Chinese action movie, and d_3 is an online video of a Chinese comedy movie. The specific tags used by each of these users to annotate each of these online documents are shown in Table 1.   □

The *personalized ranking problem* [3,25] can be fomalized as follows: given a folksonomy $\mathcal{F}(U, T, D)$ and a query q submitted by a user $u \in U$ to a search engine, it re-ranks the set of documents $d_q \in D$ that match q, in such a way that relevant documents for u are highlighted and pushed to the top for maximizing this user's satisfaction and personalizing the search results. The ranking follows an ordering $\tau = [d_1 \geq d_2 \geq \cdots \geq d_k]$ in which (i) $d_k \in D$ and (ii) $d_i \geq d_j$ iff $Rank(d_i, q, u) \geq Rank(d_j, q, u)$, where $Rank(d, q, u)$ is the result of a ranking function that quantifies similarity between q and the document d relative to u .

The vector space model (VSM) [19] is a general model used in information retrieval where the profile of a user (resp., a document) is mapped to a weighted vector in a universal term space. The terms can be tags or words. We use words when we deal with the text of the document and of the query, while tags are used when we deal with the tags of the document and of the query (each query word is considered a tag).

To calculate the similarity between two vectors, we use the well-known cosine similarity. Given two vectors $\boldsymbol{A} = (A_1, \ldots, A_n)$ and $\boldsymbol{B} = (B_1, \ldots, B_n)$, its cosine similarity $Sim(\boldsymbol{A}, \boldsymbol{B})$ is formally defined as follows, where $Sim(\boldsymbol{A}, \boldsymbol{B})$ ranges from 0 (independence) to 1 (identity):

$$Sim(\boldsymbol{A}, \boldsymbol{B}) = \frac{\boldsymbol{A} \cdot \boldsymbol{B}}{|\boldsymbol{A}||\boldsymbol{B}|} = \frac{\sum\limits_{i=1}^{n} A_i \cdot B_i}{\sqrt{\sum\limits_{i=1}^{n} A_i^2} \cdot \sqrt{\sum\limits_{i=1}^{n} B_i^2}}. \tag{1}$$

We use the textual matching score, $Score(q, d)$, to indicate how similar a query q is to the textual content of a document d using words as terms. This score is not folksonomy-based and has been widely adopted in most commercial search engines. Thus, it can be obtained directly from these search engines, when it is incorporated into a personalized ranking function.

3   Personalized Ranking Functions

In this section, we recall two state-of-the-art personalized ranking methods and illustrate their potential ordering problems in the running example. All ranking functions presented in this paper follow the widely used VSM, where the weights of tags are based on tag frequencies (*tf*), and the extension to *tf-idf* (tag frequency-inverse document frequency) [10] is trivial.

3.1 User Profile Personalized Ranking Function

Xu et al. [26] propose a ranking function to compute the ranking score $Rank(d, q, u)$ of a document d relative to a given query q issued by a user u from two aspects: (i) the textual matching score $Score(q, d)$, measuring the statistical textual quality of d relative to q; and (ii) a profile matching score $Sim(p_u, p_d)$, which estimates the interest of the user u in the document d, and which is measured by the similarity between the user profile and the general document profile. As this method uses the user's preferences that are implicitly contained in the user profile to personalize the ranking result, we call it *user profile personalized ranking (UP-PR) function*, formally defined as follows:

$$Rank(d, q, u) = \alpha \cdot Sim(p_u, p_d) + (1 - \alpha) \cdot Score(q, d), \qquad (2)$$

where p_u is the user profile indicating this user's personal preferences, and p_d is the general document profile measuring the understandings and perceptions of all users about this document. Following the VSM, p_u (resp., p_d) is a weighted vector with tags as dimensions and tag weights as values, where a tag's weight is the number of times that this tag is used by the user (resp., is used to annotate the document) for bookmarking. The following example illustrates the UP-PR function.

Example 2. Recalling Example 1, $Carl$ would like to find an interesting Chinese comedy film, so he issues a query "Interesting Chinese film" to a non-personalized search engine. Obviously, based on the knowledge in Example 1, $Carl$ would expect the ordering of the search result to be $\tau_0 = [d_3 \geq d_1 \geq d_2]$. However, the search engine computes $Score(q, d_1) = 0.6, Score(q, d_2) = 0.52$, and $Score(q, d_3) = 0.5$, i.e., the resulting ordering on the search results is $\tau_1 = [d_1 \geq d_2 \geq d_3]$, which is an unexpected ordering, as the desired document d_3 is ranked at the bottom.

    On the other hand, if we use *UP-PR* to personalize the ranking result, then we first compute the weighted vectors of the query (denoted q), the profile of $Carl$ (denoted p_{Carl}), and the profiles of the documents (denoted p_{d_1}, p_{d_2}, and p_{d_3}) as shown in Table 2. Then, the personalized *UP-PR* ranking scores of d_1, d_2, and d_3 for $Carl$ relative to this query can be computed as shown in Equation 3 with $\alpha = 0.5$. Therefore, the personalized ranking of these search results is $\tau_2 = [d_1 \geq d_3 \geq d_2]$.

$$Rank(d_1, q, Carl) = \frac{1}{2}(Sim(p_{Carl}, p_{d_1}) + Score(q, d_1))$$

$$= \frac{1}{2}(\frac{7}{\sqrt{4} \cdot \sqrt{13}} + 0.68) = \frac{1}{2}(0.97 + 0.6) = 0.79,$$

$$Rank(d_2, q, Carl) = \frac{1}{2}(Sim(p_{Carl}, p_{d_2}) + Score(q, d_2))$$

$$= \frac{1}{2}(0.57 + 0.52) = 0.55, \qquad (3)$$

$$Rank(d_3, q, Carl) = \frac{1}{2}(Sim(p_{Carl}, p_{d_3}) + Score(q, d_3))$$

$$= \frac{1}{2}(0.75 + 0.5) = 0.63.$$

However, although τ_2 obtained via *UP-PR* is better than τ_1 (promoting d_3 from the bottom to the middle), τ_2 is still not the best ordering, as d_3 is ranked lower than d_1. Specifically, we note that in Equation 3, d_1 has a higher ranking score than d_3, which is intuitively inaccurate, because (based on the knowledge in Example 1) $Sim(p_{Carl}, p_{d_3})$ should have similar value to $Sim(p_{Carl}, p_{d_1})$ (as *Carl* prefers all comedies), and $Score(q, d_3)$ should be the highest text matching score (as d_3 is a Chinese comedy film perfectly matching the query). In the following (sub)sections, we will analyze in detail the reasons for such an inaccurate ordering.   □

3.2   Social Personalized Ranking Function

We obtain a low textual matching score $Score(q, d_3)$ in Example 2, because d_3 is an online video that has little textual content to compute a proper textual matching score. This problem is common on the Social Web, and, to solve it, Bouadjenek et al. [3] propose a *social personalized ranking (SoPRa) function*, which extends the *UP-PR* function in [26] by considering a new non-personalized matching score: the social matching score $Sim(q, p_d)$ between the given query q and the social summary of document p_d. This score indicates how relevant the social summary of a document d is to q. The intuition is to use social tags to better summarize the content of a document and add further information for social resources with very little textual content (e.g., videos and images). Therefore, we have two query-related scores in *SoPRa*, which are defined as follows:

$$Rank(d, q, u) = \alpha \cdot Sim(p_u, p_d) + (1 - \alpha) \cdot [\beta \cdot Sim(q, p_d) + (1 - \beta) \cdot Score(q, d)] . \quad (4)$$

Example 3. Continuing Example 2, if we personalize the results of the search engine by *SoPRa*, the ranking scores of d_1, d_2, and d_3 for *Carl* are computed as shown in Equation 5 (with $\alpha = 0.5$ and $\beta = 0.5$).

$$Rank(d_1, q, Carl) = \frac{1}{2}(Sim(p_{Carl}, p_{d_1}) + \frac{1}{2}(Sim(q, p_{d_1}) + Score(q, d_1)))$$

$$= \frac{1}{2}(0.97 + \frac{1}{2}(\frac{4}{\sqrt{3} \cdot \sqrt{13}} + 0.6)) = \frac{1}{2}(0.97 + \frac{1}{2}(0.64 + 0.6)) = 0.8,$$

$$Rank(d_2, q, Carl) = \frac{1}{2}(Sim(p_{Carl}, p_{d_2}) + \frac{1}{2}(Sim(q, p_{d_2}) + Score(q, d_2)))$$

$$= \frac{1}{2}(0.56 + \frac{1}{2}(0.44 + 0.52)) = 0.52, \quad (5)$$

$$Rank(d_3, q, Carl) = \frac{1}{2}(Sim(p_{Carl}, p_{d_3}) + \frac{1}{2}(Sim(q, p_{d_3}) + Score(q, d_3)))$$

$$= \frac{1}{2}(0.75 + \frac{1}{2}(0.87 + 0.5)) = 0.72.$$

As we can see, the resulting ordering is the same as the one of *UP-PR*, which is not desired. Specifically, by using a social matching score, *SoPRa* narrows the gap between the ranking scores of d_1 and d_3, but the improvement is still not big enough to change the ordering of three documents.   □

4   Dual Personalized Ranking Function

The reasons for having a low profile matching score $Sim(p_{Carl}, p_{d_3})$ in the previous examples are twofold: on the one hand, the general document profile p_{d_3} does not correctly characterize $Carl$'s real perception about d_3, since tags from all users are treated equally, and the tag from Bob brings a bias; on the other hand, the user profile p_{Carl} does not properly model $Carl$'s preference, because p_{Carl} does not tag d_3, so the information used for preference modeling is not comprehensive.

Generally, the widely used general document profile, which treats tags from all users with equal importance, may not be able to summarize a special user's personal perception about a document. Similarly, the information contained in the user profile (i.e., the tags assigned by the user) is usually insufficient to comprehensively characterize the preferences of the user.

Therefore, to solve these problems, we propose a new ranking function, which will be able to better personalize search results by introducing two novel profiles: the extended user profile and the personalized document profile, to better characterize a user's preferences and better summarize his/her personal perception about a document, respectively. Specifically, instead of using the same general document profile for all users, for each of the documents, each individual user has a personalized document profile to characterize his/her perception about this document. Furthermore, we define an extended profile of user u as p'_u, which sums up all personalized document profiles of u to more comprehensively characterize u's preference. This ranking function is called *dual personalized ranking (D-PR)* function and formally defined as follows:

$$Rank(d, q, u) = \alpha \cdot Sim(p'_u, p_{u,d}) + (1 - \alpha) \cdot [\beta \cdot Sim(q, p_d) + (1 - \beta) \cdot Score(q, d)], \quad (6)$$

where the personalized profile of a document d for a user u, $p_{u,d}$, is a weighted vector of tags characterizing u's perception about d; while p'_u is an extended profile of u, obtained by summing up all personalized document profiles of u and defined as follows:

$$p'_u = \sum_{i=1}^{|D|} p_{u,d_i}. \quad (7)$$

Note that in Equation 6, we still use the general document profile p_d to compute the query-related social matching score $Sim(q, p_d)$. As defined in Section 3.2, $Sim(q, p_d)$ is a non-personalized matching score, measuring the textual similarity between q and the social summary of d, and it aims at using social tags assigned by all users to better summarize the content of a document, so here it is unreasonable to replace p_d by $p_{u,d}$.

4.1   Personalized Document Profile

It is a challenge how to obtain the personalized document profiles of a user for all online documents. The tags assigned by a user to a document may be a good outline of this user's personal perception about this document. However, it is, in fact, not practical: on the one hand, a user normally uses only a few (typically, 1 to 3) tags to annotate a document, so these tags contain too little information to comprehensively summarize

Table 2. Weighted vectors of query and profiles

	English	Chinese	Comedy	Action	Interesting	Boring
p_{Alice}	1	1	2	0	2	1
p_{Bob}	0	1	0	1	1	2
p_{Carl}	1	0	1	0	1	1
p_{d_1}	2	0	2	0	2	1
p_{d_2}	0	1	0	1	1	2
p_{d_3}	0	1	1	0	1	1
q	0	1	1	0	1	0

the document; on the other hand, only a small portion of online documents are annotated by a user, but we need a personalized document profile for each document.

Therefore, we propose to estimate the personalized document profile of a user u via using the perception similarities between u and other users as weights to sum up tags assigned to the relevant document by the users having high perception similarities with u. The underlying intuition is that users having similar perceptions about existing documents will very likely also share similar perceptions about future documents, so, for a user u, tags assigned by users having high perception similarity with u are more helpful to characterize u's personal perception about the document than tags assigned by users having low perception similarity with u. Intuitively, the higher the perception similarity between two users, the higher their tags are weighted for each other.

In this section, we first propose a method to quantify the perception similarities between users. Then, we present how to use perception similarities as weights of tags to estimate the personalized document profile. Finally, Example 4 illustrates how to apply *D-PR* in the running example.

Profile-Based Perception Similarity. Since the tags assigned by a user to a document can be treated as an outline of this user's perception about this document, it is natural to measure a user's overall perception by the weighted vector based on all the tags used by this user, i.e., his/her user profile. Thus, a perception similarity of two users can be measured by the similarity of their profiles, called *profile-based perception similarity*:

$$PerSim(u', u) = Sim(p_{u'}, p_u). \tag{8}$$

Estimate of Personalized Document Profile. For a given user, after obtaining the perception similarities between u and all other users, we first select a set of users $U_T \subseteq U$, whose perception similarity with u are higher than a predefined threshold T. Then, for a given document d, we estimate u's personalized document profile relative to d (denoted $p_{u,d}$) by using perception similarities as weights to sum up the tags assigned to d by the users belonging to U_T. Formally,

$$p_{u,d} = \sum_{i=1}^{|U_d \cap U_T|} (v_{u_i,d} \cdot PerSim(u_i, u)), \tag{9}$$

Table 3. Weighted vectors of personalized document profiles and extended user profile

	English	Chinese	Comedy	Action	Interesting	Boring
p_{Carl,d_1}	1.9	0	1.9	0	1.9	0.56
p_{Carl,d_1}	0	0.56	0	0.56	0.56	1.9
p_{Carl,d_3}	0	0.9	0.9	0	0.9	0.56
p'_{Carl}	1.9	1.46	2.8	0.56	3.36	2.16

where $v_{u_i,d}$ is also a weighted vector of tags, whose weight of a tag is the number of times that the tag is assigned by u_i to d; while $U_d \subseteq U$ is the set of users who annotate document d, and $|U_d \cap U_T|$ is the cardinality of the intersection of U_d and U_T.

Example 4. Continuing the running example, based on Equation 8, we first use p_{Alice}, p_{Bob}, and p_{Carl} as shown in Table 2 to compute the perception similarities between $Carl$ and two other users as follows:

$$PerSim(Carl, Alice) = Sim(p_{Carl}, p_{Alice}) = \frac{6}{\sqrt{4} \cdot \sqrt{11}} = 0.9,$$

$$PerSim(Carl, Bob) = Sim(p_{Carl}, p_{Bob}) = \frac{3}{\sqrt{4} \cdot \sqrt{7}} = 0.56, \quad (10)$$

$$PerSim(Carl, Carl) = Sim(p_{Carl}, p_{Carl}) = 1.$$

Then, based on Equation 9, we estimate $Carl$'s personalized document profile of d_1, d_2, and d_3 (denoted p_{Carl,d_1}, p_{Carl,d_2}, and p_{Carl,d_3}, respectively) as shown in Table 3, where the threshold T is set to 0.5, so $U_T = U$. Consequently, we further use Equation 7 to obtain the extended profile of $Carl$ (denoted p'_{Carl}) as shown in Table 3. Finally, the personalized ranking scores of d_1, d_2, and d_3 relative to $Carl$ based on the D-PR function (Equation 6) can be computed as shown in Equation 11 (with $\alpha = 0.5$ and $\beta = 0.5$), and the resulted ordering is $\tau_3 = [d_3 \geq d_1 \geq d_2]$.

$$Rank(d_1, q, Carl) = \frac{1}{2}(Sim(p'_{Carl}, p_{Carl,d_1}) + \frac{1}{2}(Sim(q, p_{d_1}) + Score(q, d_1)))$$

$$= \frac{1}{2}(\frac{17.0052}{\sqrt{11.1436} \cdot \sqrt{34.3052}} + \frac{1}{2}(0.64 + 0.6))$$

$$= \frac{1}{2}(0.87 + \frac{1}{2}(0.64 + 0.6)) = 0.75,$$

$$Rank(d_2, q, Carl) = \frac{1}{2}(Sim(p'_{Carl}, p_{Carl,d_2}) + \frac{1}{2}(Sim(q, p_{d_2}) + Score(q, d_2)))$$

$$= \frac{1}{2}(0.7 + \frac{1}{2}(0.44 + 0.52)) = 0.59, \quad (11)$$

$$Rank(d_3, q, Carl) = \frac{1}{2}(Sim(\boldsymbol{p'_{Carl}}, \boldsymbol{p_{Carl,d_3}}) + \frac{1}{2}(Sim(\boldsymbol{q}, \boldsymbol{p_{d_3}}) + Score(q, d_3)))$$

$$= \frac{1}{2}(0.88 + \frac{1}{2}(0.87 + 0.5)) = 0.78.$$

In summary, τ_3 ranked by *D-PR* is identical to the desired ordering τ_0. This is because *D-PR* solves profile modeling problems existing in the state-of-the-art approaches in the following two ways: (i) for a given user (e.g., $Carl$), *D-PR* utilizes the perception similarities to weaken the influences of tags assigned by users having different perceptions with this user (e.g., Bob) such that the resulting personalized document profiles can better capture this user's real perception about the documents; (ii) for a user u, *D-PR* obtains a personalized document profile for each document, so the extended user profile of u, computed by summing up all these personalized document profiles, contains more sufficient information to characterize u's preferences more comprehensively.   □

5   Experimental Study

In this section, we evaluate the personalization performance of our *D-PR* function by comparing it with the *SoPRa* function, which is the closest work and considered as the state-of-the-art baseline. As this experiment aims at verifying the personalization effect of introducing two novel personalized profiles, we set $\beta = 1$ to eliminate the influence of the possible non-personalized textual matching problem in $Score(q, d)$.

We conduct experimental studies based on a public real-world large scale research dataset, which is described and analyzed in [15]. This dataset gathers more than 100 000 URLs of online documents and retrieves their social annotations from Delicious.com. After removing the documents without any social annotation, the general information of the resulting dataset is as shown in Table 4. Statistically, each user assigns an average of 9.4 tags; only 0.038% of users annotate more than 100 (0.17%) online documents; and the maximum number of online documents annotated by a single user is only 442 (0.75%). These statistical results show that, for any individual user, only a very small proportion of online documents are annotated by him/her, so we need to estimate the user's personalized document profile with the help of tags assigned by others, using their perception similarities as weights.

5.1   Evaluation Methodology

Although the relevance judgment of personalized search result subjectively depends on end users, several researches [1,2,12] have already proved that the tagging behavior of a user on the Social Web is closely correlated to his/her online search behavior, i.e., if a document is annotated by a user with some tags, this document is very likely to

Table 4. Dataset information

Users	Tags	Documents
388,963	3,647,266	59,126

be visited by the same user if it appears as a search result of using the same tags as the search query. This finding provides the theoretical base of our automatic evaluation framework: if a query is issued by a user with some terms, the relevant document is the one annotated by this user using the same terms as tags.

Therefore, to generate a set of synthetic user queries, we randomly select a set of bookmarks from the dataset. For each bookmark (u, t, d), we create a query $q = t$, which is issued by user u and aims at finding document d. In this paper, we limit the size of each query to be 2 to 4 keywords, which is a typical query size issued for on-line search as studied in work [8]. Finally, we remove all selected bookmarks to avoid promoting the annotated document with bias. Furthermore, to reduce the influence of removing bookmarks, we only randomly create 100 synthetic user queries each time and conduct 10 times of evaluations independently and then report the average results.

The performance of the *D-PR* function and the *SoPRa* function are evaluated based on a widely adopted metric [3,25], called *mean reciprocal rank (MRR)*. MRR measures the performance of a personalized function by assigning a value $1/r$ for each tested personalized query answering and then computing the mean value. Formally,

$$MRR = \sum_{i=1}^{n} 1/(r_i \cdot n), \tag{12}$$

where r_i is the ranking position of the ith user query's relevant document in the personalized search result ordering, and n is the total number of tested queries.

5.2 Results

Since both the *D-PR* function and the *SoPRa* function use a parameter α to adjust the proportion of the profile matching score in the ranking score, we vary the value of the parameter α from 0 to 1.0 and report the result in each case. We set the threshold T of the perception similarity to 0.5. Recall from above that we set $\beta = 1$.

The experimental results of these two personalized functions are shown in Fig. 1. Generally, Fig. 1 shows that our *D-PR* function outperforms the *SoPRa* function in terms of MRR in almost all α cases, and its best ranking result at $\alpha = 0.3$ is about 11% better than the one of the *SoPRa* function at $\alpha = 0$. Specifically, we have the following observations in Fig. 1: (i) A continuous decline of the MRR of *SoPRa* is witnessed from $\alpha = 0$ (non-personalized) to $\alpha = 1$ (fully personalized); this observation verifies our argument that using user profiles and general document profiles cannot personalize well search results on the Social Web (here, they make ranking even worse). (ii) When the value of α rises from 0 to 0.3, the MRR of *D-PR* increases from around 0.147 to about 0.163. This indicates that the profile matching score in *D-PR* can better personalize the ranking of search results, which proves the effectiveness of the proposed personalized document profile and extended user profile. (iii) Afterwards, the MRR of *D-PR* continuously falls down to its bottom at $\alpha = 1$, which shows that excessive personalization will produce a bad ordering, because the topic matching between query and document is also critical for online search. Overall, these experimental results show that our *D-PR* function achieves better personalized search than the state-of-the-art *SoPRa* function.

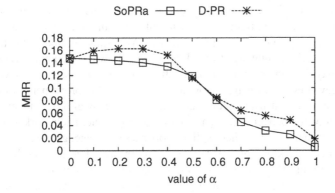

Fig. 1. *D-PR* vs *SoPRa*

6 Related Work

Personalized Web search by considering the searcher's personal attributes and preferences while evaluating a query is of great interest in information retrieval [18], since user queries are in general very short and provide an incomplete specification of the individual information need of a user. Some approaches have already been proposed to mine user preferences from both the user's explicit and implicit activities on the Web, such as query history [21], browsing history [22], the user's current task [13] or intent [23], and even eye-tracking during the search session [9]. Then, a user profile is built from the user's preferences and used for personalization by *query expansion* [5], i.e., a user's query is expanded based on the resulting profile to reflect the particular interest, or *re-ranking* [20], i.e., search results are re-ranked according to a user's profile such that personally relevant results appear higher in the search result list.

Specifically, the work in [9] shows that user preferences that are derived from click logs are reasonably reliable; so, in [16], the history of click data is used to estimate the user's hidden interests and to compute values of the *topic-sensitive PageRank* [7] for personalizing search results. Furthermore, Shen et al. [20] develop a method based on a decision-theoretic framework to convert user search histories into user profiles that are used to both expand queries and re-rank search results. As shown in [6], the benefits that can be achieved through personalization vary across queries; [13] and [23] thus propose solutions to discover the user's current tasks or intents by log analysis to help identify the queries that will benefit most from personalization.

However, mining user preferences by aggregating the user's online activities inevitably encounters a serious problem of privacy compromise [11]: due to the various online activities, Web logs usually contain some sensitive information of users, such as home address, medical record, bank account number, social security number, and so on. Therefore, as a privacy-enhanced personalization technique, folksonomy-based personalized Web search attracts more and more research efforts [3,14,25,26].

Xu et al. [26] propose to use the similarity of folksonomy-based user and document profiles to personalize search results. Then, Bouadjenek et al. [3] extend this work by introducing a social matching score to solve the textual matching problem. Instead of using *tf-idf*, Noll and Meinel [14] only use user tag frequency as the weighting of tag and normalize all document frequency to 1 to put more importance to the user profile. Vallet et al. [25] propose to use the probabilistic BM25 ranking model [17] to replace VSM. As these works weight tags from all users equally when modeling the document profile, they may encounter some personalization problems as discussed above.

In [24], Teevan et al. investigate how to use groups to improve personalized Web search and conclude that using group data collected across group members yields a significant improvement over individual personalization alone. Their work identifies the groups (or separate users) by either explicit properties (e.g., age, gender, job, location), interest groups, or desktop content; however, this information may result in privacy issues. Therefore, in our work, we propose to use the perception similarity computed from social annotations as groupization criteria to avoid such privacy problems.

7   Summary and Outlook

In this paper, we have proposed a dual personalized ranking (D-PR) function to improve personalized ranking of search on the Social Web via an extended user profile and a personalized document profile. We have formally defined the extended user profile of a user as the sum of all of his/her personalized document profiles; and we have further proposed to estimate the personalized document profile based on the perception similarities between users. Finally, a method used to quantify the perception similarity has also been presented. We have performed evaluations based on a public real-world large scale research dataset, and the results validate that our *D-PR* personalized function outperforms the state-of-the-art *SoPRa* function.

In future research, we will apply our *D-PR* ranking function to other Social Web datasets to evaluate its performance on various kinds of social resources. We will also investigate how to utilize categorical or ontology information of online documents to further enhance personalized search on the Social Web.

Acknowledgments. This work was supported by the EPSRC grant EP/J008346/1 "PrOQAW: Probabilistic Ontological Query Answering on the Web", by the European Research Council (FP7/2007–2013/ERC) grant 246858 ("DIADEM"), by a Google European Doctoral Fellowship, and by a Yahoo! Research Fellowship. We also want to thank Mohamed Reda for providing us with the SoPRa implementation and dataset.

References

1. Benz, D., Hotho, A., Jäschke, R., Krause, B., Stumme, G.: Query logs as folksonomies. Datenbank-Spektrum 10(1), 15–24 (2010)
2. Bischoff, K., Firan, C.S., Nejdl, W., Paiu, R.: Can all tags be used for search? In: Proceedings of CIKM, pp. 193–202 (2008)
3. Bouadjenek, M.R., Hacid, H., Bouzeghoub, M.: Sopra: A new social personalized ranking function for improving Web search. In: Proceedings of SIGIR, pp. 861–864 (2013)
4. Carmel, D., Zwerdling, N., Guy, I., Ofek-Koifman, S., Har'el, N., Ronen, I., Uziel, E., Yogev, S., Chernov, S.: Personalized social search based on the user's social network. In: Proceedings of CIKM, pp. 1227–1236 (2009)

5. Chirita, P.A., Firan, C.S., Nejdl, W.: Personalized query expansion for the Web. In: Proceedings of SIGIR, pp. 7–14 (2007)
6. Dou, Z., Song, R., Wen, J.: A large-scale evaluation and analysis of personalized search strategies. In: Proceedings of WWW, pp. 581–590 (2007)
7. Haveliwala, T.H.: Topic-sensitive pagerank. In: Proceedings of WWW, pp. 517–526 (2002)
8. Jansen, B.J., Spink, A., Bateman, J., Saracevic, T.: Real life information retrieval: A study of user queries on the Web. In: SIGIR Forum, pp. 5–17 (1998)
9. Joachims, T., Granka, L., Pan, B., Hembrooke, H., Gay, G.: Accurately interpreting click-through data as implicit feedback. In: Proceedings of SIGIR, pp. 154–161 (2005)
10. Jones, K.S.: A statistical interpretation of term specificity and its application in retrieval. Journal of Documentation 28, 11–21 (1972)
11. Kobsa, A.: Privacy-enhanced personalization. Commun. ACM 50(8), 24–33 (2007)
12. Krause, B., Hotho, A., Stumme, G.: A comparison of social bookmarking with traditional search. In: Macdonald, C., Ounis, I., Plachouras, V., Ruthven, I., White, R.W. (eds.) ECIR 2008. LNCS, vol. 4956, pp. 101–113. Springer, Heidelberg (2008)
13. Luxenburger, J., Elbassuoni, S., Weikum, G.: Task-aware search personalization. In: Proceedings of SIGIR, pp. 721–722 (2008)
14. Noll, M.G., Meinel, C.: Web search personalization via social bookmarking and tagging. In: Aberer, K., Choi, K.-S., Noy, N., Allemang, D., Lee, K.-I., Nixon, L.J.B., Golbeck, J., Mika, P., Maynard, D., Mizoguchi, R., Schreiber, G., Cudré-Mauroux, P. (eds.) ISWC/ASWC 2007. LNCS, vol. 4825, pp. 367–380. Springer, Heidelberg (2007)
15. Noll, M.G., Meinel, C.: The metadata triumvirate: Social annotations, anchor texts and search queries. In: Proceedings of WI-IAT, pp. 640–647 (2008)
16. Qiu, F., Cho, J.: Automatic identification of user interest for personalized search. In: Proceedings of WWW, pp. 727–736 (2006)
17. Robertson, S.E., Walker, S.: Some simple effective approximations to the 2-Poisson model for probabilistic weighted retrieval. In: Proceedings of SIGIR, pp. 232–241 (1994)
18. Salton, G., McGill, M.J.: Introduction to Modern Information Retrieval. McGraw-Hill, Inc., New York (1986)
19. Salton, G., Wong, A., Yang, C.S.: A vector space model for automatic indexing. Commun. ACM 18(11), 613–620 (1975)
20. Shen, X., Tan, B., Zhai, C.: Implicit user modeling for personalized search. In: Proceedings of CIKM, pp. 824–831 (2005)
21. Shen, X., Zhai, C.X.: Exploiting query history for document ranking in interactive information retrieval. In: Proceedings of SIGIR, pp. 377–378 (2003)
22. Sugiyama, K., Hatano, K., Yoshikawa, M.: Adaptive Web search based on user profile constructed without any effort from users. In: Proceedings of WWW, pp. 675–684 (2004)
23. Teevan, J., Dumais, S.T., Liebling, D.J.: To personalize or not to personalize: modeling queries with variation in user intent. In: Proceedings of SIGIR, pp. 163–170 (2008)
24. Teevan, J., Morris, M.R., Bush, S.: Discovering and using groups to improve personalized search. In: Proceedings of WSDM, pp. 15–24. ACM (2009)
25. Vallet, D., Cantador, I., Jose, J.M.: Personalizing Web search with folksonomy-based user and document profiles. In: Gurrin, C., He, Y., Kazai, G., Kruschwitz, U., Little, S., Roelleke, T., Rüger, S., van Rijsbergen, K. (eds.) ECIR 2010. LNCS, vol. 5993, pp. 420–431. Springer, Heidelberg (2010)
26. Xu, S., Bao, S., Fei, B., Su, Z., Yu, Y.: Exploring folksonomy for personalized search. In: Proceedings of SIGIR, pp. 155–162 (2008)

Author Index